Environmental Mycology in Public Health

Environmental Mycology in Public Health

Fungi and Mycotoxins Risk Assessment and Management

Edited by

Carla Viegas

Ana Catarina Pinheiro

Raquel Sabino

Susana Viegas

João Brandão

Cristina Veríssimo

ELSEVIER

AMSTERDAM • BOSTON • HEIDELBERG • LONDON
NEW YORK • OXFORD • PARIS • SAN DIEGO
SAN FRANCISCO • SINGAPORE • SYDNEY • TOKYO

Academic Press is an imprint of Elsevier

AP

Academic Press is an imprint of Elsevier
125 London Wall, London EC2Y 5AS, UK
525 B Street, Suite 1800, San Diego, CA 92101-4495, USA
225 Wyman Street, Waltham, MA 02451, USA
The Boulevard, Langford Lane, Kidlington, Oxford OX5 1GB, UK

ISBN: 978-0-12-411471-5

British Library Cataloguing-in-Publication Data
A catalogue record for this book is available from the British Library

Library of Congress Cataloging-in-Publication Data
A catalog record for this book is available from the Library of Congress

Typeset by TNQ Books and Journals
www.tnq.co.in

Dedication

The editors wish to acknowledge the vision of Dr. Laura Rosado, who recognized environmental exposure to fungi as part of a wider medical mycology approach. Her enthusiasm, passed on at the National Institute of Health Dr. Ricardo Jorge in Portugal, was an inspiration to the editors of this book.

Contents

Section II
Outline of Fungal Pathologies

Section III

Part II
Environmental Mycology in Public Health

Section I

Section II
Occupational Settings

Section III
Nonoccupational Exposure

Contributors

Anna Błajet-Kosicka Kazimierz Wielki University, Faculty of Natural Science, Institute of Experimental Biology, Department of Physiology and Toxicology, Bydgoszcz, Poland

Kaitlin Benedict Mycotic Diseases Branch, Centers for Disease Control and Prevention, Atlanta, GA, USA

Luís Miguel Borrego Allergy Center, CUF Descobertas Hospital, Lisboa, Portugal; Centro de Estudos de Doenças Crónicas, CEDOC, NOVA Medical School/ Faculdade de Ciências Médicas, Universidade Nova de Lisboa, Portugal

João Brandão Reference Unit for Parasitic and Fungal Infections, Department of Infectious Diseases, National Institute of Health Doutor Ricardo Jorge, Lisboa, Portugal

Mary E. Brandt Mycotic Diseases Branch, Centers for Disease Control and Prevention, Atlanta, GA, USA

Carlo Brera Istituto Superiore di Sanità – Dipartimento di Sanità Pubblica Veterinaria e Sicurezza Alimentare – Reparto OGM e Xenobiotici di origine fungina - Viale Regina Elena, Rome, Italy

Karl V. Clemons California Institute for Medical Research, Infectious Diseases Research Laboratory, San Jose, CA, USA

Sonia Colicchia Istituto Superiore di Sanità – Dipartimento di Sanità Pubblica Veterinaria e Sicurezza Alimentare – Reparto OGM e Xenobiotici di origine fungina - Viale Regina Elena, Rome, Italy

Rodrigo de Almeida Paes Laboratório de Micologia do Instituto Nacional de Infectologia Evandro Chagas, Fundação Oswaldo Cruz, Rio de Janeiro, Brazil

Barbara De Santis Istituto Superiore di Sanità – Dipartimento di Sanità Pubblica Veterinaria e Sicurezza Alimentare – Reparto OGM e Xenobiotici di origine fungina - Viale Regina Elena, Rome, Italy

Francesca Debegnach Istituto Superiore di Sanità – Dipartimento di Sanità Pubblica Veterinaria e Sicurezza Alimentare – Reparto OGM e Xenobiotici di origine fungina - Viale Regina Elena, Rome, Italy

Philippe Duquenne INRS, Laboratoire de Métrologie des AérosolsVandoeuvre-les-Nancy, France

Alesia Ferguson Department of Environmental and Occupational Health, University of Arkansas for Medical Sciences, Little Rock, AR, United States

Xi Fu Occupational and Environmental Medicine, Department of Medical Sciences, Uppsala University, University Hospital, Uppsala, Sweden

Bari Gordon Department of Civil, Architectural, and Environmental Engineering, University of Miami, Coral Gables, FL, USA

Jan Grajewski Kazimierz Wielki University, Faculty of Natural Science, Institute of Experimental Biology, Department of Physiology and Toxicology, Bydgoszcz, Poland

Emanuela Gregori Istituto Superiore di Sanità – Dipartimento di Sanità Pubblica Veterinaria e Sicurezza Alimentare – Reparto OGM e Xenobiotici di origine fungina - Viale Regina Elena, Rome, Italy

Ferry Hagen Department of Medical Microbiology & Infectious Diseases, Canisius-Wilhelmina Hospital, Nijmegen, The Netherlands

Valerie J. Harwood Department of Integrative Biology, University of South Florida, Tampa, FL, USA

Anne Hyvärinen National Institute for Health and Welfare, Department of Health Protection, Living Environment and Health Unit, Kuopio, Finland

Robert Kosicki Kazimierz Wielki University, Faculty of Natural Science, Institute of Experimental Biology, Department of Physiology and Toxicology, Bydgoszcz, Poland

Carina Ladeira Environmental Health Research Group & Genetic and Metabolism Research Group, Escola Superior de Tecnologia da Saúde de Lisboa, IPL, Portugal

Adela López de Cerain Department of Pharmacology and Toxicology, School of Pharmacy, University of Navarra, Spain

Maria Chiara Magri Istituto Superiore di Sanità – Dipartimento di Sanità Pubblica Veterinaria e Sicurezza Alimentare – Reparto OGM e Xenobiotici di origine fungina - Viale Regina Elena, Rome, Italy

Stefan Mayer Berufsgenossenschaft Handel und Warendistribution, Department of Prevention, Mannheim, Germany

Brunella Miano Istituto Superiore di Sanità – Dipartimento di Sanità Pubblica Veterinaria e Sicurezza Alimentare – Reparto OGM e Xenobiotici di origine fungina - Viale Regina Elena, Rome, Italy

Inês Andrade Mota Allergy Center, CUF Descobertas Hospital, Lisboa, Portugal

Aino Nevalainen National Institute for Health and Welfare, Department of Health Protection, Living Environment and Health Unit, Kuopio, Finland

Anne Oppliger University of Lausanne and Geneva, Switzerland

Maja Peraica Unit of Toxicology, Institute for Medical Research and Occupational Health, Zagreb, Croatia

Elena Piecková Slovak Medical University, Limbová, Bratislava, Slovakia

Ana Catarina Pinheiro Pharmacy Department, Centro Hospitalar do Algarve, Faro Unit, Portugal

Guillermo Quindós-Andrés Laboratorio de Micología Médica, Departamento de Inmunología, Microbiología y Parasitología (UFI 11/25 «Microbios y Salud»), Facultad de Medicina y Odontología, Universidad del País Vasco/Euskal Herriko Unibertsitatea, Apartado, Bilbao, Spain

Malcolm D. Richardson Mycology Reference Centre, University Hospital of South Manchester, and Centre for Respiratory Medicine and Allergy, Institute of Inflammation and Repair, University of Manchester, Manchester, UK

Raquel Sabino National Institute of Health Dr. Ricardo Jorge, Infectious Diseases Department, Lisbon, Portugal

Robert A. Samson CBS-KNAW Fungal Biodiversity Centre, Uppsalalaan, Utrecht, The Netherlands

Helena Solo-Gabriele Department of Civil, Architectural, and Environmental Engineering, University of Miami, Coral Gables, FL, USA

Sabina Soricelli Istituto Superiore di Sanità – Dipartimento di Sanità Pubblica Veterinaria e Sicurezza Alimentare – Reparto OGM e Xenobiotici di origine fungina - Viale Regina Elena, Rome, Italy

Martin Täubel National Institute for Health and Welfare, Living Environment and Health Unit, Kuopio, Finland

Magdalena Twarużek Kazimierz Wielki University, Faculty of Natural Science, Institute of Experimental Biology, Department of Physiology and Toxicology, Bydgoszcz, Poland

Cristina Veríssimo Reference Laboratory for Parasites and Fungal Infections, Department of Infectious Diseases, National Institute of Health, Lisbon, Portugal

Ariane Vettorazzi Department of Pharmacology and Toxicology, School of Pharmacy, University of Navarra, Spain

Carla Viegas Environment & Health RG, Lisbon School of Health Technology, Polytechnic Institute of Lisbon, Environmental Health Institute, Faculty of Medicine from Lisbon University, Lisbon, Portugal

Rosely Maria Zancopé Oliveira Laboratório de Micologia do Instituto Nacional de Infectologia Evandro Chagas, Fundação Oswaldo Cruz, Rio de Janeiro, Brazil

Foreword

Exposure to airborne fungal spores and other propagules in the environment, whether this occurs indoors or outdoors, in the workplace or in the home, is an everyday occurrence that may lead to a wide range of disease manifestations in humans and animals. These include mycoses, mycotoxicoses, and allergies. Improving our understanding of the role of the environment in the causation of these diseases is a critical need in the formulation and evaluation of intervention and prevention strategies to reduce their impact on global public health and medical care.

Recognition of the central importance of the environment as a source of human infection has come about, at least in part, as a result of the emergence of an unprecedented number of ubiquitous environmental fungi as major causes of disease. These hitherto "harmless" organisms have come to constitute the predominant group of life-threatening fungal pathogens seen in individuals whose immunity is impaired as a result of either an underlying disorder or its treatment. Most of these infections follow inhalation of spores from the air, and the lungs are the most common site of initial damage. Exposure to contaminated air may occur during hospitalization, especially if there is ongoing construction or renovation work, but infection arising from exposure to airborne fungal spores in the home or workplace may be more frequent.

It has long been established that some larger fungi are poisonous (or hallucinogenic), but many microfungi also produce mycotoxins in animal feed and human food. Contamination can occur throughout the entire food chain, from the crop in the field, through storage and shipping, to processed foods. Depending on dosage and duration of exposure, these nonvolatile metabolites can induce acute or chronic disease in farm animals and humans. Mycotoxicoses were first recognized and studied in the developed world, but many countries have adopted regulations to limit mycotoxin exposure. As a consequence, these diseases now mostly occur in developing countries in the tropics where environmental conditions, such as high temperatures and humidity, favor mold growth and toxin production.

Much progress has been made in improving analytical procedures for mycotoxins and in developing safer production chains in the animal feed and human food industries, at least in developed countries. However, as the production and distribution of our food shift from localized production and consumption toward an increasingly globalized network of distribution, mycotoxins in animal feed or human food now have the potential to pose a serious health risk to consumers throughout the world.

That there is a causal relationship between exposure to mold-contaminated indoor and outdoor environments and adverse health effects has been difficult to prove, owing to coexposure to many other components of bioaerosols. The only exceptions to this are the various infections and allergies, such as asthma, rhinitis, and sinusitis, for which a specific link between the outcome and the causal agent has been well established. Excessive indoor dampness is not by itself a cause of ill health, but damp indoor environments can favor mold growth in homes, offices, schools, and other buildings. However, it is frustrating that, despite extensive research, there is still insufficient or inadequate epidemiological or toxicological evidence to allow us to determine whether an association exists between the presence of molds and/or mycotoxins in damp indoor environments and adverse health effects in otherwise healthy adults.

This book provides a valuable service to all who are concerned with fungal diseases, whether with mycoses, mycotoxicoses, allergies, or other potential adverse health effects. By bringing together what is currently known about these conditions, together with the latest information on their detection, monitoring, and control, the authors have provided a comprehensive resource for all those concerned with this increasingly important and diverse field of mycology. Increased awareness of this field will be critical if the resources needed to develop successful intervention and prevention strategies are to be acquired. Only then will we be able to reduce the substantial public health burden of these diseases.

David W. Warnock
Faculty of Medical and Human Sciences,
University of Manchester, Manchester,
United Kingdom.

Part I

Fungal Specificities in Environmental Mycology

Section I

General Fungal Characteristics

Chapter 1

Cellular Constitution, Water and Nutritional Needs, and Secondary Metabolites

Robert A. Samson
CBS-KNAW Fungal Biodiversity Centre, Uppsalalaan, Utrecht, The Netherlands

Recent taxonomic treatments show that fungi and animals both belong to the group Opisthokonta.[1,2] Fungi are considered the sister group of animals and part of the eukaryotic crown group that appeared about a billion years ago. Fungi share with animals the ability to export hydrolytic enzymes that break down biopolymers, which then can be absorbed for nutrition. Fungi live in their own food supply and simply grow into new food as the local environment becomes exhausted of nutrients. The organisms traditionally regarded as "fungi" belong to three unrelated groups: the true fungi in Kingdom Fungi (Eumycota), the Oomycetes, and the slime molds.

Our current knowledge shows that there are approximately 100,000 described species, but a conservative estimate of the total number of fungal species thought to exist is 1.5 million.[3,4] However, Blackwell[5] has indicated that until recently, estimates of numbers of fungi did not include results from large-scale environmental sequencing methods. Newer estimates based on data acquired from several molecular methods have predicted that as many as 5.1 million species of fungi may exist.[6,7]

In this chapter a summary of important fungal structures and characters is given, with emphasis on the fungi that play an important role in environmental mycology. For more detailed information on fungi the reader should consult books on introductory mycology.[8–13]

FUNGAL STRUCTURES

Mycelium

The mycelium consists of hyphae, and the type of hyphae is characteristic of specific groups of fungi. Fungi that lack cross walls (nonseptate; aseptate;

Environmental Mycology in Public Health. http://dx.doi.org/10.1016/B978-0-12-411471-5.00001-6

coenocytic) are found, for example, in the Zygomycetes. In Ascomycetes and Basidiomycetes, species form septate hyphae, with perforations at the septa, called septal pores. These allow the movement of cytoplasm and organelles from one compartment to the next. The type and complexity of the septal pore are characteristic of specific groups of fungi. Yeasts are unicellular, although some species with yeast forms may become multicellular, in the majority of the cases through the formation of strings of connected budding cells known as pseudohyphae. Hyphae elongate almost exclusively at the tips, growing outward from the point of establishment. As a result of apical growth, hyphae are relatively uniform in diameter, and mycelium that grows in an unimpeded manner forms a circular colony on solid substrates that support fungal growth.

Sporangiospores

The asexual propagules that form inside a sporangium, which can be mostly spherical or cylindrical, through a process involving cleavage of the cytoplasm are named sporangiospores. These spores are thin walled, one celled, hyaline, or pale in color, and usually globose or ellipsoid in shape. One to 50,000 sporangiospores may be formed in a single sporangium. When mature, sporangiospores are released by breakdown of the sporangial wall, or the entire sporangium may be dispersed as a unit. Sporangiospores are produced by fungi of the Chytridiomycetes and Zygomycetes groups, as well the Oomycetes, a group of fungi that is phylogenetically unrelated to the true fungi. The sexual propagation of the fungi that produce sporangiospores occurs via the zygospore. The zygospores serve as resting and survival propagules and are found rarely in cultures of common fungi.

Conidiophores and Conidia

Many species that are relevant to environmental mycology are "anamorphic fungi." This is the current terminology for those fungi that used to be called Fungi imperfecti, Deuteromyces, Hyphomycetes, Coelomycetes, etc. These names were used for fungi of the Ascomycetes or Basidiomycetes that lack a sexual state, but phylogenetic studies have shown that within many genera, sexual and asexual species are closely related. Hence there is now a change in the nomenclature of fungi, which is based on the one fungus–one name concept.[14–16] In some genera, such as *Aspergillus* and *Penicillium,* with teleomorph connections (*Eurotium*, *Neosartorya*, *Eupenicillium,* etc.), the selection of the current nomenclature of the species follows the anamorphic name, for example, *Aspergillus* and *Penicillium.*[17,18]

Anamorphic fungi were also artificially grouped based on their morphological structures, such as the presence of solitary conidiophores, synnemata, or conidiophores produced within pycnidia. Phylogenetic studies have also shown that within genera or even species these different structures may occur and therefore these, sometimes distinct, morphological structures cannot be used for distinguishing genera or even species.

Among the anamorphic fungi various types of conidiogenesis can be seen. The patterns of conidiogenesis are described in detail by Cole and Samson.[19] How conidiogenesis takes place in a fungus is relevant to the mode of sporulation, number of conidia produced, and distribution of these propagules. A common type of conidiogenesis is through the phialide, which can produce masses of conidia in dry chains or conglomerates (e.g., *Aspergillus, Penicillium*) or in so-called slimy heads (e.g., *Stachybotrys, Fusarium*). Other fungi are characterized by thallic, blastic, or poroconidia (e.g., *Geotrichum, Cladosporium, Alternaria*).[20]

In addition to the phialide, conidia can be formed from different types of conidiogenous cells, which can be formed singly on hyphae, on the surface of aggregated hyphal structures, or within various types of fruiting bodies. Pycnidia and acervuli are fruiting bodies inside which conidia are formed. Sporodochia and synnemata are other examples of fruiting bodies on which conidia are formed. Conidium-forming fungi are primarily Ascomycetes, although they can also be found as anamorphic Basidiomycete species. A good example is *Wallemia sebi,* which belongs to a separate family, *Wallemiomycetes,* and is very common in indoor environments and on low water activity food.

For many years it was assumed that the spores/conidia and perhaps large mycelial fragments were the source of exposure to fungi[21] and that spore counting could be used for exposure assessment. However, it has been demonstrated that fragments significantly smaller than spores (down to 0.1 μm) are released from the mycelia of infested materials.[22–25] These fragments can be liberated in numbers hundreds of times higher than the number of spores, with no correlation between the numbers of released fragments and spores.[23] It is important to consider exposure to the small fungal fragments when assessing exposure to fungal allergens.

Ascomata and Basidiomata

The ascoma (plural: ascomata) is the fruiting body of an Ascomycete and mostly consists of very tightly interwoven hyphae and may contain asci, each of which typically contains four, eight, or more ascospores. These fruiting bodies are most commonly bowl-shaped (apothecia), spherical (cleistothecia), or flask-like (perithecia), closed or with an opening. Genera such as *Byssochlamys* are characterized by naked asci, which lack an ascoma wall. A basidioma (plural: basidiomata) is the fruiting body of a Basidiomycete and consists of a multicellular structure that bears the spore-producing hymenium. Basidiomata are characteristic of the hymenomycetes; rusts and smuts do not produce such structures. Epigeous (aboveground) basidiomata that are visible to the naked eye are commonly referred to as mushrooms, while hypogeous (underground) basidiomata are usually called false truffles.

Chlamydospores

Chlamydospores are survival structures formed from an existing hyphal cell or a conidium that develops a thickened wall and cytoplasm packed with lipid reserves. The thickened cell walls may be pigmented or hyaline, and chlamydospores develop singly or in clusters, depending upon the fungus. Chlamydospores are passively dispersed, in most instances when the mycelium breaks down. Chlamydospores are formed by many different groups of fungi and are often found in aging cultures.

Sclerotia

Compact aggregations of hyphae differentiated into an outer, pigmented rind and an inner mass of hyaline cells, called a medulla, are called sclerotia. Such fungal structures contain food reserves and are a type of survival propagule produced by a number of fungi in the Ascomycetes and Basidiomycetes. Sclerotia are mostly neglected as important fungal structures, because they are not distinctly present in most fungal isolations.

Sclerotial development has a role in dormancy and is also considered an important condition for sexual development.[26] Asci and ascospores can be found in sclerotia in species in the *Aspergillus* sections *Flavi*[27–29] and *Circumdati*,[30,31] showing that these structures are important for propagation. In these fungi, ripe asci can be obtained by mating or are produced after an extended time of incubation.

Sclerotia are also regarded as important in view of the production of specific compounds. Metabolites from the sclerotia of a non-aflatoxigenic strain of *Aspergillus flavus* showed substantial antifeedant activity against insects.[32] Arthropod predation is recognized as a selective force that has shaped the chemical defense systems of *A. flavus* and other sclerotium-producing fungi.

Wicklow and Shotwell[33] examined the distribution of aflatoxins among the conidia and sclerotia of toxigenic strains of *A. flavus* and *Aspergillus parasiticus* and found that the substantial aflatoxin levels in conidia could place agricultural workers exposed to dust containing large numbers of *A. flavus* conidia at risk. Cellular ratios of aflatoxin B_1 to aflatoxin G_1 were nearly identical in conidia and sclerotia even though levels of total aflatoxins in these propagule types may have differed greatly. Aflatoxin G_1 was detected in the sclerotia of all *A. flavus* strains but in the conidia of only one strain. Each of the *A. parasiticus* strains examined accumulated aflatoxin G_1 in both sclerotia and conidia.

Frisvad et al.[34] could induce the production of sclerotia by certain strains of *Aspergillus niger* when grown on Czapek yeast agar with raisins, on other fruits, or on rice. In strains in which sclerotia were found, up to 11 apolar indoloterpenes of the aflavinine type were detected, which had not been reported before for strains of *A. niger*. The induction of sclerotium formation can thus be a way of inducing the production of new secondary metabolites from previously silent gene clusters.

GROWTH CONDITIONS

Some fungi are symbionts or parasites on other organisms, but most species grow on land and obtain their nutrients from dead organic matter. The majority of species obtain their food by secreting enzymes, which partially digest the food extracellularly, and then absorbing the partially digested food to complete digestion internally.

Nutrients

Unlike plants, which use carbon dioxide and light as sources of carbon and energy, respectively, fungi meet these two requirements by assimilating preformed organic matter; carbohydrates are generally the preferred carbon source. Fungi can readily absorb and metabolize a variety of soluble carbohydrates, such as glucose, xylose, sucrose, and fructose. Fungi are also characteristically well suited to using insoluble carbohydrates such as starches, cellulose, and hemicelluloses, as well as very complex hydrocarbons such as lignin. Many fungi can also use proteins as a source of carbon and nitrogen. To use insoluble carbohydrates and proteins, fungi must first digest these polymers extracellularly. Saprobic fungi obtain their food from dead organic material; parasitic fungi do so by feeding on living organisms (usually plants), thus causing disease.

Water

Most fungi require very high water availability and rapidly dry out, or senesce, under dry conditions. However, fungi are also able to tolerate much lower water availability than other organisms. Survival at low water activity level (extremely low osmotic potential) has been studied in relation to food spoilage. Spoilage by xerophilic molds has proved to be a very common food contamination problem. The most xerophilic organism, *Xeromyces bisporus*, can grow at a water activity (a_w) of 0.62,[35] while many other species such as *Eurotium, Aspergillus, Wallemia,* and *Penicillium* commonly contaminate low water activity products (a_w 0.75–0.90). Xerophilic molds are also extremely common in indoor environments but are often neglected and not found if the detection method uses high water activity isolation media.

The mechanisms that enable functionality under osmotic stress are related to the presence of compatible solutes. Compatible solutes such as glycerol and other polyols are stored in high concentrations in the cell, which counters the effects of water loss. Glycerol appears to protect enzymes from the accumulation of sodium ions and loss of water, both of which may denature them. Polyols may also protect membranes. Xerophilic fungi use compatible solutes to maintain water potential in the cell, though their rates of metabolism and thus growth are extremely slow.

pH

For most fungi, a pH range of 5.5–6.5 seems to be suitable for their maximum growth and sporulation, but the hydrogen environment of fungi is difficult to study because they change the pH of their environment as they grow. A typical example is *A. niger*, which produces citric and other organic acids and thus lowers the pH of the substrate. Some species increase and others decrease the pH of their medium. The pH of the medium is important because it influences mineral availability, enzyme activity, and membrane function. Generally speaking, fungi can tolerate a wide range of pH.

Oxygen

In general, fungi require oxygen to survive, but they are also able to use fermentation when they lack oxygen. The fungi include species that are obligately aerobic or obligately anaerobic (e.g., rumen fungi). However, many fungi lie between these extremes, with the capacity to function facultatively under aerobic and anaerobic conditions. Oxygen is used for oxidative metabolism, to generate energy. However, it is also essential for the biosynthesis of sterols, unsaturated fatty acids, and some vitamins. Thus, while many fungi can exist under anaerobic conditions and respire fermentatively, they also have the capacity to transport oxygen or the products of respiration through their cytoplasm.

Temperature

Fungi can normally tolerate the range of temperature of the environment from which they are taken. Their response to temperature is quite varied, however. Active growth will usually be associated with a limited range of temperatures. There are different definitions of temperature requirements, but those fungi that grow between 15 and 35 °C are usually called mesophilic, and those growing above this range are termed thermophilic. Those that grow at low temperatures (<5 °C) are called psychrophilic. Many fungi remain alive for extended periods at temperatures unsuitable for growth. Temperature affects lag time, specific growth rate, and yield in quite different ways for each fungus. High or low temperatures may cause the fungi to enter dormancy, and reversion to original temperatures may be insufficient to restore metabolic activity.

The spores of some fungi also survive exposure to extreme temperatures when they are dry. This capacity is referred to as thermostability, and it is found widely among the fungi.

The fungi that function in extreme aridity, extreme temperatures, and saline conditions are stress-tolerant species. Ascospores of *Byssochlamys, Talaromyces, Neosartorya*, and other *Trichocomaceae* are known to be heat resistant and can cause major spoilage problems in heat-treated food products and beverages. Ascospores can survive temperatures of up to 120 °C.[36–38]

Light

Light has an important influence on fungal growth in specific cases. The effect of UV (ultraviolet) radiation on spore and fruiting body formation and phototropic release is a clear example of the importance of light. Overall, light does not play a major part in the metabolism and growth of fungi. For the cultivation and sporulation of common species, light seems not to be a limiting factor, and most anamorphic fungi on food and in indoor environments develop well in the dark. However, light might have an impact on the production of metabolites. It has previously been shown that the biosynthesis of the mycotoxins ochratoxin A and B and of citrinin by *Penicillium* is regulated by light. In wheat that was contaminated with an ochratoxin A-producing culture of *Penicillium verrucosum* and treated with blue light after a preincubation by the fungus, the concentration of the preformed ochratoxin A was reduced by roughly 50% compared to the control and differed by >90% compared to the sample that was incubated further in the dark.[39]

UV radiation can reduce the viability of conidia, especially in air. Park et al.[40] found that 600 mWs/cm^2 of UV at 260 nm could potentially be used for the inactivation of *A. niger, Penicillium citrinum*, and *Cladosporium cladosporioides* in dried fishery food products. Levels of fungi growing on insulation within air-handling units (AHUs) in an office building and levels of airborne fungi within AHUs were measured before the use of germicidal UV light and again after 4 months of operation. The fungal levels following UV operation were significantly lower than the levels in control AHUs.[41]

FUNGAL METABOLITES

The diversity of fungi is reflected in the variety of fungal metabolites, but it seems that certain groups are able to produce more metabolites than others. For example, Frisvad[42] showed that species of *Aspergillus, Penicillium,* and *Talaromyces* are particularly productive organisms for secondary metabolites. A comparison with other genera shows that most secondary metabolites have been reported from *Aspergillus* (1984), from *Penicillium* (1338), and from *Talaromyces*, (316). Two other common genera, *Fusarium* (507) and *Trichoderma* (438), produce fewer secondary metabolites.

Frisvad[42] preferred the term exometabolites for secondary metabolites and defined this term as small molecules produced during morphological and chemical differentiation that are outwardly directed, that is, secreted or deposited in or on the cell wall, and accumulated. This contrasts with endometabolites (primary metabolites), which fluctuate in concentration and are either transformed into other endometabolites or feed into exometabolites, exoproteins, exopolysaccharides, or morphological structures. While endometabolites can be found for almost all species of fungi, exometabolites, exoproteins, and exopolysaccharides are taxonomically delimited and produced in species-specific profiles. Some metabolites can occur as both endo- and exometabolites, for example, citric acid.

The biosynthetic pathways involved are also diverse, including polyketides, sesquiterpenes and diterpenes, diketopiperazines, cyclic peptides, β-lactams, and combinations of these pathways. Many of these compounds have biological activity that may be harmful, such as mycotoxins and phytotoxins, or beneficial, such as antibiotics and other pharmaceuticals.

Toxins in Food

There is a vast literature on mycotoxins, and numerous monographs have been published.[43–49] A wealth of information is available about the fungal toxins produced in food. Many books and papers have been published on the occurrence, toxicity, and detection of these compounds. Wu et al.[50] recently reviewed the public health impacts of food-borne mycotoxins. Although there are approximately 400 compounds described and considered to be toxic, the most important mycotoxins known today are (1) aflatoxins, which cause liver cancer and have also been implicated in child growth impairment and acute toxicoses; (2) fumonisins, which have been associated with esophageal cancer and neural tube defects; (3) deoxynivalenol and other trichothecenes, which are immunotoxic and cause gastroenteritis; and (4) ochratoxin A, which has been associated with renal diseases.

Toxins in Indoor Environments

There are many reports on the occurrence of mycotoxins in the indoor environment. Although species of the indoor mycobiota have the potential to produce toxic metabolites, much of the information in many publications or on the internet is not correct. The reported data mostly refer to species that grow on food (and can produce toxins on specific substrates), but it is important to know, however, whether the same species can produce toxic metabolites when grown on building material. Nielsen and Frisvad[51] have reported that the number of species producing toxins in the indoor environment is actually small. They also explained that mycotoxin production on materials occurs at high water activity (a_w>0.9 on the material surface), but significant mycotoxin production will occur only above an a_w of 0.95.

Sorensen et al.[52] found that the conidia of *Stachybotris chartarum* contain trichothecene mycotoxins. In view of the potent toxicity of the trichothecenes, the inhalation of aerosols containing high concentrations of these conidia is considered to be a potential hazard to health. However, exposure is highest from dry materials and decaying biomass. Therefore the worst case scenario is consecutive water damage, in which large quantities of biomass and mycotoxins are formed, followed by desiccation of the biomass. In such a situation, many conidia and small fungal fragments will become aerosolized and will be deposited all over the building, including the building envelope.

Xerophilic species are common indoor fungi,[20] and these molds are not known to produce important toxins in food. However, the metabolites they produce when growing in indoor environments have not been thoroughly investigated.

Slack et al.[53] reported that *Eurotium* species could produce neoechinulin A and B, epiheveadride, flavoglaucin, auroglaucin, and isotetrahydroauroglaucin as major metabolites. These compounds possess toxic properties, but the relevance to human exposure is not yet known. Furthermore Desroches et al.[54] have found that *Wallemia* strains from the built environment in Canada can produce a number of metabolites, including the known compound walleminone and a new compound, wallimidione (1-benzylhexahydroimidazo[1,5-*a*]pyridine-3,5-dione). Based on an in silico analysis, wallimidione is likely to be the most toxic of the metabolites reported to date from *W. sebi*.

REFERENCES

1. http:/tolweb.org/tree/ [accessed 27.02.15].
2. Hibbett DS, Binder M, Bischoff JF, et al. A higher-level phylogenetic classification of the Fungi. *Mycol Res* 2007;**111**:509–47.
3. Hawksworth DL. The magnitude of fungal diversity: the 1.5 million species estimate revisited. *Mycol Res* 2001;**105**:1422–32.
4. Hawksworth DL, Rossman AY. Where are all the undescribed fungi? *Phytopathology* 1997;**87**:888–91.
5. Blackwell M. The Fungi: 1, 2, 3…5.1 million species? *Am J Bot* 2011;**98**:426–38.
6. O'Brien BL, Parrent JL, Jackson JA, et al. Fungal community analysis by large-scale sequencing of environmental samples. *Appl Environ Microbiol* 2005;**71**:5544–50.
7. Taylor DL, Herriott IC, Stone KE, et al. Structure and resilience of fungal communities in Alaskan boreal forest soils. *Can J For Res* 2010;**40**:1288–301.
8. Alexopoulos CJ, Mims CW, Blackwell M. *Introductory mycology*. 4th ed. New York: John Wiley & Sons; 1996.
9. Deacon J. *Fungal biology*. 4th ed. Malden, MA: Blackwell Publishing; 2006.
10. Kendrick B. *The fifth kingdom*. 3rd ed. Newburyport, MA: Focus Publishing; 2000.
11. Moore D, Robson GD, Trinci APJ. *21st century guidebook to fungi*. New York: Cambridge University Press; 2011.
12. Webster J, Weber RWS. *Introduction to fungi*. New York: Cambridge University Press; 2007.
13. Crous PW, Verkley GJM, Groenewald JZ, et al. Fungal biodiversity. In: *CBS Laboratory Manual Series 1*. Utrecht: CBS-KNAW Fungal Biodiversity Centre; 2009. 269 pp.
14. Norvell LL. Melbourne approves a new CODE. *Mycotaxon* 2011;**116**:481–90.
15. Hawksworth DL. A new dawn for the naming of fungi: impacts of decisions made in Melbourne in July 2011 on the future publication and regulation of fungal names. Also IMA Fungus 2011; **2**: 155–162. *MycoKeys* 2011;**1**:7–20.
16. Hawksworth DL, Crous PW, Redhead SA, Reynolds DR, Samson RA, Seifert KA, et al. The Amsterdam declaration on fungal nomenclature. *Mycotaxon* 2011;**116**:491–500.
17. Samson RA, Visagie CM, et al. Phylogeny, identification and nomenclature of the genus *Aspergillus*. *Stud Mycol* 2014;**78**:141–73.
18. Visagie CM, Houbraken J, et al. Identification and nomenclature of the genus *Penicillium*. *Stud Mycol* 2014;**78**:343–71.
19. Cole GT, Samson RA. *Patterns of development in conidial fungi*. Pitman Publishing Limited; 1979. 190 pp.
20. Samson RA, Houbraken J, Frisvad JC, et al. Food and indoor fungi. In: *CBS Laboratory Manual Series 2*. Utrecht: CBS-KNAW Fungal Biodiversity Centre; 2010. 390 pp.

21. Sorenson WG. Fungal spores: hazardous to health. *Environ Health Perspect* 1999;**107**:469–72.
22. Kildesø J, Würtz H, Nielsen KF, et al. The release of fungal spores from water damaged building materials. In: Seppänen O, Säteri J, editors. *Proceedings of healthy buildings 2000, Espoo, Finland. SYI indoor air information, Helsinki*. 2000. p. 313–8.
23. Gorny RL, Reponen T, Willeke K, et al. Fungal fragments as indoor air biocontaminants. *Appl Environ Microbiol* 2002;**68**:3522–31.
24. Green BJ, Sercombe JK, Tovey ER. Fungal fragments and undocumented conidia function as new aeroallergen sources. *J Allergy Clin Immunol* 2005;**115**:1043–8.
25. Green BJ, Tovey ER, Sercombe JK, et al. Airborne fungal fragments and allergenicity. *Med Mycol* 2006;**44**:S245–55.
26. Dyer PS, O'Gorman CM. Sexual development and cryptic sexuality in fungi: insights from *Aspergillus* species. *FEMS Microbiol Rev* 2013;**36**:165–92.
27. Horn BW, Moore GG, Carbone I. Sexual reproduction in *Aspergillus flavus*. *Mycologia* 2009;**101**:423–49.
28. Horn BW, Moore GG, Carbone I. The sexual state of *Aspergillus parasiticus*. *Mycologia* 2009;**101**:275–80.
29. Horn BW, Moore GG, Carbone I. Sexual reproduction in aflatoxin-producing *Aspergillus nomius*. *Mycologia* 2011;**103**:174–83.
30. Udagawa S, Uchiyama S, Kamiya S. *Petromyces muricatus*, a new species with an *Aspergillus* anamorph. *Mycotaxon* 1994;**52**:207–14.
31. Frisvad JC, Samson RA. *Neopetromyces* gen. nov. and an overview of teleomorphs of *Aspergillus* subg. *Circumdati*. *Stud Mycol* 2000;**45**:201–7.
32. Wicklow DT, Dowd PF, Tepaske MR, et al. Sclerotial metabolites of *Aspergillus flavus* toxic to a detritivorous maize insect (*Carpophilus Hemipterus, Nitidulidae*). *Trans Br Mycol Soc* 1988;**91**:433–8.
33. Wicklow DT, Shotwell OL. Intrafungal distribution of aflatoxins among conidia and sclerotia of Asper*gillus flavus* and *Aspergillus parasiticus*. *Can J Microbiol* 1983;**29**:1–5.
34. Frisvad JC, Petersen LM, Lyhne EK, et al. Formation of sclerotia and production of indoloterpenes by *Aspergillus niger* and other species in section *Nigri*. *PLoS One* 2014;**9**:e94857.
35. Pettersson OV, Leong SL, Lantz H, et al. Phylogeny of the extreme xerophile, *Xeromyces bisporus*. *Fungal Biol* 2011;**115**:1100–11.
36. Houbraken J, Samson RA, Frisvad JC. *Byssochlamys*: significance of heat resistance and mycotoxin production. *Adv Exp Med Biol* 2006;**571**:211–24.
37. Dijksterhuis J, Samson RA. *Food mycology. A multifaceted approach to fungi and food*. London: CRC Press; 2007. 403 pp.
38. Houbraken J, Dijksterhuis J, Samson RA. Diversity and biology of heat resistant fungi. In: Wong H-C, editor. *Stress response of foodborne microorganisms*. New York: Nova Science Publishers, Inc; 2011. p. 331–53.
39. Schmidt-Heydt M, Cramer B, Graf I, et al. Wavelength-dependent degradation of ochratoxin and citrinin by light in vitro and in vivo and its implications. *Toxins* 2012;**4**:1535–51.
40. Park SY, Lee NY, Kim SH, et al. Effect of ultraviolet radiation on the reduction of major food spoilage molds and sensory quality of the surface of dried filefish (*Stephanolepis cirrhifer*) fillets. *Food Res Int* August 2014;**62**:1108–12.
41. Levetin E, Shaughnessy R, Rogers CA, et al. Effectiveness of germicidal UV radiation for reducing fungal contamination within air-handling units. *Appl Environ Microbiol* 2001;**67**:3712–5.
42. Frisvad JC. Taxonomy, chemodiversity and chemoconsistency of *Aspergillus, Penicillium* and *Talaromyces* species. Front Microbiol, in press.
43. Bennett JW, Klich M. Mycotoxins. *Clin Microbiol Rev* 2003;**16**:497–516.

44. Keller NP, Turner G, Bennett JW. Fungal secondary metabolism from biochemistry to genomics. *Nat Rev Microbiol* 2005;**3**:937–47.
45. Frisvad JC, Thrane U, Samson RA. Mycotoxin producers. In: Dijksterhuis J, Samson R, editors. *A food mycology. A multifaceted approach to fungi and food*. Boca Raton, FL: CRC Press; 2007 403 pp.
46. Frisvad JC, Thrane U, Samson RA, et al. Important mycotoxins and the fungi with produce them. *Adv Exp Med Biol* 2006;**571**:3–31.
47. Rai M, Varma A. *Mycotoxins in food, feed and bioweapons*. Germany: Springer; 2010. 405 pp.
48. Varga J, Frisvad JC, Samson RA. A reappraisal of fungi producing aflatoxins. *World Mycotoxin J* 2009;**2**:263–77.
49. Weidenbörner M. *Encyclopedia of food mycotoxins*. Germany: Springer; 2001. 294 pp.
50. Wu F, Groopman JD, Pestka JJ. Public health impacts of foodborne mycotoxins. *Annu Rev Food Sci Technol* 2014;**5**:351–72.
51. Nielsen K, Frisvad JC. Mycotoxins on building materials. In: Adan O, Samson RA, editors. *Fundamentals of mold growth in indoor environments and strategies for healthy living*. Wageningen: Wageningen Academic Press; 2011. p. 245–75.
52. Sorenson WG, Frazer DG, Jarvis BB, et al. Trichothecene mycotoxins in aerosolized conidia of *Stachybotrys atra*. *Appl Environ Microbiol* 1987;**53**:1370–5.
53. Slack GJ, Puniani E, Frisvad JC. Secondary metabolites from *Eurotium* species, *Aspergillus calidoustus* and *A. insuetus* common in Canadian homes with a review of their chemistry and biological activities. *Mycol Res* 2009;**113**:480–90.
54. Desroches TC, McMullin DR, Miller JD. Extrolites of *Wallemia sebi*, a very common fungus in the built environment. *Indoor Air* 2014;**24**:533–42.

Chapter 2

Dispersion Forms

Carla Viegas[1], **João Brandão**[2]
[1]*Environment & Health RG, Lisbon School of Health Technology, Polytechnic Institute of Lisbon, Environmental Health Institute, Faculty of Medicine from Lisbon University, Lisbon, Portugal;* [2]*Reference Unit for Parasitic and Fungal Infections, Department of Infectious Diseases, National Institute of Health Doutor Ricardo Jorge, Lisboa, Portugal*

Fungi are very common in buildings affected by moisture problems. While bacteria need relative humidity (RH) levels higher than 90% and the constant presence of water, fungi thrive at RH levels between 75% and 85%.[1]

Fungi colonize a given material when their spores germinate and their hyphae and mycelia grow in or over the material. This growth usually ends with a massive spore production. Spore formation is important for survival, replication, and fungal dispersion and spores are often regarded as the most prevalent airborne fungal particles.[2]

Based on their small size and large number, fungal spores are classified as a bioaerosol.[3] Always present in the atmosphere, their concentration changes with environmental conditions. Production and spore release vary drastically from species to species, which influences dissemination in the air and on surfaces.[3] *Aspergillus* and *Penicillium* spores may remain indoors for long periods of time, while *Stachybotrys* spores soon diminish their concentration based on the interpretation of results from air samples analyzed by culture media.[4]

Spores can be actively or passively released; their further fate depends on the wind.[5–8] Owing to their size, fungal spores can cover enormous distances using air currents[9] and remain in the atmosphere for a long time.

The distance transported is unimportant for the development of fungi unless the viability of the spores plays a role[5,7]; even a dead spore can contain an allergen for humans.[10] Spore sizes range typically between 2 and 56 μm among species and may vary even within a species. However, airborne spores may have a smaller size than quoted in manuals, since these data usually refer to the size of hydrated spores, whereas spores rapidly desiccate when dispersed in air.[10]

The main exposure route of fungal spores is inhalation. The size, shape, and surface structure of the individual or aggregated spores are of major importance for their aerodynamic behavior and define where they are deposited in the airways. A substantial finer fraction may reach the alveoli, whereas the larger spore types (>7–10 μm) do not. Owing to their size, aggregates and bigger spores such

Environmental Mycology in Public Health. http://dx.doi.org/10.1016/B978-0-12-411471-5.00002-8

as those from *Alternaria* genera[11] may deposit to a greater extent in the lower and upper airways rather than in the alveoli.

The specific gravity may have some influence on the aerodynamic diameter (AED), since the density of spores varies from 0.4 to 1.5 g/cm^3. Air humidity also has an impact, as the AED is increased by 11–27% at 100% RH. No changes have been observed at RH ranging from 30% to 90%.[12] This means that inhaled spores may increase in size in the airways, which has some effect on where they are deposited, but it is not known how rapid fungal spores absorb water. Many spores are actually hydrophobic.[13]

Spores may be released in aggregated chains or clumps that have larger AEDs than single spores. However, the correlation between increase in AED and aggregate size is relatively small.[13] Spores from genera such as *Fusarium* and *Stachybotrys* are produced in slimy aggregates, which are dispersed outdoors by raindrops (splash dispersal). The sticky spores may adhere and infect nearby plants or fall to the ground. Spores from these species can also be found airborne, probably released from plant material by mechanical friction during harvest and further handling of grain and straw.[11]

Experimental studies have demonstrated that smaller fragments can also be released from fungal cultures by an air stream. This has been shown for several species, for example, *Aspergillus versicolor, Cladosporium cladosporoides, Penicillium melinii, Streptomyces albus, Trichoderma harzianum,* and *Ulocladium* sp., but not *Penicillium chrysogenum*.[14–18] Previously, Sorenson et al. (1987)[19] reported the liberation of hyphal fragments from cultures of *Stachybotrys chartarum*. The fungal origin of these fragments is supported by the detection of fungal antigens in the particle size fraction that contained only fragments[16] and by the staining of fragments with a DNA/RNA fluorochrome.[18] Environmental and occupational studies indicate that even large hyphal fragments may have been overlooked.[20,21]

Dispersion of viable spores and fungal structures is very dependent not only on their dimension,[22] but also on their biological features,[23,24] air temperature, oxygen availability, presence of nutrients, and texture[25] and also vibrations on the surface where they land.[26]

Fungi release their spores into air streams or as a reaction to unfavorable situations such as a sudden shift in RH or even to reach new sources of nutrients. Some spores possess a thick wall and are kept aggregated by a slimy substance, which makes them heavy and difficult to be transported by air. Such is the case of *Fusarium*, *Acremonium,* and *Exophiala*. On the other hand, fungi such as *Penicillium* and *Cladosporium* have dry-walled spores, easy to dissociate and very light, making them very easy to disperse.[27] In the specific case of the *Cladosporium* genera, the small spore's size also contributes to dispersion and makes it very common in air samples[28] and on moist surfaces.[29] Thus, it is possible to separate spores into two distinct groups: the wet spores of fungi such as *Stachybotrys*, *Acremonium,* and *Trichoderma* and the dry spores represented by fungi such as *Penicillium, Aspergillus,* and *Cladosporium*. Dry spores can exist in large amounts and are easily airborne and inhaled; wet spores, produced in slimy aggregates, are not. The type of spore does not depend on water consumption by the fungi—it is an intrinsic feature of the fungal species.[30]

Wet spores are frequently associated with visible colonies that exist on walls and surfaces. This feature also influences the methodology chosen for determining their presence. When it is necessary to assess fungal contamination in a case of health complaints it is justifiable to take only surface samples, as the most relevant pathogenic and toxigenic fungi (as *S. chartarum*) are more likely to be found on walls and other surfaces than in air samples.[30] In several studies, *Stachybotrys* spores were not identified through air collection methodologies but were isolated in surface samples from the same locations.[31]

Water requirements are also determinants for the preferential location of fungi. *Aspergillus fumigatus*, being more demanding in terms of water than *A. versicolor*, for instance, is more likely to be found on moist surfaces than in air samples[30] and is, therefore, a good indicator of indoor fungal growth[11].

In addition to *A. fumigatus, A. flavus* and *A. niger* have also been detected in the water systems of hospital settings, at both hot and cold water sources.[32] When *Aspergillus* spores are disseminated they stay in the air for quite a long time and they can contaminate all the surrounding surfaces.[33,34] The spores are very light, desiccation-resistant, and easily airborne.[35] Inhalation can occur directly or by colonization of the intermediate region of the nasophararinx.[34]

Compared to *Aspergillus* and *Penicillium*, *Alternaria* spores are produced in much smaller quantities and, as such, are not as frequently found in the air as they are on surfaces.[30] Several fungal species, especially those belonging to the *Cladosporium* and *Aspergillus* genera, can be disseminated through particles. While these genera are usually isolated in indoor environments and are, often, a result of indoor contamination, *Cladosporium* species originate mostly in outdoor air.[36] *Acremonium* is rarely found in air samples and is most frequently isolated from surface samples.[37]

Fungal spores present in outdoor air can also be found indoors. Indoor spore levels may, however, be increased by unintended fungal growth or by handling of moldy materials.[11]

Considering that the main source of fungal spores indoor may be the outdoor air, one must be aware that the content of fungal spores of every taxon in outdoor air is characterized mainly by a specific seasonal and daylight cycle.[38] Among other things, these cycles depend on climate and weather conditions,[39] the accessibility of fresh substrates for the development of the fungus, the circadian daylight and darkness cycle, and other hard-to-define environmental factors.[40] Airborne fungal spores occur throughout the year, but the seasonal cadence and their spectrum of composing species depend on the climate.[41] In a moderate climate the maximum concentration of most spores occurs in summer or early autumn.[42] In Europe, differences in the pattern of these phenomena are fairly inconspicuous.[43] In tropical and subtropical regions the greatest abundance of spores is noted in cold months (November–February) and the smallest in the warm ones (May–September).[44] Daily concentrations may differ significantly in the following seasons and in different habitats. The differences may concern seasonal sums of total spores, the concentrations, and the duration of seasons, as well as the time of maximum occurrence.[45] Weather conditions affect sporulation, dispersal, and deposition of spores and these elements correlate with one another. Conditions that prevailed a few days earlier frequently influence current concentrations.[41]

In some professional activities that cause dust formation, one must consider dust particles as a vehicle for fungi and their respective metabolites, as exemplified in the following.

When feeding occurs in swine production systems, abundant organic dust particles are dispersed into the air and kept suspended for a long time or deposited on the floor owing to gravitational sedimentation.[46] Consequently, fungi and their metabolites, such as Volatile organic compounds (VOCs) and mycotoxins, will probably display the same behavior.[47–49] In addition, dust deposition and humidity on the walls and ceilings of poorly maintained buildings from swine production systems may facilitate microbial growth and proliferation. If these sources become aerosolized, an increase in fungal contamination will take place.[50]

Some of the tasks developed in the poultry industry, such as bringing the animals to the slaughterhouse, present an increased occupational risk. This is a manual activity and the poultry naturally spread their wings to resist the movement, which, in turn, is a very effective way of generating litter dust and consequently increasing the exposure to fungi and mycotoxins.[51–53] Although mycotoxins are not volatile, they are associated with the presence of mold spores or particles when present in the air's breathable fraction.[54] In these settings, other tasks, such as spreading new litter, and the litter turnover present also increase occupational risks owing to dust aerosolization—fungi and mycotoxins included.[55,56]

Particles also play a crucial role in fungal/metabolite dispersion in the cork industry. Various stages of cork transformation influence the type of fungal contamination owing to changes in environmental parameters.[57–60] Other activities with higher exposure to particles also contribute to a higher exposure to fungi and their metabolites. In the waste industry, for instance, which is prone to particles,[61–64] a high simultaneous presence of various mycotoxins has been reported[65] and, more recently, high values of aflatoxin B1 were found in the serum of waste workers.[66] Previously, other research also provided important data concerning the presence of ochratoxin A in workers' blood.[67]

In short, there are many ways for fungi to disperse: by air, but also by other dispersion methods. This chapter outlined the most obvious and relevant contexts, but did not cover the singularities of specific occupational contexts. Those will be addressed in the chapter concerning occupational exposure.

REFERENCES

1. Singh J. Occupational exposure to moulds in buildings. *Indoor Built Environ* 2001;**10**(3–4):172–8.
2. Horner WE. Assessment of the indoor environment: evaluation of mold growth indoors. *Immunol Allergy Clin N Am* 2003;**23**:519–31.

3. Buttner MP, Willeke K, Grinshpun SA. In: Hurst CJ, editor. *Sampling and analysis of airborne microorganisms. Manual of environmental microbiology*. Washington, D.C: ASM Press; 1997. p. 629–40.
4. Flannigan B, Miller JD. Health implications of fungi in indoor environments: an overview. In: Samson RA, Flannigan B, Flannigan ME, editors. lit. *Health implications of fungi in indoor environments*. Amsterdam: Elsevier; 1994. p. 1–28.
5. Edmonds RL. *Aerobiology. The ecological systems approach*. Stroudsburg, PA: Dowden, Hutchinson & Ross, Inc; 1979.
6. Hirst JM. Aerobiology in plant pathology. *Grana* 1991;**30**:25–9.
7. Ingold CT. *Fungal spores: their liberation and dispersal*. Oxford: Clarendon Press; 1971.
8. McCartney A. Dispersal of spores and pollen from crops. *Grana* 1994;**33**:76–80.
9. Hjelmroos M. Relationship between airborne fungal spore presence and weather variables. *Grana* 1993;**32**:40–7.
10. Kasprzyk I. Aeromycology—main research fields of interest during the last 25 years. *Ann Agric Environ Med* 2008;**15**:1–7.
11. Eduard W, Halstensen A. Quantitative exposure assessment of organic dust. *SJWEH* 2009;**7**:30–5.
12. Reponen T, Willeke K, Ulevicius V, Reponen A, Grinshpun SA. Effect of relative humidity on the aerodynamic diameter and respiratory deposition of fungal spores. *Atmos Environ* 1996;**30**:3967–74.
13. Reponen T. Aerodynamic diameters and respiratory deposition estimates of viable fungal particles in mold problem dwellings. *Aerosol Sci Technol* 1995;**1**:11–23.
14. Kildesø J, Würtz H, Nielsen KF, Wilkins CK, Gravesen S, Nielsen PA, et al. The release of fungal spores from water damaged building materials. In: *Healthy buildings 2000: exposure human responses and building investigations. Proceedings, Indoor air information*, **vol. 1**. 2000. p. 313–8.
15. Kildesø J, Wurtz H, Nielsen KF, Kruse P, Wilkins K, Thrane U, et al. Determination of fungal spore release from wet building materials. *Indoor Air* 2003;**2**:148–55.
16. Górny RL, Reponen T, Willeke K. Fungal fragments as indoor air biocontaminants. *Appl Environ Microbiol* 2002;**68**:3522–31.
17. Górny RL, Mainelis G, Grinshpun SA, Willeke K, Dutkiewicz J, Reponen T. Release of *Streptomyces albus* propagules from contaminated surfaces. *Environ Res* 2003;**91**:45–53.
18. Madsen AM, Würtz H. Fungal enzymes in indoor dust. In: Eckardt J, editor. *Bioaerosols, fungi, bacteria, mycotoxins and human health*. New York: Fungal Research Group Foundation, Inc; 2005. p. 384–91.
19. Sorenson WG, Frazer DG, Jarvis BB, Simpson J, Robinson VA. Trichothecene mycotoxins in aerosolized conidia of Stachybotrys atra. *Appl Environ Microbiol* 1987;**53**:1370–5.
20. Green BJ, Tovey ER, Sercombe JK. Airborne fungal fragments and allergenicity. *Med Mycol* 2006;**1**:S245–55.
21. Halstensen AS, Nordby KC, Eduard W, Klemsdal SS. Real-time PCR detection of toxigenic Fusarium in airborne and settled dust and associations with trichothecene mycotoxins. *J Environ Monit* 2006;**8**:1235–41.
22. Aydogdu H, Asan A, Otkun MT. Monitoring of fungi and bacteria in the indoor air of primary schools in Edirne city, Turkey. *Indoor Built Environ* 2005;**14**(5):411–25.
23. Gomes JF. Contaminação do ar interior por bioaerossóis. *Rev Port Pneumol VIII* 2002;**6**:689–94.
24. Lugauskas A, Krikstaponis A. Microscopic fungi found in the libraries of Vilnius and factors affecting their development. *Indoor Built Environ* 2004;**13**(3):169–82.
25. Becker R. Fungal disfigurement of constructions analysis of the effects of various factors. *Air Qual Monogr* 1994;**2**:361–80.

26. Górny RL, Reponen T, Grinshpun SA. Source strength of fungal spore aerosolization from moldy building material. *Atmos Environ* 2001;**28**:4853–62.
27. Goyer N, Lavoie J, Lazure L. *Bioaerosols in the workplace: evaluation, control and prevention guide*. Institut de Recherche en Santé et en Sécurité du Travail du Québec; 2001.
28. Ruiz SS, de la Guardia Guerrero CD, Sánchez FA. Estudio aerobiológico de los conidios de Alternaria y Cladosporium en la atmósfera de la ciudad de Almería (SE de España). *Rev Iberoam Micol* 2004;**21**:121–7.
29. Grant C, Hunter CA, Flannigan B. Water activity requirements of moulds isolated from domestic dwellings. *Int Biodeterior Biodegrad* 1989;**25**:259–84.
30. Duchaine C, Mériaux A. The importance of combining air sampling and surface analysis when studying problematic houses for mold biodiversity determination. *Aerobiologia* 2001;**17**:121–5.
31. Cooley JD, Wong WC, Jumper CA. Correlation between the prevalence of certain fungi and sick building syndrome. *Occup Environ Med* 1998;**9**:579–84.
32. Borrego C, Carvalho A, Miranda AI. *Investigação em ambiente e saúde: desafios e estratégias*. Universidade de Aveiro; 2009.
33. Bex V, Mouilleseaux A, Causse R. A survey of Aspergillus contamination in a hospital during renovation. *Healthy Build* 2000;**1**:359–64.
34. Van Den Bergh MF, Verweij PE, Voss A. Epidemiology of nosocomial fungal infections: invasive aspergillosis and the environment. *Diagnostic Microbiol Infect Dis* 1999;**34**:221–7.
35. Rodrigues AG, Araújo R. Comparison of Andersen and honey jar methods for monitoring hospital wards. *Indoor Built Environ* 2007;**16**:71–8.
36. Sarica S, Asa A, Otkun MT. Monitoring indoor airborne fungi and bacteria in the different areas of Trakya University Hospital, Edirne, Turkey. *Indoor Built Environ* 2002;**11**:285–92.
37. Lugauskas A, Krikstaponis A, Seskausas V. Species of conditionally pathogenic micromycetes in the air of dwellings and occupational premises. *Indoor Built Environ* 2003;**12**:167–77.
38. Nayak B, Nanda A, Behera N. Airborne fungal spores in an industrial area: seasonal and diurnal periodicity. *Aerobiologia* 1998;**14**:59–67.
39. Śpiewak R, Skórska C, Góra A, Horoch A, Dutkiewicz J. Young farmers with cellular reactivity to airborne microbes suffer more frequently from work-related skin symptoms and allergic dermatitis. *Ann Agric Environ Med* 2001;**8**:255–9.
40. Gravesen S. Fungi as a cause of allergic disease. *Allergy* 1979;**34**:135–54.
41. Šegvić Klarić M, Pepeljnjak S. A year-round aeromycological study in Zagreb area, Croatia. *Ann Agric Environ Med* 2006;**13**:55–64.
42. Corden JM, Millington WM, Mullins J. Long-term trends and regional variation in the aeroallergen Alternaria in Cardiff and Derby UK—are differences in climate and cereal production having an effect? *Aerobiologia* 2003;**19**:191–9.
43. D'Amato G. Spieksma FThM. Aerobiologic and clinical aspects of mould allergy in Europe. *Allergy* 1995;**50**:870–7.
44. Lim SK, Chew FT, Mohd Dali SDB, Tan HTW, Lee BW, Tan TK. Outdoor airborne fungal spores in Singapore. *Grana* 1998;**37**:246–52.
45. Kasprzyk I, Worek M. Airborne fungal spores in urban and rural environments in Poland. *Aerobiologia* 2006;**22**:169–76.
46. Kim K, Ko H, Kim Y, Kim C. Assessment of Korean farmer's exposure level to dust in pig buildings. *Ann Agric Environ Med* 2008;**15**:51–8.
47. Millner P. Bioaerosols associated with animal production operations. *Bioresour Technol* 2009;**100**:5379–85.
48. Tsapko V, Chudnovets A, Sterenbogen M, Papach V, Dutkiewicz J, Skórska C, et al. Exposure to bioaerosols in the selected agricultural facilities of the Ukraine and Poland—A review. *Ann Agric Environ Med* 2011;**18**:19–27.

49. Viegas S, Veiga L, Verissimo C, Sabino R, Figueredo P, Almeida A, et al. Occupational exposure to aflatoxin B1 in swine production and possible contamination sources. *J Toxicol Environ Health Part A Curr Issues* 2013;**76**(15):944–51.
50. Duchaine C, Grimard Y. Cormier, Y. Influence of building maintenance, environmental factors, and seasons on airborne contaminants of swine confinement buildings. *Am Indust Hyg Assoc J* 2000;**61**:56–63.
51. Health and Safety Executive. *Statement of evidence: respiratory hazards of poultry dust.* Health and Safety Executive; 2008.
52. Oppliger A, Charrie N, Droz P, Rinsoz T. Exposure to bioaerosols in poultry houses at different stages of fattening; use of real-time PCR for airborne bacterial quantification. *Ann Occup Hyg* 2008;**52**:405–12.
53. Viegas C, Carolino E, Malta-Vacas J, Sabino R, Viegas S, Veríssimo C. Fungal contamination of poultry litter: a public health problem. *J Toxicol Environ Health Part A* 2012;**75**:1341–50.
54. Bush R, Portnoy J, Saxon A, Terr A, Wood R. The medical effects of mold exposure. *J Allergy Clin Immunol* 2006;**17**:326–33.
55. Whyte R. Occupational exposure of poultry stockmen in current barn systems for egg production in the United Kingdom. *Br Poult Sci* 2002;**43**:364–73.
56. Viegas S, Veiga L, Malta-Vacas J, Sabino R, Figueredo P, Almeida A, et al. Occupational exposure to aflatoxin (AFB1) in poultry production. *J Toxicol Environ Health Part A* 2012;**75**:1330–40.
57. Pestana M, Tinoco I. A Indústria e o Comércio da Cortiça em Portugal Durante o Século XX. *Silva Lus* 2009;**17**:1–26.
58. Francuz B, Year H, Geraut L, Bensefa-Colas L, Nghiem ZH, Choudat D. Occupational asthma induced by Chrysonilia sitophila in a worker exposed to coffee grounds. *Clin Vaccine Immunol* 2010;**17**:1645–6.
59. Danesh P, Caldas F, Marques J, San-Romão M. Mycobiota in Portuguese "normal" and "green" cork throughout the manufacturing process of stoppers. *J Appl Microbiol* 1997;**82**:689–94.
60. Guedes A. *Agentes biológicos no trabalho: perigos ocultos.* Lisboa: ACT; 2008. [cited 2012 Jul]. Available from http://www.act.gov.pt/SiteCollectionDocuments/BolsaTextosSHST/AgentesBiologicosTrabalho.pdf.
61. Tolvanen O, Nykänen J, Nivukoski U, Himanen M, Veijanen A, Hänninen K. Occupational hygiene in a Finnish drum composting plant. *Waste Manag* 2005;**25**:427–33.
62. Wouters IM, Spaan S, Douwes J, Doekes G, Heederik D. Overview of personal occupational exposure levels to inhalable dust, endotoxin, β(1→3)-glucan and fungal extracellular polysaccharides in the waste management chain. *Ann Occup Hyg* 2006;**50**:39–53.
63. Domingo JL, Nadal M. Domestic waste composting facilities. A review of human health risks. *Environ Int* 2009;**35**:382–9.
64. Viegas S, Almeida-Silva M, Viegas C. Occupational exposure to particles in 2 Portuguese waste-sorting units. *Int J Occup Med Environ Health* 2014:854–62.
65. Mayer S, Vishwanath V, Sulyok M. *Airborne workplace exposure to microbial metabolites in waste sorting plants.* Albany, NY: Fungal Research Group Foundation Inc; 2012.
66. Viegas S, Veiga L, Figueiredo P, Almeida A, Carolino E, Viegas C. Assessment of Worker's exposure to aflatoxin B_1 in a portuguese waste industry. *Ann Occup Hyg* 2014;**76**:944–51.
67. Degen GH, Blaskewicz M, Lektarau Y, Grüner C. Ochratoxin A Analysen im Blut von Arbeitnehmern in der Abfallwirtschaft. *Mycotoxin Res* 2003;**19**:3–7.

Section II

Outline of Fungal Pathologies

Chapter 3

Fungal Infections

Cristina Veríssimo
Reference Laboratory for Parasites and Fungal Infections, Department of Infectious Diseases, National Institute of Health, Lisbon, Portugal

A variety of data support the concept that certain opportunistic fungal infections are increasing in frequency over time. Several reasons have been proposed for the increase in invasive fungal infections, including the use of antineoplastic and immunosuppressive agents, broad-spectrum antibiotics, prosthetic devices, grafts and more aggressive surgery. Patients with burns, neutropenia, acquired human immunodeficiency syndrome (AIDS) and pancreatitis are also predisposed to fungal infection.[1–5]

Fungal infections can be classified in several ways; the most widely used classifies them according to the initial site of the infection.[6]

SUPERFICIAL FUNGAL INFECTIONS

These infections arise from a pathogen restricted to the stratum corneum, with little or no tissue reaction.[7] According to several authors, dermatophytoses are among the most common human infections.[8–10] The term is used to describe infections of the skin and nails caused by a restricted group of related filamentous and keratinophylic fungi, the dermatophytes, which are also known as the ringworm fungi.[6] The prevalence of these infections varies greatly but it is estimated that at least 10–15% of the world's population may be infected by dermatophytes.[8]

The term *tinea* can be used to define superficial infection caused by dermatophytic fungi. According to the infected body site, these may be the following: *tinea corporis* (torso), *tinea manum* (hands), *tinea pedis* (feet), *tinea cruris* (groin), *tinea capitis* (scalp), and *tinea unguinum* (nails). The etiologic agents of the dermatophytoses are classified into three anamorphic (asexual or imperfect) genera of the anamorphic class Hyphomycetes (formerly Deuteromycota, or *Fungi imperfecti*)—*Epidermophyton*, *Microsporum*, and *Trichophyton*.[11] This group is able to cause infections in humans and can be classified according to their natural reservoir as anthropophilic, zoophilic, and geophilic, as they are adapted mainly to humans, warm-blooded animals, or soils, respectively (Table 1).[11–13]

Environmental Mycology in Public Health. http://dx.doi.org/10.1016/B978-0-12-411471-5.00003-X

TABLE 1 Classification of Dermatophyte Fungi According to Their Natural Reservoir

Genus	Anthropophilic species (area of endemicity)	Zoophilic species (typical host)	Geophilic species
Epidermophyton spp	*E. floccosum* (worldwide)		*E. stockdaleae*
Trichophyton spp	*T. concentricum* (Southeast Asia, Melanesia, Amazon area, Central America, Mexico) *T. gourvilii* (Central Africa) *T. kanei* *T. megninii* (Portugal , Sardinia)	*T. equinum* (horse) *T. mentagrophytes* (rodents, rabbit, hedgehog) *T. sarkisorii* (Bactrian camel) *T. simii* (monkey, fowl) *T. verrucosum* (cattle, sheep, dromedary)	*T. ajelloi* *T. flavescens* *T. gloriae, T. longifusum* *T. phaseoliforme, T. terrestre* (complex) *T. vanbreuseghemii*
Microsporum spp	*M. audouinii* (Africa) *M. ferrugineum* (East Asia, East Europe)	*M. canis* (cat, dog) *M. equinum* (horse) *M. gallinae* (fowl) *M. persicolor* (fowl)	*M. amazonicum* *Microsporum* anamorph of *A. cookiellum* *M. boullardii* *M. cookei* *M. gypseum* (complex of three species) *M. nanum* *M. praecox* *M. racemosum* *M. ripariae* *M. vanbreuseghemii*

Adapted from Weitzman and Summerbell.[11]

Tinea capitis is the most common dermatophytic infection of the scalp, affecting mainly children. *Microsporum canis*, a zoophilic dermatophyte, is still the most common causative agent of *tinea capitis* reported in Europe.[14,15] *Trichophyton rubrum* and *Trichophyton mentagrophytes* are the most common dermatophytes species isolated from nails; these results are corroborated by several studies,[15–17] although molecular techniques have revealed that several *T. mentagrophytes* isolates were in fact *Trichophyton interdigitale*.

Other nondermatophyte fungi can cause superficial infections of skin and nails, because of their ability to degrade keratin,[6,9,18,19] namely fungi such as *Chrysosporium* spp., *Neoscytallidium dimidiatum*, *Scopulariopsis* spp., *Fusarium* spp., and *Aspergillus* spp., which are described by several authors as being agents of dermatomycosis.[6,9] The incidence of these infections is difficult to assess from published work but it is estimated that *Scopulariopsis brevicaulis* may account for about 3% of cases of onychomycosis.[6] Araújo et al.[9] report that 4.5% of the total cases of onychomycosis infections are caused by emerging fungi, namely *Fusarium* sp., *Neoscytallidium* sp., *Trichosporon beigelii*, and *Curvularia* sp.[9]

SUBCUTANEOUS INFECTIONS

The intact skin surface presents an effective barrier to many potentially pathogenic agents, but superficial trauma resulting from maceration or minor abrasion may provide an entry route for a range of pathogenic fungi, causing subcutaneous infections.[13]

Subcutaneous infections involve the dermis, subcutaneous tissues, and bone. These infections are heterogeneous but all are caused by penetrating trauma of the skin and are acquired as a result of the implantation of organisms that grow as saprobes in soil and decomposing vegetation.[1] The disease may remain localized or spread to adjacent tissues, dissemination may occur, but it is a rare condition in immune-competent hosts.

These mycoses are rarely observed in Europe, where most cases are observed in returning travelers, aid workers, archaeologists, and immigrants.[20]

Few studies report the occurrence of deep mycosis in Africa, where the number of subcutaneous infections may thus be underestimated and where this problem requires an increasing awareness.[21]

The main subcutaneous fungal infections include sporotrichosis, chromoblastomycosis, mycetoma, lobomycosis, rhinosporidiosis, subcutaneous zygomycosis, and subcutaneous phaeohyphomycosis.[22]

Sporotrichosis is a subcutaneous infection caused by a dimorphic fungus: the species complex of *Sporothrix schenckii*,[23] which commonly occurs in nature on dead plant material as a saprophyte. Although worldwide in distribution, it grows primarily in warm temperate and tropical climates.[24] The predominant clinical manifestation of the disease is lymphocutaneous sporotrichosis, which occurs chiefly in those whose occupation or vocation involves contact with vegetation and soil. This form of the disease is generally associated with immune-competent individuals.

Mycetoma is a syndrome involving cutaneous and subcutaneous tissues, fascia, and bone and is caused by soil-inhabiting bacteria (actinomycotic mycetoma) or fungi (eumycetoma). The disease is known as "Madura foot" or maduromycosis.[25,26]

Regions above the equator have the highest prevalence and incidence for these infections. The disease is also seen in Central America, India, and all across Africa, especially in Sudan.[21,26]

FIGURE 1 Lesion on left hand of liver transplant recipient caused by *Alternaria infectoria* (Susana Brás, Dermatology Service, Central Hospital, Lisbon, Portugal).

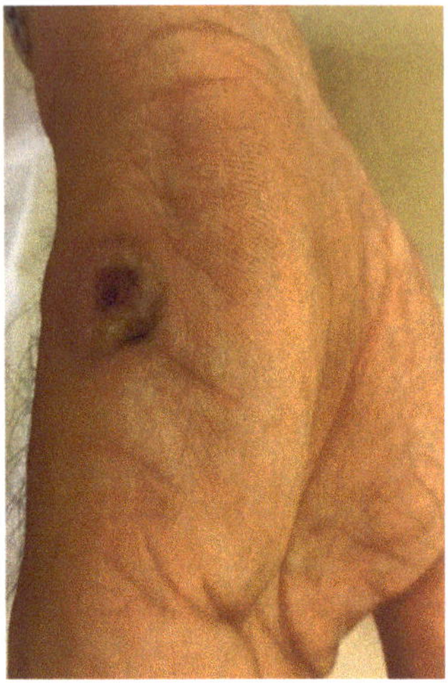

Approximately 50% of all cases of mycetoma are caused by true fungi, the most common of which are *Madurella mycetomatis* and *Trematosphaeria grisea.* Disease caused by *Madurella* is underestimated but socioeconomically important.[25] Less frequent etiologic agents of mycetoma include *Acremonium kiliense*, *Exophiala jeanselmei*, *Leptosphaeria senegalensis*, *Pseudallescheria boydii*, the teleomorph of *Scedosporium apiospermum*, and species included in more than 10 other fungal genera. The other 50% of infections called actinomycosis or actinomycotic mycetoma are caused by aerobic Actinomycetes, primarily *Actinomadura madurae* and *Nocardia brasiliensis*.

Phaeohyphomycosis is a subcutaneous infection generally found in tropical and subtropical zones, mainly in populations that do not routinely wear shoes. It is caused by darkly pigmented (dematiaceous) molds, which form thick-walled, dark-colored, multicelled structures called muriform cells or sclerotic bodies in tissue. Several dematiaceous molds have been associated with subcutaneous lesions. The most common etiologic agents are *Cladophialophora carrionii* and *Fonsecaea pedrosoii*. Less common pathogens include *Fonsecaea compactum*, *Phialophora verrucosa*, *Rhinocladiella aquaspersa*, *E. jeanselmei*, *Exophiala spinifera*, *Exophiala dermatitidis*,[22] and *Alternaria infectoria* (Figure 1).

INVASIVE FUNGAL INFECTIONS

Invasive fungal infections can be divided in two distinct groups: the endemic or dimorphic mycoses that are caused by true pathogenic fungi such as *Histoplasma*, *Coccidioides*, *Penicillium marneffei* and opportunistic infections caused by molds and yeast, that are saprophytes, which invade only immune-compromised hosts.[3,6,27] In approximately 90% of cases the lungs are the portal of entry, with nasal sinuses and the skin accounting for the remainder of cases.[4,5,27]

Over the past 30 years, there have been significant changes in the incidence and epidemiology of invasive fungal infections as a result of growing numbers of patients with a variety of risk factors (e.g., transplantation, chemotherapy, HIV infection, and use of corticosteroids or new immunosuppressive agents).[1–3,28–31]

Yeasts and molds are now among the 10 most frequently isolated pathogens among patients in intensive care units.[1,27]

Candida spp. and *Cryptococcus* spp. are the yeasts most frequently isolated in clinical practice.[3,30,33] Invasive *Candida* infections originate most frequently from endogenous reservoirs in patients with lowered host defense; however, exogenous infections can also occur and in hospitalized patients are frequently transmitted via the hands of health care workers.[27,32]

Aspergillus spp. are ubiquitous fungi with a worldwide distribution in nature. The most common species of *Aspergillus* causing human disease are *Aspergillus fumigatus*, *Aspergillus flavus*, *Aspergillus niger*, and *Aspergillus terreus*. Other species can cause disease, but *A. fumigatus* accounts for approximately 90% of all cases of life-threatening infections.[1,31] Invasive aspergillosis has a higher prevalence among bone marrow-transplanted patients and is frequently life threatening with high rates of mortality and morbidity associated.[2,31–33]

Non-*Aspergillus* molds such as Zygomycetes can generate a clinical picture indistinguishable from other fungi.[34–36] The molds of the class Mucormicotina (former Zygomycetes) have worldwide distribution and can be found in soil and decaying organic matter. Some species of this group are responsible for acute and relatively rapid, aggressive, fatal, angioinvasive diseases called zygomycosis or mucormycosis, especially in immune-compromised hosts.[37] The most common species causing angioinvasive zygomycosis is *Rhizopus arrhizus* (*Rhizopus oryzae*). Sinopulmonary involvement is predominant among those with infection; however, in diabetic individuals rhino-orbital and rhinocerebral zygomycosis is often described.[35]

Fusarium, a soil fungus, can enter the body through the respiratory system or via severe onychomycosis but has also been connected with venous access devices, its prevalence being clearly higher in patients who carry a central venous catheter[27,38] because of its ability to form biofilms.

Scedosporium, another emerging pathogen, has become recognized as a potent etiologic agent of severe infections in immune-compromised and occasionally also in immune-competent patients. Currently *Scedosporium* infections are among the most common deep mold infections.[39]

Most of the fungi have global distribution; however, some pathogenic fungi such as *Histoplasma capsulatum*, *Histoplasma duboisii*, *Blastomyces dermatitidis*, *Paracoccidioides brasiliensis*, *Coccidioides immitis*, *Coccidioides posadasi*, *P. marneffei*, and *S. schenckii* have a restricted geographical distribution.[6,23,24,40,41] All of them are dimorphic fungi, since they have the ability to change their morphology from mycelial fungi (saprophytic form) in nature to the yeast form (parasitic form) at 37 °C. Dimorphism is in fact considered a virulence factor, since it gives to fungi the ability to adapt to the human body environment. Table 2 summarizes the geographical distribution and natural habitat of these fungi.

TABLE 2 Geographical Distribution and Natural Habitat of Dimorphic Fungi

Dimorphic Fungi	Geographical Distribution	Natural Habitat	Mode of Infection
Penicillium marneffei	East and southeast Asia, including southern China, Hong Kong, and Thailand	Bamboo rats and soils	Inhalation of conidia
Histoplasma capsulatum	North, Central, and South America; east Asia; Australia; and the African continent	Soil enriched with bird or bat droppings	Inhalation of conidia
Histoplasma duboisii	Central Africa	Unknown (probably soil and plants)	Traumatic introduction with contaminated material
Sporothrix schenckii	North and South America, South Africa, and East Asia	Soil and plants and sphagnum moss	Inoculation by trauma
Blastomyces dermatitidis	South, central, and southeastern regions of North America but also Central and South America and parts of Africa	Wet soils with high organic content and acid pH	Inhalation of spores
Paracoccidioides brasiliensis	South and Central America	Soil, but little is known about its natural habitat	Inhalation of spores
Coccidioides immitis	Southwestern United States, Central and parts of central South America	Soil dust from hot arid regions	Inhalation of arthrospores

Inhalation of spores is the most common pathway for these infections. The organisms gain access to the respiratory tract and, depending on the individual immune system status, they can cause mild to severe lung infections or even widespread infections that can be fatal if not treated.[3,6]

REFERENCES

1. Binder U, Lass-Flörl C. Epidemiology of invasive fungal infections in the mediterranean area. *Mediterr J Hematol Infect Dis* 2011;**3**(1):e20110016. Open Journal System.
2. Richardson M, Lass-Florl C. Changing epidemiology of systemic fungal infections. *Clin Microbiol Infect* 2008;**14**(Suppl. 4):5–24.

3. Enoch DA, Ludlam HA, Brown NM. Invasive fungal infections: a review of epidemiology and management options. *J Med Microbiol* 2006;**55**:809–18.
4. Limper AH. The changing spectrum of fungal infections in pulmonary and critical care practice, clinical approach to diagnosis. *Proc Am Thorac Soc* 2010;**7**:163–8.
5. Zmeili OS, Soubani AO. Pulmonary aspergillosis: a clinical update. *Q J Med* 2007;**100**:317–34.
6. Richardson M, Warnock D. *Fungal infection diagnosis's and management*. 2nd ed. Blackwell Scientific Publications; 1994.
7. Schwartz RA. Superficial fungal infections. *Lancet* 2004;**364**(9440):1173–82.
8. Matsumoto T. In: Kibler CC, Mackenzie DER, Odds FC, editors. *Pathogenesis of fungal diseases, principles and practice of clinical mycology*. John and Wiley & Sons Ltd; 1996.
9. Araújo A, Bastos O, Souza MA, Oliveira O. Onychomycosis caused by emergent fungi: clinical analysis, diagnosis and revision. *An Bras Dermatol* 2003;**78**(4):445–55.
10. Havlickova B, Czaika VA, Friedrich M. Epidemiological trends in skin mycoses worldwide. *Mycoses* 2008;**51**(4):2–15.
11. Weitzman I, Summerbell RC. The dermatophytes. *Clin Microbiol Rev* 1995;**8**(2):240–59.
12. Badillet G. In: Varia, editor. *Dermatophyties et Dermatophytes—Atlas clinique et biologique*. 3eme ed.
13. MacKenzie DWR. In: Kibler CC, Mackenzie DER, Odds FC, editors. *Pathogenesis of fungal diseases, principles and practice of clinical mycology*. John and Wiley & Sons Ltd; 1996.
14. Ginter-Hanselmayer G, Weger W, Ilkit M, Smolle J. Epidemiology of tinea capitis in Europe: current state and changing patterns. *Mycoses* 2007;**50**:6–13.
15. Seebacher C, Bouchara JP, Mignon B. Updates on the epidemiology of dermatophyte infections. *Mycopathologia* 2008;**166**(5–6):335–52.
16. Ghannoum MA, Hajjeh RA, Scher R, Konnikov N, Gupta AK, Summerbell R, et al. A large-scale North American study of fungal isolates from nails: the frequency of onychomycosis, fungal distribution, and antifungal susceptibility patterns. *J Am Acad Dermatol* 2000;**43**:641–8.
17. Gupta AK, Jain C, Lynde CW, MacDonald P, Cooper EA, Summerbell RC. Prevalence and epidemiology of onychomycosis in patients visiting physicians' offices: a multicenter Canadian survey of 15,000 patients. *J Am Acad Dermatol* 2000;**43**(2, Part 1):245.
18. Torres-Rodríguez Josep M, López-Jodra Olga. Epidemiology of nail infection due to keratinophilic fungi. In: Kushwaha RKS, Guarro J, editors. *Biology of dermatophytes and other keratinophilic fungi*. Bilbao: Revista Iberoamericana de Micología; 2000.
19. Ali-Shtayeh Mohamed S, Rana MF Jamous. Keratinophilic fungi and related dermatophytes in polluted soil and water habitats. In: Kushwaha RKS, Guarro J, editors. *Biology of dermatophytes and other keratinophilic fungi*. Bilbao: Revista Iberoamericana de Micología; 2000.
20. Bonifaz A, Vázquez-González D, Perusquía-Ortiz AM. Subcutaneous mycoses: chromoblastomycosis, sporotrichosis and mycetoma. *J Dtsch Dermatol Ges* 2010;**8**(8):619–27.
21. Jacyk WK, Lawande RV, Tulpule SS. Deep mycoses in West Africa: a report of 13 cases and review of the Nigerian literature. *J Natl Med Assoc* 1981;**73**(3):251–6.
22. Queiroz-Telles F, McGinnis MR, Salkin I, Graybill JR. Subcutaneous mycoses. *Infect Dis Clin North Am* 2003;**17**:59–85.
23. Zancope-Oliveira RM, Almeida-Paes R, Evangelista de Oliveira MM, Freitas DF, Gutierrez Galhardo MC. New diagnostic applications in sporotrichosis. In: Khopkar U, editor. *Skin biopsy—perspectives*. InTech; 2011. http://dx.doi.org/10.5772/23590. Available from http://www.intechopen.com/books/skin-biopsy-perspectives/new-diagnostic-applications-in-sporotrichosis. ISBN: 978-953-307-290-6.
24. Barros MB, de Almeida Paes R, Schubach AO. *Sporothrix schenckii* and sporotrichosis. *Clin Microbiol Rev* 2011;**24**:633–54.

25. Ahmed AO, van Leeuwen W, Fahal A, van de Sande W, Verbrugh H, van Belkum A. Mycetoma caused by *Madurella mycetomatis*: a neglected infectious burden. *Lancet Infect Dis* 2004;**4**(9):566–74.
26. van Belkum A, Fahal A, van de Sande WW. Mycetoma caused by *Madurella mycetomatis*: a completely neglected medico-social dilemma. *Adv Exp Med Biol* 2013;**764**:179–89.
27. de Pauw Ben E. What are fungal infections? *Mediterr J Hematol Infect Dis* 2011;**3**. Open Journal System.
28. Mengoli C, Cruciani M, Barnes RA, Loeffler J, Donnelly JP. Use of PCR for diagnosis of invasive aspergillosis: systematic review and meta-analysis. *Lancet Infect Dis* 2009;**9**:89–96.
29. Cuenca-Estrella M, Bernal-Martinez L, Buitrago MJ, Castelli MV, Gomez-Lopez A, Zaragoza O, et al. Update on the epidemiology and diagnosis of invasive fungal infection. *Inter J Antimicrob Agents* 2008;**32**(Suppl. 2):143–7.
30. Warnock D. Trends in epidemiology of invasive fungal infection. *Jpn J Med Mycol* 2007;**48**: 1–12.
31. Barton Richard C. Laboratory diagnosis of invasive aspergillosis: from diagnosis to prediction of outcome. *Scientifica* 2013;**2013**:29. Article ID 459405. http://dx.doi.org/10.1155/2013/459405.
32. Sabino R, Sampaio P, Carneiro C, Rosado L, Pais C. Isolates from hospital environments are the most virulent of the *Candida parapsilosis* complex. *BMC Microbiol* 2011;**11**:180. http://dx.doi.org/10.1186/1471-2180-11-180.
33. Thornton CR. Detection of invasive aspergillosis. *Adv Appl Microbiol* 2010;**70**:187–216.
34. De Pauw B, Walsh TJ, Donnelly JP, et al. Revised definitions of invasive fungal disease from the European Organization for Research and Treatment of Cancer/Invasive Fungal Infections Cooperative Group and the National Institute of Allergy and Infectious Diseases Mycoses Study Group (EORTC/MSG) Consensus Group. *Clin Infect Dis* 2008;**46**:1813–21.
35. Pagano L, Valentini C, Caira M, Fianchi L. Zygomycosis: current approaches to the management of patients with haematological malignancies. *Brit J Haematol* 2009;**146**(6):597–606.
36. Chayakulkeeree M, Ghannoum MA, Perfect JR. Zygomycosis: the re-emerging fungal infection. *Eur J Microbiol Infect Dis* 2006;**25**:215–29.
37. Kontoyiannis DP, Lionakis MS, Lewis RE, Chamilos G, Healy M, Perego C, et al. Zygomycosis in a tertiary-care cancer center in era of *Aspergillus*-active antifungal therapy: case-control observational study of 27 recent cases. *J Infect Dis* 2005;**191**(8):1350–60.
38. Nucci M, Anaissie E. Cutaneous infection by *Fusarium* species in healthy and immunocompromised hosts: implications for diagnosis and management. *Clin Infect Dis* 2002;**35**:909–20.
39. Guarro J, Kantarcioglu AS, Horré R, et al. *Scedosporium apiospermum*: changing clinical spectrum of a therapy-refractory opportunist. *Med Mycol* 2006;**44**(4):295–327.
40. Brown J, Benedict K, Park BJ, Thompson GR. Coccidioidomycosis: epidemiology. *Clin Epidemiol* 2013;**5**:185–97.
41. Bagagli E, Sano A, Coelho KI, Alquati S, Miyaji M, de Camargo ZP, et al. Isolation of *Paracoccidioides brasiliensis* from armadillos (*Dasypus noveminctus*) captured in an endemic area of paracoccidioidomycosis. *Am J Trop Med Hyg* 1998;**58**(4):505–12.

Chapter 4

Allergic Response to Fungal Exposure

Inês Andrade Mota[1], Luís Miguel Borrego[1,2]
[1]Allergy Center, CUF Descobertas Hospital, Lisboa, Portugal; [2]Centro de Estudos de Doenças Crónicas, CEDOC, NOVA Medical School/Faculdade de Ciências Médicas, Universidade Nova de Lisboa, Portugal

ALLERGIC DISEASE AND FUNGAL SENSITIZATION

Allergic diseases constitute a leading cause of chronic disease in both children and adults. According to the World Health Organization, hundreds of millions of people (10–30% of population) suffer from rhinitis, and 300 million from asthma, with a negative impact on quality of life and growing socioeconomic costs. Epidemiological research has shown a dramatic increase in allergic disease prevalence, about 30–40%, especially in children.[1] The International Study of Asthma and Allergies in Childhood (ISAAC) Phase III, a 7-year cross-sectional study, reports sharp increases in their prevalence, especially in the 6- to 7-year-old age group compared to the 13- to 14-year-old age group.[2]

Allergic diseases are associated with a predisposition to producing IgE-mediated responses to widespread and harmless environmental substances (allergens), resulting from a complex gene–environmental interaction. These disorders include allergic rhinitis (AR), asthma, atopic dermatitis (AD), and food allergies, which frequently present together. Clinical manifestations can be diverse, in both location and severity, ranging from uncomfortable symptoms to systemic reactions. Mites, animal dander, pollens, foods, latex, drugs, and insect venoms are considered the major allergens.[3]

The current global environmental changes will be affecting pollen counts, stinging insects, and molds related to allergic diseases.[1]

The prevalence of sensitization to fungi among the general population is fairly unknown. In an epidemiological survey on a cohort of 4962 respiratory subjects, the prevalence of fungal sensitization was estimated to be about 19.1% by means of skin-prick test, with positive tests to at least one of the seven selected fungal species (*Alternaria*, *Aspergillus*, *Candida*, *Cladosporium*, *Penicillium*, *Saccharomyces*, and *Trichophyton*).[4]

Environmental Mycology in Public Health. http://dx.doi.org/10.1016/B978-0-12-411471-5.00004-1

Spores released by fungi can be found in both indoor and outdoor environments. Exposure can occur through contact with saprophytic species or by eating mushrooms (even those not poisonous).[4]

Unlike other allergenic sources (pollens, mites, animal dander), fungi can cause direct infections, either localized or systemic, and toxic-irritant effects due to secondary metabolites, beyond their role as allergens in severe respiratory diseases and hypersensitivity disorders.[4,5]

IMMUNE RESPONSE AND HYPERSENSITIVITY TO FUNGAL EXPOSURE

Fungi are responsible for a broad spectrum of clinical manifestations arising from hypersensitivity reactions in addition to IgE-mediated response. Although the old Gell and Coomb's classification divides the reactions into four types, more than one mechanism can be involved in the pathogenesis of hypersensitivity reactions, particularly in fungal diseases. Type I, or immediate, reactions (IgE-mediated) are implicated in AR, asthma, allergic bronchopulmonary aspergillosis (ABPA), and allergic bronchopulmonary mycosis (ABPM). Type II, or antibody-mediated, cytotoxic reactions are related to autoimmune disorders. Type III, or immune complex-mediated, reactions are responsible for hypersensitivity pneumonitis (HP) and aspergilloma development. Type IV, or delayed hypersensitivity, T-cell-mediated, reactions occur in contact dermatitis, ABPA, and HP. As a result, fungal colonization either in the sinuses (allergic fungal sinusitis) or in the bronchi (ABPA), as well as HP, may induce both T and B cell responses.[6]

Among the over 100,000 fungal species recognized, about 80 genera have been shown to induce Type I reactions in atopic individuals. The most relevant fungi, as allergenic sources, belong to the Ascomycota, mainly the genera *Alternaria*, *Aspergillus*, and *Cladosporium*. The other genera, belonging to the *Basidiomycota*, seem to be less important, with the exception of *Malassezia*, which plays an important role in patients with AD.[6]

ALLERGIC RHINITIS

AR is an extremely prevalent condition associated with exposure to inhaled allergens, causing a combination of sneezing, rhinorrhea, nasal congestion, and itching.

There is no conclusive evidence of a direct relationship between fungal exposure and clinical manifestations of AR. Nevertheless, symptoms seem to be correlated with IgE-mediated hypersensitivity, by positive skin tests or serum levels, to the common outdoor fungi (*Alternaria* and *Cladosporium*). Children with rhinoconjunctivitis sensitized to *Alternaria* have mainly nocturnal nasal symptoms, mostly in the summer, when environmental concentrations of *Alternaria* are higher.[7] In a French children cohort, integrated in

ISAAC Phase II, sensitization to *Alternaria* was associated with AR independent of asthma.[8]

Indoor spore levels of *Cladosporium* and *Aspergillus* were associated with increased risk of allergic sensitization, and sensitized children who are exposed to high levels of these fungi are prone to having symptoms of rhinoconjunctivitis.[9]

ALLERGIC FUNGAL RHINOSINUSITIS

Allergic fungal rhinosinusitis (AFRS) is a localized hypersensitivity condition resulting from fungal growth in an area of abnormal tissue drainage, mostly caused by *Aspergillus* species. The characteristics of AFRS are equivalent to those of ABPA. The mucus of these patients can contain fungal hyphae, and even though commonly associated with *Aspergillus fumigatus*, other fungi can be implicated, such as *Curvularia*.[5]

AFRS has specific diagnostic criteria that distinguish it from typical chronic sinusitis and include mucus with eosinophils (demonstrating noninvasive fungi-allergic mucin), skin hypersensitivity to specific fungal allergens and increased total serum IgE level, nasal polyposis, and characteristic radiographic findings.[5] Computed tomography can show complete opacification of at least one paranasal sinus, a typical heterogeneity of the signal within involved sinuses, and the expansion and attenuation/erosion of the bone with displacement of adjacent anatomic compartments.[10]

Treatment includes systemic corticosteroids and surgery to remove allergic mucin, but recurrences are common, unless therapy to control inflammation is maintained.

Chronic rhinosinusitis can also be attributed to fungal sensitivity, particularly to *Alternaria* species, although these organisms might be found in healthy individuals.

ASTHMA

Asthma is an inflammatory airways disease that may be worsened by numerous factors. Among the allergic causes or triggers, fungi have a predominant role. Sensitization to fungal allergens has been recognized as a risk factor for its development, persistence, and severity, particularly that to *Alternaria alternata*.[11] Exposure to fungal spores (mainly in the outdoor environment) is associated with asthma exacerbation in children.[12] Home dampness has the potential to increase the proliferation of fungi. Several studies have demonstrated a consistent association between indoor fungal exposure and asthma symptoms, suggesting that children living in damp houses or those with visible fungal growth have an increased risk of having lower respiratory tract symptoms (particularly cough and wheezing) and airway hyperresponsiveness.[13–15] Prospective studies have shown that infants at high risk for developing asthma who are exposed to high levels of *Penicillium* are at significant risk for wheeze and persistent cough.[16]

Studies have suggested that outdoor fungal exposure is as important as indoor exposure, if not more so, in increasing asthma symptoms and risk of exacerbations.[17] Children with higher outdoor exposure to spores in the first 3 months of life are at increased risk of early wheezing.[18] Despite this evidence, the dampness promotes development of other organisms such as bacteria and dust mites along with fungi, so its impact on respiratory symptoms is not completely unequivocal. Studies have been performed to evaluate the levels of indoor and outdoor fungi during the months that followed Hurricanes Katrina and Rita. Homes more damaged by flooding showed higher fungal growth compared to those less affected. Nevertheless, no increase in adverse health outcomes has been observed.[19] The HEAL Study characterized post-Katrina exposures to mold and allergens in children with asthma and identified the presence of *Alternaria* antigens in 98% of homes, with elevated concentrations in half of them. It was previously known that *Alternaria* at high exposure levels was associated with the onset of childhood asthma, but to assess if *Alternaria* in these situations is a significant asthma trigger will require further investigation.[20]

ALLERGIC BRONCHOPULMONARY ASPERGILLOSIS

ABPA is an inflammatory disease that arises in patients with asthma or cystic fibrosis. It is caused by an immunologic reaction to *Aspergillus* antigens, mainly *A. fumigatus*, a ubiquitous indoor and outdoor fungus, which colonizes the tracheobronchial tree of asthmatic patients as a result of poor clearance of airway secretions.

This disorder is characterized by frequent asthma exacerbations, recurrent transient chest radiographic infiltrates, peripheral blood eosinophilia, and development of bronchiectasis. The presence of pulmonary infiltrates along with eosinophilia distinguishes this condition from the underlying asthma or even cystic fibrosis, but the differential diagnosis includes other symptoms known as PIE (pulmonary infiltrates with eosinophilia) syndrome.

Patients may experience chronic wheezing, chest discomfort, cough with sputum production of brownish mucous plugs (containing fungal hyphae), and systemic symptoms such as fever and malaise.

ABPA may affect around 13% of asthmatic patients.[21] The bronchial colonization by *A. fumigatus* can induce a mix of hypersensitivity reactions (Types I, III, and IV) that are responsible for causing symptoms from asthmatic exacerbations to fatal pulmonary damage. The pulmonary impairment can lead to irreversible end-stage fibrosis.

Although several clinical diagnostic criteria have been established, serological findings confirming sensitization to *A. fumigatus* are essential to confirm or exclude the disease. Screening of sensitization to *A. fumigatus* should be routinely performed by skin-prick test and/or serum-specific IgE. The most sensitive diagnostic test is *A. fumigatus*-specific IgE level, so all patients with poor controlled asthma should be screened with this test to allow early diagnosis.[19] Furthermore, recombinant allergens may contribute to a more accurate diagnosis.[6]

The diagnosis of ABPA does not require positive cultures for *A. fumigatus*, because the presence of serological and radiological findings is enough. However, multiple fungi can colonize the airways of patient with ABPA and frequently there is a dissociation between colonizing and sensitizing species. This should be taken into account in treatment, particularly regarding resistance to fungicides.[22]

The management of ABPA depends on the control of asthma, to prevent exacerbation and particularly the occurrence or enlargement of bronchiectasis. Corticosteroids are the treatment of choice, being frequently required prolonged courses of oral therapy. Antifungals may be used in selected cases. The use of fungicides, namely azoles, such as itraconazole, is reserved for patients with recurrent exacerbations and glucocorticoid-dependent disease.[19]

Allergen immunotherapy with *A. fumigatus* is normally avoided because of potential development of immune complex disease, but this has not been systematically studied. Omalizumab may eventually be a therapeutic option in the near future.[23]

ABPM resembles ABPA, because this syndrome also includes severe asthma, total and pulmonary eosinophilia, increased levels of total and specific IgE, and bronchiectasis. Airway fungal colonization may be induced by other fungi (*Candida*, *Penicillium*, and *Curvularia*).

A new phenotype of asthma has been described as "severe asthma with fungal sensitization" (SAFS), which is considered a diagnosis of exclusion.[24] This condition seems to respond to an oral antifungal (itraconazole), as a clinical trial demonstrated improvement in quality of life in about 60% of patients.[25] Further research will be needed to assess the value of these options.

ATOPIC DERMATITIS

AD is a chronic inflammatory and extremely pruritic skin disorder with a complex pathophysiology. *Malassezia* yeasts stimulate keratinocytes to produce several cytokines, in a species-dependent manner. The role of these organisms in AD is still under intense investigation. Current data have shown that at least *Malassezia sympodialis* has the ability to induce AD in susceptible individuals. Along the evolution of the disease, sensitization to *Malassezia* allergens can lead to the production of *Malassezia*-specific IgE. This sensitization is specific for the skin manifestations and does not occur in atopic patients with mainly respiratory symptoms. IgE-binding allergens have been identified for *M. sympodialis*, *Malassezia furfur*, and, lately, *Malassezia globosa*.[26] In a cohort of patients with AD patients who reacted positively to an intradermal skin test using *M. furfur* extract, 83.3% of them had elevated serum levels of IgE to purified Mal f 4, a major *M. furfur* allergen.[27] *Malassezia sympodialis*, which colonizes the skin of both AD and healthy individuals, has been associated with IgE-mediated sensitization exclusively in patients with AD. The disrupted skin barrier facilitating allergen uptake may be the reason for this specific sensitization.[6] Cultures of skin samples to assess the distribution of *Malassezia* species on the skin of patients with AD or psoriasis (PS) and healthy volunteers showed that AD patients yielded exclusively *M. sympodialis* isolates, whereas *M. furfur* isolates were observed only in PS patients. The isolation of *M. sympodialis* was more frequent among AD patients and healthy volunteers than among PS patients.[28]

OTHER DISEASES

There are other diseases related to occupational fungal exposure. This group includes HP, farmer's lung, bagassosis, and mushroom worker's lung, which result from exposure to thermophilic *Actinomycetes* found in hay, bagasse, and mushroom compost, respectively.[6]

HP, also referred as extrinsic allergic alveolitis, can present as three far overlapping forms—acute, subacute, and chronic. It is an uncommon but relevant allergic disease that can be related to high-level fungal exposure, usually in an occupational setting, although it can be associated with other inhaled sources (organic dust of bacterial, vegetable, or avian origin). This disease seems to be a cell-mediated (delayed) hypersensitivity. Allergen-specific precipitins are related to exposure and might contribute to the acute noninfectious episodes of interstitial pneumonitis (alveolitis), which may produce restrictive irreversible lung disease. Exposure to domestic specific indoor fungal spores is an extremely unlikely cause of HP that occurs only in rare situations of workplace exposure or resulting from some contamination at home.

DIAGNOSIS OF FUNGAL-RELATED ALLERGIC DISEASES

The diagnosis of fungal sensitization can be made with skin-prick tests and/or by evaluating specific IgE levels, including *A. alternata*, *A. fumigatus*, *Cladosporium* species, and *Penicillium chrysogenum*.

Fungal sensitization is significant and clinically relevant, but the lack of standardized fungal allergenic extracts compromises the diagnosis. There are many problems with the standardization of fungal extracts, including commercial and technical issues.

Several tests have been developed to detect circulating antibodies against allergens in sensitized patients. In vitro methods, such as enzyme-linked immunosorbent assay or ImmunoCAP®, provide evidence for specific IgE. Discrepancies between results of skin test and allergen-specific IgE determinations have been reported.[29] The reliability of in vitro tests is uncertain, because it is not possible to get the extracts used to produce the in vitro test system as a skin test solution, and comparison of skin reactivity with specific IgE levels is of paramount importance to exclude cross-reactivity.[6] Significant improvements should be made in in vivo and in vitro diagnostic tools.

TREATMENT

Avoidance is the first measure to put into practice. Nevertheless, fungal exposure mainly occurs in the outdoor environment, so complete avoidance is quite impossible. Otherwise, indoor exposure can be better controlled by improving indoor air quality and cleaning visible fungal growth with chlorine-containing bleach solutions and detergent.

Although there is little evidence of immunotherapy for fungal allergens and, overall, it is not recommended, trials with *A. alternata* and *Cladosporium herbarum* extracts have suggested some clinical benefit.[6] Subcutaneous immunotherapy has been used, but additional studies are required. In a 3-year prospective clinical trial, double-blind and placebo-controlled, with children and adolescents who were allergic to only *A. alternata*, immunotherapy using standardized *A. alternata* extract has reduced symptoms of and medication for asthma and rhinoconjunctivitis without serious side effects. The active therapy was generally well tolerated, the most common side effect being local edema at the site of injection in four patients after 11 injections (1.1% of all injections).[30]

A 3-year randomized controlled study using sublingual immunotherapy in patients with *A. alternata* respiratory allergies (AR with or without mild-to-moderate asthma) showed it to be efficacious and well tolerated (17% reported side effects, in general mild and transient).[31]

In a Spanish prospective study, 36 patients with respiratory allergy (rhinitis alone or associated with asthma), monosensitized to fungi, received specific immunotherapy, which was administered mostly by the subcutaneous route (80%) and in the remaining cases by the sublingual route. The profile of sensitization included *Alternaria*, *Aspergillus*, or *Penicillium* and combinations of two or more, mainly *Alternaria* and *Cladosporium*. Treatment was well tolerated, without severe adverse effects either during the induction phase or during the first 12 months.[32]

Considering that most patients sensitized to fungi have severe asthma, either inhaled or oral corticosteroids are commonly used to control symptoms. Exacerbation of ABPA, ABPM, or SAFS is also treated with corticosteroids, orally for a few weeks, added to inhaled corticosteroids. The role of inhaled corticosteroids is particularly relevant in reducing exacerbation and controlling asthma. Antifungal agents can be used to eradicate fungal colonization, although their effects should be further investigated.[6]

REFERENCES

1. Pawankar R, Canonica GW, Holgate ST, Lockey RF. *White Book on Allergy 2011—2012 Executive Summary.* WAO—World Allergy Organization; 2011.
2. Asher MI, Montefort S, Björkstén B, Lai CK, Strachan DP, Weiland SK, et al. Worldwide time trends in the prevalence of symptoms of asthma, allergic rhinoconjunctivitis, and eczema in childhood: ISAAC phase one and three repeat multicountry cross-sectional surveys. *Lancet* 2006;**368**:733–43.
3. Galli SJ, Tsai M, Piliponsky AM. The development of allergic inflammation. *Nature* 2008;**9**:445–54. http://dx.doi.org/10.1038/nature07204.

4. Mari A, Schneider P, Wally V, Breitenbach M, Simon-Nobbe B. Sensitization to fungi: epidemiology, comparative skin tests, and IgE reactivity of fungal extracts. *Clin Exp Allergy* 2003;**33**:1429–38.
5. Bush RK, Portnoy JM, Saxon A, Terr AI, Wood RA. The medical effects of mold exposure. *J Allergy Clin Immunol* 2006;**117**:326.
6. Crameri R, Garbani M, Rhyner C, Huitema C. Fungi: the neglected allergenic sources. *Allergy* 2013. http://dx.doi.org/10.1111/all.12325.
7. Andersson M, Downs S, Mitakakis T, Leuppi J, Marks G. Natural exposure to *Alternaria* spores induces allergic rhinitis symptoms in sensitized children. *Pediatr Allergy Immunol* 2003;**14**(2):100.
8. Randriamanantany ZA, Annesi-Maesano I, Moreau D, Raherison C, Charpin D, Kopferschmitt C, et al. *Alternaria* sensitization and allergic rhinitis with or without asthma in the French Six Cities study. *Allergy* 2010;**65**(3):368.
9. Jacob B, Ritz B, Gehring U, Koch A, Bischof W, Wichmann HE, et al. Indoor exposure to molds and allergic sensitization. *Environ Health Perspect* 2002;**110**(7):647.
10. Dutre T, Al Dousary S, Zhang N, Bachert C. Allergic fungal rhinosinusitis-more than a fungal disease? *J Allergy Clin Immunol* August 2013;**132**(2):487–9.
11. Bush RK, Prochnau JJ. *Alternaria*-induced asthma. *J Allergy Clin Immunol* 2004;**113**:227–34.
12. Delfino RJ, Caote BD, Zeiger RS, Seltzer JM, Street DH, Koutrakis P. Daily asthma severity in relation to personal ozone exposure and outdoor fungal spores. *Am J Respir Crit Care Med* 1996;**154**:633–41.
13. Peat J, Dickerson J, Li J. Effects of damp and mould in the home on respiratory health: a review of the literature. *Allergy* 1998;**53**:120–8.
14. Hagmolen of Ten Have W, van den Berg NJ, van der Palen J, van Aalderen WM, Bindels PJ. Residential exposure to mould and dampness is associated with adverse respiratory health. *Clin Exp Allergy* 2007;**37**(12):1827.
15. Antova T, Pattenden S, Brunekreef B, Heinrich J, Rudnai P, Forastiere F, et al. Exposure to indoor mould and children's respiratory health in the PATY study. *J Epidemiol Community Health* August 2008;**62**(8):708–14.
16. Gent JF, Ren P, Bealanger K, Triche E, Bracken MB, Holford TR, et al. Levels of household mold associated with respiratory symptoms in the first year of life in a cohort at risk for asthma. *Environ Health Perspect* 2002;**110**:A781–6.
17. Pongracic JA, O'Connor GT, Muilenberg ML, Vaughn B, Gold DR, Kattan M, et al. Differential effects of outdoor versus indoor fungal spores on asthma morbidity in inner-city children. *J Allergy Clin Immunol* 2010;**125**(3):593.
18. Harley KG, Macher JM, Lipsett M, Duramad P, Holland NT, Prager SS, et al. Fungi and pollen exposure in the first months of life and risk of early childhood wheezing. *Thorax* 2009;**64**(4):353.
19. Barbeau DN, Grimsley LF, White LE, El-Dahr JM, Lichtveld M. Mold exposure and health effects following hurricanes Katrina and Rita. *Annu Rev Public Health* 2010;**31**:165–78.
20. Grimsley LF, Chulada PC, Kennedy S, White L, Wildfire J, Cohn RD, et al. Indoor environmental exposures for children with asthma enrolled in the HEAL study, post-Katrina New Orleans. *Environ Health Perspect* November 2012;**120**(11):1600–6.
21. Agarwal R, Chakrabarti A. Allergic bronchopulmonary aspergillosis in asthma: epidemiological, clinical and therapeutic issues. *Future Microbiol* 2013;**8**(11):1463–74.
22. Matsuse H, Tsuchida T, Fukahori S, Kawano T, Nishino T, Fukushima C, et al. Dissociation between sensitizing and colonizing fungi in patients with allergic bronchopulmonary aspergillosis. *Ann Allergy Asthma Immunol* September 2013;**111**(3):190–3.

23. Bush RK. Fungal allergy. In: Pawankar Ruby, Holgate Stephen T, Rosenwasser Lanny J, editors. *Allergy frontiers: clinical manifestations*. Springer; 2009. p. 472–81. [Google books. Fri. December 27, 2013. 2010]. http://books.google.pt/.
24. Agarwal R. Severe asthma with fungal sensitization. *Curr Allergy Asthma Rep* 2011;**11**:403–13.
25. Denning DW, O'Driscoll BR, Powell G, Chew F, Atherton GT, Vyas A, et al. Randomized controlled trial of oral antifungal treatment for severe asthma with fungal sensitization: the Fungal Asthma Sensitization Trial (FAST) study. *Am J Respir Crit Care Med* 2009;**179**(1):11.
26. Gaitanis G, Magiatis P, Hantschke M, Bassukas ID, Velegraki A. The *Malassezia* genus in skin and systemic diseases. *Clin Microbiol Rev* January 2012;**25**(1):106–41.
27. Onishi Y, Kuroda M, Yasueda H, Saito A, Sono-Koyama E, Tunasawa S, et al. Two-dimensional electrophoresis of *Malassezia* allergens for atopic dermatitis and isolation of Mal f 4 homologs with mitochondrial malate dehydrogenase. *Eur J Biochem* April 1999;**261**(1):148–54.
28. Jagielski T, Rup E, Ziółkowska A, Roeske K, Macura AB, Bielecki J. Distribution of *Malassezia* species on the skin of patients with atopic dermatitis, psoriasis, and healthy volunteers assessed by conventional and molecular identification methods. *BMC Dermatol* March 7, 2014;**14**(1):3.
29. Crameri R. Fungal allergy and pathogenicity. In: Breitenbach M, Crameri R, Lehrer SB, editors. *Chemical immunology and allergy*, vol. 81. 2002. [Google books].
30. Kuna P, Kaczmarek J, Kupczyk M. Efficacy and safety of immunotherapy for allergies to *Alternaria alternata* in children. *J Allergy Clin Immunol* 2011;**127**:502–8. e1–6.
31. Pozzan M, Milani M. Efficacy of sublingual specific immunotherapy in patients with respiratory allergy to *Alternaria alternata*: a randomised, assessor-blinded, patient-reported outcome, controlled 3-year trial. *Curr Med Res Opin* December 2010;**26**(12):2801–6.
32. Torres-Rodríguez JM, Pulido-Marrero Z, Vera-García Y. Respiratory allergy to fungi in Barcelona, Spain: clinical aspects, diagnosis and specific treatment in a general allergy unit. *Allergol Immunopathol (Madr)* September–October 2012;**40**(5):295–300.

Chapter 5

Mycotoxicoses

Maja Peraica
Unit of Toxicology, Institute for Medical Research and Occupational Health, Zagreb, Croatia

Exposure to mycotoxins may cause toxic syndromes called mycotoxicoses, which may affect various systems according to the target organs of the mycotoxin. Mycotoxicoses are much more frequent in tropical than in moderate regions because high levels of humidity and temperature favor the growth of molds and mycotoxin production. This effect is enhanced by the lack of an appropriate control of mycotoxins throughout the whole chain of food production, but their toxicity is more pronounced in malnourished persons and children. Mycotoxicosis occurring in a single individual is usually not recognized because its diagnosis requires sophisticated equipment, and medical doctors are not educated to recognize them. Mycotoxins as causative agents of disease should be taken into consideration when the disease appears in several persons and is not connected with some other causative agent, such as microorganisms.

Mycotoxicoses may be acute and chronic. Acute mycotoxicoses are caused by exposure to high amounts of mycotoxins that have acute toxicity properties. Historically, acute mycotoxicoses were common even in moderate temperature zones, causing epidemics that devastated entire regions, sometimes influencing the course of human history.[1] Nowadays, they still appear in tropical countries with equal severity and high mortality. In acute mycotoxicoses, symptoms appear quickly and, if exposure continues, the outcome may be fatal. Chronic mycotoxicosis is the consequence of long-term exposure to smaller amounts of mycotoxins. The initial phase of exposure is frequently insidious and without obvious initial symptoms.

Owing to the introduction of good agricultural practices, ergotism (mycotoxicosis caused by ergot alkaloids), which has caused the deaths of large numbers of people, has not been seen in temperate zones for centuries.[2] Some mycotoxicoses, such as the serious neurological disease called sugarcane disease, which affects children in China and is caused by 3-nitropropionc acid, have not been recently reported, although they probably still appear. Kwashiorkor and Reye syndrome were erroneously considered to be mycotoxicoses caused by aflatoxins. This chapter means to provide a short overview of symptoms, prognosis, and outcome of human mycotoxicoses.

Environmental Mycology in Public Health. http://dx.doi.org/10.1016/B978-0-12-411471-5.00005-3

AFLATOXINS

Aflatoxins are a group of mycotoxins mostly produced by *Aspergillus flavus* and *Aspergillus parasiticus*. Among them, the most frequently found and the most toxic is the highly hepatotoxic AFB1. In India and Kenya, AFB1 has caused several outbreaks of acute aflatoxicosis with high mortality rates (10–60%),[3,4] the most recent occurring in Kenya in 2004. The initial symptoms are high temperature, vomiting, and abdominal pain followed by anorexia, depression, jaundice, diarrhea, and photosensitivity. If exposure continues, jaundice progresses rapidly with hepatosplenomegaly, ascites, and coma. At postmortem the main lesions are consistent with acute liver injury. The consequences of long-term exposure to lower amounts of AFB1 are chronic liver injury, cirrhosis, ascites, and primary hepatocellular carcinoma (HCC). The risk of HCC increases 25- to 30-fold when exposure to AFB1 is combined with hepatitis B or hepatitis C virus. In Africa, children are exposed to aflatoxins already in utero, with the peak of exposure after complete weaning. Early exposure to aflatoxins is connected with impaired growth, decreased immunity, and interference with protein metabolism and nutrition, which probably causes a higher mortality rate in young adulthood.[5]

The International Agency for Cancer Research classified the mixture of natural aflatoxins (AFB1, AFB2, AFG1, and AFG2) as a group A1 carcinogen (sufficient evidence for carcinogenicity in humans), whereas AFM1 (metabolite of AFB1 with carcinogenic activity 10 times lower than that of AFB1) was classified as possibly carcinogenic to humans (group 2B).[6,7]

FUMONISINS

Fumonisins are group of 15 mycotoxins produced by *Fusarium* molds (mostly *Fusarium verticilloides* and *Fusarium moniliforme*). The most frequently found member of this group is fumonisin B1. It interferes with the metabolism of sphingolipids.[8] In a single FB1-caused acute intoxication, only gastrointestinal symptoms (diarrhea, vomiting, borborygmus) were reported. Chronic exposure to FB1 is connected to a high rate of neural tube defects (brain and spinal cord malformations) in regions where the main food is maize, which is frequently contaminated with FB1 (South Texas, USA; Mexico; Guatemala; China; and South Africa).[9] FB1 is also supposed to be the causative agent of esophageal cancer in some parts of Africa (Transkei region) and primary liver tumors in China. FB1 is classified as possibly carcinogenic to humans (group 2B).[6]

OCHRATOXINS

Ochratoxins are a group of mycotoxins produced by *Penicillium verrucosum* and various species of *Aspergillus* molds (*Aspergillus alliaceus, Aspergillus auricomus*, *Aspergillus carbonarius*, *Aspergillus glaucus*, *Aspergillus melleus*, *Aspergillus niger*) that contaminate crops in the field, leading to field and storage ochratoxins contamination. There is a single reported case of acute ochratoxicosis in humans that resulted in acute renal failure. Chronic exposure to the most toxic ochratoxin A (OTA) is suspected to be the chief causative agent of Balkan endemic nephropathy (BEN) and urothelial tumors found very frequently in the BEN region.[10]

ZEARALENONE

Zearalenone (ZEA) is produced by some *Fusarium* species. This is a mycotoxin with low acute toxicity in experimental animals and there is no report of acute toxicity in humans.[11] Symptoms of chronic exposure are caused by interactions of ZEA and its metabolites with estrogen receptors. Premature telarche (development of breasts in young girls) with precocious pseudopuberty has been seen in young girls in Costa Rica exposed to residues of ZEA in meat (used as a growth-promoting compound). Premature telarche in girls and male gynecomastia were reported in Hungarian children consuming naturally ZEA-contaminated "healthy" food.

TRICHOTHECENES

Trichothecenes are a group of about 170 mycotoxins produced mostly by molds of *Fusarium* strains, although some strains from *Trichoderma*, *Trichothecium*, *Myrothecium*, and *Stachybotrys* may also produce them. In the former Soviet Union in the 1930s, epidemics of T-2 mycotoxicosis called alimentary toxic aleukia appeared with severe gastrointestinal symptoms that lasted 3–9 days.[12] In the second stage, the symptoms improved but the anemia, thrombocytopenia, and leukopenia increased during several weeks. If exposure continued, the third stage with necrotic lesions in airways and gastrointestinal tract with infections and hemorrhages developed and was fatal in 60% of patients. The fourth is the stage of recovery. There are still cases of trichothecenes mycotoxicosis with mild symptoms in small communities in tropical countries.

A summary of the mycotoxins discussed in this chapter is presented in Table 1.

Acute mycotoxicoses are serious, sometimes fatal diseases, that appear only sporadically in tropical countries. In developed countries scientists are more concerned about the continuous human exposure to various mycotoxins with genotoxic and carcinogenic properties in experimental animals. There is no single, simple, and cheap method for decontamination of several mycotoxins in food and therefore human exposure could be avoided only by prevention of food contamination with molds.

TABLE 1 Most Common Producers of Mycotoxins, Target Organs of Toxicity, and Symptoms of Mycotoxicoses

			Symptoms	
Mycotoxin	**Producers**	**Target Organ**	**Acute Toxicity**	**Chronic Toxicity**
Aflatoxins (AFB1)	*Aspergillus flavus* *Aspergillus parasiticus*	Liver	Gastrointestinal disturbance, jaundice, photosensitivity, hepatosplenomegaly, ascites, coma, death	Chronic liver injury, cirrhosis, ascites, liver carcinoma
Fumonisins (FB1)	*Fusarium verticilloides* *Fusarium moniliforme*	–	Diarrhea, vomiting, borborygmus	Brain and spinal cord malformations Esophageal cancer
Ochratoxins (OTA)	*Penicillium verrucosum* *Aspergillus alliaceus* *Aspergillus auricomus* *Aspergillus carbonarius* *Aspergillus melleus* *Aspergillus niger*	Kidney	Acute renal failure	Chronic renal injury in Balkan endemic nephropathy
Zearalenone	*Fusarium* spp.	Endocrine system	–	Girls: Premature telarche, precocious pseudopuberty Boys: Gynecomastia
Trichothecenes (T-2 toxin)	*Fusarium* spp. *Trichoderma* spp. *Trichothecium* spp. *Myrothecium* spp. *Stachybotrys* spp.	Hematopoietic tissues	Signs of myelotoxicity	–

REFERENCES

1. Peraica M, Rašić D. The impact of mycotoxins on human history. *Arh Hig Rada Toksikol* 2012;**63**:511–6.
2. Peraica M, Radić B, Lucić A, Pavlović M. Toxic effects of mycotoxins in humans. *Bull WHO* 1999;**77**:711–86.
3. Ngindu A, Johnson BK, Kenya PR, et al. Outbreak of acute hepatitis caused by aflatoxin poisoning in Kenya. *Lancet* June 12, 1982:1346–7.
4. Tandon BN, Krishnamurthy L, Koshy A, Tandon HD, et al. Study of epidemic of jaundice presumably due to toxic hepatitis, in Northwest India. *Gastroenterology* 1977;**72**:488–94.
5. Williams JH, Phillips TD, Jolly PE, et al. Human aflatoxicosis in developing countries: a review of toxicology, exposure, potential health consequences, and interventions. *Am J Clin Nutr* 2004;**80**:1106–22.
6. IARC (International Agency for Research on Cancer). Some naturally occurring substances: food items and constitutents, heterocyclic aromatic amines and mycotocins. In: *Report of an IARC working group on the evaluation of carcinogenic risks to humans*. Lyon: IARC; 1993. [IARC Monographs on the Evaluation of Carcinogenic Risks to Humans, vol. 56].
7. *Evaluation of certain mycotoxins in food (fifty-sixth report of the joint FAO/WHO expert committee on food additives)*. WHO Technical Report Series, No. 906. 2002.
8. Voss KA, Smith GW, Haschek WM. Fumonisins: toxicokinetics, mechanism of action and toxicity. *Anim Feed Sci Technol* 2007;**137**:299–325.
9. Marasas WFO, Riley RT, Hendricks KA, et al. Fumonisins disrupt sphingolipid metabolism, folate transport and neural tube development in embryo culture and in vivo: a potential risk factor for human neural tube defects among populations consuming fumonisin-contaminated maize. *J Nutr* 2004;**134**:711–6.
10. Pfohl-Leszkowicz A, Tozlovanu M, Manderville R, et al. New molecular and field evidences for the implication of mycotoxins but not aristolochic acid in human nephropathy and urinary tract tumor. *Mol Nutr Food Res* 2007;**51**:1131–46.
11. Zinedine A, Soriano JM, Molto JC, et al. Review on the toxicity, occurrence, metabolism, detoxification, regulation and intake of zearalenone: an oestrogenic mycotoxin. *Food Chem Toxicol* 2007;**45**:1–18.
12. Sudakin DL. Trichothecenes in the environment: relevance to human health. *Toxicol Lett* 2003;**143**:97–107.

Section III

Chapter 6

Risk Groups for Acquiring Fungal Infections

Guillermo Quindós-Andrés
Laboratorio de Micología Médica, Departamento de Inmunología, Microbiología y Parasitología (UFI 11/25 «Microbios y Salud»), Facultad de Medicina y Odontología, Universidad del País Vasco/Euskal Herriko Unibertsitatea, Apartado, Bilbao, Spain

The epidemiology of invasive mycoses or invasive fungal diseases (IFDs) is in continuous transformation. Social changes together with medical and surgical advances have improved the hope and quality of life in many countries. However, these advances also have brought with them the onerous toll of an increase in the number of people at risk for an IFD.

Fungi probably appeared a billion and a half years ago and it is estimated that the number of fungal species far surpasses a million, the vast majority still unknown. These microorganisms have been able to adapt and survive in extensive ecological niches with the presence of decomposing organic matter. Most people in regular contact with fungi do not suffer any IFD, despite their ubiquity. However, there is a growing population with diseases or underlying factors that predispose to IFD. Despite the current progress in diagnostic methods and the availability of new antifungal drugs there has not been a decrease in the incidence and mortality of IFDs.[1] IFDs can be caused by more than 100 different species of fungi, although 20 species are the most common etiologies. Invasive candidiasis is the most frequent mycosis in most patients and hospitals.[2] Invasive aspergillosis is more frequent than candidiasis in bone marrow transplant (BMT) recipients, especially if they receive prophylaxis with an antifungal, such as fluconazole, without activity against *Aspergillus*.[3,4] In persons infected with the human immunodeficiency virus (HIV) or suffering from acquired immunodeficiency syndrome (AIDS), especially those who do not receive antiretroviral treatment, cryptococcosis is more frequent than candidiasis.[5] Other fungi such as *Pneumocystis*, *Fusarium*, and *Rhizopus* can cause devastating illnesses. In addition, the list of emerging fungi grows continuously, including yeasts and pseudoyeasts (*Trichosporon*, *Saccharomyces*, *Rhodotorula*, *Saprochaete*, etc.), mucorales (*Mucor*, *Rhizomucor*, *Absidia*, *Cunninghamella*, etc.), hyaline molds (*Acremonium*, *Lomentospora*,

Environmental Mycology in Public Health. http://dx.doi.org/10.1016/B978-0-12-411471-5.00006-5

Scedosporium, *Scopulariopsis*, *Paecilomyces*, *Trichoderma*, etc.), and dematiaceous fungi (*Alternaria*, *Bipolaris*, *Curvularia*, *Exophiala*, *Exserohilum*, etc.).[3]

A small group of fungi, belonging to the genera *Candida* and *Malassezia*, is part of the human microbiota and can cause endogenous superficial or invasive mycoses depending on the patient's immune status. Invasive candidiasis occurs when *Candida* enters the bloodstream, usually through the intestinal mucosa. Exogenous fungal infections are acquired by inhalation of airborne conidia that can multiply and spread from lung to other organs. The high number of conidia present in the air and the low incidence of invasive mycoses reflect the high efficiency of human defensive mechanisms, eliminating the large number of conidia to which most people are exposed. Other major routes of entry for fungi in the human body are through contact with fungal propagules in soil or present in some animals (dermatophytes), fungal implantation by a trauma or punctures with thorns or splinters (Mucorales, *Lomentospora*, *Scedosporium*, or *Sporothrix*), ingestion and colonization of the digestive tract (*Candida*, Mucorales, and *Microsporidium*), and entering the bloodstream through needles or catheters used for intravenous administration of drugs or nutrition, prostheses, or other biomedical devices or surgical procedures (*Candida*, *Geotrichum*, *Malassezia*, *Saprochaete*, etc.). Some human activities favor superficial and subcutaneous mycoses. For example, *tinea pedis* is more common in athletic, military, and other professions that wear shoes that keep a constant humidity in the feet. Conversely, these dermatomycoses are less frequent in people who wear open shoes, such as sandals. Other superficial mycoses that affect inguinal and crural skin are more common in sedentary people, such as office clerks or taxi or bus drivers. Cutaneous mycoses are more common in regions with warm and humid climates. Mycetoma is observed more frequently in the feet of people with professional activities related to agriculture or forestry work, and the entry of the fungus is facilitated by penetrating injuries caused by thorns, splinters, nails, stones, bites, etc. In many of these mycoses, such as sporotrichosis, malnutrition, chronic alcoholism, debilitating diseases, and/or immunodeficiency can be involved and facilitate the spread and severity of fungal infection. From the door of entry, fungal invasion of contiguous tissues or spread to other organs through the blood or lymph depends largely on the health status of the patient and is more frequent in those individuals with significant immunodeficiency. In some mycoses, usually endogenous ones such as candidiasis or essentially exogenous ones such as mucormycosis or fusariosis, it may be difficult to specify the time and place of acquisition of infection because the incubation periods of most opportunistic mycoses are long, vague, or unknown.[1–4]

The annual incidence of IFD estimated per 100,000 people is 7–20 for invasive candidiasis, 3–6 for cryptococcosis, and 1–3 for invasive aspergillosis. However, there are significant epidemiological variations between countries and between hospitals in the same country that must be related both to local characteristics of the disease and its risk factors and to differences in the medical practices (speed of diagnosis, guidelines for treatment, prophylaxis, etc.) of the medical and surgical services. The annual incidence of less frequent fungal infections has been estimated to be 0.2 annual cases of mucormycosis per 100,000 inhabitants, 0.12 cases of other hyalohyphomycosis (caused by *Fusarium* and other hyaline filamentous fungi), and 0.1 cases of phaeohyphomycosis (caused by *Bipolaris*, *Exophiala*, and other dematiaceous filamentous fungi).[1–5]

Patients at higher risk are those admitted to intensive care units (ICUs); those who use prostheses, catheters, or other intravenous devices; those who receive various immunosuppressant treatments or antineoplastic chemotherapy; and the recipients of BMT or solid organ transplants (SOT).[2,3,6] In addition, there are opportunistic fungal infections due to infection by HIV and AIDS, mainly in those countries where antiretroviral treatment is not fully available. These IFDs include cryptococcosis, candidiasis, histoplasmosis, and penicilliosis.[7] At present, although IFDs in HIV-infected persons have declined dramatically in countries with better social and health care, in many countries of sub-Saharan Africa and Asia, the number of people suffering from cryptococcosis and other IFDs continues to increase.[5,7–8]

The classical division of invasive mycoses is as cosmopolitan or endemic. However, transoceanic travel, migration of people and animals, clearance of new land for human settlement, and the modification of the environment together with regional and global changes in climate are causing significant changes in the geographic distribution of IFDs. The emergence of cryptococcosis caused by *Cryptococcus gattii* on Vancouver Island and in western Canada and in the states of Washington and Oregon (USA) is considered an important infectious danger for inhabitants and tourists. These areas are far from the usual ecological niches of *C. gattii*.[1,7] Enlargement of habitat related to the high adaptive capacity of fungi such as *Coccidioides immitis* and *Coccidioides posadasii* to more extreme conditions has led to an increase in coccidioidomycosis in the southwestern United States and northwestern Mexico.[8,9]

Potentially pathogenic fungi can be isolated at all geographic latitudes and in all natural or human environments. However, it is difficult to specify clearly on many occasions whether the IFD is nosocomial or community acquired. The drift of many hospital treatments to their realization in the home of the patient is becoming more frequent. In addition, many ambulatory patients remain, during a time that can be extended, in a so-called day hospital to receive a session of chemotherapy or radiation therapy, among other treatments, which earlier forced the hospitalization of the patient. For this reason, many researchers prefer to call them generically "IFDs associated with health care." Some IFDs are usually endogenous, such as candidiasis, others are essentially exogenous, such as mucormycosis, but it may be difficult to specify when and where exactly the origin of invasive mycoses is. Attributed mortality is high, ranging from 30% in invasive candidiasis to more than 50% in invasive aspergillosis, and can reach 90–100% in mucormycosis or scedosporiasis.[1–4] Most of these patients suffer from significant immunodeficiency or are hospitalized for severe underlying diseases. Among the more predisposed are newborns of low weight (immunological immaturity); the elderly (immunosenescence); patients treated with major surgery, especially of the digestive apparatus, BMT, or SOT; patients with any kind of immunodeficiency or chronic disease who receive high or prolonged doses of corticosteroids and other immunosuppressive agents; patients with chronic inflammatory diseases or autoimmune diseases who are treated with monoclonal antibodies; critical patients admitted to the ICU; and HIV-infected individuals with advanced infection who receive no treatment with antiretroviral drugs (Tables 1–3).[1,2,6]

TABLE 1 Predisposing Factors for Invasive Fungal Diseases

Factor	Mycoses
Environment: air, earth movement with organic matter in decomposition (powder) in forests, gardens, urban works, housing repair, etc.	Histoplasmosis Coccidioidomycosis Paracoccidioidomycosis Blastomycosis Hyalohyphomycosis Phaeohyphomycosis
Food, water, and other beverages	Aspergillosis Mucormycosis Infection by *Saccharomyces, Rhodotorula, Trichosporon* Hyalohyphomycosis Phaeohyphomycosis
Iatrogenic origin: hands, clothes, tools and instruments, objects and surfaces—fomites	Candidiasis Mucormycosis
Alteration or breakage of anatomical barriers (surgery, catheters, chemotherapy, assisted ventilation, hemodialysis, peritoneal dialysis, etc.)	Candidiasis Aspergillosis
Modification of the microbiota (antimicrobial treatment)	Candidiasis Aspergillosis
Dysfunction of the phagocytic activity of neutrophil leukocytes (quantitative or qualitative neutropenia)	Candidiasis Aspergillosis Trichosporonosis Geotrichosis Hyalohyphomycosis Phaeohyphomycosis
Cellular immunodeficiency (HIV infection and AIDS)	Cryptococcosis Pneumocystosis Histoplasmosis Coccidioidomycosis
Immunological immaturity (neonates and infants) and immunosenescence (elders)	Candidiasis
Diabetes and other metabolic disorders	Mucormycosis Candidiasis

If we take as an example the oncohematological patients, one of the major at-risk populations, these patients can be grouped as regards the main factors of risk into three categories: those who belong to the host, environmental factors, and those related to antimicrobial treatments used.[2–4] Host factors include the stage of immunosuppression, dysfunction of vital organs, microbial colonization, and reactivation of latent infections. Risk factors related to the surrounding environment of the patient include the fungal load in the air, in food, and on objects. Finally, the antifungal prophylaxis schemes selected, or iatrogenic factors used as antibacterial or antiviral treatments, are important to elucidate what fungal pathogens may be involved in the etiology (Table 4).

TABLE 2 Predisposing Factors for Invasive Fungal Diseases in Hematological Patients

Clinical Condition	Factors
State of immunosuppression	Neutropenia (intensity, duration, and dynamic) Lymphopenia (intensity, duration, and dynamic) Cellular and humoral immunodeficiency Malnutrition Senescence High doses of antineoplastic chemotherapy or corticoids Immunosuppressant drugs (alemtuzumab, infliximab, anti-thymocyte globulin, etc.) Infection by immunomodulatory viruses (cytomegalovirus, Epstein–Barr virus, human herpesvirus-6, human herpesvirus-7, human herpesvirus-8)
Dysfunction of organs	Alteration of mucosa (mucositis) and skin by radiotherapy, chemotherapy, graft versus host disease, herpesvirus infections, vascular catheters Renal and/or hepatic failure Digestive tract obstruction Hyposplenia or asplenia Respiratory distress secondary to viral infections
Microbial colonization and/or reactivation of latent infections	Use of broad-spectrum antibiotics Antacid use Prolonged hospital stay Foreign objects Ciliary dysfunction Reactivation of latent infection by mycobacteria, *Toxoplasma*, cytomegalovirus, Epstein–Barr virus, herpesvirus simplex, varicella-zoster virus, hepatitis B virus, or hepatitis C virus Increased iron deposits in the reticuloendothelial system (bone marrow, liver, etc.) Previous invasive fungal infection

TABLE 3 Factors and Populations at Risk of Developing Invasive Candidiasis

General factors	Severity of acute illness (high APACHE values) Age (infants <1 year and elderly >65 years) Major surgery (primarily gastrointestinal) Prolonged stay in ICU Catheters and other intravascular devices Serious underlying diseases: diabetes mellitus, liver cirrhosis, malnutrition Multiple blood transfusions Intravenous nutrition Mechanical ventilation
Factors in high risk patients	Premature newborn with low weight (<1500 g) Significant and prolonged neutropenia Central venous catheter Antineoplastic chemotherapy Therapy with corticosteroids or immunosuppressive drugs Pancreatitis, visceral perforation, etc. Renal failure and/or hemodialysis Polytraumatism Extensive burns Broad-spectrum antibiotics Multiple colonization by the same species of *Candida* in various body locations

TABLE 4 Factors and Populations at Risk of Developing an Invasive Aspergillosis

Factor	Neutropenia Phagocytic dysfunction Cellular immune deficiency Therapy with corticosteroids and other immunosuppressive drugs Mucocutaneous barrier breaks Environmental exposure (high conidia concentrations)
Patient groups	Neutropenia (<500 neutrophils/mm^3 during >10 days); acute myeloid leukemia or allogeneic bone marrow transplant Graft versus host disease treated with immunosuppressive drugs Solid organ transplantations: lung > heart > intestine > liver > kidney HIV infection/AIDS without antiretroviral treatment and <100 CD4/μl Chronic granulomatous disease Biological therapy using monoclonal antibodies: alemtuzumab (anti-CD52) and infliximab, adalimumab and etanercept (anti-TNFα) Nonhematologic critical patients
	Chronic obstructive pulmonary disease undergoing chronic corticoid treatment Liver cirrhosis, advanced liver disease Major and/or complex surgery

IFDs adopt very different clinical presentations ranging from fungemia and peritonitis to more localized infections, such as lung infections or meningoencephalitis. Many of these are associated with the presence of catheters, prostheses, and other intravascular devices that facilitate both the access of the fungus to the bloodstream and tissues and the development of chronic septic foci. The great complexity of patients presenting important risks of IFD and the growing diversity of pathogenic fungi are major diagnostic and therapeutic challenges. Adequate knowledge of the etiology and epidemiology of these IFDs is one of the fundamental foundations for making an early and correct diagnosis and establishing the most appropriate treatment for each patient.

REFERENCES

1. Warnock DW. Fungal diseases: an evolving public health challenge. *Med Mycol* 2006;**44**:697–705.
2. Quindós G. Epidemiology of candidaemia and invasive candidiasis. A changing face. *Rev Iberoam Micol* 2014;**31**:42–8.
3. Pontón J, Quindós G, editors. *Guía de bolsillo de las micosis invasoras en los pacientes onco-hematológicos*. Bilbao: Asociación Española de Micología; 2009, ISBN: 978-84-8086-188-5.
4. Pemán J, Luque P, Nieto M, Pozo JC, Solé A, Zaragoza R. Update on invasive mycoses by filamentous fungi in critically ill patients. *Enferm Infecc Microbiol Clin* 2011;**29**(Suppl. 4):36–41.
5. Park BJ, Wannemuehler KA, Marston BJ, Govender N, Pappas PG, Chiller TM. Estimation of the current global burden of cryptococcal meningitis among persons living with HIV/AIDS. *AIDS* 2009;**23**:525–30.
6. Zaragoza R, Llinares P, Maseda E, Ferrer R, Rodríguez A, Épico Project Group. Épico project. Development of educational recommendations using the DELPHI technique on invasive candidiasis in non-neutropenic critically ill adult patients. *Rev Iberoam Micol* 2013;**30**:135–49.
7. MacDougall L, Fyfe M, Romney M, Starr M, Galanis E. Risk factors for *Cryptococcus gattii* infection, British Columbia, Canada. *Emerg Infect Dis* 2011;**17**:193–9.
8. Brown J, Benedict K, Park BJ, Thompson 3rd GR. Coccidioidomycosis: epidemiology. *Clin Epidemiol* 2013;**5**:185–97.
9. Laniado-Laborín R. Coccidioidomycosis and other endemic mycoses in Mexico. *Rev Iberoam Micol* 2007;**24**:249–58.
10. Myint T, Anderson AM, Sanchez A, Farabi A, Hage C, Baddley JW, et al. Histoplasmosis in patients with human immunodeficiency virus/acquired immunodeficiency syndrome (HIV/AIDS): multicenter study of outcomes and factors associated with relapse. *Medicine (Baltimore)* 2014;**93**:11–8.

Part II

Environmental Mycology in Public Health

Section I

Chapter 7

Pathways and Routes of Natural Exposure to Fungal Infection

Karl V. Clemons[1], Malcolm D. Richardson[2]
[1]*California Institute for Medical Research, Infectious Diseases Research Laboratory, San Jose, CA, USA;* [2]*Mycology Reference Centre, University Hospital of South Manchester, and Centre for Respiratory Medicine and Allergy, Institute of Inflammation and Repair, University of Manchester, Manchester, UK*

EXPOSURE PATHWAYS AND ROUTES OF INFECTION IN HUMANS

Fungi are ubiquitous in the outdoor environment and grow on a wide variety of natural substrates such as wood, vegetation, soil, water, and dust during heavy excavation and construction.[1] The elevated temperatures found in composting vegetation are selective for thermophilic species. In general these fungi are unable to utilize cellulose and lignin. They are characterized by rapid germination. There are approximately 3300 species of currently known soil fungi. For many years there has been the view that most soil fungi were cosmopolitan and that species at a particular site were selected only by various soil parameters. It has also been the view that most fungal species would potentially spend part of their life in the soil. This view has now been modified considerably, as very many plant-parasitic species are never isolated from the soil. This is particularly pertinent when considering pathogenic fungi. These properties of wide ecologic distribution, rapid growth, and thermotolerance are of particular importance in developing human disease.

Fungal growth in indoor environments such as water-damaged homes, schools, children's day care centers, offices, and hospitals poses a severe hygienic problem and is a potential human health risk. Inhalation of spores, hyphal fragments, and allergens produced by filamentous fungi from indoor air leads to irritative and nonspecific symptoms in sensitive persons.[2,3] Long-term exposure to these fungal propagules and allergens may cause severe, debilitating disease, and fatal infections, such as asthma, allergic diseases, alveolitis, and invasive pulmonary disease, and have an impact on other chronic pulmonary diseases, for example, chronic obstructive pulmonary disease.

Most commonly, pathogenic fungi gain entry to the body through the respiratory tract. Conidia are deposited in the nasal turbinates and may be inhaled

Environmental Mycology in Public Health. http://dx.doi.org/10.1016/B978-0-12-411471-5.00007-7

into the pulmonary alveoli. In the case of primary cutaneous fungal disease, for example, primary cutaneous aspergillosis, conidia are introduced directly into abraded or occluded skin. They then proliferate and can disseminate.

Nosocomial outbreaks of fungal disease, such as hospital-related *Aspergillus* infections, have been linked to construction or renovation work, as well as to contaminated ventilation systems. However, most cases of aspergillosis in hospitalized patients are sporadic in nature and it is much more difficult to determine whether these infections are acquired inside or outside the hospital setting.

Superficial Infections

These are infections limited to the outermost layers of the skin, the nails and hair, and the mucous membranes.[1] The principal infections in this group are the dermatophytoses and superficial forms of candidosis (also known as candidiasis). These diseases affect millions of individuals worldwide, but there are regional variations.

DERMATOPHYTOSIS

Unlike most other fungal infections, dermatophytosis is a contagious, host-to-host transmissible disease of humans and animals. Infection with an anthropophilic dermatophyte is acquired by direct or indirect contact with an infected individual. Indirect transfer may occur via the floors of communal bathing facilities or on shoes, clothing, brushes, towels, bedding, and other fomites. Dermatophyte arthroconidia can remain viable in the environment for long periods of time and the interval between deposition and transfer to another host may be considerable.

Infection with a zoophilic *Microsporum* or *Trichophyton* species is often the result of direct animal-to-human contact. These dermatophytes may then be transmitted among humans to a limited extent. Infection can also occur as a result of environmental contamination by an infected domestic pet or when buildings, gates, fence posts, or implements are contaminated by farm animals. Infection with geophilic dermatophytes, such as *Microsporum gypseum*, is uncommon, and involves transmission of a soil-borne inoculum to humans or other animals. Occasional outbreaks of human disease, originating from contaminated soil, have been reported.

SUPERFICIAL CANDIDOSIS

In most cases, superficial *Candida* infection is derived from the individual's own endogenous reservoir in the mouth, gastrointestinal tract, lower genital tract, or skin. In some cases, however, transmission of organisms from person to person can occur. For instance, neonatal oral candidosis is more common in infants born of mothers with vaginal candidosis, which suggests that infection occurs when the infant takes in some of the vaginal contents during parturition. The hands of mothers and health care workers are another potential source of neonatal infection.

MYCOTIC KERATITIS

The etiological agents of keratomycosis are widespread in the environment, being commonly found in indoor and outdoor air, in the soil and dust, and on decomposing plant matter. Human infection usually follows the traumatic implantation of spores into the corneal epithelium. Much less commonly, it results from their inadvertent introduction during surgical procedures, such as corneal transplantation.

OTOMYCOSIS

The etiological agents of otomycosis are commonly found in indoor and outdoor air, in the soil and dust, and on decomposing plant matter. Their prevalence varies with climatic conditions, but warm humid environments support their growth, and the human ear canal is ideal for their proliferation.

Opportunistic and Systemic Infections

The organisms that cause systemic fungal infection can be divided into two distinct groups: the true pathogens and the opportunists.[1] The first of these groups comprises a handful of organisms, mostly dimorphic fungi, that are able to invade and develop in the tissues of a normal host with no recognizable predisposition. The principal diseases are blastomycosis, coccidioidomycosis, histoplasmosis, and paracoccidioidomycosis. The second group, the opportunists, consists of less virulent and less well-adapted organisms that are able to invade the tissues of only immunocompromised hosts. Although new species of fungi are regularly being identified as causes of disease in immunocompromised patients, five diseases still account for most reported infections: aspergillosis, candidosis, cryptococcosis, mucormycosis, and pneumocystosis.

ASPERGILLOSIS

Inhalation of *Aspergillus* spores is the usual mode of infection in humans. The incubation period is unknown. Less frequently, infection follows the traumatic implantation of spores, as in corneal infection, or direct inoculation from contaminated dressings. Inhalation of contaminated water aerosols during patient showering has been suggested as an additional potential source of infection.

The etiological agents of aspergillosis are ubiquitous in the environment and the likelihood that infection will occur following inhalation or implantation of spores largely depends on host factors. Invasive aspergillosis has emerged as a major problem in several groups of immunocompromised patients. The likelihood of aspergillosis developing in these individuals depends on a number of host factors, the most important of which is the level of immunosuppression, whether this manifests as profound or prolonged neutropenia, as in graft versus host disease in hematopoietic stem cell transplant (HSCT) recipients, or as rejection in solid organ transplant (SOT) recipients.

SYSTEMIC CANDIDOSIS

In most cases invasive *Candida* infection is endogenous in origin, but transmission of organisms from person to person can also occur. In the health care setting, the main mechanisms of cross infection include direct patient-to-patient transmission; transmission from a colonized or infected person to a recipient via a third person, often a health care worker; transmission from a colonized or infected person to another individual via a medical device; and more or less simultaneous transmission to two or more patients from a common source, such as contaminated intravenous infusions. In most reports of outbreaks or cross infections due to *Candida albicans* in intensive care unit (ICU) patients, the suspected or proven source has been the hands of health care workers. Outbreaks of *Candida parapsilosis* infection among ICU patients have been linked to nosocomial acquisition via medical devices, such as vascular catheters, and/or parenteral nutrition. There are also a number of reports of *C. parapsilosis* cross infection related to hand carriage by health care workers.

CRYPTOCOCCOSIS

Inhalation of airborne propagules from an environmental source is the usual mode of infection with *Cryptococcus neoformans* and *Cryptococcus gattii* in humans and animals. However, it remains unclear whether the infectious particles are small desiccated acapsular yeast cells or basidiospores of the teleomorph stage of the fungus. It is thought that the encapsulated yeast cells are too large to penetrate the upper respiratory tract.

Outbreaks, investigations of which have helped to elucidate the sources, incubation periods, and risk factors for other fungal infections such as histoplasmosis, have not been described for *C. neoformans* infection. Thus, although bird roosting sites have been clearly implicated in outbreaks of histoplasmosis, the association of *C. neoformans* infection with exposure to pigeon droppings, the best recognized of the putative environmental risk factors for this disease, is mostly derived from information on the isolation of the fungus from the environment, not from studies that attribute disease to this exposure. The incubation period for *C. neoformans* is unknown and could be weeks, months, or even longer. Estimated incubation periods for outbreak cases of *C. gattii* infection among residents of Vancouver Island, BC, Canada, and elsewhere have ranged from 2 to 13 months (median 6–7 months).

MUCORMYCOSIS

Most human infections follow inhalation of spores that have been released into the air, and the lungs and nasal sinuses are the most common initial sites of infection. Infection may also follow cutaneous or percutaneous inoculation of organisms, either through traumatic disruption of skin barriers or as a result of catheter insertion or injections. Less frequently, infection follows ingestion of contaminated food. Many of the etiological agents of mucormycosis are ubiquitous in the environment and the likelihood that infection will occur following inhalation, implantation, or ingestion of spores largely depends on host factors.

PNEUMOCYSTIS JIROVECII PNEUMONIA

Pneumocystis infection is acquired by inhalation, but it remains unclear whether the source of infection is environmental or person-to-person spread. No environmental form or cycle of *Pneumocystis* has been found, but animal models and nosocomial case clusters among immunosuppressed patients suggest that airborne transmission from one host to another is important. Epidemiological studies showing the clustering of *Pneumocystis jirovecii*-specific genotypes with patients' place of residence are consistent with the hypothesis of airborne person-to-person transmission of the organism.

BLASTOMYCOSIS

Inhalation of *Blastomyces dermatitidis* spores is the usual mode of infection in humans. The incubation period, which has been estimated from outbreaks in which exposure occurred over a limited time period, is 4–6 weeks. Occasional cases have followed traumatic cutaneous inoculation.

Outbreaks of blastomycosis are uncommon, but have provided us with important information on risk factors for development of the disease. Most reported point-source outbreaks have been associated with occupational and recreational activities, often along streams or rivers, and have resulted from exposures to moist soil enriched with organic matter. Apart from outbreaks, blastomycosis is more commonly seen in adults than in adolescents or children. Those at greatest risk for blastomycosis are middle-aged men with outdoor occupations, such as construction or farming, or recreational interests, such as hunting or fishing. As with outbreaks, sporadic cases often involve exposures that occur near freshwater. In some reports, the incidence of sporadic infection has been reported to be much higher among African Americans, although it is unclear whether this is due to genetic risk factors or increased exposure.

COCCIDIOIDOMYCOSIS

Infection is thought to occur when *Coccidioides* arthroconidia from disrupted soil are inhaled. The incubation period, which has been estimated from point-source outbreaks in which exposure occurred over a limited time period, ranges from 1 to 3 weeks. In contrast to histoplasmosis, once individuals have recovered from *Coccidioides* infection, they are usually immune to reinfection. However, later reactivation of quiescent lesions can occur in persons who become immunocompromised. Coccidioidomycosis is not contagious but person-to-person spread has occurred by transmission from an organ donor to a recipient.

The major risk factor for infection is environmental exposure. The risk of infection depends on a number of factors including the nature of the environmental site, the activities performed, and the duration and degree of dust or soil exposure. Longer and more intense exposures may lead to a shorter incubation time and usually result in more severe pulmonary disease. Human infection has been associated with ground-disturbing activities, such as building construction, landscaping, farming, archaeological excavation, and numerous recreational pursuits. Natural events that result in the generation of dust clouds, such as earthquakes and windstorms, have been associated with an increased risk of infection and have resulted in large outbreaks.

HISTOPLASMOSIS

Inhalation of *Histoplasma capsulatum* spores is the usual mode of infection in humans. The incubation period, which has been estimated from point-source outbreaks in which exposure occurred over a limited time period, is 1–3 weeks. In cases of reinfection, the incubation period appears to be shorter (4–7 days after exposure). Histoplasmosis is not contagious but person-to-person spread has occurred by transmission from an organ donor to a recipient.

The major risk factor for development of infection with *H. capsulatum* is environmental exposure. The risk of infection depends on a number of factors including the nature of the environmental site, the activities performed, and the duration and degree of dust or soil exposure. Longer and more intense exposures usually result in more severe pulmonary disease. Most reported outbreaks have been associated with exposures to sites contaminated with *H. capsulatum* or have followed activities that disturbed accumulations of bird or bat droppings. These include building construction, cleaning, renovation and demolition, soil excavation, spelunking, and clearing sites harboring the fungus. In many cases, however, infections with *H. capsulatum* do not involve major disturbance of infected soil. Mild dust exposure can also result in outbreaks or sporadic cases of histoplasmosis.

PARACOCCIDIOIDOMYCOSIS

Inhalation of *Paracoccidioides brasiliensis* spores is believed to be the usual mode of infection in humans. However, in a few cases infection may be the result of traumatic inoculation. The infection is thought usually to be acquired during the first two decades of life, but clinical disease is uncommon in this age range. The incubation period is unknown, but it is clear that the fungus can remain dormant for very long periods in the lymph nodes following asymptomatic primary infection. Among more than 70 reported cases of imported paracoccidioidomycosis diagnosed in nonendemic countries, the latency period ranged from 4 months to 60 years with an average duration of asymptomatic infection of 14 years.

Diseases of Implantation—Subcutaneous Infections

These are infections of the dermis, subcutaneous tissues, and adjacent bones that generally show slow localized spread.[1] They usually result from the traumatic implantation of saprophytic fungi from soil or vegetation. More widespread dissemination of the infection, through the blood or lymphatics, is uncommon and usually occurs only if the host is in some way debilitated or immunocompromised. The principal subcutaneous mycoses are mycetoma, sporotrichosis, phaeohyphomycosis, and chromoblastomycosis. These infections are most frequently encountered among the rural populations of the tropical and subtropical regions of the world, where individuals go barefoot and wear the minimum of clothing.

CHROMOBLASTOMYCOSIS

Minor trauma, such as abrasions or wounds due to thorns or wood splinters, is often sufficient to introduce the many environmental fungi associated with chromoblastomycosis. The incubation period is unknown, but the disease is rare in children and adolescents exposed to the same environmental conditions as adults with the disease. This suggests a long latency period before signs of disease begin to appear.

ENTOMOPHTHOROMYCOSIS

It is still uncertain how entomophthoromycosis is acquired, but subcutaneous infection is thought to follow traumatic inoculation. Minor trauma, such as a thorn prick or an insect bite, is often sufficient to introduce the organism. Basidiobolomycosis is unusual among subcutaneous fungal infections in that it is more common in children and adolescents than in adults. As with other subcutaneous fungal infections, most affected individuals are otherwise normal, and there does not appear to be any predisposition apart from exposure. It is thought that gastrointestinal disease is acquired through ingestion of soil, animal feces, or food contaminated by either. Potential risk factors include prior ranitidine use.

MYCETOMA

Infection results from the traumatic implantation of the pathogen into the skin or subcutaneous tissue. In some cases the organisms are introduced on thorns or wood splinters; in others infection is due to later contamination of the wound with soil organisms. The initial lesion appears several months after the traumatic incident, but patients most commonly present for medical attention with long-standing infection.

SPOROTRICHOSIS

Infection usually follows the traumatic implantation of *Sporothrix schenckii* into the skin or subcutaneous tissue. Minor trauma, such as abrasions or wounds due to thorns or wood splinters, is often sufficient to introduce the organism. Zoonotic sporotrichosis occurs with exposure to infected animals, most commonly cats. Infection has also been transmitted from uninfected animals that transfer the organisms from soil through scratching or biting. The initial lesion usually appears 1–4 weeks after inoculation. Pulmonary and disseminated forms of infection, although uncommon, follow spore inhalation.

PHAEOHYPHOMYCOSIS

Subcutaneous infection is the most frequently reported form of phaeohyphomycosis. It is thought to result from the traumatic implantation of the causative fungus into the subcutaneous tissue. Minor trauma, such as abrasions or wounds due to thorns or wood splinters, is often sufficient to introduce the organism. Less commonly, it may occur as a result of hematogenous dissemination of invasive disease in an immunocompromised individual. The incubation period is unknown. The principal etiological agents include *Bipolaris* species, *Exophiala jeanselmei*, *Exophiala* (*Wangiella*) *dermatitidis*, and *Phialophora* species, but many less common mould have also been reported to cause this condition. Most cases occur sporadically in otherwise healthy individuals, often in older men with outdoor occupations, but infections have also been reported in immunocompromised individuals, such as SOT recipients. These patients may be at increased risk of subsequent dissemination.

HYALOHYPHOMYCOSIS

The mechanisms by which human *Fusarium* infection is acquired are not well understood. The incubation period is unknown. There are several suggested routes of transmission, including inhalation, implantation following trauma, and acquisition via contaminated intravascular devices. In some cases, it has been found that the source of disseminated *Fusarium* infection in an immunocompromised individual was a preexisting nail infection or a localized skin infection. The isolation of *Fusarium* species from hospital water supply and distribution systems has led some experts to suggest that inhalation of bioaerosols generated during showering could be an important source of invasive infection in HSCT recipients and other immunocompromised individuals.

Species of *Scedosporium* are ubiquitous in the environment and the likelihood that infection will occur following inhalation from the air, inhalation or aspiration of contaminated water, or implantation of spores largely depends on host factors. There are also several reports of acquisition via contaminated intravascular devices. The incubation period is unknown. Trauma is the major risk factor for localized *Scedosporium* infection in the immunocompetent individual. In addition to infection of the subcutaneous tissue and bone (mycetoma), cases of endophthalmitis, osteomyelitis, and arthritis have been reported following the traumatic introduction of the organism. Unlike other opportunistic mould pathogens, *Scedosporium* species can cause pneumonia in immunocompetent individuals following aspiration of polluted water. Hematogenous dissemination to the brain is a frequent complication.

Infection with *Talaromyces marneffei* (formerly known as *Penicillium marneffei*) is thought to follow inhalation of spores that have been released from thus far unidentified environmental sources into the air. The incubation period is variable, but it is clear that the fungus can sometimes remain dormant for long periods following asymptomatic primary infection. In contrast, cases have been reported in infants and children with the acquired immunodeficiency syndrome (AIDS) in which disseminated infection occurred within a few weeks of exposure to the organism.

Animal Models and Routes of Infection

Animal models are important tools in the study of infections caused by fungi and have been comprehensively reviewed.[4] With the primary aim of mimicking clinical disease, there are a number of choices that an investigator needs to make before beginning a study. These include the choice of animal species, fungal species, and route of infection to emulate the type of clinical infection being modeled. The ethics of using laboratory animals and a more detailed discussion of the various aspects of designing a model have been discussed at length previously.[4] Animal models can be designed for specific studies on pathogenesis, host response, and therapeutic intervention, with the model differing possibly in host species or fungal species based on the questions being asked by the investigator. Thus, defining what it is the investigator is trying to accomplish will determine how the model needs to be designed. For most studies, the mouse is the species of choice for a number of reasons. However, other animals such as rats, rabbits, hamsters, dogs, cats, and birds have been used.

As noted in the first part of this chapter, there are several routes by which humans and animals are exposed to potential fungal pathogens. However, a prospective study of these exposures is almost impossible because outbreaks are usually unexpected and often not all exposed individuals develop clinical disease. It is desirable to understand the events of a natural exposure to a fungal pathogen, and many animals are equally or more susceptible to the infections than are humans. Although the use of sentinel animals might be possible, it is not practical, in that no disease may develop or few animals become ill, making statistically valid studies almost impossible. Thus, to allow a practical and valid study, laboratory animals are used and exposed to the fungal pathogens via specific routes of infection to mimic natural exposures. Exposure to potentially pathogenic fungi from the environment occurs in several ways, including inhalation, ingestion, and direct contact through skin, mucous membranes, or cornea. In addition, exposure to a fungal pathogen can occur by translocation of normal flora from the skin or intestinal tract under the right circumstances.

INHALATIONAL MODELS OF INFECTION

A majority of the serious infections caused the primary fungal pathogens (i.e., *Coccidioides, Histoplasma, Blastomyces,* and *Paracoccidioides*), as well as more opportunistic pathogens (e.g., *Aspergillus, Cryptococcus, Scedosporium, Penicillium,* etc.), are acquired via inhalation. To study infections acquired through exposure by inhalation, the infecting organism is inoculated via the respiratory tract. There are three methods used to infect the animals by this route and several species of laboratory animals have been used to study a variety of infections.[4–8]

The first method is the simplest to perform and is the intranasal administration of the infectious inoculum to the animal. If done in mice, the animal is anesthetized to a plane of anesthesia that eliminates the swallow reflex using isoflurane fumes or with injectables such as a cocktail of ketamine and xylazine. The animal is held upright and the mouth held shut and a small volume of fungal suspension (10–50 μl) placed on the nares. The animal will inhale the droplets into the lungs. A critical aspect is the plane of anesthesia. If it is too deep, the animal may not inhale deeply and if it is too light the animal will tend to swallow much of the inoculum.

A second method for inducing pulmonary exposure is the use of intratracheal administration. Again, this method requires anesthesia and most often a minor surgical incision above the trachea. The fungal inoculum is injected using a tuberculin syringe directly into the trachea, which has the advantage of no loss of organisms. After injection the skin incision is closed with sutures or staples. Similar to this method is the use of a catheter inserted into the trachea or an atomizer cannula.

The third method of inhalational exposure is through the use of a chamber to disperse the infecting conidia and allowing the animal to breathe naturally. This method has been used very successfully in studies of pulmonary aspergillosis done in mice.[4,6] Exposure by this method results in deeper penetration of the conidia into the lungs and an even distribution of infection, whereas the intratracheal and intranasal methods can result in uneven distribution of organisms. However, this method also requires the proper chamber and mechanism to disperse the conidia.[9,10]

MUCOSAL MODELS OF INFECTION

Although acquisition of fungal infection through ingestion is not common, mucosal *C. albicans* infection is common in immunosuppressed patients and in untreated human immunodeficiency virus-infected patients. *C. albicans* is part of the normal skin and gut flora of many individuals and causes orogastric mucosal infection when the immune system is compromised or the balance of normal flora has been altered by high-dose antibacterial therapy or the mucosal layer is damaged by anticancer therapies.[11] Two models of mucosal disease are relevant.[4,12] The first utilizes severe combined immunodeficiency mice that are allowed to drink an inoculum of *C. albicans* for a period of 24h; *C. albicans* is not part of the normal flora of mice. A mucosal infection is established in the superficial layers of the mucosa throughout the gastrointestinal tract from the tongue to the colon. However, in this model the organisms do not translocate into the bloodstream, mimicking disease in AIDS patients. In a second model the mucosal epithelium is damaged by high-dose 5-fluorouracil treatment and, after infection, the animals not only have the orogastric infection but also develop systemic disease in the kidneys and liver because the organism has translocated across the gut and into the bloodstream. This model is similar to the disease in cancer chemotherapy patients.

Vaginitis due to *C. albicans* is also a mucosal infection and can be modeled in estrogenized female mice or rats.[13,14] The animals are given estradiol to induce estrus and then inoculated intravaginally with a suspension of yeast. Maintenance of the infection over more than 1 week requires periodic dosing with the estradiol.

DIRECT INFECTION

Dermatomycosis is one of the most frequent fungal infections worldwide and acquired by exposure of the skin to the organisms in the soil or on hard surfaces in, for instance, shower facilities. The organism enters the skin through small breaks in the keratinized surface and establishes a superficial infection. Both mice and guinea pigs have been used for modeling these infections, which are initiated by applying a suspension of conidia to lightly abraded skin. The infections are acute to the surface of the skin, but both will self-resolve within weeks.[15]

Other fungal infections, such as eumycetoma and sporotrichosis, can occur after puncture of the skin to inoculate the organisms into the subcutaneous tissue. Models of these diseases are lacking with respect to the subcutaneous part of the infection, and animals are often infected parenterally. More work is needed in these diseases to develop models that more readily mimic human disease.

CONCLUSIONS

As should be evident from the information presented in this chapter, natural exposure to various fungal infections is a common occurrence. The study of these exposures and their epidemiology is a very important aspect of human health care and for the prevention of these diseases where possible by avoidance or by hygiene practices. It should be remembered that these practices apply to both humans and animals. The use of laboratory animals to model the infections resulting from natural exposures is critical to our better understanding of the pathogenesis and progression of these infections, as well as for studies on the development of preventative or curative therapies.

REFERENCES

1. Richardson MD, Warnock D. *Fungal infection – diagnosis and management*. 4th ed. Chichester: Wiley-Blackwell; 2012.
2. McGinnis MR. Indoor mould development and dispersal. *Med Mycol* 2007;**45**(1):1–9.
3. Meheust D, Le Cann P, Reboux G, Millon L, Gangneux JP. Indoor fungal contamination: health risks and measurement methods in hospitals, homes and workplaces. *Crit Rev Microbiol* 2014;**40**(3):248–60.
4. Capilla J, Clemons KV, Stevens DA. Animal models: an important tool in mycology. *Med Mycol* 2007;**45**(8):657–84.
5. Brummer E, Clemons KV. Animal models of systemic mycoses. In: Miyaji M, editor. *Animal models in medical mycology*. Boca Raton, FL: CRC Press; 1987. p. 79–95.
6. Clemons KV, Stevens DA. The contribution of animal models of aspergillosis to understanding pathogenesis, therapy and virulence. *Med Mycol* 2005;**43**(Suppl. 1):S101–10.
7. Clemons KV, Capilla J, Stevens DA. Experimental animal models of coccidioidomycosis. *Ann NY Acad Sci* 2007;**1111**:208–24.
8. Clemons KV, Stevens DA. Cryptococcosis in experimental animals, lessons learned. In: Heitman J, Kozel T, Kwon-Chung J, Perfect J, Casadevall A, editors. *Cryptococcus from human pathogen to model yeast*. Washington, DC: American Society for Microbiology; 2011. p. 473–88.
9. Steinbach WJ, Benjamin Jr DK, Trasi SA, Miller JL, Schell WA, Zaas AK, et al. Value of an inhalational model of invasive aspergillosis. *Med Mycol* 2004;**42**(5):417–25.
10. Sheppard DC, Rieg G, Chiang LY, Filler SG, Edwards Jr JE, Ibrahim AS. Novel inhalational murine model of invasive pulmonary aspergillosis. *Antimicrob Agents Chemother* 2004;**48**(5):1908–11.
11. Samaranayake YH, Samaranayake LP. Experimental oral candidiasis in animal models. *Clin Microbiol Rev* 2001;**14**(2):398–429.
12. Clemons KV, Stevens DA. Orogastrointestinal model of mucosal and disseminated candidiasis. *Methods Mol Biol* 2012;**845**:557–67.
13. Fidel Jr PL, Cutright J, Steele C. Effects of reproductive hormones on experimental vaginal candidiasis. *Infect Immun* 2000;**68**(2):651–7.
14. Sobel JD, Muller G, McCormick JF. Experimental chronic vaginal candidosis in rats. *Sabouraudia* 1985;**23**(3):199–206.
15. Shimamura T, Kubota N, Shibuya K. Animal model of dermatophytosis. *J Biomed Biotechnol* 2012;**2012**:125384.

Section II

Occupational Settings

Chapter 8

Highly Contaminated Workplaces

Anne Oppliger[1], Philippe Duquenne[2]
[1]*University of Lausanne and Geneva, Switzerland;* [2]*INRS, Laboratoire de Métrologie des AérosolsVandoeuvre-les-Nancy, France*

INTRODUCTION

Fungi are eukaryotic microorganisms that are ubiquitous in the environment. Most of them produce large quantities of spores and fungal material can easily be released into the air both through intrinsic/natural mechanisms and through external events such as human activities. Thus, airborne fungi are found at low concentrations in the air of environments, considered as lowly contaminated, such as outdoor air, office buildings, and homes.[1–3] High levels of airborne fungi are also found in various occupational sectors in which materials containing fungi are handled.[4] For example, in the food industry, some specific fungus species are essential for production processes and are deliberately added, and their growth is generally controlled and confined. In the agricultural or waste management sectors, environmental fungi colonize and grow on organic matter without any human control. Fungi play a major role in the decomposition of organic matters and are greatly involved in waste and composting. They are also present in agriculture (grain harvesting and handling, vegetable cultivation, animal farming) and industries (wood processing), sometimes in very high concentrations, but here they are considered as contaminants. Indeed, all workers who handle organic matter (vegetal or animal) are potentially exposed to a high airborne fungal load. The incidence and abundance of fungal species depends on the natural selection acting on these very complex microbiological communities, and several parameters are involved.

Airborne fungal particles found at occupational settings consist of spores, mycelium fragments, and debris that are present as single particles or complex aggregates. Once in suspension in the air, fungal particles can be inhaled by exposed workers and cause diverse symptoms including allergies, irritation, and opportunistic infections.[5,6] Irritation of the eyes, nose, and throat, as well as cough, are often reported as short-term effects of exposure to airborne fungi, whereas longer exposure is associated with increased risks of chronic diseases.[6]

Environmental Mycology in Public Health. http://dx.doi.org/10.1016/B978-0-12-411471-5.00008-9

The link between exposure to fungi and occupational diseases is often difficult to prove owing to coexposure to other components of bioaerosols.

To prevent these adverse health effects, characterization of occupational airborne fungal exposure is essential. In most studies, the presence and quantification of airborne fungi is assessed either by the culture of viable and culturable fungal particles (called colony-forming units or cfu) on nutrient agar or by counting the total number of spores collected on a filter using a microscope. It is also possible to assess fungal contamination by measuring β-D-glucans, which are components of the fungi's cell walls. Levels of exposure depend on various factors, which are important to identify. The purpose of this chapter is to review the fungal concentrations found in highly contaminated work environments and to highlight the main determinants. The review focuses on scientific studies published between 2000 and 2014. It will also review the most frequently encountered fungal species and the more exposing tasks.

FUNGAL AEROSOLS IN ANIMAL CONFINEMENT BUILDINGS

Modern animal farming seems to require bigger and bigger installations housing more and more animals. For example, it is no longer rare to find pig or poultry farms housing more than 1000 animals, and installations with more than 10,000 animals are common in certain countries. Most of the time, these animals are kept in enclosed buildings for their entire lives and the accumulation of organic dust is difficult to control. Aerosolization of settled dust occurs easily as animals move about and workers do their jobs. Organic dust is composed of both nonviable particles (originating from food, litter, feces, skin squames, etc.) and viable particles such as viruses, bacteria, and fungi. Many studies have quantified and/or identified airborne fungi in livestock farms. Most fungi samples were taken at stationary points and reflected airborne concentrations, not workers' personal exposure. This is an important point because it is known that concentrations based on personal samples measured directly in a worker's breathing zone are usually higher than ambient concentrations measured by stationary sampling in the workplace.[7] However, this has not always been substantiated.[8] Direct impact onto nutrient agar is by the far the most widely used procedure for assessing the fungal content of bioaerosols. This usually involves sampling air for short periods (between 0.5 and 5 min), but daily variations are not taken into account.

Pig Farms

The intensity and variability of personal exposure to organic dust among pig farmers depends on the farm's characteristics and the farmer's work activities. For instance, activities related to feed handling, as well as high-pressure water cleaning, increase exposure to organic dust.[9] The use of dry feed, slatted floor coverings, and neutral ventilation systems is important factors related to increases in dust exposure.[9] Dust exposure is also increased during specific work tasks involving intense animal handling (castration, teeth cutting, handling sick animals, etc.) and pig load-out.[10,11] In temperate regions, a seasonal effect is also reported, with an increase in airborne dust and/or bioaerosol concentrations during the winter period.[9,12,13] However, the inverse has also been found.[14,15] Basinas et al.[9] demonstrated that in Denmark, a 10°C increase in outdoor temperature was associated with a decrease in dust exposure as high as 30%. The authors suggested that this might be attributable to the higher ventilation rates used in higher temperatures. Table 1 shows a data synthesis of 14 studies carried out in 12 different countries and published between 2000 and 2013. Results using culture-dependent methods (direct impact of culturable fungi on nutrient agar) vary between studies: from $<10^2$ to 48×10^3 cfu/m^3. When counting by microscope (total fungal particles collected on filters), sampling time is higher (usually between 60 and 240 min) and results vary between 5.8×10^3 and 8760×10^3 cells/m^3. One study,[16] which sampled air over a very long period (48 h) on a glass fiber filter and counted fungal particles in a counting chamber, revealed a very high fungal concentration of 12×10^5 cell/m^3. This study's culture-based count showed that only 0.4% of the fungi on the nutrient agar used were viable. As discussed above, the seasonal effects observed in some studies leave us with contradictory results, which deserve further investigation.

One study[17] reported no differences in airborne fungi concentrations between different types of piggeries (breeding, farrowing, nursery, growing, and finishing). In Poland, Sowiak et al.[18] showed that fungal concentrations were higher in pig farms using a litter bed system than in those using a pit manure system (18×10^3 vs 0.9×10^3 cfu/m^3) and that manual feed distribution led to higher fungal concentrations than mechanical feeding (8.4×10^3 vs 0.6×10^3 cfu/m^3). The influence of litter was also shown in Canada,[19] where pig farms with sawdust beds generated higher fungal concentrations than conventional barns with slatted floors (48×10^3 vs 0.86×10^3 cfu/m^3).

The diversity of fungal genera found in pig farms is extremely variable from one study to another. It seems that each pig farm has its own specific and dynamic fungal community. However, Table 1 shows that *Aspergillus* sp. and *Penicillium* sp. were present in all the pig farms studied. *Cladosporium* sp. and *Scopulariopsis* sp. were also very present.

Poultry Farms

Large-scale production has also led to increased bird densities in poultry houses.[20] As in pig farms, concentrations of airborne organic dust in poultry houses depend on work activities and farm characteristics. Types of work activities in poultry farms are fewer than in pig farms because less animal care is needed. The stage at which birds are growing rapidly (4–6 weeks) is important because the biomass of feces and feather dandruff increases sharply during the fattening period. For workers, the load-out for slaughter also generates a

TABLE 1 Methods, Arithmetic Mean, Geometric Mean (or Median), Range (Min–Max), and Fungal Species Identified in Pig Farms in Various Countries

Country	Method	AM (10^3 cfu/m^3)	GM or Median (10^3 cfu/m^3)	Min–Max (10^3 cfu/m^3)	Species (Most Frequent)	Remarks	References
Portugal	Culture	3.21 2.26 2.16 7.38	ND		*Aspergillus versicolor* *Scopulariopsis brevicaulis* *Penicillium* sp.		Viegas et al.[21] Sabino et al.[22]
Switzerland	Culture	5.7	0.77	(0.02–52)	ND		Masclaux et al.[13]
Denmark	Culture Counting Chamber PCR and sequencing	1200			*Penicillium* sp. *S. brevicaulis* *Stachybotrys chartarum* *Chrysosporium* *Aspergillus eurotium* (65%) *Wallenia* *Mucorales* *Russulales*		Kristiansen et al.[16]
Poland	Culture	15.5 (Total fungi) 10.7 (Respirable fungi)	2.7 1.5	(0.2–108) (0.02–70)	ND		Sowiak et al.[23]
Korea	Culture	0.631 (Winter) 9.2 (Summer)	0.454 7.1	(0.07–2.5) (0.9–22.7)	*Aspergillus* sp. *Cladosporium* sp. *Penicillium* sp.		Jo and Kang[14]

Finland	Culture	NDs	2.0–1100	(0.2–3500)	*Aspergillus* sp. *Penicillium* sp. *Cladosporium* *Scopulariopsis* sp.	Traditional and composting Swineries	Rautiala et al.[24]
Canada	Culture	48 (Litter = sawdust) 0.86 (Slatted floor)	ND		*Aspergillus* sp. *Acremonium* *Beauveria* *Cladosporium* sp. *Muco* sp. *Scedosporium* sp. *Scopulariopsis* sp. *Penicillium* sp. *Paecilomyces* *Petriella*		Létourneau et al.[19]
United States Iowa	Culture	ND	28.3 (Hoop building) 20.5 (Conventional)	(2.1–428) (2.1–209)	ND		Thorne et al.[25]
United States Ohio	Culture Counting on cellulose ester membrane	4.1/2.0 (Summer/winter) 14.2/5.8 (Summer/winter)	ND		*Aspergillus* sp. *Penicillium* sp. *Basidiospore* *Cladosporium*	Personal exposure	Lee et al.[26]
Korea	Culture	3.14 log (cfu/m^3)	ND		ND		Kim et al.[27]

Continued

TABLE 1 Methods, Arithmetic Mean, Geometric Mean (or Median), Range (Min–Max), and Fungal Species Identified in Pig Farms in Various Countries—cont'd

Country	Method	AM (10^3 cfu/m^3)	GM or Median (10^3 cfu/m^3)	Min–Max (10^3 cfu/m^3)	Species (Most Frequent)	Remarks	References
Denmark	Counting/CAMNEA Culture	ND	8760 (Total cell/m^3) 380 (Viable fungi)	(0–140,000) (0–4300)	*Aspergillus* sp. *Cladosporium* *Penicillium* sp.		Radon et al.[28]
Taiwan	Culture	1.7 (Andersen microbial sampler) 3.3 (All-glass impinger) 3.8 (Polycarbonate filters)	ND	(0.248–4.3) (1.0–6.4) (0.8–10.1)	*Cladosporium* sp. *Aspergillus* sp. *Alternaria* *Penicillium* sp. *Fusarium*	Comparison of three sampling methods	Chang et al.[17]
Germany	Culture Counting/CAMNEA	0.22 540	0.13 440	(0.06–0.62) (170–1200)	ND	Test of biofilter efficacy	Martens et al.[29]

ND = not done.

significant increase in exposure to organic dust.[30] There are two main types of poultry production: farms housing laying hens for egg production and farms with broilers for meat production. Animals are kept either directly on the floor or in cages. Table 2 shows the airborne fungal concentrations measured in poultry farms in nine countries (11 studies in total). Results range between 97 and 440×10^3 cfu/m^3. In contrast to pig farms, seasonal influences have rarely been investigated in poultry farms. Monthly variations were found in Croatia, with the highest concentrations measured in June and the lowest in September (85.6×10^3 vs 0.75×10^3 cfu/m^3). The influences of other parameters were not reported. As in pig farms, the same four genera were predominant: *Aspergillus* sp., *Penicillium* sp., *Cladosporium* sp., and *Scopulariopsis* sp. However, a study performed in Texas[31] using pyrosequencing of airborne fungal DNA showed that the predominant species was *Sagenomella sclerotialis*, which represented 37% of the 14 identified fungi species. *Aspergillus ochraceus* and *Penicillium janthinellum* represented only 14% and 13%, respectively.

Dairy Farms

In dairy farms, animals do not stay in enclosed buildings throughout their entire life cycle and animal population densities (animals per square meter), as well as the total number of animals per farm, are usually lower than those observed in pig and poultry farms. Milking activities represent an important daily task during which animals are very active and thus generate a lot of organic dust. A 2014 study[32] showed that exposure to organic dust was higher when fully automatic milking was used, as well as during repenning of animals, handling of feed and seeds, and handling of silage and when spreading bedding. In big Californian dairies (more than 1000 cows), the time spent rebedding is associated with higher inhalable dust concentrations than the time spent milking.[33] In the Netherlands, personal exposure to airborne dust is higher in barns utilizing compost bedding compared with those with sawdust bedding.[34] Table 3 presents the results of four studies that measured airborne concentrations of fungi in enclosed barns. In France, seasonal variations in levels of several fungal species were observed, and peaks of fungal contamination were observed during straw handling.[35] This study also detected airborne aflatoxins and showed that farmers could possibly be exposed to *Stachybotrys chartarum* during routine barn work. In Romania, the level of fungi was higher in barns with bedding than in those without (19.1×10^3 vs 3.2×10^3 cfu/m^3).[36] In India, a monthly survey carried out over 2 years showed seasonal variations in the levels of total airborne and culturable fungi, with higher measurements in winter, late summer, and the rainy season.[37] Lee et al.[15] compared fungal exposure in various animal farms by using personal sampling. They showed that, on average, dairy farmers were not exposed to levels very different from those of pig and poultry farmers. *Aspergillus* sp., *Penicillium* sp., and *Cladosporium* sp. were the most frequently encountered species in dairy farms.

TABLE 2 Methods, Arithmetic Mean (AM), Geometric Mean (GM) or Median, Range (Min–Max), and Fungal Species Identified in Poultry Farms in Various Countries

Country	Method	AM (10^3 cfu/m^3)	GM or Median (10^3 cfu/m^3)	Min–Max (10^3 cfu/m^3)	Species (Most Frequent)	Remarks	References
Croatia	Culture	0.8–85.6	ND		ND	Sampling twice a month during 1 year	Matkovic et al.[38]
Poland	Culture Culture	ND	1.3–23.49 (Stationary) 7.6–30.44 (Personal)		17 Genera identified but without indication of frequencies	Stationary and personal sampling	Lawniczek-Walcyk et al.[39]
Portugal	Culture	ND	ND		*Scopulariopsis brevicaulis* *Rhizopus* sp. *Penicillium* sp. *Aspergillus* sp.		Viegas et al.[40]
United States North Carolina	Culture	3.0 7.4–17.0		(0.123–10.8) (0.131–82.9)	*Cladosporium* *Fusarium* Yeast *Fusarium* *Acremonium*	Inside the poultry house Outside the poultry house	Wang-Li et al.[41]
France	Culture	0.161 (CIP 10–M sampler 0.097 (Airport MD8 sampler)	ND		*Aspergillus* sp. *Penicillium* sp. *Alternaria* sp. *Scopulariopsis* sp. *Cladosporium*	Comparison of samplers and culture growth temperature	Nieguitsila et al.[42]

Poland	Culture	4.1–3.6	ND		*Aspergillus* sp. *Penicillium* sp. *Alternaria* sp. *Cladosporium* *Fusarium* sp.		Plewa and Lonc[43]
Croatia	Culture	12.7–25.9	ND	(4.9–68.4)	*Penicillium* sp. *Fusarium* sp. *Aspergillus* sp. *Mucor* sp. *Rhizopus* sp. *Scopulariopsis* sp.	Measurement of IgG to various molds	Rimac et al.[44]
United States Texas	DNA pyrosequencing		689 cells/m³ 255 cells/m³ 248 cells/m³ 1810 cells/m³		*Saganomella* *Aspergillus ochraceus* *Penicillium janthinellum* Total	Sampling on a mannequin	Nonnenmann et al.[31]
United States Ohio	Culture Counting on cellulose ester membrane	28 18	ND		*Cladosporium* *Aspergillus* sp. *Penicillium* sp. *Basidiospore*	Personal exposure	Lee et al.[26]
South Africa	Culture	0.7		(0–4.3)	ND	Comparison inside/outside	Venter et al.[45]
Switzerland	Counting/ CAMNEA Culture	ND	20,000 cells/ m³ (total fungi) 440 (Viable fungi)	(0–1,100,000) (14–11 × 10^7)	*Eurotium* sp. Thermophilic fungi		Radon et al.[28]

ND = not done.

TABLE 3 Methods, Arithmetic Mean, Geometric Mean (or Median), Range (Min–Max), and Fungal Species Identified in Dairy Farms in Various Countries

Country	Method	AM (10^3 cfu/m^3)	GM or Median (10^3 cfu/m^3)	Min–Max (10^3 cfu/m^3)	Species (Most Frequent)	Remarks	References
France	Culture	ND	0.0648–0.076	(0.002–0.59)	*Aspergillus fumigates* *Aspergillus glaucus* *Cladosporium cladosporioides* *Penicillium chrysogenum* *Stachybotrys chartarum* *Ulocladium chartarum*	Mycotoxin measurement	Lanier et al.[35]
Romania	Culture	16.1 24.1	15.4 21.8	(2.7–41) (5.8–78.5)	*Aspergillus* sp. *Penicillium* sp. *Cladosporium* sp. Yeast	Morning Evening	Popescu et al.[36]
United States Ohio	Culture Counting on cellulose ester membrane	39/0.3 (Summer/winter) 36/0.9 (Summer/winter)	ND	ND	*Aspergillus* sp. *Penicillium* *Basidiospore* *Cladosporium*	Personal exposure	Lee et al.[26]
India	Culture Count	0.165–2.22 (Viable fungi) 0.233–2.98 (Total fungi)	ND	ND	*Aspergillus* sp. *Penicillium* sp. *Cladosporium* sp. *Alternaria* *Nigrospora* sp. *Periconia* sp.	Monthly evolution of presence of each species	Adhikari et al.[8]

ND = not done.

FUNGAL AEROSOLS IN SAWMILLS

In sawmills, exposure to organic dust is high.[46–48] Before sawing, timber is very often kept outdoors under damp conditions that favor mold and bacterial growth. When timber is handled and sawn, fungi are aerosolized. Fungal concentrations up to 10^6–10^7 spores/m^3 have been reported.[49–51] Table 4 presents the airborne fungus concentrations measured in sawmills in various countries. A Korean study[52] comparing the levels of airborne fungi in various work sectors reported that levels in sawmills were higher than in either the livestock feed industry or metalworking fluid handling plants, with levels reaching up to 30×10^3 cfu/m^3. Very similar airborne concentrations of fungi were also observed in Switzerland,[53] where all of the 37 sawmills studied exceeded the occupational exposure guideline value of 1000 cfu/m^3. In Croatia, Klaric et al.[54] observed seasonal differences in fungal levels in two sawmills. Average concentrations were significantly higher in April, May, and July than in other months (September, November, December, and February), with a minimum of 160 cfu/m^3 measured during the winter period and a maximum of 14×10^3 cfu/m^3 measured in July. In Norway, *Rhizopus* has been shown to be the most abundant mold,[50,54] with a maximum count of 10×10^6 spores/m^3.[56] Wood trimmers in those high-exposure sawmills had significantly higher levels of *Rhizopus microsporus*-specific IgG and IgA antibodies. *Rhizopus* and *Penicillium* were the most abundant in Finnish sawmills,[49] whereas *Penicillium* predominated in Canadian[50] and Swiss sawmills.[53] In Poland, *Penicillium citrinum* was predominant in sawmills processing deciduous wood, whereas *Aspergillus fumigatus* predominated in sawmills processing coniferous wood.[57] In Canada and Poland, debarking work stations were the most highly contaminated by mold (up to 15×10^5 cfu/m^3 and mean of 15.3×10^3 cfu/m^3, respectively), in comparison to planning, sawing, and sorting.[51,57] In Poland[58] airborne concentrations of fungi were higher in fiberboard and chipboard factories than in sawmills (33.5×10^3 and 10.1×10^3 cfu/m^3 vs 4.2×10^3 cfu/m^3, respectively).

FUNGAL AEROSOLS IN WASTE SECTORS

Fungal Exposure during Collection, Sorting, and Recycling of Waste

Household waste represents a mix of components including biodegradable matter (kitchen and food waste), often called "biowaste," and nonbiodegradable matter (paper, cardboard, glass, plastics, cans, textiles, etc.), often called "dry waste." Wastes sorted at the source and nonseparated wastes are collected and transferred to specialized centers to be recycled and recovered. These waste recycling centers (WRCs) also treat similar wastes from municipal organizations, offices, hospitals, industry, etc. The nonrecoverable fraction of waste is generally sent to landfills or incinerators for ultimate disposal. The recycling process embraces several steps and involves both mechanical and manual operations for waste collection, transfer loading, crushing, and sorting.

TABLE 4 Methods, Arithmetic Mean, Geometric Mean (or Median), Range (Min–Max), and Fungal Species Identified in Sawmills in Various Countries

Country	Method	AM (10^3 cfu/m^3)	GM or Median (10^3 cfu/m^3)	Min–Max (10^3 cfu/m^3)	Species (Most Frequent)	Remarks	References
Croatia	Culture	1.6–7.3		(0.16–14)	*Penicillium* sp. *Paecilomyces* spp. *Chrysonilia sitophilia*		Klaric et al.[54]
Korea	Culture	ND	2.2	(0.108–30.9)	ND	Comparison with work sectors	Park et al.[52]
Canada	Culture	83		(0–1700)	*Penicillium spinulosum* *Penicillium myczinski* *Penicillium fellutanum* *Eupenicillium* sp.		Duchaine and Mériaux[51]
Norway	Microscopic spore observation			(1.0–10,000)		Only *Rhizopus microsporus*	Rydjord et al.[56]
Switzerland	Culture	14.776		(9.5–30)	*Penicillium* sp. *Aspergillus* sp.		Oppliger et al.[53]
Poland	Culture	33.5 (Fiber board factory) 10.1 (Chipboard factories)			*Aspergillus fumigatus* *Penicillium* sp. *Cladosporium*		Dutkiewicz et al.[58]
Poland	Culture	4.2 (Deciduous wood) 3.9 (Coniferous wood)			*A. fumigatus* *Penicillium* sp.		Dutkiewicz et al.[57]

ND = not done.

Domestic waste usually contains high quantities of fungi that are able to develop on organic matter and can be aerosolized when wastes are handled. The emission of fungal aerosols during waste recycling operations has been reported in several studies. The mean ambient concentration of airborne fungi in Korean recycling centers was measured at 1.8×10^4 cfu/m^3.[59] A study carried out in two municipal solid waste treatment plants in Finland revealed ambient concentrations from 470 to 2.9×10^5 cfu/m^3 for airborne culturable fungi and from less than the limit of detection (LOD) to 2.7 mg/m^3 for airborne dust.[60] Similar ambient levels were reported in another study carried out in Finland,[61] and lower levels were measured during glass bottle recycling in Canada.[62] Exposure of workers handling waste has also been reported. Thus, individual exposure to airborne culturable fungi among waste collectors in Quebec was found to be between 4.8×10^3 and 1.0×10^5 cfu/m^3.[63] Lower levels of exposure to fungi ($<1.8 \times 10^4$ cfu/m^3) were found among waste collectors in Germany.[64] In Finland, individual exposure to bioaerosols of workers handling waste (including waste collectors and compost workers) was from <LOD to 2.0×10^6 spores/m^3 for fungi (microscopic counts), from 2 to 220 ng/m^3 for (1,3)-β-D-glucans, and from <LOD to 6.6 mg/m^3 for inhalable dust.[65] Exposures of domestic waste collectors in the Netherlands were from 260 to 3.1×10^4 ng/m^3 for (1,3)-β-D-glucans and from 0.2 to 9.1 mg/m^3 for inhalable dust.[66] In Germany, individual exposures to airborne fungi were measured up to 6.2×10^6 cfu/m^3 during recycling of textiles, up to 1.8×10^6 cfu/m^3 during sorting and recycling of paper and paperboard, and up to 3.2×10^6 cfu/m^3 during sorting and recycling of plastic wastes.[67] Gladding et al.[68] reported bioaerosol measurements in WRCs in England and Wales treating a mixture of household and commercial waste materials. They found individual exposure levels from <LOD to 137 ng/m^3 for (1,3)-β-D-glucans and from <LOD to 62.6 mg/m^3 for inhalable dust.[68] Waste handlers working in a paper sorting plant in Denmark were found to be exposed to airborne fungi at levels between 1.5×10^4 and 4.5×10^5 cfu/m^3.[69] Exposure levels between 2.4×10^4 and 1.1×10^5 cfu/m^3 were recorded during waste recycling and sorting in Korea.[70]

The work tasks at greatest risk of exposure included unloading, crushing, sorting, and maintenance operations. The main airborne fungi isolated in recent studies have belonged to *Penicillium* sp., *Aspergillus niger*, *A. fumigatus*, *Cladosporium* sp., and *Chrysonilia sitophilia*.[60,61,69,71] Few studies suggested seasonal variation in exposure.[64,72] Exposure to bioaerosols in WRCs is associated with a range of adverse respiratory health effects, including upper airway inflammation, allergic asthma, and allergic rhinitis.[65,66,68] However, data on exposure and health are scarce, making it difficult to determine the precise role of fungi in these reported symptoms.[73]

Fungal Exposure Associated with Waste Composting

Composting is a natural process involving the biological degradation of organic wastes (green waste, household waste, sewage sludge, etc.) under aerobic conditions. It leads to a reduction in waste volumes and produces a valuable end-product for agriculture and gardening. The process is based on the proliferation of high concentrations of microorganisms and the development of complex microflora as temperatures increase (up to 60–70 °C). Thus, thermophilic bacteria and fungal microorganisms in compost have been reported at levels of up to 10^5–10^7 cfu/g of matter.[74,75] Composting plants vary greatly in size, type of waste composted, design, and degree of enclosure.[75] Published molecular biology studies have brought new perspectives on the biodiversity of the microbial community involved in the composting process. These studies revealed that the fungal taxa that appeared most abundantly fluctuated according to such factors as type of waste, type of process, and the stage of that process. They included genera such as *Chaetomium*, *Cladosporium*, *Scytalidium*, *Thermomyces*, *Arthrographis*, *Fusarium*, *Aspergillus*, and *Penicillium*.[76–78]

Waste and compost handling during the process (compost turning, shredding, screening, movement of vehicles, etc.) has been shown to release airborne microorganisms into the ambient air of composting facilities. Thus, a wide range of concentrations of airborne fungi have been reported in composting facilities treating household wastes, green wastes, wood chips, sewage, biowastes, or the products of anaerobic digestion. Concentrations of mesophilic fungi were measured from 1.0×10^2 to 5.0×10^6 cfu/m^3 in France,[79–81] from 480 to 1.7×10^4 cfu/m^3 in the United States[82] and from 1.2×10^2 to 3.0×10^5 cfu/m^3 in Austria.[83] A mean concentration of 4.0×10^6 cfu/m^3 was reported in a study carried out in Germany.[84] For airborne thermophilic fungi, ambient concentrations were measured from $<$LOD to 3.4×10^7 cfu/m^3 in France[79,81] and at a mean concentration of 3.0×10^5 cfu/m^3 in Germany.[84] Airborne levels of *A. fumigatus* were found between the LOD and 1.9×10^5 cfu/m^3 in various countries.[83,85–87] One specific study was carried out in the vehicle cabins of different composting facilities, with reported ambient concentrations from 7.5×10^2 to 5.7×10^5 cfu/m^3.[88] High personal exposure levels to airborne fungi (up to 10^7 cfu/m^3), to inhalable dust, and to (1,3)-β-D-glucans were measured in Canada, France, Germany, and the Netherlands.[63,89–91] The dispersion of bioaerosols to populations neighboring composting plants is also a concern.[92]

Several studies have identified shredding, pile turning, and compost screening as specific activities that emit heavy loads of bioaerosols.[79,92,93] The abundance of *Aspergillus*, *Thermomyces*, and *Penicillium* in bioaerosols generated by composting activities has been revealed by several studies based on both culture methods[80,94] and molecular methods.[95,96] Jobs that have already been described as subject to exposure include driving in vehicles with cabins[89,90,97,98] as well as clearing up spilled soil and performing maintenance.[90] Indeed, bioaerosol exposure in composting facilities has been associated with increased respiratory disorders and dermal pathologies among compost workers.[89,90,100] Thus, preventive measures are required in composting plants to improve employee protection from exposure to fungal aerosols.

FUNGAL AEROSOLS IN THE FOOD INDUSTRY

Dry Meat

Fungi are involved in the production of dry meat products such as dry-cured sausages, salami, and ham. For example, the production process for dry sausages begins with grinding the fresh meat pieces, followed by blending the ground meat with the curing ingredients and microbial inoculants. The subsequent meat paste is then stuffed into natural or synthetic casings to form sausages that are hung on trolleys. In some processes the sausages are watered with a commercial starter culture containing fungi (*Penicillium nalgiovense* or *Penicillium chrysogenum*). In other processes, the surface inoculation step is not carried out and products directly undergo the next processing stage. The sausages are incubated at optimal temperature and humidity for the time necessary for the fermentation and transformation of the meat. Once the fermentation step is complete, the sausages undergo a drying cycle (by lowering air humidity and enhancing air circulation) to remove excess moisture from the products. Fungal growth occurs on the surface of dry meat products during this process. Specific operations such as curing can be included in the process. Products can also be brushed manually or mechanically to remove fungi from their surfaces. They are then handled for packaging and shipment. Operations are usually performed in workshops dedicated to more than one step.

Typically, dry meat products are spontaneously colonized by a heterogeneous mycoflora including several genera and species. However, the diversity of the mycoflora that develops depends on the type of meat products, the processes, the configuration of workshops, and the season. The fungi most frequently isolated from dry sausage casings are *Penicillium* sp. (*P. nalgiovense*, *P. chrysogenum*, *Penicillium olsonii*, *Penicillium verrucosum*, *Penicillium viridicatum*, *Penicillium nordicum*, *Penicillium solitum*, *Penicillium oxalicum*), *Eupenicillium crustaceum*, and *Eurotium amstelodami*.[101–103] Ochratoxin A has also been reported on such products.[101] These genera, and others, have been found on salami and cured-meat products.[104] The surface mycobiota contributes to the product's appearance, and the development of its organoleptic properties is driven by microbial metabolic activity (proteolysis, lipolysis, etc.). The use of commercial starter cultures of *P. nalgiovense* or *P. chrysogenum* usually ensures standardized final products and prevents the proliferation of undesired microorganisms such as pathogenic bacteria and mycotoxins that produce molds.[105]

Whatever the production process used, the handling of moldy products may lead to the release of fungal spores, fragments, debris, products, and compounds into workshop air. Data regarding concentrations of airborne fungi during the production of fermented and dried meat and the associated occupational exposure are scarce. In a dry-sausage production plant in Italy, concentrations of airborne fungi were found at 10^5 and 10^8 cfu/m^3 during the maturation–drying step and the brushing step, respectively.[106] The airborne fungi most isolated were *P. chrysogenum*, *Penicillium notatum*, *Aspergillus versicolor*, *Mucor mucedo*, and *Cryptococcus laurentii*. Another Italian study reported concentrations up to 1.1×10^9 cfu/m^3 in the production of salami, the dominant fungi being *Penicillium camemberti*.[107] Lower concentrations ($<1.0\times10^3$ cfu/m^3) were measured in other studies.[108] Qualitative bioaerosol assessments were also done by collecting the air samples by sedimentation directly on petri dishes. The results from those samples can be considered, but this method is not suitable for bioaerosol studies carried out for risks purposes.[102,109]

Cases of hypersensitivity pneumonitis have been reported among workers in dry meat product workshops. Some of these cases were ascribed to occupational exposure to fungi from the surface mycoflora of dry sausages, especially *P. nalgiovense*.[110,111] Fungi were also incriminated in allergies observed among workers carrying out a similar activity.[112] In a factory producing salami, similar symptoms were attributed to *P. camemberti*.[107] *Penicillium camemberti* was also involved in asthma among workers packaging dry sausages.[113] Fungi belonging to *Penicillium* sp. produce large amounts of conidia that can be easily aerosolized during such processes and several allergenic species have been identified in the genera.[114]

Fungi in Wineries and Wine Cellars

Fungi are well-known components of the bioaerosols in wine production and storage cellars. They are usually found on the surfaces of walls and wine bottles[115] and may be released into the air during work operations and air movements in cellars. Published studies revealed moderate ambient concentrations ranging from 55 to 1.6×10^4 cfu/m^3 in Austria[77,116] and from 57 to 2.5×10^3 cfu/m^3 in France.[117] Airborne culturable fungi from wine cellars are also characterized by a high biodiversity, with dominant genera belonging to *Penicillium*, *Aspergillus*, *Cladosporium*, *Phialophora*, *Phoma*, *Trichoderma*, and *Ulocladium*. The occurrence of allergenic fungi suggests possible health hazards among workers and risks for the alteration of wine quality during production and storage.

Fungal Aerosols during Cheese Production

Cheeses are food goods produced by curdling milk using bacteria or enzymes. The resulting curds are drained, processed, and cured and ripened in a range of ways and using numerous techniques that lead to thousands of varieties around the world. Fungi also play a fundamental role during cheese production, especially during the ripening of washed rind cheeses (Reblochon, Saint-Nectaire, etc.), blue-veined cheeses (Roquefort, etc.), or mold rind cheeses (Camembert, Brie, etc.). They rapidly grow over the cheese's surface, preventing the growth of contaminants as well as promoting the development of flavor and other organoleptic features.[117] The biodiversity of fungi on cheeses depends on the type of cheese and on the ripening conditions; the most cited fungi belong to *Penicillium*, *Acremonium*, *Alternaria*, *Aspergillus*, *Cladosporium*, *Geotrichum*, *Mucor*, *Rhizopus*, and *Trichoderma*.[60,116,118–120]

These fungi may be aerosolized when cheeses are handled during ripening or packaging, but data regarding occupational exposure of cheese workers to fungi are scarce. One study reported ambient concentrations of airborne fungi in ripening cellars of up to 2.5×10^4 cfu/m^3 when pressed, cooked cheese (Comté) was brushed; up to 2.0×10^3 cfu/m^3 during the storage of similar cheeses (Emmental); and up to 4.0×10^4 cfu/m^3 during the storage of soft cheese (Mont d'Or).[121] Lower concentrations, from 90 to 6.1×10^2 cfu/m^3, were found in a Brazilian cheese production plant,[122] and below 7.5×10^2 cfu/m^3 during the production of pressed, uncooked cheeses.[123,124] A 2014 study carried out in a French cellar during the ripening of natural-rind cheeses revealed high exposure to fungi, from 1.0×10^4 to 2.0×10^8 cfu/m^3, with the dominant fungi being *Mucor fuscus* and *Penicillium* sp.[125]

That study suggested that cheese workers (brushers, selectors, packagers, etc.) may be exposed to high microbial levels and that additional studies are required for a better characterization of exposure in the sector. Several cases of the development of hypersensitivity pneumonitis to fungi among cheese workers have been described.[126,127]

Fungi in Other Food Production

Very few data have been published about other occupational environments in which fungi could be aerosolized from food products. For example, concentrations from the LOD to 1.5×10^5 cfu/m^3 were recorded during grape stemming and crushing.[128] Moderate ambient concentrations of airborne fungi (mean values between 23 and 3.0×10^3 cfu/m^3) were found in a noodle factory.[121] The main fungi present were yeasts belonging to the *Cladosporium*, *Penicillium*, and *Aspergillus* taxa, and noodle factory atmospheres also contain flour, flavorings, and other ingredients. Furthermore, fungi and mycotoxins have been found in spices and herbs used as food industry ingredients,[129,130] and bioaerosol exposure of workers handling such products may be an issue. Further investigations may be required in such occupational environments.

FUNGAL AEROSOLS DURING PLANT AND GRAIN HANDLING

Exposure to Airborne Fungi during Grain and Cereal Handling

Grain and cereal harvesting, handling, and shredding generally release large amounts of airborne particles including dust and fungi. High ambient fungal levels have been reported during threshing.[131] Studies carried out in Norway revealed high exposures to fungal spores (microscopic analysis) from the LOD to 6.4×10^6 spores/m^3 (Ref. 131) and up to 5.2×10^9 spores/m^3 (Ref. 132) during threshing and storage of different grains. Mycotoxin-producing fungi were also detected.[133] Exposures to airborne fungi were found between 1.8×10^3 and

1.3×10^7 cfu/m^3 during grain handling in the United States.[15] Lower levels were reported during grain handling in eastern Poland.[131] Grain grinding and flour processing were also shown to be activities that expose workers to dust, fungi, and mycotoxins.[134–136]

Exposure to Airborne Fungi during Herb Processing

Processing of herbs has also been associated with a high concentration of airborne fungi. Ambient concentrations up to 6.3×10^5 cfu/m^3 were measured during cleaning, cutting, grinding, sieving, sorting, and packing of various herbs used for production of spices, medications, and cosmetics.[137] Ambient levels between 10 and 1.0×10^5 cfu/m^3 were found in farm air during processing of peppermint and chamomile plants, with *Alternaria alternata* as the dominant fungus.[131,138] Airborne fungi were found at a concentration of 9.7×10^4 cfu/m^3 during thyme threshing and associated with a case of allergic alveolitis and occupational dermatitis.[139,140] Health effects among valerian growers exposed to herb dust were also reported.[141]

Exposure to Airborne Fungi in Greenhouses

A growing interest has been paid to bioaerosols produced during work in greenhouses. Indeed, high individual exposure levels to airborne dust (up to 15 mg/m^3) and fungi (up to 8.7×10^7 cfu/m^3) were measured in greenhouses producing and packaging vegetables (tomatoes, cucumbers, cabbage, etc.) as well as plants and flowers.[142–146] Fungi were also found, using a nonquantitative method, in the air of a botanical garden.[147] Exposure was found to depend on the work task, the type of crop, and its growth stage and open-field versus closed working spaces.

Exposure to Airborne Fungi during Peat Moss Processing

Peat moss is a production mainly composed of organic matter colonized by a large population of microorganisms. Ambient concentrations were measured up to 441.7 mg/m^3 for inhalable dust and up to 1.0×10^8 cfu/m^3 for mesophilic fungi during sieving, mixing, and bagging.[148] A high incidence of mold sensitization of workers is suggested in peat moss processing plants.[149]

CONCLUSIONS

Fungi are ubiquitous in the environment and are able to colonize a wide range of ecological niches including plants, grains, and animals, as well as anthropogenic environments such as wastes. Their metabolic properties are also exploited in the food industry. This review of 14 years' literature shows that workers in many occupational sectors are exposed to moderate, high, or very high levels of airborne fungal particles. Such exposures are reported in a wide variety of occupational settings, including animal confinement buildings; swine, dairy, and poultry farms; sawmills; recycling centers; composting facilities; cheese-ripening cellars; greenhouses; flour mills; etc. The exposures generally occur during specific work tasks involving the handling of contaminated materials as well as operations promoting the release of large amounts of fungal particles into the air (brushing, shredding, turning, etc.).

Results of different studies are difficult to compare. Indeed, many different methods are used to collect and analyze airborne fungi and results can be highly variable depending on the methodology. Collection of airborne fungi is based on impaction, impingement, or filtration. Several media are used (different nutrient agars, different types of filters). Air is sampled at different volumes, at different flow rates, and for different durations. Culture-dependent analyses differ in temperature and length of incubation, whereas non-culture-dependent methods use different techniques (counting chamber, coloration with fluorescent or nonfluorescent dyes, DNA amplification, pyrosequencing, etc.). Each method has its advantages, disadvantages, and limitations. Methods for the quantitative assessment of airborne levels of noninfectious microorganisms in highly contaminated work environments were comprehensively discussed by Eduard and Heederik.[150] The development of international standards would be a very important step forward because it would allow the reliable comparison of exposures between different occupational environments around the world.

Fungal particles measured in occupational environments include spores, hyphal fragments, debris from these entities, and associated compounds such as glucans and mycotoxins.[2] Inhaled particles may be deposited in the respiratory tract and exposures have been associated with health disorders.[5] There are no occupational exposure limit values admitted at an international level. Guidelines exist, but they are currently based on culture-dependent methods, which are now recognized to underestimate true concentrations. Moreover, it is globally accepted that fragments of fungi or dead fungal cells can still have allergenic, toxic, or irritating properties, and thus they must be taken into account for bioaerosol exposure characterization in occupational hygiene situations. In the future, those characterizations should include the measurement of airborne mycotoxins[21,40] or the detection of mycotoxin metabolites in biological samples using biomarkers.[151]

Fungal exposures found in highly contaminated environments often exceed the guideline values recommended by the pertinent local authorities. Technical protective measures are not always easy to put in place, and wearing individual respiratory protection is not always possible. Harvesting grain, washing cheese, and handling salami seem to be the occupational situations with the most potential for exposure, with levels of up to 10^8 to 10^9 cfu/m^3.

REFERENCES

1. Fischer G, Dott W. Relevance of airborne fungi and their secondary metabolites for environmental, occupational and indoor hygiene. *Arch Microbiol* 2003;**179**(2):75–82.
2. Gorny LR. Filamentous microorganisms and their fragments in indoor air—a review. *Ann Agric Environ Med* 2004;**11**:185–97.
3. Mandal J, Brandl H. Bioaerosols in indoor environment—a review with special reference to residential and occupational locations. *Open Environ Biol Monit J* 2011;**4**:83–96.
4. Eduard W. The Nordic Expert Group for criteria documentation of health risks from chemicals. 139. Fungal spores. *Arb och Hälsa* 2006;**21**:1–145.
5. Fung F, Hughson WG. Health effects of indoor fungal bioaerosol exposure. *Appl Occup Environ Hyg* 2003;**18**(7):535–44.
6. Eduard W. A health-based criteria document on fungal spore exposure in the working population. Is it relevant for the general population? *Indoor Air* 2008;**18**(3):257–8.
7. Simon X, Duquenne P. Assessment of workers' exposure to bioaerosols in a French cheese factory. *Ann Occup Hyg* 2014;**58**(6):677–93.
8. Adhikari A, Reponen T, Lee SA, Grinshpun SA. Assessment of human exposure to airborne fungi in agricultural confinements: personal inhalable sampling versus stationary sampling. *Ann Agric Environ Med* 2004;**11**(2):269–77.
9. Basinas I, Schlunssen V, Takai H, Heederik D, Omland O, Wouters IM, et al. Exposure to inhalable dust and endotoxin among Danish pig farmers affected by work tasks and stable characteristics. *Ann Occup Hyg* 2013;**57**(8):1005–19.
10. O'Shaughnessy PT, Donham KJ, Peters TM, Taylor C, Altmaier R, Kelly KM. A task-specific assessment of swine worker exposure to airborne dust. *J Occup Environ Hyg* 2010;**7**(1): 7–13.
11. O'Shaughnessy P, Peters T, Donham K, Taylor C, Altmaier R, Kelly K. Assessment of swine worker exposures to dust and endotoxin during hog load-out and power washing. *Ann Occup Hyg* 2012;**56**(7):843–51.
12. Duchaine C, Grimard Y, Cormier Y. Influence of building maintenance, environmental factors, and seasons on airborne contaminants of swine confinement buildings. *Am Ind Hyg Assoc J* 2000;**61**(1):56–63.
13. Masclaux FG, Sakwinska O, Charriere N, Semaani E, Oppliger A. Concentration of airborne *Staphylococcus aureus* (MRSA and MSSA), total bacteria, and endotoxins in pig farms. *Ann Occup Hyg* 2013;**57**(5):550–7.
14. Jo W-K, Kang J-H. Exposure levels of airborne bacteria and fungi in Korean swine and poultry sheds. *Arch Environ Occup Health* 2005;**60**(3):140–6.
15. Lee SA, Adhikari A, Grinshpun SA, McKay R, Shukla R, Reponen T. Personal exposure to airborne dust and microorganisms in agricultural environments. *J Occup Environ Hyg* 2006;**3**(3):118–30.
16. Kristiansen A, Saunders AM, Hansen AA, Nielsen PH, Nielsen JL. Community structure of bacteria and fungi in aerosols of a pig confinement building. *FEMS Microbiol Ecol* 2012;**80**(2):390–401.
17. Chang CW, Chung H, Huang CF, Su HJJ. Exposure of workers to airborne microorganisms in open-air swine houses. *Appl Environ Microbiol* 2001;**67**(1):155–61.
18. Sowiak M, Brodka K, Buczynska A, Cyprowski M, Kozajda A, Sobala W, et al. An assessment of potential exposure to bioaerosols among swine farm workers with particular reference to airborne microorganisms in the respirable fraction under various breeding conditions. *Aerobiologia* 2012;**28**(2):121–33.

19. Letourneau V, Nehme B, Meriaux A, Masse D, Duchaine C. Impact of production systems on swine confinement buildings bioaerosols. *J Occup Environ Hyg* 2010;**7**(2):94–102.
20. Sauter EA, Petersen CF, Steele EE, Parkinson JF, Dixon JE, Stroh RC. The airborne microflora of poultry houses. *Poult Sci* 1981;**60**(3):569–74.
21. Viegas S, Veiga L, Verissimo C, Sabino R, Figueiredo P, Almeida A, et al. Occupational exposure to aflatoxin B1: the case of poultry and swine production. *World Mycotoxin J* 2013;**6**(3):309–15.
22. Sabino R, Faisca VM, Carolino E, Verissimo C, Viegas C. Occupational exposure to aspergillus by swine and poultry farm workers in Portugal. *Toxicol Environ Health A* 2012;**75**(22–23):1381–91.
23. Sowiak M, Brodka K, Kozajda A, Buczynska A, Szadkowska-Stanczyk I. Fungal aerosol in the process of poultry breeding – quantitative and qualitative analysis. *Medyc Pracy* 2012;**63**(1):1–10.
24. Rautiala S, Kangas J, Louhelainen K, Reiman M. Farmers' exposure to airborne microorganisms in composting swine confinement buildings. *AIHAJ* 2003;**64**(5):673–7.
25. Thorne PS, Ansley AC, Perry SS. Concentrations of bioaerosols, odors, and hydrogen sulfide inside and downwind from two types of swine livestock operations. *Occup Environ Hyg* 2009;**6**(4):211–20.
26. Lee JA, Thorne PS, Reynolds SJ, O'Shaughnessy PT. Monitoring risks in association with exposure levels among wastewater treatment plant workers. *Occup Environ Med* 2007; **49**(11):1235–48.
27. Kim KY, Ko HJ, Kim HT, Kim CN, Kim YS. Assessment of airborne bacteria and fungi in pig buildings in Korea. *Biosys Eng* 2008;**99**(4):565–72.
28. Radon K, Danuser B, Iversen M, Monso E, Weber C, Hartung J, et al. Air contaminants in different European farming environments. *Ann Agric Environ Med* 2002;**9**(1):41–8.
29. Martens W, Martinec M, Zapirain R, Stark M, Hartung E, Palmgren U. Reduction potential of microbial, odour and ammonia emissions from a pig facility by biofilters. *Int J Hyg Environ Health* 2001;**203**(4):335–45.
30. Oppliger A, Charriere N, Droz P-O, Rinsoz T. Exposure to bioaerosols in poultry houses at different stages of fattening; use of real-time PCR for airborne bacterial quantification. *Ann Occup Hyg* 2008;**52**(5):405–12.
31. Nonnenmann MW, Bextine B, Dowd SE, Gilmore K, Levin JL. Culture-independent characterization of bacteria and fungi in a poultry bioaerosol using pyrosequencing: a new approach. *J Occup Environ Hyg* 2010;**7**(12):693–9.
32. Basinas I, Sigsgaard T, Erlandsen M, Andersen N, Takai H, Heederik D, et al. Exposure-affecting factors of dairy farmers' exposure to inhalable dust and endotoxin. *Ann Occup Hyg* 2014;**58**(6):707–24.
33. Garcia J, Bennett DH, Tancredi D, Schenker MB, Mitchell D, Reynolds SJ, et al. Occupational exposure to particulate matter and endotoxin for California dairy workers. *Int J Hyg Environ Health* 2013;**216**:56–63.
34. Samadi S, van Eerdenburg FJCM, Jamshidifard A-R, Otten GP, Droppert M, Heederik DJJ, et al. The influence of bedding materials on bio-aerosol exposure in dairy barns. *J Expo Sci Environ Epidemiol* 2012;**22**(4):361–8.
35. Lanier C, Andre V, Seguin V, Heutte N, El Kaddoumi A, Bouchart V, et al. Recurrence of *Stachybotrys chartarum* during mycological and toxicological study of bioaerosols collected in a dairy cattle shed. *Ann Agric Environ Med* 2012;**19**(1):61–7.
36. Popescu S, Borda C, Diugan EA. Microbiological air quality in tie-stall dairy barns and some factors that influence it. *Afr J Agric Res* 2011;**6**(32):6726–34.
37. Adhikari A, Sen MM, Gupta-Bhattacharya S, Chanda S. Volumetric assessment of airborne fungi in two sections of a rural indoor dairy cattle shed. *Environ Int* 2004;**29**(8):1071–8.

38. Matkovic K, Vucemilo M, Stokovic I, Simic R, Marusic D, Vinkovic B, et al. Concentrations of airborne bacteria and fungi in a livestock building with caged laying hens. *Vet Arhiv* 2013;**83**(4):413–24.
39. Lawniczek-Walczyk A, Gorny RL, Golofit-Szymczak M, Niesler A, Wlazlo A. Occupational exposure to airborne microorganisms, endotoxins and beta-glucans in poultry houses at different stages of the production cycle. *Ann Agric Environ Med* 2013;**20**(2):259–68.
40. Viegas S, Veiga L, Malta-Vacas J, Sabino R, Figueiredo P, Almeida A, et al. Occupational exposure to aflatoxin (AFB1) in poultry production. *J Toxicol Environ Health Part A* 2012;**75**:1330–40.
41. Wang-Li L, Li Q, Byfield GE. Identification of bioaerosols released from an egg production facility in the Southeast United States. *Environ Eng Sc* 2013;**30**(1):2–10.
42. Nieguitsila A, Arne P, Durand B, Deville M, Benoit-Valiergue H, Chermette R, et al. Relative efficiencies of two air sampling methods and three culture conditions for the assessment of airborne culturable fungi in a poultry farmhouse in France. *Environ Res* 2011;**111**(2):248–53.
43. Plewa K, Lonc E. Analysis of airborne contamination with bacteria and moulds in poultry farming: a case study. *Pol J Environ Stud* 2011;**20**(3):725–31.
44. Rimac D, Macan J, Varnai VM, Vucemilo M, Matkovic K, Prester L, et al. Exposure to poultry dust and health effects in poultry workers: impact of mould and mite allergens. *Int Arch Occup Environ Health* 2010;**83**(1):9–19.
45. Venter P, Lues JFR, Theron H. Quantification of bioaerosols in automated chicken egg production plants. *Poult Sc* 2004;**83**(7):1226–31.
46. Alwis U, Mandryk J, Hocking AD, Lee J, Mayhew T, Baker W. Dust exposures in the wood processing industry. *Am Ind Hyg Assoc J* 1999;**60**(5):641–6.
47. Cormier Y, Merlaux A, Duchaine C. Respiratory health impact of working in sawmills in eastern Canada. *Arch Environ Health* 2000;**55**(6):424–30.
48. Demers PA, Teschke K, Davies HW, Kennedy SM, Leung V. Exposure to dust, resin acids, and monoterpenes in softwood lumber mills. *AIHAJ* 2000;**61**(4):521–8.
49. Roponen M, Seuri M, Nevalainen A, Hirvonen MR. Fungal spores as such do not cause nasal inflammation in mold exposure. *Inhal Toxicol* 2002;**14**(5):541–9.
50. Eduard W, Sandven P, Levy F. Serum IgG antibodies to mold spores in two Norwegian sawmill populations: relationship to respiratory and other work-related symptoms. *Am J Ind Med* 1993;**24**(2):207–22.
51. Duchaine C, Meriaux A, Thorne PS, Cormier Y. Assessment of particulates and bioaerosols in eastern Canadian sawmills. *AIHAJ* 2000;**61**(5):727–32.
52. Park H, Park H, Lee I. Microbial exposure assessment in sawmill, livestock feed industry, and metal working fluids handling industry. *Saf Health Work* 2010;**1**(2):183–91.
53. Oppliger A, Rusca S, Charriere N, Duc TV, Droz PO. Assessment of bioaerosols and inhalable dust exposure in Swiss sawmills. *Ann Occup Hyg* 2005;**49**(5):385–91.
54. Klaric MS, Varnai VM, Calusic AL, Macan J. Occupational exposure to airborne fungi in two Croatian sawmills and atopy in exposed workers. *Ann Agric Environ Med* 2012;**19**(2):213–9.
55. Eduard W, Sandven P, Levy F. Relationships between exposure to spores from *Rhizopus microsporus* and *Paecilomyces variotii* and serum IgG antibodies in wood trimmers. *Int Arch Allergy Immunol* 1992;**97**(4):274–82.
56. Rydjord B, Eduard W, Stensby B, Sandven P, Michaelsen TE, Wiker HG. Antibody response to long-term and high-dose mould-exposed sawmill workers. *Scand J Immunol* 2007;**66**(6):711–8.
57. Dutkiewicz J, Krysinska-Traczyk E, Prazmo Z, Skorska C, Sitkowska J. Exposure to airborne microorganisms in polish sawmills. *Ann Agric Environ Med* 2001;**8**(1):71–80.
58. Dutkiewicz J, Olenchock SA, Krysinska-Traczyk E, Skorska C, Sitkowska J, Prazmo Z. Exposure to airborne microorganisms in fiberboard and chipboard factories. *Ann Agric Environ Med* 2001;**8**(2):191–9.

59. Park D, Ryu S, Kim S, Byun H, Yoon C, Lee K. Airborne bacteria and fungi associated with waste-handling work. *Int J Occup Environ Health* 2013;**19**(4):311–8.
60. Lehtinen J, Tolvanen O, Nivukoski U, Veijanen A, Hänninen K. Occupational hygiene in terms of volatile organic compounds (VOCs) and bioaerosols at two solid waste management plants in Finland. *Waste Manag* 2013;**33**(4):964–73.
61. Tolvanen OK. Airborne bio-aerosols and noise in a dry waste treatment plant in Pietarsaari, Finland. *Waste Manag Res* 2001;**19**(2):108–14.
62. Kennedy SM, Copes R, Bartlett KH, Brauer M. Point-of-sale glass bottle recycling: indoor airborne exposures and symptoms among employees. *Occup Environ Med* 2004;**61**(7):628–35.
63. Lavoie J, Dunkerley CJ, Kosatsky T, Dufresne A. Exposure to aerosolized bacteria and fungi among collectors of commercial, mixed residential, recyclable and compostable waste. *Sci Total Environ* 2006;**370**:23–8.
64. Neumann HD, Balfanz J, Becker G, Lohmeyer M, Mathys W, Raulf-Heimsoth M. Bioaerosol exposure during refuse collection: results of field studies in the real-life situation. *Sci Total Environ* 2002;**293**(1–3):219–31.
65. Heldal KK, Halstensen AS, Thorn J, Djupesland P, Wouters I, Eduard W, et al. Upper airway inflammation in waste handlers exposed to bioaerosols. *Occup Environ Med* 2003;**60**(6):444–50.
66. Wouters IM, Hilhorst SKM, Kleppe P, Doekes G, Douwes J, Peretz C, et al. Upper airway inflammation and respiratory symptoms in domestic waste collectors. *Occup Environ Med* 2002;**59**(2):106–12.
67. Hebisch R, Linsel G. Workers' exposure to hazardous substances and biological agents in recycling enterprises. *Gefahrst Reinhalt Luft* 2012;**72**(4):163–9.
68. Gladding T, Thorn J, Stott D. Organic dust exposure and work-related effects among recycling workers. *Am J Ind Med* 2003;**43**(6):584–91.
69. Breum NO, Würtz H, Midtgaard U, Ebbehøj N. Dustiness and bio-aerosol exposure in sorting recyclable paper. *Waste Manag Res* 1999;**17**(2):100–8.
70. Park D-U, Ryu S-H, Kim S-B, Yoon C-S. An assessment of dust, endotoxin, and microorganism exposure during waste collection and sorting. *J Air Waste Manag Assoc* 2011;**61**(4):461–8.
71. Malta-Vacas J, Viegas S, Sabino R, Viegas C. Fungal and microbial volatile organic compounds exposure assessment in a waste sorting plant. *J Toxicol Environ Health Part A* 2012;**75**(22–23):1410–7.
72. Nielsen BH, Nielsen EM, Breum NO. Seasonal variation in bioaerosol exposure during biowaste collection and measurements of leaked percolate. *Waste Manag Res* 2000;**18**(1):64–72.
73. Porta D, Milani S, Lazzarino A, Perucci CA, Forastiere F. Systematic review of epidemiological studies on health effects associated with management of solid waste. *Environ Health* 2009;**8**(1):60–74.
74. Lacey J. Actinomycetes in composts. *Ann Agric Environ Med* 1997;**4**(1):113–21.
75. Swan JRM, Kesley A, Crook B, Gilbert EJ. In: Executive H-HS, editor. *Occupational and environmental exposure to bioaerosols from composts and potential health effects—a critical review of published data*. 2003. Norwich.
76. Sørensen LM, Jacobsen T, Nielsen PV, Frisvad JC, Koch AG. Mycobiota in the processing areas of two different meat products. *Int J Food Microbiol* 2008;**124**(1):58–64.
77. Haas D, Galler H, Habib J, Melkes A, Schlacher R, Buzina W, et al. Concentrations of viable airborne fungal spores and trichloroanisole in wine cellars. *Int J Food Microbiol* 2010;**144**(1):126–32.
78. Panelli S, Buffoni JN, Bonacina C, Feligini M. Identification of moulds from the Taleggio cheese environment by the use of DNA barcodes. *Food Control* 2012;**28**:385–91.

79. Duquenne P, Simon X, Koehler V, Goncalves-Machado S, Greff G, Nicot T, et al. Documentation of bioaerosol concentrations in an indoor composting facility in France. *J Environ Monit* 2012;**14**(2):409–19.
80. Persoons R, Paratb S, Stoklova M, Perdrixa A, Maitre A. Critical working tasks and determinants of exposure to bioaerosols and MVOC at composting facilities. *Int J Hyg Environ Health* 2010;**213**:338–47.
81. Poirot P, Duquenne P, Grosjean J, Monta N, Nicot T, Zimmerman F, et al. Approche des risques chimiques et microbiologiques dans le secteur du compostage. *Hygiène Sécurité du Travail* 2010. **221**(4ieme trimestre 2010-ND 2336-221-10):3–16.
82. Hryhorczuk D, Curtis L, Scheff P, Chung J, Rizzo M, Lewis C, et al. Bioaerosol emissions from a suburban yard waste composting facility. *Ann Agric Environ Med* 2001;**8**:177–85.
83. Reinthaler F, Wüst G, Haas D, Feierl G, Ruckenbauer G, Marth E. Simple emission-reducing measures in an open biological waste treatment plant. *Aerobiologia* 2004;**20**(1):83–8.
84. Fischer G, Albrecht A, Jäckel U, Kämpfer P. Analysis of airborne microorganisms, MVOC and odour in the surrounding of composting facilities and implications for future investigations. *Int J Hyg Environ Health* 2008;**211**(1–2):132–42.
85. Epstein E, Wu N, Youngberg C, Croteau G. Dust and bioaerosols at a biosolids composting facility. *Compost Sci Util* 2001;**9**(3):250–5.
86. Pankhurst LJ, Akeel U, Hewson C, Maduka I, Pham P, Saragossi J, et al. Understanding and mitigating the challenge of bioaerosol emissions from urban community composting. *Atmos Environ* 2011;**45**(1):85–93.
87. Sanchez-Monedero MA, Stentiford EI. Generation and dispersion of airborne microorganisms from composting facilities. *Process Saf Environ Prot* 2003;**81**(3):166–70.
88. Schlosser O, Huyard A, Rybacki D, Do Quang Z. Protection of the vehicle cab environment against bacteria, fungi and endotoxins in composting facilities. *Waste Manag* 2012;**32**(6):1106–15.
89. Bünger J, Schappler-Scheele B, Hilgers R, Hallier E. A 5-year follow-up study on respiratory disorders and lung function in workers exposed to organic dust from composting plants. *Int Arch Occup Environ Health* 2007;**80**(4):306–12.
90. Schlosser O, Huyard A, Cartnick K, Yaez A, Catan V, Quang Z Do. Bioaerosol in composting facilities: occupational health risk assessment. *Water Environ Res* 2009;**81**:866–77.
91. Wouters IM, Spaan S, Douwes J, Doekes G, Heederik D. Overview of personal occupational exposure levels to inhalable dust, endotoxin, beta(1->3)-glucan and fungal extracellular polysaccharides in the waste management chain. *Ann Occup Hyg* 2005;**50**(1):39–53.
92. Taha MPM, Drew GH, Longhurst PJ, Smith R, Pollard SJT. Bioaerosol releases from compost facilities: evaluating passive and active source terms at a green waste facility for improved risk assessments. *Atmos Environ* 2006;**40**(6):1159–69.
93. Sanchez-Monedero MA, Stentiford EI, Urpilainen ST. Bioaerosol generation at large-scale green waste composting plants. *J Air Waste Manag Assoc* 2005;**55**(5):612–8.
94. Byeon JH, Park CW, Yoon KY, Park JH, Hwang J. Size distributions of total airborne particles and bioaerosols in a municipal composting facility. *Bioresour Technol* 2008;**99**(11):5150–4.
95. Le Goff O, Bru-Adan V, Bacheley H, Godon JJ, Wery N. The microbial signature of aerosols produced during the thermophilic phase of composting. *J Appl Microbiol* 2010;**108**(1):325–40.
96. Adams JDW, Frostick LE. Analysis of bacterial activity, biomass and diversity during windrow composting. *Waste Manag* 2009;**29**(2):598–605.
97. Herbert-Nielsen B, Würtz H, Breum NO, Otto MP. Microorganisms and endotoxin in experimentally generated bioaerosols from composting household waste. *Ann Agric Environ Med* 1997;**4**:159–68.

98. Fracchia L, Pietronave S, Rinaldi M, Martinotti MG. The assessment of airborne bacterial contamination in three composting plants revealed site-related biological hazard and seasonal variations. *J Appl Microbiol* 2006;**100**:973–84.
99. Bünger J, Antlauf-Lammers M, Schulz TG, Westphal GA, Muller MM, Ruhnau P, et al. Health complaints and immunological markers of exposure to bioaerosols among biowaste collectors and compost workers. *Occup Environ Med* 2000;**57**(7):458–64.
100. Douwes J, Heederik D. Epidemiologic investigations of endotoxins. *Int J Occup Environ Health* 1997;**3**(1):S26–31.
101. Lacumin L, Chiesa L, Boscolo D, Manzano M, Cantoni C, Orlic S, et al. Moulds and ochratoxin A on surfaces of artisanal and industrial dry sausages. *Food Microbiol* 2009;**26**(1):65–70.
102. Castellari C, Quadrelli AM, Laich F. Surface mycobiota on Argentinean dry fermented sausages. *Int J Food Microbiol* 2010;**142**(1–2):149–55.
103. Canel RS, Wagner JR, Stenglein SA, Ludemann V. Indigenous filamentous fungi on the surface of Argentinean dry fermented sausages produced in Colonia Caroya (Córdoba). *Int J Food Microbiol* 2013;**164**(1):81–6.
104. Galvalisi U, Lupo S, Piccini J, Bettucci L. *Penicillium* species present in Uruguayan salami. *Rev Argent Microbiol* 2013;**44**(1):36–42.
105. Bernáldez V, Córdoba JJ, Rodríguez M, Cordero M, Polo L, Rodríguez A. Effect of *Penicillium nalgiovense* as protective culture in processing of dry-fermented sausage “salchichón”. *Food Control* 2013;**32**(1):69–76.
106. Palmas F, Meloni V. Fungi as an occupational health hazard in seasoned-food-industry workers. *Environ Monit Assess* 1997;**48**(3):273–84.
107. Marchisio VF, Sulotto F, Botta GC, Chiesa A, Airaudi D, Anastasi A. Aerobiological analysis in a salami factory: a possible case of extrinsic allergic alveolitis by *Penicillium camemberti*. *Med Mycol* 1999;**37**(4):285–9.
108. Asefa DT, Kure CF, Gjerde RO, Omer MK, Langsrud S, Nesbakken T, et al. Fungal growth pattern, sources and factors of mould contamination in a dry-cured meat production facility. *Int J Food Microbiol* 2010;**140**(2–3):131–5.
109. Asefa DT, Gjerde RO, Sidhu MS, Langsrud S, Kure CF, Nesbakken T, et al. Moulds contaminants on Norwegian dry-cured meat products. *Int J Food Microbiol* 2009;**128**(3):435–9.
110. Guillot M, Bertoletti L, Deygas N, Raberin H, Faure O, Vergnon J-M. Pneumopathie à la fleur de saucisson: trois observations. *Rev Mal Respir* 2008;**25**(5):596–600.
111. Rouzaud P, Soulat JM, Trela C, Fraysse P, Recco P, Carles P, et al. Symptoms and serum precipitins in workers exposed to dry sausage mould: consequences of exposure to sausage mould. *Int Arch Occup Environ Health* 2001;**74**(5):371–4.
112. Farber P, Geisen R. Antagonistic activity of the food-related filamentous fungus *Penicillium nalgiovense* by the production of penicillin. *Appl Environ Microbiol* 1994;**60**(9):3401–4.
113. Merget R, Sander I, Rozynek P, Heinze E, Imoehl M, Raulf-Heimsoth M, et al. Occupational immunoglobulin E-mediated asthma due to *Penicillium camemberti* in a dry-sausage packer. *Respiration* 2006;**622**:1–3.
114. Recco P, Muris M, Mallea M, Linas MD. Allergènes fongiques. *Rev Franc Lab* 1996;**281**:21–8.
115. Goto S, Takayama K, Shinohara T. Occurrence of molds in wine storage cellars. *J Ferment Bioeng* 1989;**68**(4):230–2.
116. Clemenz A, Sterflinger K, Kneifel W, Mandl K. Airborne fungal microflora of selected types of wine-cellars in Austria. *Mittl Klosterneubg* 2008;**1**:17–22.
117. Simeray J, Mandin D, Mercier M, Chaumont J-P. Survey of viable airborne fungal propagules in French wine cellars. *Aerobiologia* 2001;**17**(1):19–24.
118. Tiwari KL, Jadhav SK, Kumar A. Morphological and molecular study of different *Penicillium* species. *Middle-East J Sci Res* 2011;**7**(2):203–10.

119. Haleem Khan AA, Mohan Karuppayil S. Fungal pollution of indoor environments and its management. *Saudi J Biol Sci* 2012;**19**(4):405–26.
120. Bokulich NA, Mills DA. Facility-specific "house" microbiome drives microbial landscapes of artisan cheesemaking plants. *Appl Environ Microbiol* 2013;**79**(17):5214–23.
121. Tsai M-Y, Liu H-M. Exposure to culturable airborne bioaerosols during noodle manufacturing in central Taiwan. *Sci Total Environ* 2009;**407**(5):1536–46.
122. Salustiano VC, Andrade NJ, Brandao SCC, Azeredo RMC, Lima SAK. Microbiological air quality of processing areas in a dairy plant as evaluated by the sedimentation technique and a one-stage air sampler. *Braz J Microbiol* 2003;**34**(3):255–9.
123. Kure CF, Borch E, Karlsson I, Homleid JP, Langsrud S. Use of the selective agar medium CREAD for monitoring the level of airborne spoilage moulds in cheese production. *Int J Food Microbiol* 2008;**122**(1–2):29–34.
124. Grinshpun SA, Adhikari A, Li C, Yermakov M, Reponen L, Johansson E, et al. Inactivation of aerosolized viruses in continuous air flow with axial heating. *Aerosol Sci Technol* 2010;**44**(11):1042–8.
125. Simon X, Duquenne P. Assessment of workers' exposure to bioaerosols in a French cheese factory. *J Occup Environ Hyg* 2014;**58**(6):16.
126. Campbell JA, Kryda MJ, Treuhaft MW, Marx JJ, Roberts RC. Cheese worker's hypersensitivity pneumonitis. *Am Rev Respir Dis* 1983;**127**(4):495–6.
127. Guglielminetti M, Valoti E, Cassini P, Taino G, Caretta G. Respiratory syndrome very similar to extrinsic allergic alveolitis due to *Penicillium verrucosum* in workers in a cheese factory. *Mycopathologia* 2000;**149**(3):123–9.
128. Zollinger M, Krebs W, Brandl H. Bioaerosol formation during grape stemming and crushing. *Sci Total Environ* 2006;**363**(1–3):253–9.
129. Al-Sohaibani S, Murugan K, Lakshimi G, Anandraj K. Xerophilic aflatoxigenic black tea fungi and their inhibition by *Elettaria cardamomum* and *Syzygium aromaticum* extracts. *Saudi J Biol Sci* 2011;**18**(4):387–94.
130. Bokhari FM. Spices mycobiota and mycotoxins available in Saudi Arabia and their abilities to inhibit growth of some toxigenic fungi. *Mycobiology* 2007;**35**(2):47–53.
131. Halstensen AS, Heldal KK, Wouters IM, Skogstad M, Ellingsen DG, Eduard W. Exposure to grain dust and microbial components in the Norwegian grain and compound feed industry. *Ann Occup Hyg* 2013;**57**(9):1105–14.
132. Halstensen AS, Nordby KC, Wouters IM, Eduard W. Determinants of microbial exposure in grain farming. *Ann Occup Hyg* 2007;**51**(7):581–92.
133. Halstensen AS, Nordby KC, Eduard W, Klemsdal SS. Real-time PCR detection of toxigenic *Fusarium* in airborne and settled grain dust and associations with trichothecene mycotoxins. *J Environ Monit* 2006;**8**(12):1235–41.
134. Abdel Hameed AA. Airborne dust, bacteria, actinomyceteKs and fungi at a flourmill. *Aerobiologia* 2007;**23**(1):59–69.
135. Oluwafemi F, Odebiyi T, Kolapo A. Occupational aflatoxin exposure among feed mill workers in Nigeria. *World Mycotoxin J* 2012;**5**(4):385–9.
136. Desai MR, Ghosh S. Occupational exposure to airborne fungi among rice mill workers with special reference to aflatoxin producing *A. flavus* strains. *Ann Agric Environ Med* 2003;**10**(2):159–62.
137. Dutkiewicz J, Krysińska-Traczyk E, Skórska C, Sitkowska J, Prazmo Z, Golec M. Exposure to airborne microorganisms and endotoxin in herb processing plants. *Ann Agric Environ Med* 2001;**8**(2):201–11.

138. Skorska C, Sitkowska J, Krysinska-Traczyk E, Cholewa G, Dutkiewicz J. Exposure to airborne microorganisms, dust and endotoxin during processing of peppermint and chamomile herbs on farms. *Ann Agric Environ Med* 2005;**12**:281–8.
139. Spiewak R, Skorska C, Dutkiewicz J. Occupational airborne contact dermatitis caused by thyme dust. *Contact Dermat* 2001;**44**(4):235–9.
140. Mackiewicz B, Skórska C, Dutkiewicz J, Michnar M, Milanowski J, Prazmo Z, et al. Allergic alveolitis due to herb dust exposure. *Ann Agric Environ Med* 1999;**6**:167–70.
141. Skorska C, Golec M, Mackiewicz B, Gora A, Dutkiewicz J. Health effects of exposure to herb dust in valerian growing farmers. *Ann Agric Environ Med* 2005;**12**:247–52.
142. Madsen AM, Tendal K, Frederiksen MW. Attempts to reduce exposure to fungi, β-glucan, bacteria, endotoxin and dust in vegetable greenhouses and a packaging unit. *Sci Total Environ* 2014;**468–469**(0):1112–21.
143. Madsen AM, Hansen VM, Nielsen SH, Olsen TT. Exposure to dust and endotoxin of employees in cucumber and tomato nurseries. *Ann Occup Hyg* 2009;**53**(2):129–38.
144. Adhikari A, Gupta J, Wilkins JR, Olds RL, Indugula R, Cho KJ, et al. Airborne microorganisms, endotoxin, and (1–3)-beta-D-Glucan exposure in greenhouses and assessment of respiratory symptoms among workers. *Ann Occup Hyg* 2011;**55**(3):272–85.
145. Hansen VM, Meyling NV, Winding A, Eilenberg J, Madsen AM. Factors affecting vegetable growers' exposure to fungal bioaerosols and airborne dust. *Ann Occup Hyg* 2012;**56**(2):170–81.
146. Li D-W, LaMondia J. Airborne fungi associated with ornamental plant propagation in greenhouses. *Aerobiologia* 2010;**26**(1):15–28.
147. Rodolfi M, Lorenzi E, Picco AM. Study of the occurrence of greenhouse microfungi in a botanical garden. *J Phytopathol* 2003;**151**(11–12):591–9.
148. Mériaux A, Pageau P, Cormier Y, Goyer N, Duchaine C. Bioaerosols in peat moss processing plants. *J Occup Environ Hyg* 2006;**3**(8):408–17.
149. Mériaux A, Cormier Y, Pageau P, Israël-Assayag E, Duchaine C. Sensitization to airborne molds and its health effects in peat moss processing plant workers. *J Occup Environ Hyg* 2006;**3**(8):442–7.
150. Eduard W, Heederik D. Methods for quantitative assessment of airborne levels of noninfectious microorganisms in highly contaminated work environments. *Am Ind Hyg Assoc J* 1998;**59**(2):113–27.
151. Hooper D, Bolton V, Gray MR. *Fungal mycotoxins can be detected in tissue and body fluids of patients with a history of exposure to toxin producing molds*2008. Available at: http://www.realtimelab.com/documents/MycotoxinPosterMay16207.pdf.

Chapter 9

Fungi in Low-contamination Occupational Environments

Aino Nevalainen, Anne Hyvärinen

National Institute for Health and Welfare, Department of Health Protection, Living Environment and Health Unit, Kuopio, Finland

INTRODUCTION

There are no health-based, generally accepted threshold levels for indoor or outdoor fungal concentrations and, therefore, the concept of "high" and "low" fungal levels has no precise administrational or regulatory content. However, there are a lot of data presenting observed concentrations of various indoor and outdoor locations, and based on this knowledge, one can draw conclusions and make a rational assessment of the quantities of concentrations that are observed.

Most occupational environments with definitely large concentrations of fungi are those where organic or biological material is processed or treated.[1] They almost always contain other biological materials as well, such as bacteria, amoebae, and plant- and animal-derived materials. Examples of such environments are agricultural settings, solid waste, and wastewater treatment facilities, and many food processing industries. How do the "low-contamination" settings differ from these environments? In the occupational environments with low contamination, no professional, large-scale treatment of such material is being done. This does not mean that the low-level settings would be free of fungal contamination; it is only that their sources of contamination are different, such as outdoor air, building-related factors, and occupant-related factors.

Whereas measured fungal concentrations are strongly linked with the methods used, and fungal concentrations vary greatly within a location depending on, e.g., season and climate, one can perhaps use a rule of thumb that what is meant by a typical "low" concentration is between the orders of magnitude of 10^0 and 10^3 cfu/m^3. The concentration ranges in highly loaded environments often exceed these levels by several orders of magnitude. It must be kept in mind, however, that fungal concentrations have great variation in time and space. Even in highly contaminated environments, the measured levels may be low for long times and increase only during certain phases of work, and in low-concentration environments, sporadic local peaks of concentration can be observed.

Environmental Mycology in Public Health. http://dx.doi.org/10.1016/B978-0-12-411471-5.00009-0

Indoor occupational environments with low fungal contamination are offices, schools, day care centers, hospitals and other institutions, shops, museums, and recreational facilities. Also, industrial environments where no organic materials are processed may be low in their fungal concentrations. In all these, those who work in the environment in question are exposed to the fungal contamination of that building, whether caused by outdoor contamination, building-related factors, or activities carried out in the building.

In indoor environments with low fungal contamination, occupants may have many adverse health effects that are linked to fungal and other microbial contamination. Major risk factors for such outcomes are dampness or moisture problems of the building that lead to mold growth, which are strongly associated with respiratory and other health outcomes.[2] This phenomenon, extensively documented in the literature, concerns not only residences but also other buildings that are the focus of this chapter. Many allergic, immunotoxic, and toxic reactions have been reported in connection with building dampness, as summarized by, e.g., the World Health Organization[2] and the Institute of Medicine[3] and as reviewed by, e.g., Mendell et al.[4] After the publication of these extensive documents, further evidence on mold and dampness as risk factors for occupants' health have been reported by, e.g., Norbäck et al.[5] (concerning residences).

It is important to realize, however, that even if adverse health effects are strongly linked to dampness, moisture, and fungal growth (mold), the causal connections between the outcome and the agent exposed to are not yet clear. This is repeatedly brought forward in the documents summarizing the present knowledge of dampness and mold. Therefore, we cannot present the health outcomes as something "caused by fungi," but rather as symptoms and diseases for which dampness, moisture, and consequent microbial growth are risk factors. The approach used in this chapter deals with the aspects of fungal contamination and behavior of fungi in indoor environments, and less emphasis is put on the health aspects that are presented elsewhere in this book.

MEASUREMENT ASPECTS

When discussing fungal concentrations or exposure assessment or observing a contamination, it is fundamentally important to understand the complexity of fungal measurements, i.e., the quantification of fungi and the qualitative aspect of fungal measurements. The approaches used may vary greatly depending on the purpose of the study. Fungal measurements always include two phases: sampling and analysis.

Samples of fungi in an indoor environment can be taken using many different methods. The selection of the method should be done by planning the purpose and approach of the sampling so that the result would best give the answer to the question that is asked before the sampling. Different approaches are needed, for example, whether the aim is to measure, e.g., peak exposures, personal exposure of a single employee, or long-term average exposure; to locate and verify fungal growth on surfaces and building materials; or to investigate a whole school building versus a single classroom.

Airborne exposure is assessed with air samples can be taken, e.g., with cascade impactors, with impingers, or on filters. Passive collection of settling particles or collection of house dust may also be relevant techniques for certain purposes. Collection methods of airborne microbial samples have been extensively reviewed by Reponen et al.[6] and Méheust et al.[7] among others.

Fungal concentrations in indoor air vary constantly in time and space (e.g., Hyvärinen et al.,[8] Frankel et al.,[9] and Mentese et al.[10]) for many reasons, including climate and weather, building-related factors, and the activities of the occupants. Therefore, a sufficient number of samples is critical to obtain a relevant estimate of the concentration. To obtain a reliable estimate of the indoor air fungal concentration, 11 samples during a minimum of two periods are needed[8] or, in another estimate, 16 samples over four time periods.[11]

Repeated sampling campaigns are especially needed when using short time sampling. Only when the investigator has a good knowledge of the "normal" levels in the indoor environment in question, it is possible to check whether those levels are exceeded using a smaller number of samples. However, usually a minimum of three samples are needed for a good assessment of air concentration. The detection and enumeration of fungi by filter samples have been described in standard 16000–16 of the International Organization for Standardization.[12]

Settling plates that are left open for a defined period of time can be used for a rough assessment, but the method is only semiquantitative at its best. Another possible way to assess airborne exposure is to sample house dust by passive collection of settled dust,[13,14] enabling longer sampling times. Dust samples can also be collected by vacuuming from, e.g., surfaces and floors.[15,16] Sampling can also be targeted to the whole indoor environment including various surfaces, which gives more alternatives for describing the fungal content of the location in question. Surfaces can be sampled with cotton swabs, adhesive tape, electrostatic wipes, or vacuuming.[17]

Another approach in microbial sampling, used especially in cases of suspected or observed dampness- or moisture-related mold growth, is to take samples of the damaged materials. This is a way to verify the location of the source of fungal contamination.[18,19]

Any time a sample is taken for fungal assessment, one must have something to compare the result with. In practical situations of hygienic monitoring of indoor spaces, adequate reference materials should be collected to be able to interpret the "meaning" of the results. The idea is that the result is informative from a contamination point of view only if the concentration is higher than "normal" or if the mycobiota of the sample differs from what is considered normal. If the investigator does not have previous reference data for comparison and assessment of the result, he or she should take reference samples from another room, another surface, or undamaged material. When air sampling is conducted, outdoor air should always be sampled in parallel with the indoor samples, as outdoor air is a major source of indoor fungi and outdoor concentrations vary greatly.

After sampling, fungal concentrations can be assessed as amounts of viable propagules, as numbers of total spore counts, or as amounts of their DNA present. Among indirect ways to quantify fungal material are measurement of ergosterol, which is a chemical marker of fungal material,[20,21] and determination of fungal extracellular polysaccharides (EPS).[22,23] Both these methods can be used to quantify the fungal material in a sample, ergosterol for total biomass and EPS for that of certain genera, e.g., *Penicillium* and *Aspergillus*.

The fungal speciation can be done from cultured samples, and certain genera and groups of fungi may also be assessed by direct microscopy.[7,18,24] Sequencing methods provide information about the microbial communities in the sample, but at present, there is not much background material produced with these methods to provide a practical solution for public health purposes. Single fungal species or genera can be quantified with quantitative polymerase chain reaction (qPCR),[25] but to benefit from these fast and specific methods, previous information is needed on what is actually being looked for.[7] Methods that aim to specifically quantify fungal components with certain biological characteristics are measurement of (1,3)-β-D-glucan, a fungal cell wall component with immunotoxic properties,[23] or fungal allergens.[26]

All these methods are used in the published literature. Since they all measure different things and the results are expressed in different units, the comparison or meta-analysis of the available data is difficult. The majority of the information currently available on fungal contamination in various low-contamination environments is culture-based data and therefore, the summaries of the various environments are mainly based on reports using culture-based methods.

It is also evident that a single number expressing the concentration does not have any universal meaning as related to health. For example, a fungal concentration of 1000 cfu/m^3 is "low" if measured outdoors in an environment with a lot of vegetation and warm weather, but it is considered very high in an office environment with mechanical ventilation in a moderate or cold climate.

FUNGAL SPECIES IN INDOOR ENVIRONMENTS

As Samson[27] summarizes in a 2011 book on fungal growth in indoor environments, "the mycobiota of indoor environments contains about 100–150 species which is a small fraction of the more than 100,000 species of described fungi." The fungi that are most commonly observed in indoor air are those that produce a lot of dry spores, for example, the genera *Penicillium, Cladosporium,* and *Aspergillus*. They occur commonly in outdoor air, and they are also present in practically all samples taken from indoor environments. In addition, a number of other genera are usually found in lower numbers.

Some fungi can be considered indicators of moisture-related mold growth.[28] They are not part of the commonly found "normal" mycobiota but appear in connection with microbial growth that acts as a source of such fungi. Well-known fungi that typically grow on wetted building materials are, e.g., *Acremonium, Aspergillus, Chaetomium, Cladosporium, Mucor, Penicillium, Paecilomyces, Stachybotrys, Trichoderma, Ulocladium,* and yeasts.[19,29,30] It is worth noting that the common fungal genera may also grow on moist materials. In such a case, the growing mold acts as an additional source of these fungi, which may be seen as air concentrations higher than normal or as an unusual rank order of the genera in indoor air.

In hospitals, species of *Aspergillus* are a special concern as it is an opportunistic pathogen for immunocompromised patients. Its occurrence in hospital air is usually considered outdoor contamination and the control strategies focus on the filtration of intake air.

Our understanding of indoor microbial communities is currently making great progress because of the increased use of noncultural, DNA-based methods such as large-scale sequencing of the microbiomes.[31,32] These new data will deepen the understanding of indoor microbiomes and provide information about the fungal sources and the behavior of fungal material in indoor and outdoor environments.

GENERAL ASPECTS OF FUNGAL CONTAMINATION

Hygienic assessment of airborne fungal contamination in a low-concentration occupational environment is strongly bound to the location. Therefore, the investigator should be aware of what are the typical, "normal" ranges of concentrations and what is the usual fungal content, "normal" mycobiota, in the building type in question in his or her own region. Most data available today have been analyzed with culture-based methods, and therefore the ideas presented in this chapter are also mainly based on experiences and studies using those methods. It should be kept in mind, however, that "total" levels of airborne microbial particles are 10–100 times higher than those observed with culture-based methods.[33,34] Furthermore, the fungal communities analyzed by sequencing the entire fungal DNA present are much more diverse than what is seen by culture-based or morphological observation.[31]

A microbial concentration in an indoor environment can be considered elevated if it exceeds the normal level by at least one order of magnitude. Signs of an unusual fungal content are both the occurrence of fungal genera or species that are not normally observed in the environment in question and an unusual rank order of the genera or species.[18,35] The rationale behind this approach is that, whereas each location has its typical, "normal" levels and speciation of fungi that originate from "normal" sources, an unusual observation tells about an unusual source that can be interpreted as contamination.

Studies on the biological quality of indoor environments have shown that each environment has its own typical levels of fungi and typical, characteristic mycobiota in indoor air, on surfaces, and in house dust. Air concentrations vary in time and space depending on, e.g., multiple sources, air currents, and activities within the space. Instead, the concentrations on surfaces and in house dust are more stable and do not vary on an hourly or daily basis. The "normal" concentrations and mycobiota are results of the following factors.

ESSENTIAL SOURCES OF INDOOR FUNGI

The main source of background fungi in indoor environments is outdoor air. Fungi are always present in outdoor air, and fungal particles enter indoors through ventilation systems, through open doors and windows, and through cracks in the construction. Fungal transport from outdoors is also caused by occupants, both on their feet and on their clothes and hair. In occupational environments, there rarely are pets such as cats and dogs that would add to the fungal transport on their feet and fur.

Fungi are monitored worldwide in outdoor locations for the purposes of allergen reports, and a lot of data on the occurrence and concentrations of major outdoor fungi have been accumulated during the long periods of such monitoring. Typically, the main sources of outdoor fungi are vegetation and soil, and consequently, fungal concentrations are higher in warmer regions than in colder climates. Fungal occurrence has seasonal variation typical to different geographical regions, and occasional episodes of unusual fungal occurrence are recorded that are due to unusual weather conditions, for example, sand storms. Indoor concentrations of fungi are usually lower than those outdoors, as the building envelope acts as a filter to a certain extent. If the building is equipped with mechanical ventilation with a filtered air intake, the indoor concentrations are much lower than those outdoors.

Indoor sources of fungi that occur in most environments are house plants. Plant leaves have fungi on their surface, and the contribution of plants to the indoor fungi may also depend on the quality of the soil and the cleanliness of the pots. Any organic material that is handled in indoor rooms may be sources of fungi. Examples of such materials in nonresidential environments are plant material, moss or soil that may be used for children's craft projects in schools, or fruit that is peeled and eaten in offices or class rooms. Food items and biological waste are handled on a larger scale in institutional kitchens and, to some extent, in coffee rooms of the employees. In a study by Andersen et al.[19] the species found in moldy materials were much the same as those generally observed in food items, suggesting the important role of food items as a source of indoor fungi.

ROLE OF VENTILATION IN FUNGAL CONTAMINATION OF INDOOR SPACES

Ventilation is necessary for any indoor environment. The ventilation system of a building provides fresh air for the occupants and removes moisture and contaminants by the particles or gaseous chemicals that are produced within the building. Ventilation may be based on gravity only, or it may be a mechanical system with either mechanical exhaust alone or mechanical intake and exhaust. As mentioned earlier, the building envelope may act as a filter for outdoor fungal particles, resulting in lower fungal concentrations indoors than outdoors. If the ventilation system has a mechanical intake, it is equipped with a filter that removes outdoor air particles from the air coming indoors. This is how ventilation may greatly decrease the levels of fungi that are present in the indoor environment. The other way in which ventilation decreases fungal material is by removal along with the exhaust air.

In large modern buildings, the ventilation system is usually combined with heating and cooling functions. This may be an additional factor contributing to fungal contamination of the system. Improperly maintained cooling systems may accumulate condensed water that becomes highly contaminated with microbial growth. Cases of hypersensitivity pneumonitis have been reported[36] in connection with contamination with *Penicillium*. Even aging of the ventilation system was a risk factor for respiratory symptoms in a case reported from Brazil, although no extreme fungal contamination could be observed.[37]

Ducts and filters of the ventilation system may be sources of fungal contamination if not maintained properly. Air filters, being made of porous materials, may be good habitats for microbial growth if enough moisture is present. In such events, a filter may be a source of indoor fungal propagules[38] or volatile compounds.[39] Fungi may also grow on the inner surface of ducts. This does not depend on duct material because fungal growth takes place if the duct surface is moist.[40]

Also infectious agents may be spread via a ventilation system. A review[41] concluded that there is strong and sufficient evidence to demonstrate an association between ventilation, air movements in buildings, and the transmission of infectious diseases in buildings. Although the review emphasized bacterial and viral infections, the risk is similar for fungal infections, for example, in hospitals. Fungal infections may be fatal for immunocompromised patients, and fungal contamination can be harmful for both patients and personnel in operating rooms and other facilities.[42]

Recommendations for maintenance and service of ventilation systems have been developed to guarantee the proper function of the systems and protection of thc building occupants.[43]

CARPETS

Wall-to-wall carpets may act both as a sink and as a source of indoor fungi. In a study by Foarde and Berry,[44] carpeted and tile flooring in schools were compared for one year. Carpet was shown to be a significant sink for fungi, as the airborne concentrations were significantly lower than in the school with tile flooring; the geometric mean (GM) and geometric standard deviation (GSD) of the fungal concentrations were 50 (2.4) and 160 (3.1) cfu/m^3, respectively.

A significant difference was also observed in total spores, 1200 (2.8) spores/m^3 versus 2700 spores/m^3, and in airborne 1,3-D-glucan concentrations, 0.2 (1.4) ng/m^3 versus 0.5 (1.3) ng/m^3, respectively. As expected, the surface dust loading of carpet floor was much higher than that of tile floor: 1,3-D-glucan concentrations were 1,200,000 and 50,000 ng/m^2, respectively. However, this difference could not be seen as culturable fungi.

Settled spores can be resuspended into the air by human activity. Normal walking on a carpet or activity such as vacuuming can significantly increase spore counts.[45] The extent of resuspension is dependent not only on flooring type but also on the fungal genus; the spores have various sizes and aerodynamic properties.[46] Also, area rugs may be similar to carpets.[21,47] The accumulation of fungal agents is derived by the activities and characteristics of the location, e.g., influence of potential sources such as cellars, food items, and level of cleaning. Frequent vacuuming has been noticed to decrease glucan, EPS, and ergosterol levels in floor dust.[23,47,48]

FUNGAL GROWTH DUE TO MOISTURE OR DAMPNESS

A major indoor source of fungal contamination is mold growth, caused by dampness or other means of excess water accumulation into the structures or surfaces of the building. Mold in all kinds of buildings is common everywhere in the world.[2] The main consequence of mold growth is the health effects that occur in mold-contaminated indoor environments. In low-contamination work environments such as offices, schools, and hospitals, it is causing occupational diseases such as asthma, fungal allergy, hypersensitivity pneumonitis, and other health outcomes.[49–51] It is not known what the agents of exposure causing these health effects are, nor are the pathophysiological mechanisms revealed.

Mold growth consists of various organisms, including fungi, bacteria, amoebae, and other organisms. It also includes metabolites of the growing organisms, for example, volatile organic compounds and microbial toxins. Mold growth may be seen by the naked eye; or its presence may be assessed as signs of water intrusion or moisture, such as condensation, leakage, or discoloring of wood; or mold may be observed by its characteristic odor. There are no commonly agreed metrics to assess the extent of mold growth, but quantitative measurements can be used to, e.g., measure the moisture content of materials or the fungal concentration of surfaces, damaged materials, or indoor air. In fungal measurements, qualitative assessment of viable fungal genera and species is also typically imperative to identify an unusual contamination. Aspects of mold growth in buildings are extensively presented in several documents.[18,52]

GENERAL OBSERVATIONS ON THE DATA FROM LOW-CONTAMINATION ENVIRONMENTS

There are quite a number of studies that provide data from various occupational environments in different regions. However, the overall picture of the concentration ranges and the many factors affecting them is still obscure. Most of the data represent a limited number of buildings, and only a few large-scale studies have been reported. Another aspect that limits the clarity of the overall conclusions is the great number of sampling methods, culture media, and culturing conditions that have been used in various studies. In the absence of generally accepted "golden standard" methods, different versions of measurement methodology mean that the results are not comparable in a precise way.

Those studies that have focused on the variation of fungal concentrations have invariably reported a remarkable variation in time and location in air concentrations of a given building. Within-building, between-building, and seasonal variations may all be of importance.[8–10,53] This large variation highlights the important implications of using a well-thought-out sampling strategy with a sufficient number of samples whenever fungal measurements of indoor air are needed.

In the following, examples of fungal concentrations in various low-contamination occupational environments are presented. The studies referred to were made in different countries in various regions of the world representing different climatic conditions.

OFFICES

Offices may be located in large, mechanically ventilated buildings or in smaller buildings with various ventilation solutions. Office work does not usually contain handling of any organic material, and therefore, there are no specific bioaerosol sources connected to this kind of work. The fungal concentrations in office environments are mostly dependent on the building and ventilation type, amount of human traffic, type of flooring and—as always—the climatic factors. Existence of dampness or moisture-related fungal sources may be a cause for unusual fungal findings.

Tsai et al.[53] studied 100 office buildings in the United States for their indoor fungal concentrations. Arithmetic means (standard deviations (SD)) of concentrations of viable fungi in indoor and outdoor air were 100/680 (230/840) cfu/m^3. Corresponding total counts of spores were 270/6540 (1190/6780) spores/m^3. Fewer groups of fungi were observed indoors than outdoors, and a seasonal variation was seen as more groups of fungi in summertime.

Similar findings of the fungal levels have been reported from other studies. Gots et al.[54] reported an average fungal concentration of 233 cfu/m^3 from 149 noncompliant commercial buildings in the United States. An example of extremely low concentrations of viable fungi can be seen in wintertime samples in northern climates. During the cold weather and snow cover, the outdoor fungal concentrations are very low and this can be seen in indoor air as well. In a study of 77 Finnish office buildings, both buildings with known mold problems and those without were investigated. In mold-free buildings, 90% of the air samples had fungal concentrations <15 cfu/m^3, and even in those with moisture-related mold growth, only 20% of the samples exceeded 100 cfu/m^3.[55]

The indoor concentrations of fungi are usually lower than those outdoors, and fewer types of fungi are usually observed. These findings support the general conclusion that some of the outdoor air fungi are filtered by the ventilation system and by the building envelope. This is, however, highly dependent on the type and efficiency of the ventilation system and on the construction of the building. Seasonal variation is usually seen as higher concentrations and as more fungal groups in summertime. This kind of seasonal variation has been widely documented.[9]

SCHOOLS AND DAY CARE CENTERS

Both in schools and in day care centers for younger children, a number of studies have focused on the connections between indoor air pollutants and the respiratory health of the children. Only a few studies have focused on the occupational health of the teachers and other personnel, but they report on occupational health problems associated with exposure to mold in school buildings.[56,57] Although no conclusive knowledge exists on the possible link between fungal exposure in these premises and the health of the occupants, the studies often report fungal concentrations that can be used in hygienic evaluation of the indoor air quality.

A multicenter study in five European countries included fungal measurements in schools using nuclepore filters as a sampling method. Mean fungal concentrations (SD) were highest in Reims, France, 685 (310) cfu/m^3, and lowest in Uppsala, Sweden, 78 (33) cfu/m^3.[58] The concentrations were almost an order of magnitude higher in all southern European schools than in those of northern Europe.

In subtropical regions such as Taiwan, the fungal concentrations are remarkably higher than in colder climates. The GM of fungal counts was 9672 cfu/m^3 in winter and 4389 cfu/m^3 in summer as measured from schools of asthmatic and nonasthmatic children.[59] In Singapore, where the authors describe the climate as being hot and humid, a large survey of day care centers was conducted. The fungal concentrations were shown to be dependent on the type of ventilation. For example, in 59 centers with natural ventilation, mean (SD) fungal concentrations were 820 (1312) cfu/m^3, whereas in the 19 centers with air conditioning, the numbers were 677 (1151) cfu/m^3.[60] Interestingly, most outdoor concentrations of fungi were relatively close to those indoors. In a study carried out in Paris, the GM (GSD) of fungal concentrations in play rooms of day care centers varied between 120 (1.8) cfu/m^3 in winter and 353 (2.4) cfu/m^3 in summer.[61]

Thus, the concentrations in warm or hot climates may be one to two orders of magnitude higher than those observed in cold or moderate climates. This shows the importance of regional and local approaches in the evaluation of indoor fungal contamination. The same guideline numbers cannot be applied throughout the world, because the effect of outdoor fungi on the normal background concentrations of indoor air is overwhelming.

HOSPITALS AND INSTITUTIONS

In hospitals, fungal contamination is a special concern owing to the risk of nosocomial fungal infections, caused especially by *Aspergillus* species.[62] Therefore, the ventilation systems with high-efficiency filtration of intake air aim to prevent the entrance of outdoor air fungi into the hospital air. However, fungal contamination from outdoor air cannot be totally eliminated, and occasional indoor sources are also possible. Particular concerns in hospitals are renovations and construction activities within the premises, as well as nearby land excavation.[63]

Air measurements are a way to monitor the efficiency of the control measures. Fungal concentrations have been reported to be mainly under 10 cfu/m^3 [64] or, in other studies, to vary but remain usually under 100 cfu/m^3.[1,65] Also, higher concentrations have been observed: in a case study of one hospital, fungal concentrations were lowest in nursing stations, 37 ± 17 cfu/m^3, and highest in the orthopedic operations room, 97 ± 217 cfu/m^3.[66] Falvey and Streifel[65] also reported on occasional bursts of fungal contamination during which the airborne counts exceeded the mean by 3 SD.

Thus, the fungal levels in hospitals appear to be generally lower than those in offices, schools, and other such public spaces. This shows the effect of more strictly controlled ventilation systems and more rigid cleaning practices. From the point of public health, hospitals are a specific type of building with unique types of activities compared with schools, offices, and public buildings, and specific practices are needed to ensure their safe indoor air quality.

Similar levels of airborne fungi have been reported from Portuguese elderly care centers.[67] Lowest levels were observed in bedrooms, 32 cfu/m^3, and highest in the storage areas, 228 cfu/m^3. On surfaces, 40 different fungal species were observed, with *Penicillium* and *Aspergillus* species being most prevalent.

OTHER LOCATIONS

The importance of tailored, location-specific hygiene programs and approaches to controlling fungal contamination was well presented in an 11-year monitoring study of a university canteen.[68] The fungal concentrations, while having seasonal variations similar to most other indoor environments, were decreased from several hundreds of cfu/m^3 to mostly <100 cfu/m^3, a result of controlled changes in construction practices and regular hygiene training of the employees.

EXPERIENCES FROM INTERVENTIONS

Intervention studies have usually been connected to cases in which dampness, moisture, and mold contamination has been eliminated by renovations of the contaminated building. Resulting from such remediations, decreases in adverse health outcomes have often been reported as reviewed by Sauni et al.[69] Decreases in indoor contamination have also been reported in many studies as well as decreases in immunotoxic properties of airborne particulate matter.[70]

These studies support the hypothesis that fungal contamination that is harmful to the health of the occupants can be eliminated or at least decreased by renovating the mold damage and that the process has a positive effect on the building occupants.

As a general principle, problems with dampness, moisture, and mold contamination cannot be solved by means of disinfection. Disinfection chemicals that are generally targeted at destroying pathogenic bacteria are usually not effective against fungi and their spores. In addition, the cause of such mold growth is the accumulation of water for some reason. If this accumulation is not stopped by proper repairs, the mold will rapidly grow again after any disinfection effort.

The success of mold renovations can be assessed by repeated measurements of the indoor air, and changes in the occupants' health by repeated questionnaires. Meklin et al.[71] monitored two school buildings going through various repairs. For both schools, a reference building was also monitored. Indoor air concentrations of a thoroughly renovated school decreased from GM 22.6 to GM 6.3 cfu/m^3, reflecting after renovations the levels of the reference school building. It was concluded that the sources of fungi that had caused the elevated levels were successfully removed. Instead, the other index school building could be only partly repaired because of insufficient funding for the repair process. In that school, concentrations were even slightly increased (from GM 18.5 to GM 23 cfu/m^3), which can be considered a sign of an "unsuccessful" mold remediation.

Similar changes in symptom profiles were also seen in the questionnaire data of the students and teaching personnel; symptoms were decreased in the thoroughly renovated building more than in the school undergoing only partial renovation. Each time, the same questionnaire was used for both index schools and the two reference schools. In another study, the health of the teachers was documented in a school that underwent a mold remediation.[56] During the follow-up of 3 years, the incidence of self-reported bronchitis and conjunctivitis decreased in the index group in parallel with the decrease in fungal concentrations to a normal level.[72]

The majority of the microbiological data from intervention studies have been obtained using culturable methods. The few studies so far that report such results using DNA-based or other nonculturing methods do not challenge the conclusions of the studies using culture-based methods. For example, while monitoring the result of a school renovation, a significant decrease in concentrations of 13 fungi was observed using qPCR as reported by Roponen et al.[70] A decrease in the airborne fungal component (1,3)-β-D-glucan was reported as a result of renovation of a day care center with microbial growth problems.[73]

The fact that a great number of methods are available for fungal analyses should not confuse the good practices of fungal assessments. It is difficult to compare results of various sampling or analyzing methods. Different methods have different capacities for collection and observation of various fungi. In practice, although ideally the best methods should be used, the most important principle is to use exactly the same method when comparing the result to results of reference measurements. Therefore, it is essential to be consistent in creating the database that is used for public health monitoring, i.e., recognizing a situation of undesired fungal contamination.

Furthermore, using a consistent methodology allows not only the recognition of fungal contamination but also the verification of the success of control measures. The importance of comparative sampling before and after remediation in assessment of the success of the remediation project was also emphasized by Kleinheinz et al.[74] According to these authors, "there should not be any abnormally high level of one organism present in a remediated sample versus its comparative samples." To ensure the relevance of a comparison of concentrations, postremediation comparative samples must be taken using the same method and from the same locations and under similar conditions.

An important conclusion of these studies is that air measurements, if carried out to a sufficient extent, can be used to monitor the microbiological quality of air in different situations and to assess the hygienic status of a building. The use of reference buildings, although it makes the measurements extensive and labor-intensive, helps in the assessment of the results, i.e., to give the answer to the question, "are these concentrations within the normal range or do they show a contamination of the indoor air?"

The use of surface samples is often a simpler and inexpensive way to monitor the fungal contamination of indoor spaces.[15,18] Whereas fungal concentrations in surface samples do not constantly vary owing to the many factors that affect air concentrations, they may serve as adequate tools for fungal monitoring in many situations. The same principle of comparative sampling applies here as well: the use of similar methods and similar techniques allows the comparison of the results.

APPLICATION OF GUIDANCE REFERENCE VALUES FOR FUNGAL CONTAMINATION

In low-contamination indoor environments, we cannot conclude that "fungi are causing" certain health effects. The only exceptions are fungal infections or allergies for which a specific link between the outcome and the causal agent can be established. Otherwise, there are no conclusive data showing that certain concentrations of fungi—assuming the concentrations are not massively high—would per se cause any specific health effect. Neither are there any dose–response data for concentrations of common airborne fungi, although links between certain levels of fungi and some health outcomes are sporadically reported.

Instead, a consistent observation is that respiratory symptoms are strongly connected with growing fungi, "mold," a phenomenon caused by building dampness or excess moisture. Dampness, moisture, and mold may occur in any type of building, including all kinds of occupational environments. Such mold growth always includes not only fungi but also bacteria and other organisms, as well as various microbial metabolites, but it is unclear what the roles of various agents may be in the development of symptoms. Also, it is very well documented that mold growth and its surrogate, signs of water intrusion, lead to a number of adverse health effects including wheeze and increased risk of asthma.[2,4] Building dampness and mold should be controlled and prevented as part of environmental public health programs.

When suspecting fungal contamination due to moisture and mold damage, the visual and technical inspection of a building forms the basis of the identification of the problem.[18] Suspected microbial growth can be verified with bulk samples from surfaces and building materials. An inspection should also include observations of potential risk structures in hidden areas. Air sampling is typically conducted to reveal potential hidden sources.

To control problems in the building in the future, remediation should always consist of both removal of the existing mold and repair of the original cause of the water accumulation, be it leakage, condensation, or another building-related problem. Technical repair measures are necessary to eliminate potential health risks. This is an area in which public health practices are linked with building maintenance and building management practices as well as occupational health services. Such links are not often traditionally very close but are necessary to ensure the proper renovations and elimination of other exposures that are related to building mold.

It is evident that in the absence of known causal links between fungal agent exposure and specific health outcomes, or their dose–response relationships, no numerical health-based reference values for indoor airborne fungi can be given. However, reference numbers may be useful in the detection of unusual fungal contamination in the indoor environment. A summary of various guideline numbers that have been given for fungal concentrations is presented in Rao et al.[75]

The examples presented in this chapter from studies around the world prompt the apparent conclusion that indoor fungal concentrations are strongly dependent on location, climate, season, and even type of building and its ventilation, as well as the methods of sampling and analysis. Therefore, the public health-related hygienic monitoring of indoor spaces must be based on reference data that are produced on a regional basis, taking into account the parameters mentioned. The data must be assessed on both a quantitative basis (concentrations) and a qualitative basis, observing the species and their mutual prevalence ratios. As soon as the reference database includes data on normal ranges in each building type, unusual contamination can be identified with routine-type sampling procedures. Having said this, one must never forget to take outdoor samples in parallel with the indoor sampling, as the outdoor air, with its greatly varying fungal concentrations, is always worth comparison.

IMPORTANCE OF INDOOR ENVIRONMENTAL INVESTIGATIONS IN PUBLIC HEALTH

Environmental investigations are needed for many kinds of purposes in public health. Hospitals are a specific group of buildings in which both the safety of patients and the well-being of employees are criteria of the environmental hygiene activities. In such environments, the contamination that must be controlled is caused by pathogens or opportunistic species. The environment must be monitored to locate a possible contamination or to verify the success of control measures. On such occasions, air sampling may be the primary means of investigation.

All kinds of buildings may have fungal problems due to dampness or moisture accumulation and subsequent mold growth. Experience from various regions of the world has taught that basic principles of action are very similar in the recognition and remediation of the problem, but in each region, these principles must be applied to the local conditions.[18,35,76]

If dampness-related mold is the target of the investigations, the first steps should be technical inspections of the building for signs of dampness and moisture and for the functioning of the ventilation system, to assess and locate the cause and site of damage. This inspection is done by a technical professional familiar with the building's functions. Using construction-relevant methods allows the finding of the causes of the moisture problems and the creation of a basis of reasonable renovations and their planning. However, the hygienic quality of indoor air or a possible microbial growth on building materials cannot be assessed by these methods, and microbiological measurements are needed. Technical inspection and measurements are then followed by surface or material sampling for microbial analysis of suspected locations. Only to complement these measures, if needed, is microbial sampling of the indoor air relevant, as in cases of damage hidden behind a surface.

Characterization of the fungal concentrations and mycobiota of the indoor environment may reveal a situation of contamination as elevated levels, as an occurrence of an unusual species, or as an unusual rank order of the species. Fungal measurements are also a useful way to show the success of remediation. Using a suitable set of samples taken before and after the repair measures, the success of the process can be verified. Such monitoring should be part of any renovation or decontamination project.

Measurements of airborne fungi may be needed to show an unusual contamination of indoor air in many types of occupational environments. The approach may also be to document the exposure of an individual employee or group of employees. Well-conducted microbial measurements may be costly and labor-intensive and this may limit their applications even if needed for the purposes of the building investigations. Whatever method is used for sampling and analysis, the main principle is to take a sample where contamination is suspected and to always take another sample from a similar but unsuspected location for comparison. For good practice in public health, the investigator should also have a reference database to which the results of an individual case are compared. Such a database should include data from various indoor environments that belong to the building stock to be monitored, including information about the range of normal concentrations and their seasonal variation and the corresponding information about the mycobiota. This kind of database is a useful tool for revealing causes of health complaints, in verifying the success of remediation measures, and in the prevention of indoor-related health problems.

REFERENCES

1. Górny R, Dutkiewicz J. Bacterial and fungal aerosols in industrial environments in Central and Eastern European countries. *Ann Agric Environ Med* 2002;**9**:17–23.
2. WHO. *Guidelines for indoor air quality: dampness and mould*. Copenhagen: WHO Regional Office for Europe; 2009.
3. IOM (Institute of Medicine). *Damp indoor spaces and health*. Washington D.C: NAS Press; 2004.
4. Mendell MJ, Mirer AG, Cheung K, Tong M, Douwes J, et al. Respiratory and allergic health effects of dampness, mould, and dampness-related agents: a review of the epidemiological evidence. *Environ Health Perspect* 2011;**119**:748–56.
5. Norbäck D, Zock J-P, Plana E, Heinrich J, Svanes C, Sunyer J, et al. Mould and dampness in dwelling places, and onset of asthma: the population-based cohort ECRHS. *Occup Environ Med* 2013;**70**:325–31.
6. Reponen T, Willeke K, Grinshpun S, Nevalainen A. Biological particle sampling. In: Kulkarni P, Baron PA, Willeke K, editors. *Aerosol measurement, principles, techniques and applications*. 3rd ed. Holbroken, New Jersey: John Wiley & Sons, Inc.; 2011.
7. Méheust D, Le Cann P, Reboux G, Millon L, Gangneux JP. Indoor fungal contamination: health risks and measurement methods in hospitals, homes and workplaces. *Crit Rev Microbiol* 2014;**40**:248–60.
8. Hyvärinen A, Vahteristo M, Meklin T, Jantunen M, Nevalainen A, Moschandreas D, et al. Temporal and spatial variation of fungal concentrations in indoor air. *Aerosol Sci Technol* 2001;**35**:688–95.
9. Frankel M, Bekö G, Timm M, Gustavsen S, Hansen EW, Madsen AM, et al. Seasonal variations of indoor microbial exposures and their relation to temperature, humidity, and air exchange rate. *Appl Environ Microbiol* 2012;**78**:8289–97.
10. Mentese S, Rad AY, Arisoy M, Gullu G. Multiple comparisons of organic, microbial, and fine particulate pollutants in typical indoor environments: diurnal and seasonal variations. *J Air Waste Manag Assoc* 2012;**62**:1380–93.
11. Burge HA. The fungi. In: *Indoor air quality handbook*. New York: McGraw Hill; 2001. p. 45.1–33.
12. ISO. *ISO 16000–16. Indoor air–Part 16: detection and enumeration of moulds–sampling by filtration*. Geneva: International *Organization for Standardization (ISO)*; 2008.
13. Würtz H, Sigsgaard T, Valbjorn O, Doekes G, Meyer HW. The dustfall collector – a simple passive tool for long-term collection of airborne dust: a project under the Danish Mould in Buildings program (DAMIB). *Indoor Air* 2005;**15**(Suppl. 9):33–40.
14. Noss I, Wouters IM, Visser M, Heederik DJ, Thorne PS, Brunekreef B, et al. Evaluation of a low-cost electrostatic dust fall collector for indoor air endotoxin exposure assessment. *Appl Environ Microbiol* 2008;**74**:5621–7.
15. Macher JM. Review of methods to collect settled dust and isolate culturable microorganisms. *Indoor Air* 2001;**11**:99–110.
16. Karvonen AM, Hyvärinen A, Gehring U, Korppi M, Doekes G, Riedler J, et al. Exposure to microbial agents in house dust and wheezing, atopic dermatitis and atopic sensitization in early childhood: a birth cohort study in rural areas. *Clin Exp Allergy* 2012;**42**:1246–56.
17. Thorne PS, Metwali N, Avol E, McConnell RS. Surface sampling for endotoxin assessment using electrostatic wiping clothes. *Ann Occup Hyg* 2005;**49**:401–6.

18. Prezant B, Weekes D, Miller JD, editors. *Recognition, evaluation and control of indoor mold.* Fairfax, USA: American Industrial Hygiene Association; 2008.
19. Andersen B, Frisvad JC, Sondergaard I, Rasmussen IS, Larsen LS. Associations between fungal species and water-damaged building materials. *Appl Environ Microbiol* 2011;**77**:4180–8.
20. Sebastian A, Larsson L. Characterization of the microbial community in indoor environments: a chemical-analytical approach. *Appl Environ Microbiol* 2003;**69**:3103–9.
21. Hyvärinen A, Sebastian A, Pekkanen J, Larsson L, Korppi M, Putus T, et al. Characterizing microbial exposure with ergosterol, 3-hydroxy fatty acids, and viable microbes in house dust: determinants and association with childhood asthma. *Arch Environ Occup Health* 2006;**61**:149–57.
22. Douwes J, van der Sluis B, Doekes G, van Leusden F, Wijnands L, van Strien R, et al. Fungal extracellular polysaccharides in house dust as a marker for exposure to fungi: relations with culturable fungi, reported home dampness, and respiratory symptoms. *J All Clin Immunol* 1999;**103**:494–500.
23. Giovannangelo ME, Gehring U, Nordling E, Oldenwening M, van Rijswijk K, de Wind S, et al. Levels and determinants of beta (1->3)-glucans and fungal extracellular polysaccharides in house dust of (pre-)school children in three European countries. *Environ Int* 2007;**33**:9–16.
24. Fairs A, Wardlaw AJ, Thompson J, Pashley CH. Guidelines on ambient intramural airborne fungal spores. *J Investig Allergol Clin Immunol* 2010;**20**:490–8.
25. Pitkäranta M, Meklin T, Hyvärinen A, Paulin L, Auvinen P, Nevalainen A, et al. Analysis of fungal flora in indoor dust by ribosomal DNA sequence analysis, quantitative PCR, and culture. *Appl Environ Microbiol* 2008;**74**:233–44.
26. Shi C, Belisle D, Miller JD. Quantification of the Aspergillus versicolor allergen in house dust. *J Immunol Methods* 2011;**372**:89–94.
27. Samson RA. Ecology and general characteristics of indoor fungi. In: Adan OCG, Samson RA, editors. *Fundamentals of mold growth in indoor environments and strategies for healthy living.* The Netherlands: Wageningen Academic Publishers; 2011.
28. Samson RA, Flannigan B, Flannigan ME, Verhoeff AP, Adan OCG, editors. *Health implications of fungi in indoor environments.* Amsterdam, The Netherlands: Elsevier; 1994.
29. McGregor HJ, Miller JD, Rand T, Solomon J. Mold ecology: recovery of fungi from certain moldy building materials. In: Prezant B, Weekes D, Miller JD, editors. *Recognition, evaluation and control of indoor mold.* Fairfax, USA: American Industrial Hygiene Association; 2008.
30. Hyvärinen A, Meklin T, Vepsäläinen A, Nevalainen A. Fungi and actinobacteria in moisture damaged building materials – concentrations and diversity. *Int Biodeterior Biodegr* 2002;**49**:27–37.
31. Pitkäranta M, Meklin T, Hyvärinen A, Nevalainen A, Paulin L, Auvinen P, et al. Molecular profiling of fungal communities in moisture-damaged buildings before and after remediation – comparison of culture-dependent and –independent methods. *BMC Microbiol* 2011;**11**:235–50.
32. Adams RI, Miletto M, Lindow SE, Taylor JW, Bruns TD. Airborne bacterial communities in residences: similarities and differences with fungi. *PLoS One* 2014;**9**:e91283.
33. Toivola M, Nevalainen A, Alm S. Personal exposures to particles and microbes in relation to microenvironmental concentrations. *Indoor Air* 2004;**14**:351–9.
34. Luoma M, Batterman SA. Characterization of particulate emissions from occupant activities in offices. *Indoor Air* 2001;**11**:35–48.
35. The Ministry of Social Affairs and Health. A guide for healthy housing. Guidance documents of the Ministry of Social Affairs and Health 1/2003. Helsinki, Finland; 2003 (in Finnish).
36. Bernstein RS, Sorenson WG, Garabrant D, Reaux C, Treitman RD. Exposures to respirable, airborne *Penicillium* from a contaminated ventilation system: clinical, environmental and epidemiological aspects. *Am Ind Hyg Assoc J* 1983;**44**:161–9.

37. Graudenz GS, Kalil J, Saldiva PH, Gambale W, Latorre Md Mdo R, Morato-Castro FF, et al. Upper respiratory symptoms associated with aging of the ventilation system in artificially ventilated offices in São Paulo, Brazil. *Chest* 2002;**122**:729–35.
38. Kemp PC, Neumeister-Kemp HG, Lysek G, Murray F. Survival and growth of microorganisms on air filtration media during initial loading. *Atm Environ* 2001;**35**:4739–49.
39. Schleibinger H, Ruden H. Air filters from HVAC systems as possible source of volatile organic compounds (VOC) – laboratory and field assays. *Atm Environ* 1999;**33**:4571–7.
40. Buttner MP, Cruz-Perez P, Garrett PJ, Stetzenbach LD. Dispersal of fungal spores from three types of air handling system duct material. *Aerobiologia* 1999;**15**:1–8.
41. Li Y, Leung GM, Tang JW, Yang X, Chao CY, Lin JZ, et al. Role of ventilation in airborne transmission of infectious agents in the built environment – a multidisciplinary systematic review. *Indoor Air* 2007;**17**:2–18.
42. Araujo R, Cabral JP, Rodrigues AG. Air filtration systems and restrictive access conditions improve indoor air quality in clinical units: Penicillium as a general indication of hospital indoor fungal levels. *Am J Infect Control* 2008;**36**:129–34.
43. Brief RS, Bernath T. Indoor pollution: guidelines for prevention and control of microbiological respiratory hazards associated with air conditioning and ventilation systems. *Appl Ind Hyg* 1988;**3**:5–10.
44. Foarde K, Berry M. Comparison of biocontaminant levels associated with hard vs. carpet floors in nonproblem schools: results from a year long study. *J Expo Sci Environ Epidemiol* 2004;**14**:S41–8.
45. Buttner MP, Stetzenbach LD. Monitoring airborne fungal spores in an experimental indoor environment to evaluate sampling methods and the effects of human activity on air sampling. *Appl Environ Microbiol* 1993;**59**:219–26.
46. Shorter C. Fungi in New Zealand homes: measurement, aerosolization & association with children's health Ph.D. Thesis. New Zealand: University of Otago; 2013.
47. Leppänen HK, Nevalainen A, Vepsäläinen A, Roponen M, Täubel M, Laine O, et al. Determinants, reproducibility, and seasonal variation of ergosterol levels in house dust. 2014. http://dx.doi.org/10.1111/ina.12078.
48. Sordillo JE, Alwis UK, Hoffman E, Gold DR, Milton DK. Home characteristics as predictors of bacterial and fungal microbial biomarkers in house dust. *Environ Health Perspect* 2011;**119**:189–95.
49. Jaakkola MS, Jaakkola JJK. Indoor molds and asthma in adults. *Adv Appl Microbiol* 2004;**55**:309–38.
50. Akpinar-Elci M, White SK, Siegel PD, Park JH, Visotcky A, Kreiss K, et al. Markers of upper airway inflammation associated with microbial exposure and symptoms in occupants of a water-damaged building. *Am J Ind Med* 2013;**56**:522–30.
51. Seuri M, Husman K, Kinnunen H, Reiman M, Kreus R, Kuronen P, et al. An outbreak of respiratory diseases among workers at a water-damaged building. *Indoor Air* 2000;**10**:138–45.
52. Adan OCG, Samson RA, editors. *Fundamentals of mold growth in indoor environments and strategies for healthy living*. The Netherlands: Wageningen Academic Publishers; 2011.
53. Tsai FC, Macher JM, Hung YY. Biodiversity and concentrations of airborne fungi in large US office buildings from the BASE study. *Atm Environ* 2007;**41**:5181–91.
54. Gots RE, Layton NJ, Pirages SW. Indoor healths: background levels of fungi. *AIHA J* 2003;**64**:427–38.
55. Salonen H, Lappalainen S, Lindroos O, Harju R, Reijula K. Fungi and bacteria in mould-damaged and non-damaged office environments in a subarctic climate. *Atmos Environ* 2007;**41**:6797–807.
56. Patovirta R-L, Husman T, Haverinen U, Vahteristo M, Uitti JA, Tukiainen H, et al. The remediation of mold damaged school – a three-year follow-up study on teachers' health. *Cent Eur J Public Health* 2004;**12**:36–42.

57. Rudblad S, Andersson K, Stridh G, Bodin L, Juto JE. Slowly decreasing mucosal hyperreactivity years after working in a school with moisture problems. *Indoor Air* 2002;**12**:138–44.
58. Simoni M, Cai GH, Norback D, Annesi-Maesano I, Lavaud F, Sigsgaard T, et al. Total viable molds and fungal DNA in classrooms and association with respiratory health and pulmonary function of European schoolchildren. *Pediatr Allergy Immunol* 2011;**22**:843–52.
59. Su HJJ, Wu PC, Lin CY. Fungal exposure of children at homes and schools: a health perspective. *Arch Environ Health* 2001;**56**:144–9.
60. Zuraimi MS, Tham KW, Chew FT, Ooi PL. The effect of ventilations strategies of child care centers on indoor air quality and respiratory health in Singapore. *Indoor Air* 2007;**17**:317–27.
61. Roda C, Barral S, Ravelomanantsoa H, Dusseaux M, Tribout M, Le Moullec Y, et al. Assessment of indoor environments in Paris child day care centers. *Environ Res* 2011;**111**:1010–7.
62. Morris G, Kokki MH, Anderson K, Richardson MD. Sampling of *Aspergillus* spores in air. *J Hosp Infect* 2000;**44**:81–92.
63. Cheng SM, Streifel AJ. Infection control considerations during construction activities: land excavation and demolition. *Am J Infect Control* 2001;**29**:321–8.
64. Sautour M, Sixt N, Dalle F, L'ollivier C, Calinon C, Fourquenet V, et al. Prospective survey of indoor fungal contamination in hospital during a period of building construction. *J Hosp Infect* 2007;**11**:367–73.
65. Falvey DG, Streifel AJ. Ten-year air sample analysis of *Aspergillus* prevalence in a university hospital. *J Hosp Infect* 2007;**67**:35–41.
66. Azimi F, Naddafi K, Nabizadeh R, Hassanvand MS, Alimohammadi M, Afhami S, et al. Fungal air quality in hospital rooms: a case study in Tehran Iran. *J Environ Health Sci Eng* 2013;**11**:30.
67. Viegas C, Almeida-Silva M, Gomes AQ, Wolterbeek HT, Almeida SM. Fungal contamination assessment in portuguese elderly care centers. *J Toxicol Environ Health A* 2014;**77**:14–23.
68. Osimani A, Aquilanti L, Tavoletti S, Clementi F. Microbiological monitoring of air quality in a university canteen: an 11-year report. *Environ Monit Assess* 2013;**185**:4765–74.
69. Sauni R, Uitti J, Jauhiainen M, Kreiss K, Sigsgaard T, Verbeek JH, et al. Remediating buildings damaged by dampness and mould for preventing or reducing repiratory tract symptoms, infections and asthma. *Cochrane Rev J* 2013;**8**:944–1000.
70. Roponen M, Meklin T, Rintala H, Hyvarinen A, Hirvonen MR. Effect of moisture-damage intervention on the immunotoxic potential and microbial content of airborne particles and on occupants' upper airway inflammatory responses. *Indoor Air* 2013;**23**:295–302.
71. Meklin T, Putus T, Pekkanen J, Hyvarinen A, Hirvonen MR, Nevalainen A, et al. Effects of moisture-damage repairs on microbial exposure and symptoms in schoolchildren. *Indoor Air* 2005;**15**(Suppl. 10):40–7.
72. Haverinen U, Husman T, Toivola M, Suonketo J, Pentti M, Lindberg R, et al. An approach to management of critical indoor air problems in school buildings. *Environ Health Perspect* 1999;**107**:509–14.
73. Airborne Rylander R. (1->3)-Beta-D-glucan and airway disease in a day-care center before and after renovation. *Arch Environ Health* 1997;**52**:281–5.
74. Kleinheinz GT, Langolf BM, Englebert E. Characterization of airborne fungal levels after mold remediation. *Microbiol Res* 2006;**161**:367–76.
75. Rao CY, Burge HA, Chang JCS. Review of quantitative standards and guidelines for fungi in indoor air. *J Air Waste Manag Assoc* 1996;**46**:899–908.
76. Howden-Chapman P, Bennett J, Siebers R, editors. *Do damp and mould matter? Health imapcts of leaky homes. He Kainga Oranga/Housing and Health Research Programme.* Wellington: University of Otago; 2010. Aotearoa, New Zealand: Steele Roberts Publishers.

Section III

Non-occupational Exposure

Chapter 10

Domestic Environment

Indoor Mycobiota As a Public Health Risk Factor

Elena Piecková
Slovak Medical University, Limbová, Bratislava, Slovakia

Depending on their lifestyle, modern people are used to spending up to 90% of their entire time in some indoor environment. Various indoor factors can contribute to many health symptoms in occupants, including the so-called building-related illnesses, i.e., allergic, infectious, toxic, or inflammatory diseases, leading even to precancer.[1] All of these might be caused by chemicals, viruses, bacteria (especially gram-negative endotoxins), or fungi. A complex mixture of health troubles and general discomfort affecting occupants of certain buildings is known as sick building syndrome (SBS).[1] SBS symptoms are nonspecific and usually depend on various indoor microclimate characteristics—temperature, relative humidity (RH), dust, cigarette smoke, ventilation, building materials, and furnishings—but also on the personality characteristics (e.g., satisfaction in the work, ability to establish and maintain reliable professional relationships, etc.) of the affected persons. The etiological role of fungi in the health troubles connected with staying in certain buildings has not been fully clarified yet. Monitoring of exposure to indoor fungi is rather complicated owing to a lack of standard practical methods for evaluating how the indoor microclimate, outdoor surroundings, and microscopic fungi affect one another.

Molds are found almost everywhere in our environment, both outdoors and indoors. They can grow on just about any substance, as long as moisture and oxygen are available. Visible or cryptic mold growth may occur when excessive moisture accumulates in buildings or on building materials, including carpets, ceiling tiles, insulation, paper, wallboard, wood, or surfaces behind wallpaper or in heating, ventilation, and air conditioning systems. Figure 1 illustrates an expert search for a source of an unpleasant moldy smell in a new-built home. Finally, *Penicillium chrysogenum* overgrowth on the back side of plasterboard ceilings and walls was identified as the causative agent.

Figure 2 shows results of tests for antifungal properties of common building materials. The tests were performed according to ISO 846: 1997 E with *Aspergillus versicolor*.[2] The fungus developed visible growth onto almost all materials tested

Environmental Mycology in Public Health. http://dx.doi.org/10.1016/B978-0-12-411471-5.00010-7

FIGURE 1 Searching for a source of indoor moldy smell, *Penicillium chrysogenum* identified as the agent.

under the experimental conditions (water activity of 0.75, 0.8, or 0.94 for a duration of 1–3 months). Wood and lime plasters revealed the highest antifungal resistance.

Figure 3 presents the results of the same tests as mentioned above, but materials covered with a sterile house dust were used. Such conditions mimic real indoor environments; the dust serves as an additional source of nutrients for the molds. Again, wood and inorganic (lime) plaster showed the best antifungal properties.

It is impossible to eliminate all molds and fungal propagules from indoor environments. However, moisture control is the most important strategy for reducing indoor mold growth. The dampness of the substrate is one of the crucial conditions for a mold's survival. Their optimal water activity in substrate ranges between 0.60 and 0.99. The ability of *Penicillium* spp. and *Aspergillus* spp., so-called first colonizers, to grow on/in common house dust under RH of 76–80% can probably explain their dominant prevalence even in healthy buildings (see also Figures 2 and 3). The secondary (*Cladosporium* spp., *Alternaria* spp., *Chaetomium* spp., 85% RH needed) and tertiary colonizers (*Fusarium* spp., *Acremonium* spp., yeasts, optimal RH above 90%) are able to biodeteriorate any building material under optimal thermal and moisture conditions.[3–5]

An example of the plots obtained from microclimate detectors placed in a moldy apartment for 10 days is given in Figure 4. Peaks of RH represent rising humidity due to cooking activity without proper exhaustion. The temperature falls when windows are opened.

Presented in Figure 5 is a 5-year mycological study carried out in 72 Slovak flats with indoor mold problems. The percentages represent the relative numbers of flats with a certain average quantity of indoor airborne mycobiota

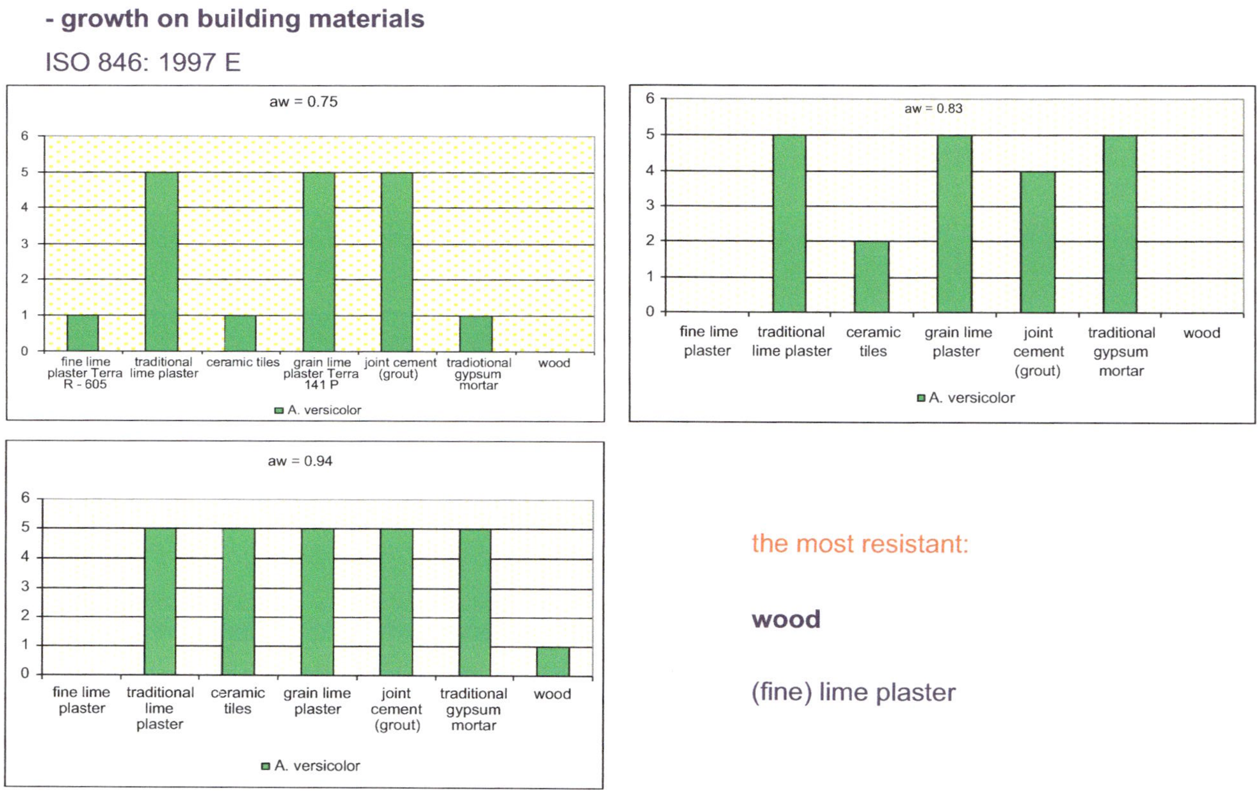

FIGURE 2 Tests of antifungal properties of building materials—model fungus *Aspergillus versicolor* and its growth on material surfaces (0, no growth; 5, total surface overgrown).

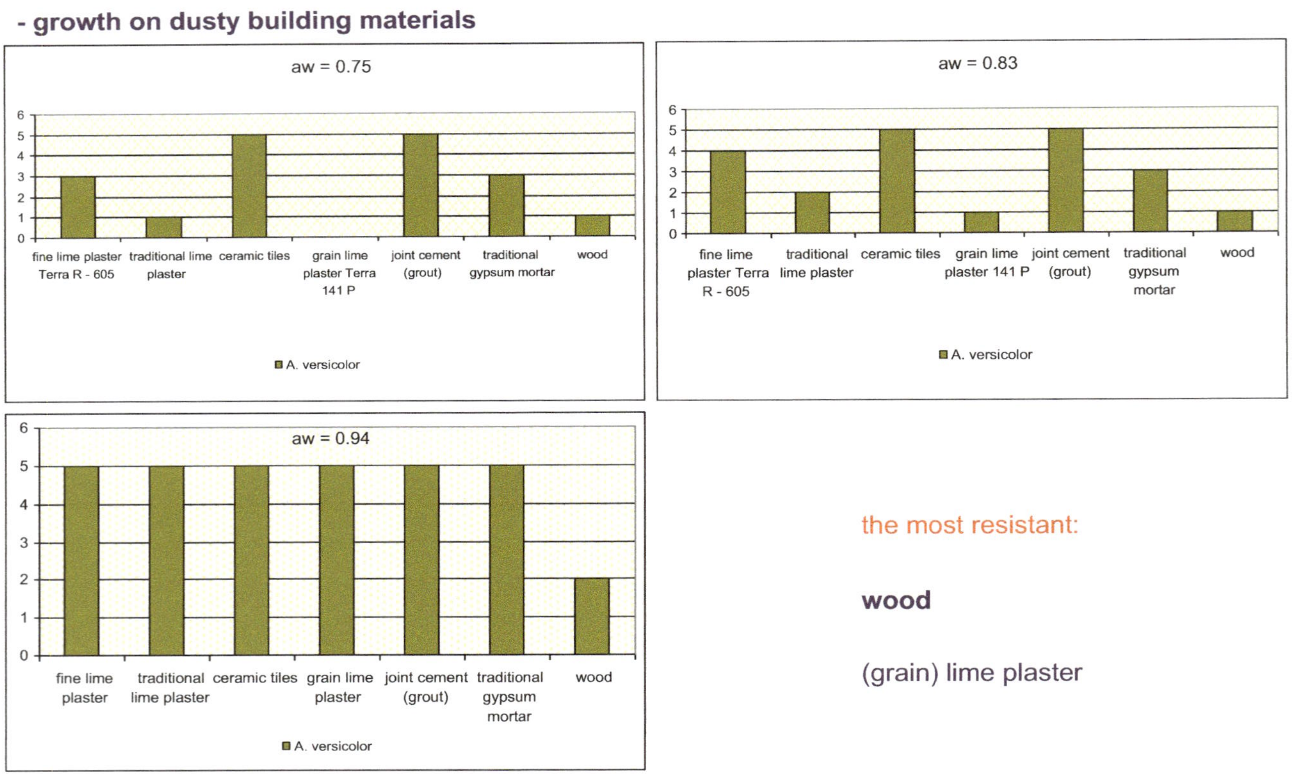

FIGURE 3 Test of building materials' susceptibility to fungal growth while being dusty. *According to ISO 846: 1997 E,* Aspergillus versicolor *as the model fungus.*

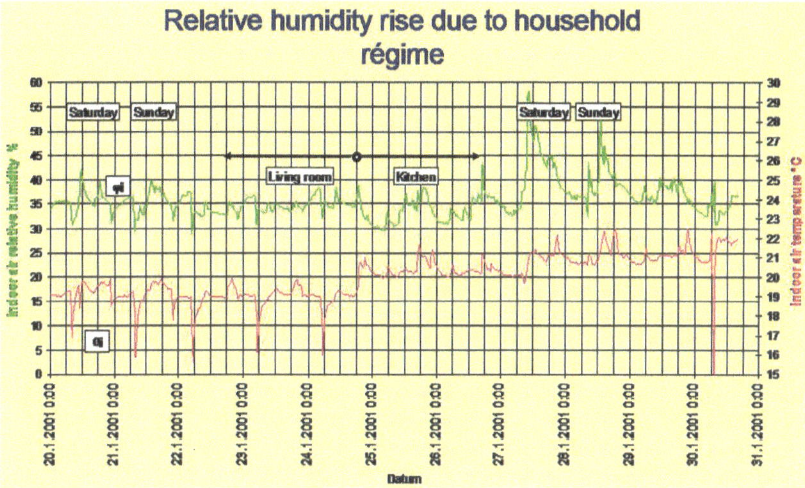

FIGURE 4 Automatic drawings from an indoor microclimate detector in a moldy apartment.

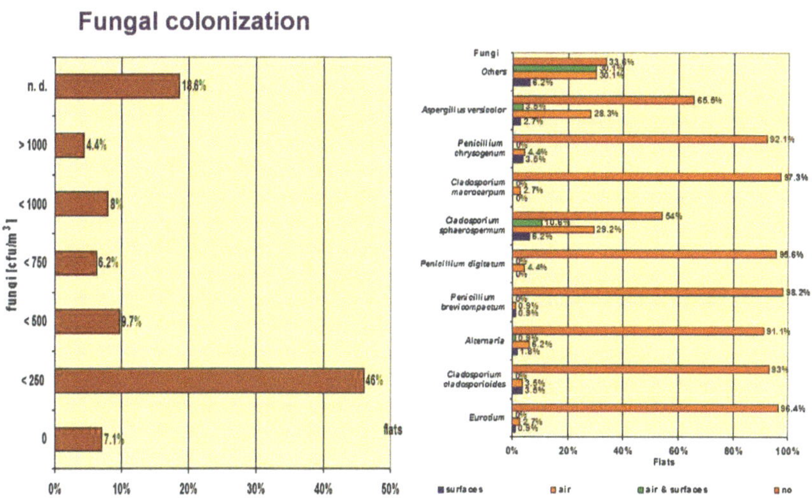

FIGURE 5 Quantitative (R) and qualitative (L) analysis of 72 Slovak moldy apartments over 5 years.

(given at the right—R) and qualitative mycological analysis (illustrated at the left—L) of the same flats (indoor air, surfaces) is shown—aspergilli, penicillia, cladosporia, and alternariae clearly dominate.

Another significant parameter related to moisture in building materials is the dew point temperature. Water condenses in cooler air around surfaces that are below the dew point. Dew points on inner wall surfaces under various thermal and humidity conditions in an apartment in Bratislava, Slovakia, are illustrated in Figures 6 and 7.

Critical surface temperature to promote fungal growth

237,3*LN(B5*D5/100/488,4)/(17,269-LN((B5*D5/100/488,4)))

STN 7305 40 -4, art. 11.2.3 p. 25

$$\theta_{dp} = \frac{236 \cdot \ln(p_d) - 1513,867}{23,59 - \ln(p_d)}$$

T(air) °C	r. h. (air) %	r. h. calculated for T(air) 20 °C (pd=2336.7 Pa)	crit. T(surface) (r. h. 80%)
18	50	44,1	10,74
19	50	47,0	11,68
20	50	50,0	12,62
21	50	53,2	13,57
22	50	56,5	14,51
23	50	60,1	15,45
24	50	63,8	16,40
25	50	67,7	17,34

FIGURE 6 Calculation of dew point (or critical surface temperature to promote fungal growth) and its source data.

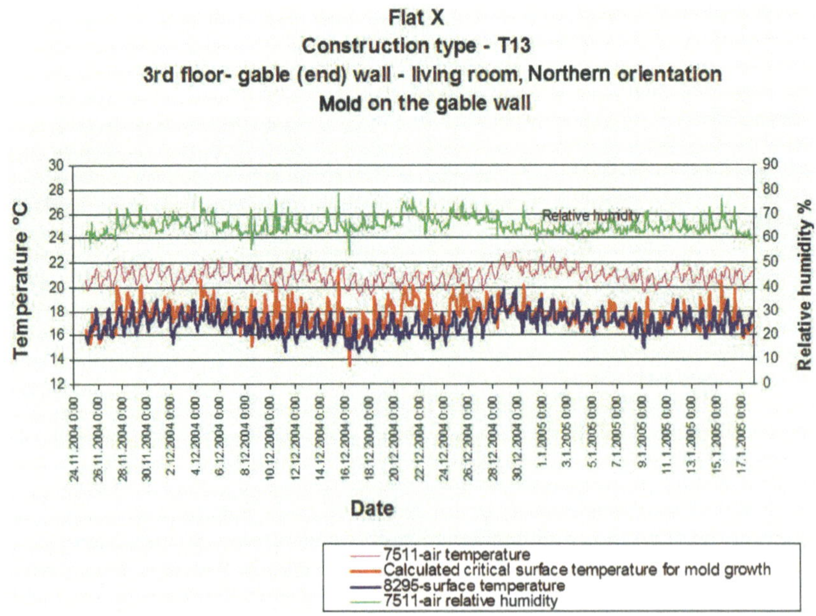

FIGURE 7 Automatic plots of indoor microclimatic conditions in a moldy flat over 3 months with the calculated dew point line (red). Measured surface temperatures of the building wall are almost always under the calculated critical ones. Thus, it is no wonder the wall had become moldy.

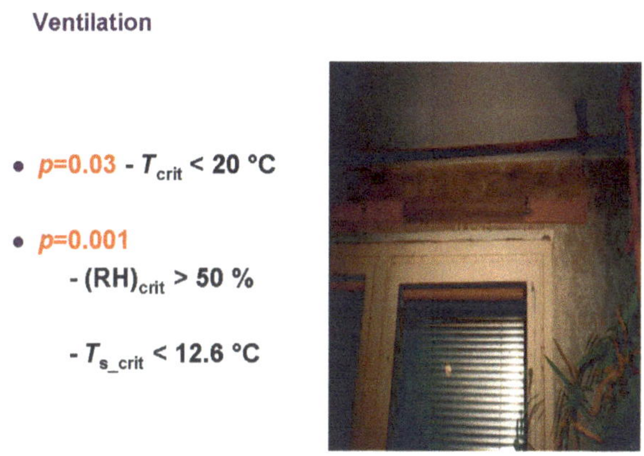

FIGURE 8 Continual vent opening causing cooling of the wall above, resulting in mold on it.

Building materials contain certain amounts of the so-called steady moisture, which is absorbed from the surrounding air. This absorbed humidity is dependent on temperature, RH, porosity of material, and presence of hygroscopic salts. The building material becomes the source of humidity when its humidity is higher than the steady moisture. In this instance, the moisture of building can be high even when the RH is low.

The main sources of moisture in buildings are:

- Capillary action of moisture: imperfections in the hydroisolation of the building subsoil and inadequate reduction of moisture causing wetness in the construction of the building subsoil;
- Improper or insufficient ventilation: continual vent opening–the cause of fungal growth on the wall above a window (statistical analysis of these household microclimate conditions: critical indoor air temperature <20 °C, $p=0.03$, critical surface temperature to promote fungi <12.6 °C at RH >50% related at the level $p=0.001$) (Figure 8);
- Rain: in roof constructions and in breaches in circumferential walls;
- Leaks in sanitary installations: at uncaulked joints of sanitary distribution systems, damaged outlet baths, washbasin, and kitchen sink;
- Condensation of water on internal surfaces of construction: in places with surface temperature lower than the dew point (Figures 9 and 10).
- Activities inside the building: an excessive amount of flowers, aquaria, vapors, laundry, cooking, and lack of "natural" (micro)ventilation owing to sealed windows, i.e., additional moisture sources. Figure 11 shows household activities and habits contributing to elevated RH (over 50%) that might promote indoor fungal growth (when the critical surface temperature drops below 12.6 °C under given microclimatic conditions). The most potent sources of extra moisture indoors seem to be babies/toddlers, pets (intensive breathing), and indoor plants (watering).

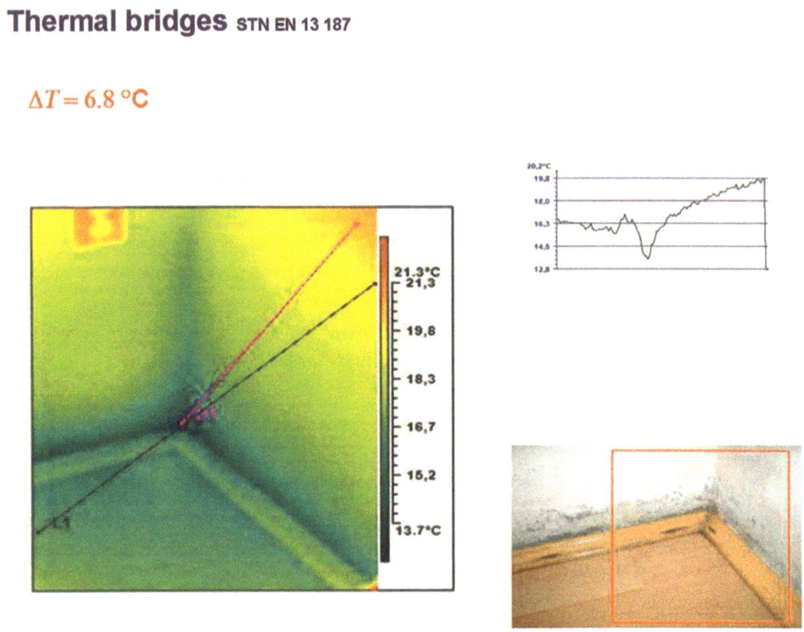

FIGURE 9 Thermovisualization of a thermal bridge with $\Delta T = 6.8\,°C$ at the low corner of a room. *According to the STN EN 13 187.*[6]

FIGURE 10 Severe mold growth on walls as a consequence of thermal bridges with the critical surface temperature <12.6 °C at RH >50%. Statistical probability of undesired fungal colonization at $p < 0.001$.

Based on the above-mentioned occupant behavior at home, it is possible to rank house rooms according to their fungal growth risk as given in Figure 12. So, kitchen and bathrooms present the highest risk.

Currently, a general approach to the study of the mechanisms of fungal effects on human beings is becoming more urgent. Such an approach includes

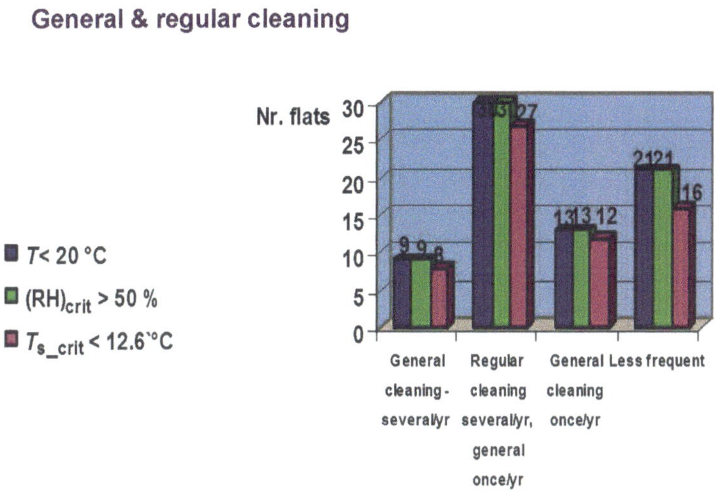

FIGURE 11 Dwelling occupants' activities elevating indoor humidity.

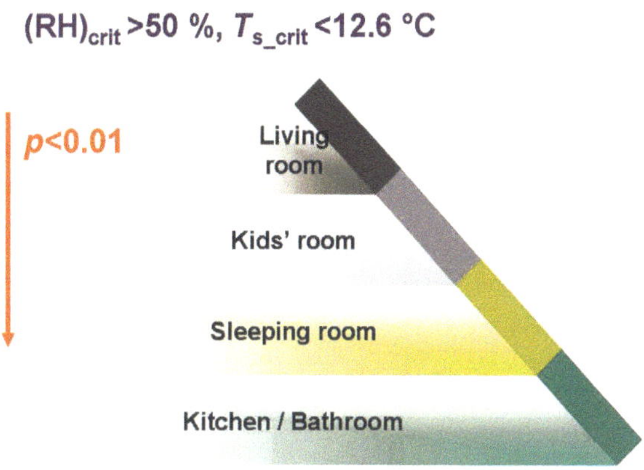

FIGURE 12 Diagram of fungal overgrowth possibility in particular house rooms (the lowest in the living room) under favorable microclimatic conditions.

the immunosuppressive influence of β-glucans from fungal cell walls as well as toxic and irritative effects of their exo- and endometabolites—mycotoxins and/or volatile organic compounds (VOCs).[7] Regarding mycotoxins with rather well-characterized toxicity (including carcinogenicity, mutagenicity, teratogenicity, cytotoxicity, and immunosuppression) after ingestion or dermal exposure, it has been found that an adverse biological effect can be caused by inhalation of a dose one-tenth the amount of one taken orally.[8]

Water-damaged building materials are often contaminated with fungi that produce detectable levels of mycotoxins, which may be aerosolized (material detritus, dust particles, most of them within respirable range) and contribute to indoor air pollution.[3,9,10] Tuomi et al.[11] analyzed 17 mycotoxins from 79 bulk building materials collected from water-damaged buildings. Their results showed that sterigmatocystin was present in 24% of the samples, trichothecenes in 19% of the samples, and citrinin in three samples. *Aspergillus versicolor* was found on most sterigmatocystin-containing samples, and *Stachybotrys chartarum* on the samples in which satratoxins were present. Indoor mycobiota, particularly *A. versicolor*—the most frequent indoor colonizer under central European housing conditions[12] (Figures 13 and 14), also have the capacity to contribute to the indoor levels of irritant VOCs. The spectrum of volatiles produced varies with microenvironmental condition (temperature, RH, substrate) and the presence of other microbes as well.[13]

Indoor *A. versicolor*

> frequency – flats > 1986 yr - ***p*=0.05**
- N orientation, living rooms
- laundry drying + exhaust – ***p*<0.05**
- general cleaning once/yr, regular cleaning less frequent
- without room plants & pets

FIGURE 13 Household characteristics enabling *Aspergillus versicolor* indoor colonization and their statistical potential.

- molecular epidemiology

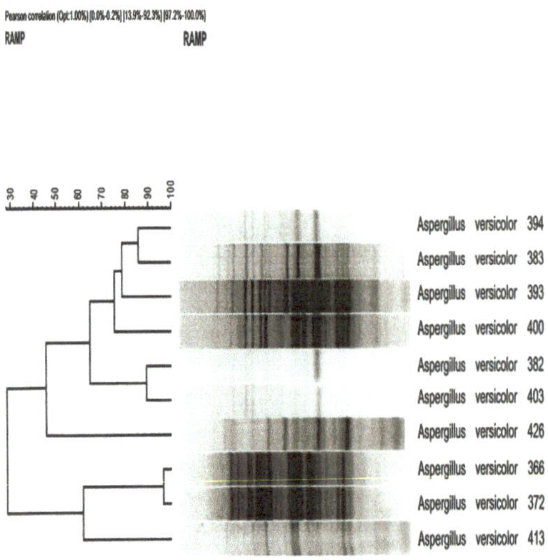

Isolate Nr.	Localization
393	indoor air (HVAC)
394	indoor air (HVAC)
382	indoor air (bath)
383	indoor air (bath)
400	indoor surface
403	indoor surface
426	indoor air (room)
366	wall (bath)
372	outdoor air
413	HVAC filter

FIGURE 14 Molecular epidemiological analysis of *Aspergillus versicolor* indoor and related outdoor isolates. HVAC, heating, ventilation, and air conditioning.

Figure 13 summarizes housing conditions favoring *A. versicolor* indoor incidence. This fungus—as a xerophilic one—does not insist on humidity extremes (room plants, pets, or improper ventilation as additional moisture sources are not necessary). It is apparent that lower indoor temperatures owing to the northern orientation of the affected rooms and dusty surroundings because of less frequent cleaning were sufficient to promote the occurrence of the fungus. Moreover, spores of *A. versicolor* and *Penicillium* sp. retained their viability on different plasters after 40 days to 3 months, i.e., until the end of the particular experiment that was being carried out.[3,4]

Molecular epidemiological analysis (RAMP) of *A. versicolor* indoor and related outdoor isolates clarified their relations. Some indoor ones were of outdoor origin, whereas others came from the same indoor source (Figure 14).

The metabolite synthesis of fungi depends on the quality of construction materials.[14,15] In our previous study on tracheal organ cultures of one-day-old chicks, varying ciliostatic activity was found in chloroform extracts of biomass from building materials (mineral wool, plasterboard, cardboard) inoculated with cultures of molds of indoor origin (*P. chrysogenum*, *Penicillium palitans*, *Trichoderma viride*, *Stachybotrys* sp., and *A. versicolor*). Generally, extracts from fungal growth on materials composed of finely divided cellulose were more active than those from growth on mineral wool.[16] The only data available on the ciliostatic activity of pure indoor mold metabolites involve sterigmatocystin, from *A. versicolor* and *Chaetomium* spp., which is one of the most active metabolites examined.[17,18] It is proven that more than 90% of the *A. versicolor* isolates are able to produce sterigmatocystin (Figure 15).

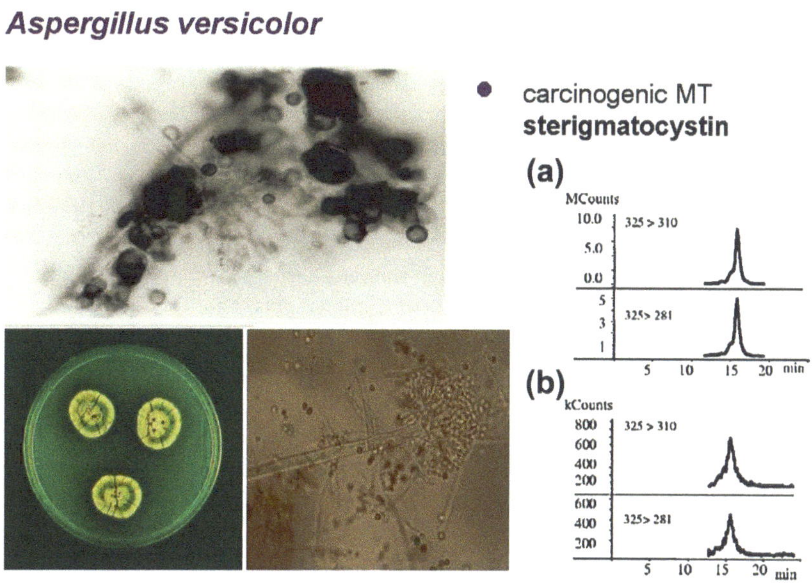

FIGURE 15 Top: direct microscopy of moldy plaster. Bottom: pure fungal culture. Right: LC–MS/MS analysis of extrolites yielded sterigmatocystin peaks.

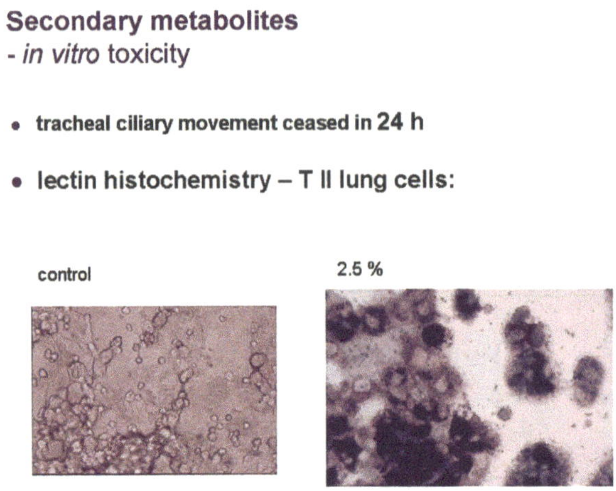

FIGURE 16 In vitro debilitating effects of crude chloroform extracts of *Aspergillus versicolor* metabolites.

There is a great need to better understand the adverse health effects of short- and long-term exposure to fungal metabolites and to determine whether health effects are reversible.[19] Studies so far have dealt with the spores, with respect to extracted solid toxicants, including mycotoxins, after respiratory exposure. According to previous experiments, it became apparent that pulmonary inflammation, or cytotoxic damage to alveolar type II cells and alveolar macrophages (AMs) (particularly sensitive) was related to methanol-soluble as well as chloroform-soluble toxins.[20–23]

Crude chloroform extracts of the *A. versicolor* secondary metabolites prevented tracheal ciliary beating in vitro within 24 h. They also caused coarsening of T II rat lung cells (analyzed by lectin histochemistry) (Figure 16). Both mechanisms are involved in the upper airways' self-cleaning processes.

When applied intratracheally in dimethyl sulfoxide, the same extracts showed concentration-dependent cytotoxic effects on lung tissue (DNA damage and oxidative stress in the cells and significantly higher proportion of free cytosolic and lysosomal enzymes cathepsin D and alkaline phosphatase). On the other hand, the phagocytic activity and viability of AMs were significantly depressed. Together with overproduction of young immature AMs, the parameters indicate the inflammatory potential of the extrolites, too (Figure 17).

The effects were statistically even more significant when extracts of coculture of *A. versicolor* and *S. chartarum* were used (Figure 18).

Why was *S. chartarum* used in the experiments? In an investigation of fatal infant idiopathic pulmonary hemorrhage outbreaks in the United States, particular fungal isolates were found to be toxic in vitro. Later studies showed that *S. chartarum* produced cytotoxic and immunosuppressive macrocyclic trichothecenes (stachybotryotoxins) and spirocyclic drimanes that caused inflammation and hemorrhages in the respiratory tract and intestines of laboratory animals. In this case, isolates of *Memnoniella echinata* (producer of

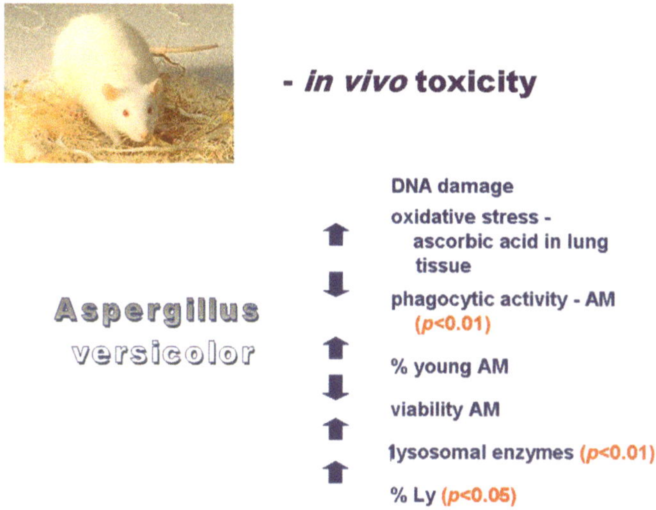

FIGURE 17 In vivo toxicity of *Aspergillus versicolor* secondary metabolites in rats—cytotoxic and inflammation-promoting effects.

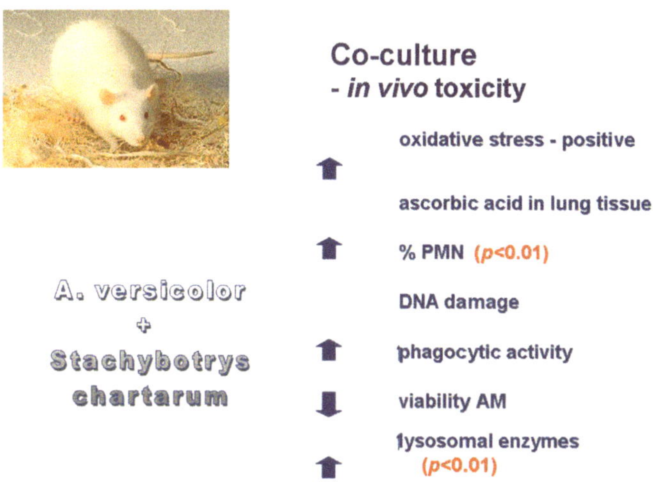

FIGURE 18 In vivo pulmonary toxicity of metabolites of coculture of *Aspergillus versicolor* and *Stachybotrys chartarum* in rats—cytotoxic and inflammation-inducing effects.

griseofulvin) and *A. versicolor* (carcinogenic sterigmatocystin) were detected as well.[24] It was this case in particular in which smoking was identified by epidemiologists as a factor in increased health risk for fungal intoxication.[25] Experimental conditions of sub- or chronic exposure to the toxins and inclusion of other environmental factors (tobacco smoke) better mimic the real indoor exposure scenario.[26]

We performed tests on the in vivo toxicity of extrolites of *A. versicolor* grown together with *S. chartarum* (intratracheally instilled) in rats subjected to tobacco smoke simultaneously and 6 weeks before the experiment. All bioindicators of cytotoxic and inflammatory injury (including inflammatory cytokines release) were then changed positively. Thus, the general toxic effect was more pronounced in this group compared with the group not exposed to smoke (Figures 19 and 20).

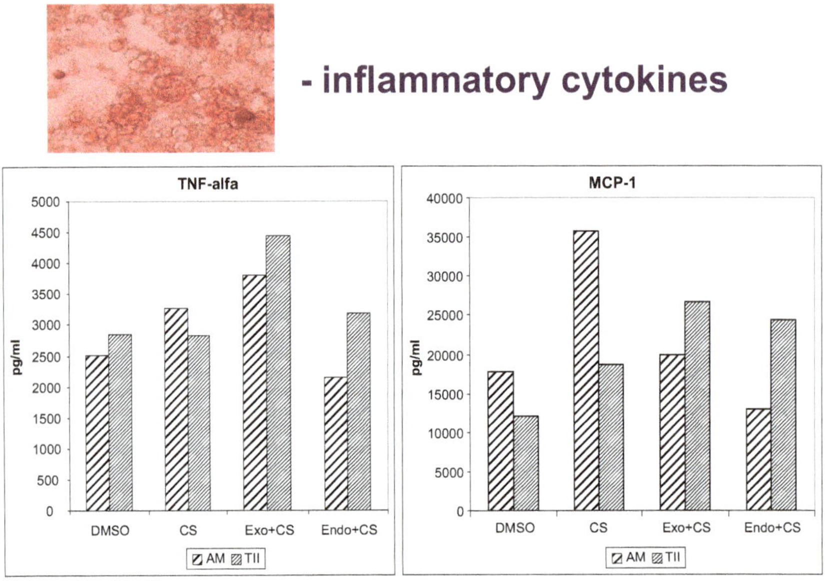

FIGURE 19 In vivo toxicity of fungal metabolites together with tobacco smoke—cytotoxic potential.

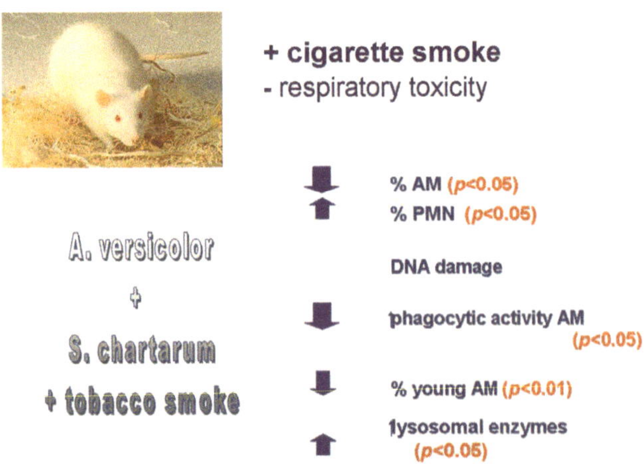

FIGURE 20 Inflammatory damage of rat lung tissue after being exposed to fungal metabolites (*Aspergillus versicolor*+*Stachybotrys chartarum*) and tobacco smoke for 3 days.

PREVENTIVE MEASURES

There is a high need to control moisture in both new and existing constructions owing to the significant health consequences resulting from dampness. Dampness and mold in buildings is undoubtedly a significant public health problem with substantial economic impact.[27]

Indoor fungal prevention consists of:

- provision of suitable air circulation by periodic ventilation (ventilation system, exchange air by opening windows);
- repairs, remodels, construction of spaces and buildings to eliminate all sources of humidity;
- periodic building maintenance;
- use of suitable impermeable materials for flooring;
- adjustment of climate conditions of buildings by correct regulation of fumigation and humidity (maintain below 60% RH, ideally 30–50%, if possible);
- adjustment for correct function of cooker hoods in kitchens and regular defrosting, washing, and disinfecting of refrigerator and freezer;
- performance of periodic cleaning of spaces (use suitable cleaners and disinfectants, regular disposal of waste) (according to our findings, the mycologically nonproblematic apartments usually undergo general deep cleaning more frequently than the moldy ones (Figure 21));
- proper storage of materials (e.g., objects should not be stored in close contact with walls);
- periodic control of occurrence of molds.[28]

Figure 22 presents the World Health Organization's recommendations for building and maintaining sustainably healthy indoor environments from the myco- and mycotoxicological point of view.[29]

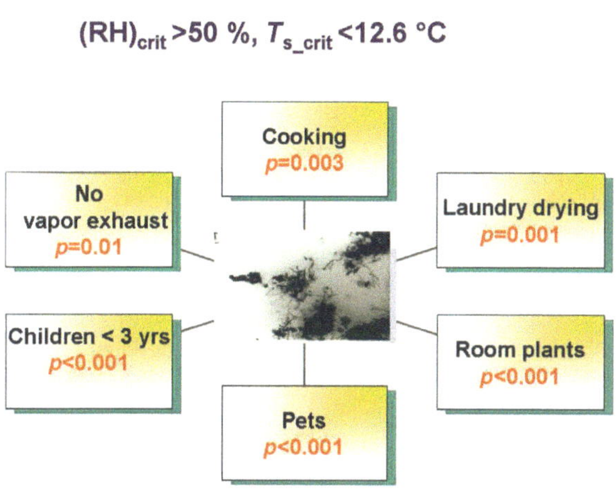

FIGURE 21 Microclimatic (including risky) parameters in the flats according to their cleaning mode.

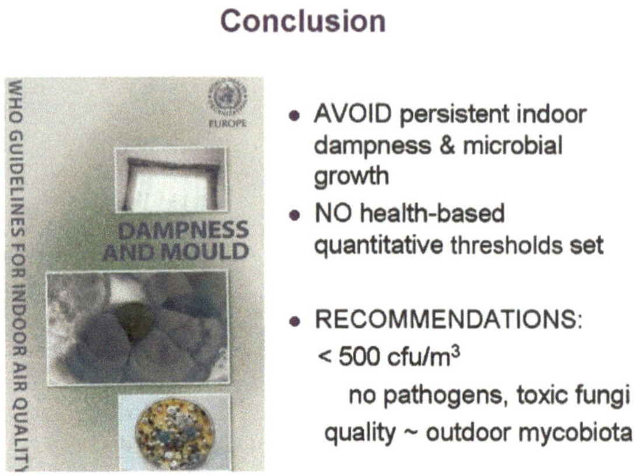

FIGURE 22 General WHO recommendations for indoor mold prevention (2009).

COLLABORATORS

- Slovak Medical University, Bratislava

  Silvia Černá, Marta Hurbánková, Zuzana Kolláriková, Zuzana Kováčiková, Miroslava Kuricová, Mária Majorošová, Soňa Wimmerová, Ladislava Wsólová

- Research & Development Institute of Building Construction, n.o., Bratislava

  Jana Bendžalová, Emil Droba, Zuzana Sternová

- NIOH, Copenhagen, Denmark

  Ken Wilkins

- NIOH Budapest, Hungary

  Elisabet Tátrai

- University of Lund, Sweden

  Erica Bloom

REFERENCES

1. Kroeling P. Sick building syndrome. *Allergol* 1998;**21**:180–91.
2. ISO 846: 1997 E. Technical committee ISO/TC 61 plastics, subcommittee SC6 aging, chemical and environmental resistance.
3. Piecková E, Wilkins K. Airway toxicity of house dust and its fungal composition. *Ann Agric Environ Med* 2004;**11**:67–73.
4. Piecková E, Jesenská Z. Evaluation of in vitro fungal contamination of plasters. (In Slovak). *Hygiena* 2000;**45**:37–41.
5. Piecková E, Kolláriková Z, Sternová Z, Droba E. Do building materials posses antifungal properties? (In Slovak). *Derma* 2009;**9**:11–6.

6. STN EN 13 187. Thermo-technical properties of buildings. *Qual Determ Therm Impair Buildings Extern Walls* 2001. [Infrared method].
7. Garrett MH, Rayment PR, Hooper MA, Abramson MJ, Hooper BM. Indoor airborne fungal spores, house dampness and associations with environmental factors and respiratory health in children. *Clin Exper Allergy* 1998;**28**:459–67.
8. Hendry KM, Cole EC. A review of mycotoxins in indoor air. *J Toxicol Environ Health* 1993;**38**:161–82.
9. Nikulin M. *Toxigenic Stachybotrys chartarum in water-damamged building materials: ecology and biology*. Microbial Publ. 46/1999. Univ. Helsinki, FI, 2000.
10. Piecková E, Kunová Z. Indoor fungi and their ciliostatic metabolites. *Ann Agric Environ Med* 2002;**9**:59–63.
11. Tuomi T, Reijula K, Johnson T, Hemminki K, Hintikka EL, Lindroos O, et al. Mycotoxins in crude building materials from water damaged buildings. *Appl Environ Microbiol* 2000;**66**:1899–904.
12. Piecková E, Majorošová M, Hurbánková M, Černá S, Pangallo D. *Aspergillus versicolor* (Vuill.) Tiraboschi—the most frequent indoor fungal colonizer under our housing conditions. *Derma* 2010;**3**:3–6.
13. Wilkins CK, Larsen ST, Hammer M, Poulsen OM, Wolkoff P, Nielsen GD. Respiratory effects in mice exposed to airborne emissions from *Stachybotrys chartarum* and implications for risk assessment. *Pharmacol Toxicol* 1998;**83**:112–9.
14. Nielsen KF, Thrane U, Larsen TO, et al. Production of mycotoxins on artificially inoculated building materials. *Int Biodeter Biodeg* 1998;**42**:9–16.
15. Gravesen S, Nielsen PA, Iversen R, Nielsen KF. Microfungal contamination of damp buildings—examples of risk constructions and risk materials. *Environ Health Perspect* 1999;**107**:505–8.
16. Wilkins K, Piecková E. Detection of ciliostatic activity in fungal growth on building materials. *Environ Sci Pollut Res* 2002;**9**:105–6.
17. Jesenská Z, Bernát D. Effects of mycotoxins on in vitro movement of tracheal cilia from one-day-old chicks. *Folia Microbiol* 1994;**39**:155–8.
18. Piecková E. In vitro toxicity of indoor Chaetomium Kunze ex Fr. *Ann Agric Environ Med* 2003;**10**:9–14.
19. Li DW, Yang CS. Fungal contamination as a major contributor to sick building syndrome. In: Straus DC, editor. *Advances in applied microbiology*, vol. 55. Academic Press, Elsevier; 2004. p. 31–112.
20. Rao CY, Brain JD, Burge HA. Reduction of pulmonary toxicity of *Stachybotrys chartarum* spores by methanol extraction of mycotoxins. *Appl Environ Microbiol* 2000a;**66**:2817–21.
21. Rao CY, Burge HA, Brain JD. The time course of responses to intratracheally instilled toxic *Stachybotrys chartarum* spores in rats. *Mycopathol* 2000b;**149**:27–34.
22. Piecková E, Hurbánková M, Černá S, Pivovarová Z, Kováčiková Z. Pulmonary cytotoxicity of secondary metabolites of *Stachybotrys chartarum* (Ehrenb.) Hughes. *Ann Agric Environ Med* 2006;**13**:259–62.
23. Piecková E, Hurbánková M, Černá S, Lišková A, Kováčiková Z, Kolláriková Z, et al. Inflammatory and haematotoxic potential of indoor *Stachybotrys chartarum* (Ehrenb.) Hughes metabolites. *Arh Hig Rada Toksikol* 2009;**60**:401–9.
24. Andersen B, Nielsen KF, Jarvis B. Characterisation of morphologically, chemically and physiologically different *Stachybotrys* species from water-damaged buildings. *Mycologia* 2002;**94**:392–403.
25. Kuhn DM, Ghannoum MA. Indoor mold, toxigenic fungi, and *Stachybotrys chartarum*: infectious disease perspective. *Clin Microbiol Rev* 2003;**16**:144–72.
26. Piecková E, Šegvič-Klarič M, Jakšič D, Majorošová M. Synergism-addition-antagonism of the indoor mycotoxins' mixtures effects onto health. (In Slovak). In: Petráš D, editor. *23rd Int. Conf. Indoor Climate 2012—Proceedings. STU Bratislava*. 2012. p. 55–8.

27. Mudarri D, Fisk WJ. Public health and economic impact of dampness and mold. *Indoor Air* 2007;**17**:226–35.
28. Warsco K, Lindsey PF. Proactive approaches for mold-free interior environments. *Arch Environ Health* 2003;**58**:512–22.
29. WHO Regional Office for Europe. *WHO guidelines for indoor air quality: dampness and mould*2009. 228 p.

Chapter 11

Urban Environment

Fungal Specificities: Nonoccupational Exposure and Urban Environment

Rosely Maria Zancopé Oliveira[1], **Rodrigo de Almeida Paes**[1], **Ferry Hagen**[2]
[1]*Laboratório de Micologia do Instituto Nacional de Infectologia Evandro Chagas, Fundação Oswaldo Cruz, Rio de Janeiro, Brazil;* [2]*Department of Medical Microbiology & Infectious Diseases, Canisius-Wilhelmina Hospital, Nijmegen, The Netherlands*

INTRODUCTION

Conceptually, occupational exposure happens after the exposure of an individual to harmful chemical, physical, or biological agents that occurs as a result of one's occupation. According to this concept, there is convincing evidence that some mycotic diseases are associated with occupational exposure to fungal agents. Among them, the respiratory mycotic diseases, such as cryptococcosis, histoplasmosis, paracoccidioidomycosis, coccidioidomycosis, and blastomycosis, result from the exposure of workers to the fungal agents, mostly in the form of particles or dust. In addition, a good example of an occupational fungal disease is sporotrichosis, which is regarded as an occupation-related disease that usually occurs in the form of isolated cases or small outbreaks. This fungal disease involves people exposed to plants or soil rich in organic matter and, occasionally, laboratory technicians.[1,2] However, there is a high proportion of mycotic cases that do not report any occupational exposure through the patient's working life. In this chapter, we focus on examples of mycotic diseases acquired from nonoccupational exposure, all of them associated with domestic and environmental exposure.

CRYPTOCOCCOSIS

Together with the clinically relevant species within the fungal genera *Aspergillus* and *Candida*, the two *Cryptococcus* species *Cryptococcus gattii* and *Cryptococcus neoformans* are major causes of fungal morbidity and mortality worldwide. The yeast genus *Cryptococcus* contains more than 100 species, but only *C. gattii* and *C. neoformans* are known as frequent sources of disease, and only a handful of case reports have described the involvement of other cryptococcal species.[3] It has been

Environmental Mycology in Public Health. http://dx.doi.org/10.1016/B978-0-12-411471-5.00011-9

estimated that annually nearly a million human immunodeficiency virus (HIV)-infected individuals develop cryptococcal meningitis, and an estimated approximately 625,000 patients will not survive this fungal infection.[4] The majority of cryptococcal infections are a result of inhalation of the disease culprit, followed by hematogenous dissemination, and if untreated it finally enters the central nervous system with fatal consequences.[5] Cutaneous infections are less often reported and are frequently related to a traumatic introduction of infectious particles.[6,7]

Until the early 1980s cryptococcal infections were rarely reported, but with the onset of the HIV pandemic the number of infections rose to high levels, especially in sub-Saharan Africa.[4,5] Based on a plethora of epidemiological studies, it has been observed that *C. neoformans* is the major cause of cryptococcal disease worldwide; this pathogen mainly infects humans and animals that have a compromised immune system.[5] On the other hand its sibling, *C. gattii,* is the major cause of cryptococcal infection among immunocompetent patients; this species was previously reported only in (sub)tropical areas but since the turn of the twenty-first century it has been on the rise in temperate climate zones.[8–10]

In addition to the difference in host predilection, both cryptococcal species differ also in terms of ecology, genetics, and physiology.[5,11] *C. neoformans* has a global distribution pattern and is mainly associated with bird excreta, especially that from pigeons. Cryptococcal cells cannot multiply within the gastrointestinal tract of pigeons but they have the ability to proliferate on nitrogen-rich excreta from which desiccated yeast cells and/or basidiospores disperse into the environment.[12,13] *C. gattii*, which was until the beginning of this century regarded as a tropical pathogen, has a distribution pattern that spans all tropical and subtropical areas, as well as regions that have a climate similar to that of Mediterranean Europe.[8,10] In 1990 the environmental niche of *C. gattii* was found to be decaying wood of *Eucalyptus* trees, a genus that originated in Australia, from where this tree was globally exported to numerous countries.[14] At that time, it was assumed that *C. gattii* was globally distributed via this route, but this was disputed by the culturing of *C. gattii* from decaying wood obtained from a *Guettarda acreana* tree in the pristine Amazon rain forest.[15] A large-scale environmental screening survey, initiated after the sudden onset of an ongoing and expanding *C. gattii* outbreak in the temperate climate zone of British Columbia, Canada, revealed that many more tree species harbor *C. gattii*.[16,17] Environmental sampling has shown that *C. gattii* is present in high concentrations in the air and soil that surrounds trees. It was also observed that *C. gattii* can survive in seawater and that a variety of sea animals have developed *C. gattii* infection, as well.

In a study performed in 2001, it was shown that in early childhood, nearly all individuals had developed antibodies against *C. neoformans* and thus that they had been in contact with this yeast.[18] However, *C. neoformans* cells can remain dormant in the human body for decades until the immune system has become attenuated. This was also observed for the primary pathogen *C. gattii,* which was found to remain dormant for months up to decades until the moment that the host's immune status became weakened owing to another illness.[9,19]

C. gattii can be present in the environment in high concentrations and it is not surprising that not only inhabitants of the affected outbreak areas develop cryptococcal diseases, but also tourists who briefly visit these localities develop infections.[9,19–21] *C. gattii* has been isolated from recreational areas, as has been shown by an environmental sampling study on the Iberian Peninsula that revealed that the strains obtained were genetically indistinguishable from those isolated from clinical and veterinary sources, including those isolated from several small outbreaks in Spain.[9,22,23]

Reports from occupational exposure and subsequent cryptococcal infections are rare and mainly describe infections after exposure of the patient to *Cryptococcus*-contaminated bird excreta, especially those patients who have been in close contact with birds.[12,24–29] Another category of individuals who are at risk for cryptococcal disease are those working in forestry, especially those working in areas known to have a high airborne concentrations of cryptococcal cells, and those subjects who work with wood and/or plant material.[7,16,30,31]

Histoplasmosis

Histoplasmosis is a serious community-acquired infection in the United States and in several Latin American countries.[32] This fungal disease is caused by the dimorphic fungal pathogen *Histoplasma capsulatum,* which displays two distinct morphological forms, filamentous and yeast forms, depending on the nutritional factors and temperature.[33] The majority of *H. capsulatum* biomass exists in the soil as a saprophytic mold, especially under alkaline conditions (pH 5–10) and temperatures between 20 and 30 °C. Usually the fungus is associated with soil enriched with organic nitrogen sources such as animal excrement, especially that from bats and chickens.[34,35] In fact, *H. capsulatum* has been isolated from several protected environments, such as caves, abandoned construction sites, and chicken coops. However, *H. capsulatum* does not cause disease in birds, so their role in the environmental spread of *H. capsulatum* is probably due to fungal carriage on their feathers, beaks, or claws. On the other hand, bats are fungal reservoirs, because they harbor the fungus in their intestinal mucosa and release *H. capsulatum* in their feces.[36–38] In addition, histoplasmosis can occur in bats, with lethal consequences. Contaminated soil has been found to be the source of infection for both humans and animals such as dogs, cats, horses, cattle, pigs, rodents, and marsupials.[32] *H. capsulatum* infection is typically acquired by the inhalation of microconidia after disturbance of contaminated soil or excreta. For this reason, histoplasmosis is generally not associated with person-to-person spread of the disease. Although rarely reported, histoplasmosis can be acquired by cutaneous inoculation of the fungus[39–41] and, even more unusually, vertical transmission can occur, given that there is one such case report known from the literature.[42] In the mold form *H. capsulatum* is composed of hyaline septated hyphae that produce the two different asexual reproduction structures of macroconidia and microconidia. The latter are the assumed infectious propagules, as their size of 2–6 μm is well suited to deposition into distal alveoli. Upon entry into a susceptible host, the microconidia rapidly convert to the pathogenic budding yeast-like form, which can also be cultivated in laboratory medium at 37 °C. Inhalation of the conidia can result in pulmonary histoplasmosis and, in some cases, severe disseminated disease and death. The risk of infection with *H. capsulatum* depends on the time of contact and the activity performed associated with soil disruption or involving bird or bat guano, such as farming, soil excavation, spelunking, construction, renovation, demolition, and cleaning sites harboring the fungus.[43] In addition, the presence of bats in caves, attics, ceilings, and roofs, as well as bird nests and abandoned chicken and poultry houses is epidemiologically important as it can increase the chance of human acquisition of *H. capsulatum.*

Histoplasmosis has long been recognized as a common recreational disease among cavers in North America, but reports indicate an increase in the number of cases in individuals who were engaged in other forms of adventure tourism and ecotourism. Outbreaks of histoplasmosis have been reported among traveling U.S. residents after trips to Central and South America.[44] Since 2010, epidemics of acute respiratory histoplasmosis have frequently been reported from several countries, the majority associated with contact with migratory bird excreta or bat guano.[45–51] Histoplasmosis is one of the most common systemic mycoses in Brazil, where epidemiological surveys have been carried out using the histoplasmin skin test, which indicated that this mycosis is endemic in all surveyed areas.[35] As of this writing, 161 cases of *H. capsulatum* infection have been associated with outbreaks that occurred in various Brazilian regions. Of these, 14 outbreaks were identified in Rio de Janeiro State. Although the epidemiology of histoplasmosis has been well studied there, cases of histoplasmosis in Brazil may be underestimated, suggesting that the endemic areas are more widespread than previously thought.

A variety of procedures have been used for decontaminating infected areas of *H. capsulatum* growth. Regular cleaning of locations with the potential to become a source of infection would be the first preventive measure in histoplasmosis. Appropriate masks should be worn by individuals at risk of exposure at suspicious sites. In a situation of high risk of exposure to *H. capsulatum* antifungal prophylaxis can be indicated. Sites that have been proven to be contaminated can be proscribed and decontaminated. The most successful method is to spray the affected sites with 3% formalin solution. However, this procedure has the side effect that it can cause environmental damage.

SPOROTRICHOSIS

Sporotrichosis is a subcutaneous mycosis that has long been recognized as the gardener's disease, because the infection is in most cases a result of the inoculation of the fungus *Sporothrix* spp. by thorns, stings, scrapes, and minor injuries. Certain occupational and leisure activities such as floriculture, horticulture, gardening, hunting, agriculture, mining, and others facilitate the exposure to the fungus and have over the past years been associated with sporotrichosis.[52]

For several years, sporotrichosis was described as a disease caused by the single species *Sporothrix schenckii*.[53] However, studies based on polyphasic taxonomy showed that this single species could be divided into several cryptic sibling species, which were named *Sporothrix brasiliensis*, *Sporothrix globosa*, *Sporothrix luriei*, *Sporothrix mexicana*, and *Sporothrix schenckii* sensu stricto.[54] These species present variations in virulence,[55] antifungal susceptibility,[56] and the clinical aspects of sporotrichosis.[57]

In 1998, in the state of Rio de Janeiro, Brazil, an epidemic of sporotrichosis was recognized and involved several cities.[58] In these endemic areas, the fungus was found not to be transmitted in the classical way, i.e., through plants, contaminated soil, or timber, but rather through scratches or bites from or other contact with domestic cats infected with *S. brasiliensis*.[57,59] This form of propagation of the disease has led to several unusual clinical manifestations of sporotrichosis.[57]

The domestic cat (*Felis catus*) is the animal species most affected by sporotrichosis. Between 1998 and 2012, around 3800 cases were diagnosed in a reference center for this mycotic infection in Rio de Janeiro, Brazil. In this epizootic occurring in Rio de Janeiro, male, unsterilized, young adult cats with access to the extradomiciliary environment constitute the animal group most susceptible to this mycosis. This occurs because the feline transmission of *S. brasiliensis* occurs mainly after fights involving an infected cat.[60] Different from humans and dogs, the skin lesions of cats have a high fungal burden, which is one of the most important factors that make the cat an important source of *Sporothrix* spp. infection of humans and other animals.[61] Respiratory signs are most frequently seen, mainly sneezing, which was observed in approximately 40% of the cases.[62]

Human sporotrichosis can be classified into the following clinical forms: fixed cutaneous, lymphocutaneous, disseminated cutaneous, and extracutaneous.[52] Nonoccupational sporotrichosis in Rio de Janeiro, Brazil, is associated with a high number of disseminated cutaneous cases of the infection. In this scenario, 9% of the patients present this clinical form, which is rarely seen in other areas. This can be explained by multiple inoculations (bites, scratches) of the fungus by naturally infected cats.[63] In addition, sporothrichosis is a serious problem in patients co-infected with HIV.[64] An HIV infection aggravates sporotrichosis, with a higher incidence of severe disseminated cases and a higher number of hospitalizations and deaths.[65]

China represents another important endemic area for sporotrichosis.[66] Although occupational sporotrichosis prevails in this area, there are some reports of nonoccupational infection by *S. globosa*, the main agent of sporotrichosis in Asia.[67] In a study conducted at Jilin province (northeast China), activities such as stacking cornstalks or reeds at home and burning them for heating the house and cooking food in colder seasons were risk factors related to sporotrichosis. Because of this epidemiological scenario, women are the group most affected by this subcutaneous mycosis in China, unlike other countries in which occupational sporotrichosis occurs.[66]

Under epidemic or endemic nonoccupational situations, the medium- and long-term control of sporotrichosis requires basic educational measures that emphasize the responsible ownership of animals and reproductive control programs for the cat population.[60] The owners of infected cats should be advised about the risks of transmission and appropriate prevention measures, particularly the use of gloves when handling the cat, and cleaning the environment with sodium hypochlorite. In addition, the carcasses of dead animals should be incinerated to avoid environmental contamination.[68] Human patients suffering from a trauma caused by scratches and/or bites of infected cats should seek medical advice immediately. Correct diagnosis and early treatment are the key to reducing the transmission of *Sporothrix* spp. between animals and humans.[52]

REFERENCES

1. Copper CR, Dixon D, Salkin IF. Laboratory-acquired sporotrichosis. *J Med Vet Mycol* 1992;**30**:169–71.
2. Hajjeh R, McDonnel S, Reef S, Licitra C, Hankins M, Toth B, et al. Outbreak of sporotrichosis among three nursery workers. *J Infect Dis* 1997;**176**:499–504.
3. Arendrup MC, Boekhout T, Akova M, Meis JF, Cornely OA, Lortholary O. European Society of Clinical Microbiology and Infectious Diseases Fungal Infection Study Group, European Confederation of Medical Mycology. ESCMID and ECMM joint clinical guidelines for the diagnosis and management of rare invasive yeast infections. *Clin Microbiol Infect* 2014;**20**(3):76–98.
4. Park BJ, Wannemuehler KA, Marston BJ, Govender N, Pappas PG, Chiller TM. Estimation of the current global burden of cryptococcal meningitis among persons living with HIV/AIDS. *AIDS* 2009;**23**(4):525–30.
5. Bovers M, Hagen F, Boekhout T. Diversity of the *Cryptococcus neoformans-Cryptococcus gattii* species complex. *Rev Iberoam Micol* 2008a;**25**(1):S4–12.
6. Christianson JC, Engber W, Andes D. Primary cutaneous cryptococcosis in immunocompetent and immunocompromised hosts. *Med Mycol* 2003;**41**(3):177–88.
7. Lingegowda BP, Koh TH, Ong HS, Tan TT. Primary cutaneous cryptococcosis due to *Cryptococcus gattii* in Singapore. *Singap Med J* 2011;**52**(7):e160–162.
8. Hagen F, Ceresini PC, Polacheck I, Ma H, van Nieuwerburgh F, Gabaldón T, et al. Ancient dispersal of the human fungal pathogen *Cryptococcus gattii* from the Amazon rainforest. *PLoS One* 2013;**8**(8):e71148.
9. Hagen F, Colom MF, Swinne D, Tintelnot K, Iatta R, Montagna MT, et al. Autochthonous and dormant *Cryptococcus* gattii infections in Europe. *Emerg Infect Dis* 2012;**18**(10):1618–24.
10. Springer DJ, Chaturvedi V. Projecting global occurrence of *Cryptococcus gattii*. *Emerg Infect Dis* 2010;**16**(1):14–20.
11. Bovers M, Hagen F, Kuramae EE, Boekhout T. Six monophyletic lineages identified within *Cryptococcus neoformans* and *Cryptococcus gattii* by multi-locus sequence typing. *Fungal Genet Biol* 2008b;**45**(4):400–21.
12. Lagrou K, Van Eldere J, Keuleers S, Hagen F, Merckx R, Verhaegen J, et al. Zoonotic transmission of *Cryptococcus neoformans* from a magpie to an immunocompetent patient. *J Intern Med* 2005;**257**(4):385–8.

13. Nielsen K, De Obaldia AL, Heitman J. *Cryptococcus neoformans* mates on pigeon guano: implications for the realized ecological niche and globalization. *Eukaryot Cell* 2007;**6**(6):949–59.
14. Ellis DH, Pfeiffer TJ. Natural habitat of *Cryptococcus neoformans* var. *gattii*. *J Clin Microbiol* 1990;**28**(7):1642–4.
15. Fortes ST, Lazéra MS, Nishikawa MM, Macedo RC, Wanke B. First isolation of *Cryptococcus neoformans* var. *gattii* from a native jungle tree in the Brazilian Amazon rainforest. *Mycoses* 2001;**44**(5):137–40.
16. Kidd SE, Chow Y, Mak S, Bach PJ, Chen H, Hingston AO, et al. Characterization of environmental sources of the human and animal pathogen *Cryptococcus gattii* in British Columbia, Canada, and the Pacific Northwest of the United States. *Appl Environ Microbiol* 2007b;**73**(5):1433–43.
17. Randhawa HS, Kowshik T, Chowdhary A, Preeti Sinha K, Khan ZU, Sun S, et al. The expanding host tree species spectrum of *Cryptococcus gattii* and *Cryptococcus neoformans* and their isolations from surrounding soil in India. *Med Mycol* 2008;**46**(8):823–33.
18. Goldman DL, Khine H, Abadi J, Lindenberg DJ, Pirofski La, Niang R, et al. Serologic evidence for *Cryptococcus neoformans* infection in early childhood. *Pediatrics* 2001;**107**(5):E66.
19. Hagen F, van Assen S, Luijckx GJ, Boekhout T, Kampinga GA. Activated dormant *Cryptococcus gattii* infection in a Dutch tourist who visited Vancouver Island (Canada): a molecular epidemiological approach. *Med Mycol* 2010;**48**(3):528–31.
20. Chambers C, MacDougall L, Li M, Galanis E. Tourism and specific risk areas for *Cryptococcus gattii*, Vancouver Island, Canada. *Emerg Infect Dis* 2008;**14**(11):1781–3.
21. Lindberg J, Hagen F, Laursen A, Stenderup J, Boekhout T. *Cryptococcus gattii* risk for tourists visiting Vancouver Island, Canada. *Emerg Infect Dis* 2007;**13**(1):178–9.
22. Colom MF, Hagen F, Gonzalez A, Mellado A, Morera N, Linares C, et al. *Ceratonia siliqua* (carob) trees as natural habitat and source of infection by *Cryptococcus gattii* in the Mediterranean environment. *Med Mycol* 2012;**50**(1):67–73.
23. Morera N, Hagen F, Juan-Sallés C, Artigas C, Patricio R, Serra JI, et al. Ferrets as sentinels of the presence of pathogenic *Cryptococcus* species in the Mediterranean environment. *Mycopathologia* 2014;**178**(1–2):145–51.
24. Baes H, Van Cutsem J. Primary cutaneous cryptococcosis. *Dermatologica* 1985;**171**(5):357–61.
25. Fessel WJ. Cryptococcal meningitis after unusual exposures to birds. *N Engl J Med* 1993;**328**(18):1354–5.
26. Goonetilleke AK, Krause K, Slater DN, Dev D, Wood ML, Basran GS. Primary cutaneous cryptococcosis in an immunocompromized pigeon keeper. *Br J Dermatol* 1995;**133**(4):650–2.
27. Nosanchuk JD, Shoham S, Fries BC, Shapiro DS, Levitz SM, Casadevall A. Evidence of zoonotic transmission of *Cryptococcus neoformans* from a pet cockatoo to an immunocompromised patient. *Ann Intern Med* 2000;**132**(3):205–8.
28. Tanphaichitra D, Sahaphongs S, Srimuang S. Cryptococcal antigen survey among racing pigeon workers and patients with cryptococcosis, pythiosis, histoplasmosis and penicilliosis. *Int J Clin Pharmacol Res* 1988;**8**(6):433–9.
29. Wegener HH, Staib F. Fatal cryptococcosis in a bird fancier. A clinical case report on pathology, diagnosis and epidemiology of cryptococcosis. *Zentralbl Bakteriol Mikrobiol Hyg A* 1983;**256**(2):231–8.
30. Kidd SE, Bach PJ, Hingston AO, Mak S, Chow Y, MacDougall L, et al. *Cryptococcus gattii* dispersal mechanisms, British Columbia, Canada. *Emerg Infect Dis* 2007a;**13**(1):51–7.
31. Leão CA, Ferreira-Paim K, Andrade-Silva L, Mora DJ, da Silva PR, Machado AS, et al. Primary cutaneous cryptococcosis caused by *Cryptococcus gattii* in an immunocompetent host. *Med Mycol* 2011;**49**(4):352–5.

32. Cano MV, Hajjeh RA. The epidemiology of histoplasmosis: a review. *Semin Respir Infect* 2001;**16**:109–18.
33. Maresca B, Kobayashi GS. Dimorphism in *Histoplasma capsulatum*: a model of study of cell differentiation in pathogenic fungi. *Infect Imm* 1989;**64**:3333–40.
34. Guimarães AJ, Nosanchuk JD, Zancopé-Oliveira RM. Diagnosis of histoplasmosis. *Braz J Microbiol* 2006;**37**:1–13.
35. Zancopé-Oliveira RM, Muniz MM, Wanke B. Histoplasmose. In: Coura JR, editor. *Dinâmica das Doenças Infecciosas e Parasitárias*. 2nd ed. Rio de Janeiro: Guanabara Koogan; 2013. p. 1238–49.
36. Klite PD, Diercks FH. *Histoplasma capsulatum* in fecal contents and organs of bats in the Canal Zone. *Am J Trop Med Hyg* 1965;**14**:433–9.
37. Galvão-Dias MA, Zancope-Oliveira RM, Giudice MC, Montenegro Neto H, Jordão LRJ, Grigorio IM, et al. Isolation of *Histoplasma capsulatum* from bats in the urban área of Sao Paulo State, Brazil. *Epidemiol Infect* 2011;**139**:1642–4.
38. González-González AE, Aliouat-Denis CM, Ramírez-Bárcenas JA, Demanche C, Pottier M, Carreto-Binaghi LE, et al. *Histoplasma capsulatum* and *Pneumocystis* spp. co-infection in wild bats from Argentina, French Guyana, and Mexico. *BMC Microbiol* 2014;**5**:14–23.
39. Vasudevan B, Ashish B, Amitabh S, Mohanty AP. Primary cutaneous histoplasmosis in a HIV-positive individual. *J Glob Infec Dis* 2010;**2**:112–5.
40. Buitrago MJ, Gonzalo-Jimenez N, Navarro M, Rodriguez-Tudela JL, Cuenca-Estrella M. A case of primary cutaneous histoplasmosis acquired in the laboratory. *Mycoses* 2011;**54**:e859–61.
41. Gomez-Moyano E, Crespo-Erchiga V, Vera-Casaño A. Probable primary cutaneous histoplasmosis in a patient infected with HIV. *J Mycol Med* 2011;**21**(3):210–3.
42. Alverson B, Alexander N, LeGolvan MP, Dunlap W, Levy C. A human immunodeficiency virus-positive infant with probable congenital histoplasmosis in a nonendemic area. *Pediatr Infect Dis J* 2010;**29**(11):1055–7.
43. Hage CA, Knox KS, Wheat LJ. Endemic mycoses: overlooked causes of community acquired pneumonia. *Resp Med* 2012;**106**:769–76.
44. Panackal AA, Hajjeh RA, Cetron MS, Warnock DW. Fungal infections among returning travelers. *Clin Infect Dis* November 1, 2002;**35**(9):1088–95.
45. Muñoz B, Martínez MA, Palma G, Ramírez A, Frías MG, Reys MR, et al. Molecular characterization of *Histoplasma capsulatum* isolated from a outbreak in treasure hunters. *BMC Infect Dis* 2010;**10**:264–70.
46. Ohno H, Ogata Y, Suguro H, Yokota S, Watanabe A, Kamei K, et al. An outbreak of histoplasmosis among healthy young Japanese women after traveling to Southeast Asia. *Intern Med* 2010;**49**:491–5.
47. Corcho-Berdugo A, Muñoz-Hernández B, Palma-Cortés G, Ramírez-Hernándrz A, Martínez-Rivera M, Frías-de León M, et al. An unusual outbreak of histoplasmosis in residents of the state of Mexico. *Gac Med Mex* 2011;**147**:377–84.
48. Centers for Disease Control. Histoplsmosis outbreak among day camp attendees-Nebraska. *MMWR Morb Mortal Wkly Rep* June 2012;**61**:747–8.
49. Kajfasz P, Basiak W. Outbreak of pulmonary histoplasmosis involving a group of four Polish travelers returning from Ecuador. *Int Marit Health* 2012;**63**:59–62.
50. Cottle LE, Gkrania-Klotsas E, Willians HJ, Brindle HE, Carmichael AJ, Fry G, et al. A multinational outbreak of histoplsmosis following a biology field trip in the Uganda rainforest. *J Travel Med* 2013;**20**:83–7.
51. Haselow DT, Safi H, Holcomb D, Smith N, Wagner KD, Bolden BB, et al. Arkansas Department of Health and Center for Disease Control and Prevention (CDC). Histoplasmosis associated with bamboo bonfire-Arkansas, October 2011. *MMWR Morb Mortal Wkly* 2014;**63**:165–8.

52. Barros MBL, Almeida-Paes R, Schubach AO. *Sporothrix schenckii* and sporotrichosis. *Clin Microbiol Rev* 2011;**24**:633–54.
53. Morris-Jones R. Sporotrichosis. *Clin Exp Dermatol* 2002;**27**:427–31.
54. Marimon R, Gene J, Cano J, Guarro J. *Sporothrix luriei*: a rare fungus from clinical origin. *Med Mycol* 2008a;**46**:621–5.
55. Arrillaga-Moncrieff I, Capilla J, Mayayo E, Marimon R, Marine M, Gene J, et al. Different virulence levels of the species of *Sporothrix* in a murine model. *Clin Microbiol Infect* 2009;**15**:651–5.
56. Marimon R, Serena C, Gené J, Cano J, Guarro J. In vitro antifungal susceptibility of five species of *Sporothrix*. *Antimicrob Agents Chemother* 2008b;**52**:732–4.
57. Almeida-Paes R, Oliveira MME, Freitas DFS, Valle ACF, Zancopé-Oliveira RM, Gutierrez-Galhardo MC. Sporotrichosis in Rio de Janeiro, Brazil: *Sporothrix brasiliensis* is associated with atypical clinical presentations. *PLoS Negl Trop Dis* 2014;**8**:e3094.
58. Barros MBL, Schubach TMP, Gutierrez-Galhardo MC, Schubach AO, Monteiro PCF, Reis RS, et al. Sporotrichosis: an emergent zoonosis in Rio de Janeiro. *Mem Inst Oswaldo Cruz* 2001;**96**:777–9.
59. Barros MBL, Schubach AO, Valle ACF, Gutierrez-Galhardo MC, Conceição-Silva F, Schubach TMP, et al. Cat-transmitted sporotrichosis epidemic in Rio de Janeiro, Brazil: description of a series of cases. *Clin Infect Dis* 2004;**38**:529–35.
60. Pereira SA, Gremiao IDF, Kitada AAB, Boechat JS, Viana PG, Schubach TMP. The epidemiological scenario of feline sporotrichosis in Rio de Janeiro, State of Rio de Janeiro, Brazil. *Rev Soc Bras Med Trop* 2014;**47**:392–3.
61. Schubach TMP, Schubach AO, Okamoto T, Barros MBL, Figueiredo FB, Cuzzi T, et al. Evaluation of an epidemic of sporotrichosis in cats: 347 cases (1988–2001). *J Am Vet Med Assoc* 2004;**224**:1623–9.
62. Pereira SA, Passos SRL, Silva JN, Gremião ID, Figueiredo FB, Teixeira JL, et al. Response to azolic antifungal agents for treating feline sporotrichosis. *Vet Rec* 2010;**166**:290–4.
63. Freitas DFS, Valle ACF, Almeida-Paes R, Bastos FI, Gutierrez-Galhardo MC. Zoonotic sporotrichosis in Rio de Janeiro, Brazil: a protracted epidemic yet to be curbed. *Clin Infect Dis* 2010;**50**:453.
64. Freitas DFS, Hoagland BS, Valle ACF, Fraga BB, Barros MBL, Schubach AO, et al. Sporotrichosis in HIV-infected patients: report of 21 cases of endemic sporotrichosis in Rio de Janeiro, Brazil. *Med Mycol* 2012;**50**:170–8.
65. Freitas DFS, Valle ACF, Silva MB, Campos DP, Lyra MR, Souza RV, et al. Sporotrichosis: an emerging neglected opportunistic infection in HIV-infected patients in Rio de Janeiro, Brazil. *PLoS Negl Trop Dis* 2014;**8**:e3110.
66. Song Y, Li SS, Zhong SX, Liu YY, Yao L, Huo SS. Report of 457 sporotrichosis cases from Jilin province, northeast China, a serious endemic region. *J Eur Acad Dermatol Venereol* 2013;**27**:313–8.
67. Liu TT, Zhang K, Zhou X. Molecular identification of *Sporothrix* clinical isolates in China. *J Zhejiang Univ Sci B* 2014;**15**:100–8.
68. Schubach TMP, Menezes RC, Wanke B. Sporotrichosis. In: Greene CE, editor. *Infectious diseases of the dog and cat*. 4th ed Saint Louis: Elsevier; 2012. p. 645–50.

Chapter 12

Urban Settings

Fungi in Archives: A Double Concern

Ana Catarina Pinheiro
Pharmacy Department, Cèntro Hospitalar do Algarve, Faro Unit, Portugal

Composed of thousands of genera and species, fungi are ubiquitous and particularly relevant because of their strong adaptability and resilience to less favorable environments. Although they are incredibly beneficial in some areas, the world is more accustomed to see fungi as threats to health, and it is not wrong to think so. However, it is not just human health that is in jeopardy when close contact is established with some species of fungi or their presence is maintained in contaminated environments. Fungi can hydrolyze a wide variety of polymers as a result of their efficient degrading enzymes.[1,2] As an organic material, paper is susceptible to biodeterioration by microfungi, especially by those known to be cellulolytic, or cellulose degrading.[2,3]

In archives and libraries, the presence of fungi can therefore be detrimental to both the health of the individuals who work there and the documents and book safe kept on the premises. This duality makes it of the utmost importance to know the environment where books are kept and where people consult them or tend to their care.

The subject of the impact of fungal growth on the health of workers and attendees of libraries and archives has been mentioned for almost a century, but despite its relevance there is still a lack of information concerning the fungal population in libraries and archives (either quantitatively or qualitatively) and its relation to health issues.

One of the first records on this concern dates back to the 1915 August 23 issue of *Every Week*, in which Rose Murray, one of the six women referred to in the article "Women Who Hold Down Unusual Jobs," is described as "the only woman in the world who holds the position of physician and surgeon to 'sick' books [...]. She is the 'doctor' for all the volumes in the New York Free Public Library. There is a very lively element of danger in her position, because books, like people, derive their sickness largely from germs and microbes. That is why Miss Murray goes about her work dressed just like a surgeon at an operation. Her equipment consists

Environmental Mycology in Public Health. http://dx.doi.org/10.1016/B978-0-12-411471-5.00012-0

of a huge apron and a veil of cheesecloth." The image that illustrates the text shows a woman nearly enveloped in a white gown, with only her eyes and hands and the bottom of her dress showing.[4] Was she right in her choice to protect herself heavily?

According to some authors, the most common malaises reported by staff working in libraries, archives, or book-containing premises are dermatitis, rhinitis, allergies, asthma, allergic bronchopulmonary aspergillosis, and hypersensitivity pneumonitis.[5] In a museum environment, Wiszniewska et al.[6] concluded that 30% of museum employees were sensitized to at least one of the fungal allergens tested and that the prevalence of allergic symptoms among subjects was relatively high and frequently related to specific sensitization. High-level exposure to airborne viable fungi (10^6 colony-forming units (CFU)/m^3) was determined to be the cause in a case of organic dust toxic syndrome in a museum staff handling moldy books.[7]

In most studies on indoor fungal contamination, four main measures are applied: total viable mold counts, total mold counts (viable and nonviable), specific mold species (qualitative or quantitative), and β-(1,3)-*d*-glucan level (β-1 to -3) glucans are components of the fungus cell wall and are considered potent inflammatory agents[8] responsible for hypersensitivity reactions.[9] For the first three options, several methods can be chosen: gravitational deposition (Durham sampler, Tauber trap, or simple Petri dishes); and impact (of which the Andersen sampler is the most popular), suction, filtration, electrostatic precipitation, thermal precipitation, and impingement.[10]

The resulting samples can be analyzed using the traditional culturing methods of incubation and microscopic analysis or by molecular biology techniques, which are capable of going further in terms of fungal identification and sorting out the nonviable fraction. Whenever possible, the best option is to perform both analyses.

To determine the atmospheric allergen load (β-[1, 3]-*d*-glucan level included), immunological techniques such as enzyme immunoassay and enzyme-linked immunoassay are frequently used.[11] Some authors[12] consider air sampling to be the best method to determine fungal contamination.

In 1997, Zyska[13] performed an extensive review of fungi encountered and identified in archives. The results presented in the 1997 compilation refer to an analysis on books, documents, and air dust/samples identified by conventional culturing methods. Some of the fungi encountered are common air contaminants, such as *Penicillium* sp. or *Cladosporium* sp.; others can be considered detrimental to human health, such as *Stachybotrys chartarum* or *Aspergillus fumigatus*.

Figure 1 summarizes the air sample results compiled by Zyska[13] and Gallo et al.[14] to which new data were added by Aira,[15] Borrego et al.,[16,17] Bueno et al.,[18] Gambale et al.,[19] Harkawy et al.,[20] Jain,[21] Lugauskas and Krikštaponis,[22] Pinheiro et al.,[23–27] Rakotonirainy et al.,[28] Ruga et al.,[29] Valentin,[5] and Wlazlo et al.[30]

Posing a risk for documents preservation, many of the fungi most commonly encountered in archives and libraries are capable of degrading cellulose: the genera *Aspergillus*, *Chaetomium*, and *Alternaria* are included in this group.[2]

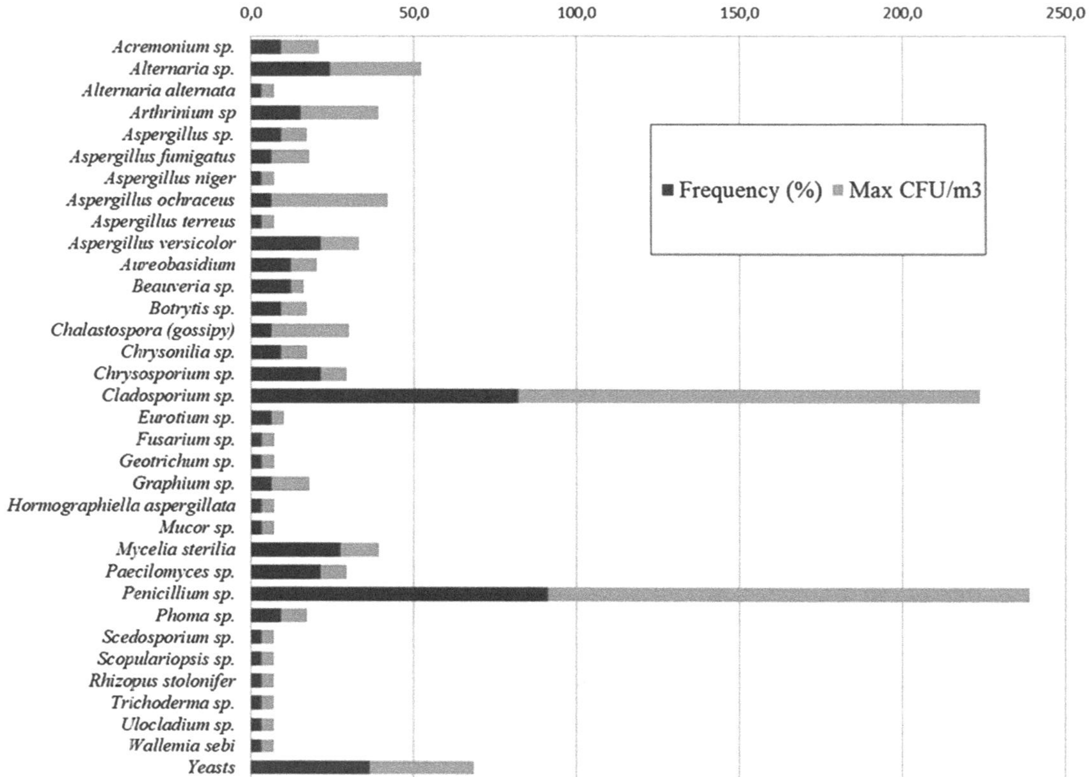

FIGURE 1 Air samples performed in four Portuguese archives and analyzed in terms of frequency (presence per total number of samples) and maximum number of CFU per cubic meter.

Capitelli et al.[31] performed a quantitative aerial study; average counts in all locations varied between 350 and 1416 CFU/m^3. For conservation purposes, Dutch guidelines defend the need for awareness when the level crosses 100 CFU.[32]

Health-risk levels suggested in the literature vary greatly among authors and through the years. For the World Health Organization, the maximum concentration is 150 CFU/m^3.[33] According to Holmberg,[34] 2200 CFU/m^3 sets the limit for a healthy environment; for Ohgke et al.,[35] levels higher than 100 CFU/m^3 are a sign of internal contamination; and for Reynolds et al.,[36] 500 CFU/m^3 is the maximum acceptable concentration. More recently, Klánová[37] recommended treating higher than 2000 CFU/m^3 as a health threat.

In Portugal, Ordinance 353A/2013,[38] December 4, which regulate a building's climatic certification, stipulates that the fungal load should never exceed the 500 CFU/m^3; the presence of a mixture of relatively uncommon species should not exceed 150 CFU/m^3, and one considered uncommon species (*Acremonium* sp., *Chrysonilia* sp., *Trichothecium* sp., *Curvularia* sp. and *Nigrospora* sp.) should not exceed 50 CFU/m^3. The presence of potentially toxinogenic fungi such as *Stachybotrys chartarum (Stachybotrys atra)*, *Fusarium moniliforme* or *Fusarium culmorum*, *Aspergillus versicolor*, *A. flavus*, *A. fumigatus*, *A. niger*, *A. ochraceus*, *A. terreus*, and *Trichoderma viridae*, and potentially pathogenic fungi such as *Histoplasma capsulatum*, *Cryptococcus neoformans*, *Blastomyces dermatidis*, and *Coccidioides immitis* are a sign of an environment of low(er) quality. Only air samples are accounted for in these guidelines, but the information highlights the need for fungal identification and not just mere quantification. The established relation between the minimum amount of spores that can cause serious allergic reactions and the fungal species has been reported by Valentin et al.[5] as between 100/m^3 for *Alternaria alternata* and 3000/m^3 for *Cladosporium herbarum*, which reinforces the idea of identification rather than quantification only.[39]

Table 1 shows only the most common genera; the potentially problematic species (Valentin et al.[8]; Ordinance 353A/2013[38]) *A. alternata*, *Aspergillus glaucus*, *Aspergillus niger*, *A. fumigatus*, and *Aspergillus versicolor*, *Chaetomium globosum*, *Cladosporium herbarum*, *Penicillium brevicompactum*, and *Stachybotrys atra* have all been identified.

In archives and libraries, some studies were performed using just an analysis of surfaces. In Maggi et al.,[40] the most represented genera were *Aspergillus* (*A. fumigatus* included), *Cladosporium* sp., *Penicillium* sp., *Chaetomium* sp., and *Alternaria* sp. An average of 10 CFU was determined per sampled area (24 cm^2).

In some cases, no difference in exposure–effect relations is observed when exposure assessment is based on air or dust samples,[7] but it can be useful to couple both methods because some fungi such as *Stachybotrys* sp. can be difficult to isolate in an air sample even when present in the studied environment.[41,42] A study performed by Zielinska-Jankiewicz[43] in 2008 included both air and surface samples. *Cladosporium* and *Penicillium* were the most

TABLE 1 Most Identified Fungal Genera in Air Sampling Studies. Nomenclature Changes Have Deemed Obsolete Some of the Fungi Mentioned in the Consulted Sources: *Cephalotrichum* (Now *Doratomyces* sp.), Cephalosporium (Now *Acremonium* sp.); Chloridium (Now *Ramichloridium* sp.), and Hormiscium (Now *Torula* sp.). In the Torula Genera Some Species Are Now Classified as Belonging to the *Exophialia, Candida, Rhodotorula, Cladophialophora,* or *Cryptococcus* Genera

Number of Studies Where Fungus Was Identified	Fungal Genera
0	Mentioned as common by the literature: *Acrothecium* sp., *Chaetomella* sp., *Coprinus* sp., *Gymnoascus* sp., *Pellicularia* sp., *Ramichloridium* sp., *Rhizoctonia* sp., *Serpula* sp., *Thielavia* sp., *Torula* sp.
1	*Acremoniella* sp., *Basipetospora* sp., *Beauveria* sp., *Bispora* sp., *Ceratosporum* sp., *Chalastospora* sp., *Cryptococcus* sp., *Dactylella, Doratomyces* sp., *Dreschelera* sp., *Emericella* sp., *Gonytrichum* sp., *Graphium* sp., *Harpographium* sp., *Heterocephalum* sp., *Hormographiella* sp., *Microsporum* sp., *Monascus* sp., *Monosporium* sp., *Mortiriella* sp., *Pestalotia* sp., *Prohytroma* sp., *Scedosporium* sp., *Scolecobasidium* sp., *Sepedonium* sp., *Sporodiniella* sp., *Sporotrichum* sp., *Sporothrix* sp., *Syncephalastrum* sp., *Trycophyton* sp., *Verticillium* sp., *yeasts*
2 to 4	*Arthrinium* sp., *Aureobasidium* sp., *Candida* sp., *Chrysonilia* sp., *Chrysosporium* sp., *Epicoccum* sp., *Eurotium* sp., *Gliocladium* sp., *Helminthosporium* sp., *Humicola* sp., *Mucar* sp., *Mycelia sterilia, Myrothecium* sp., *Neurospora* sp., *Nigrospora* sp., *Oidiodendrum* sp., *Phoma* sp., *Rhodotorula* sp., *Rhizomucor* sp., *Stachybotrys* sp., *Stemphylium* sp., *Trichocladium* sp., *Trichothecium* sp., *Ulocladium* sp., *Wallemia* sp.
5 to 7	*Acremonium* sp., *Botrytis* sp., *Chaetomium* sp., *Curvularia* sp., *Fusarium* sp., *Paecilomyces* sp., *Trichoderma* sp., *Rhizopus* sp., *Scopulariopsis* sp.
8 to 10	*Alternaria* sp., *Aspergillus* sp., *Cladosporium* sp., *Geotrichum* sp., *Penicillium* sp.

prevalent genera. Twelve species were regarded as potentially pathogenic for humans: eight had allergic properties, and 11, toxic properties. Quantitatively, the levels reached from 2.3×10^3 CFU/m^3 for air samples to 8–10 CFU/100 cm^2 for surface samples.

Sampling surfaces is also the best choice when looking for dermatophytes. These are fungi capable of degrading the keratin existing in soil; hence, the name keratinophylic.[44] These fungi can contaminate and invade living tissues, causing dermatomycosis. None of the three genera (*Microsporum*, *Trichophyton*, and *Epidermophyton*) were found in previous studies. However, this does not mean that special handling procedures should be dismissed, because Mesquita et al.[45] identified in a paper sample *Toxicocladosporium irritans*, a fungi that produces ample amounts of volatile metabolites and causes a skin rash within minutes of opening an inoculated dish.

At the Jasna Góra monastery library, Harkawi et al.[46] concluded that maximum viable fungal aerosol concentrations did not exceed 100 CFU/m^3 but *A. niger* and *A. versicolor* were present in the sampled air. The settled dust showed a different microbial structure with only five fungal genera and/or species encountered and a different prevalence distribution. The analysis performed on the dust samples achieved a maximum of 10,000/m^2.

Pinheiro et al.[24–26] performed a study on four Portuguese archives and sampled both surfaces and air samples (quantitatively and qualitatively). Regarding air samples, the number of CFU per cubic meter never exceeded 500 CFU/m^3. Indoor contamination (fungi not present in the control outdoor samples) was present in all of the studied settings. The indoor/outdoor ratio, another parameter considered in Ordinance 353A/2013, was above 1 in 21% of samples.

Compared with the global results presented in Table 1, yeasts stand out as one of the most commonly present in air samples retrieved from archives. The other most represented fungi follow the trend depicted in Table 1, with *Penicillium* sp. and *Cladosporium* sp. taking the lead. From the *Aspergillus* group, *A. versicolor* is the most represented; this fungus is known to be able to produce mycotoxins in an indoor environment.[9] Some of the fungi identified and recognizably involved in the degradation of paper can produce mycotoxins, which may pose a health problem to staff and visitors of libraries, galleries, and museums.[45] From the initial compilation performed by Zyska (1997),[13] 19% could generate health issues.[45] Also, many of the sampled locations are not ventilated and conditions for the accumulation of metabolites are ideal.

Many studies focused on indoor health issues, and the production of mycotoxins is directed at trichothecenes, although this group of metabolites is interesting only when growth of *Stachybotrys* occurs.[47] *Stachybotrys* was absent from the air samples but was identified in the swab samples of one of the studied settings. Moreover, when comparing the air sample results with the surface samples (Figure 2), most of the fungal spectra would be lost without both studies.

As in other settings, although recognizably important, it is still not clear the effect biological agents have on the human health and heritage conservation in archives.

In these particular institutions, there is still a long way to go because two different populations must be protected. Interactions established between fungi and written heritage are still confusing and, similar to what has been found for

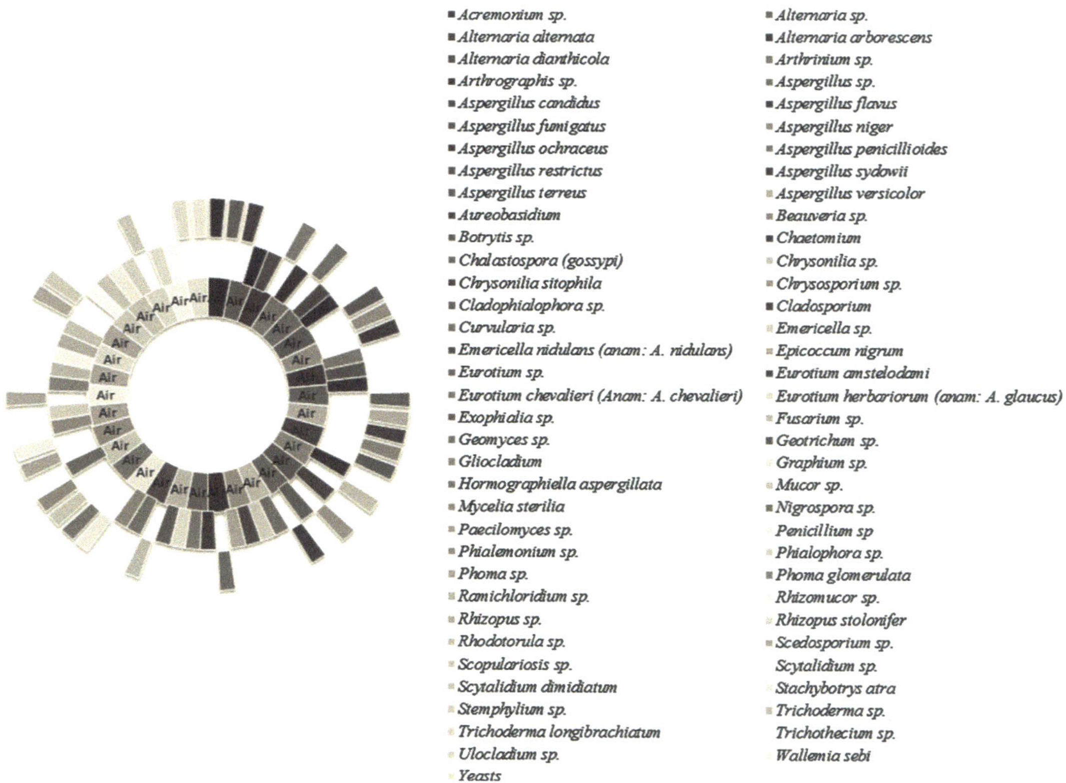

FIGURE 2 Total fungi genera/species found in all four Portuguese archives. The inner circle represents the results obtained with the air samples and the outer circles correspond to the spectra found in the surface samples. The outermost circle depicts fungi that are common to both air and surface samples. Four fungi were solely present in the air surfaces: *Geotrichum*, *Chalastospora*, *Wallemia sebi*, and *Rhizopus stolonifer*. *Eurotium amstelodamii* is still under study to ascertain the corresponding anamorph (ICPA, http://www.aspergilluspenicillium.org/index.php/eurotium).

humans, probably complex. Despite the existence of health issues and symptoms, data on human exposure and effects on staff and visitors of archives and libraries are scarce.

The fact that there is international recognition of the importance of indoor fungal communities has led to major developments in the area. The scientific community is now closer to running relevant epidemiological studies to establish a dose–response exposure level and design robust standards.

REFERENCES

1. Da Silva M, Moraes AML, Nishikawa MM, Gatti MJA, de Lencar MA Vallim, Brandão LE, et al. Inactivation of fungi from deteriorated paper materials by radiation. *Int Biodeterior Biodegrad* 2006;**57**:163–7.
2. Caneva G, Maggi O, Nugari MP, Pietrini AM, Piervittori R, Ricci S, et al. The biological aerosol as a factor of biodeterioration. In: Mandrioli P, Caneva G, Sabbioni C, editors. *Cultural heritage and aerobiology—methods and measurement techniques for biodeterioration monitoring*. Kluwer Academic Publishers; 2003. p. 3–29.
3. Adamo M, Magaudda G, Nisini PT, Tronelli G. Susceptibility of cellulose to attack by cellulolytic microfungi after gamma irradiation and ageing. *Restaurator* 2003;**24**:145–51.
4. *Abbey Newsl* 1994;**18**(6). http://cool.conservation-us.org/byorg/abbey/an/an18/an18-6/an18-602.html.
5. Valentin N. *Microbial contamination in archives and museums: health hazards and preventive strategies using air ventilation systems*. Tenerife, Spain: Contribution to the Experts' Roundtable on Sustainable Climate Management Strategies; 2007.
6. Wiszniewska M, Walusiak-Skorupa J, Pannenko I, Draniak M, Palczynski C. Occupational exposure and sensitization to fungi among museum workers. *Occup Med* 2009;**59**(4):237–42.
7. Kolstad HA, Brauer C, Iversen M, Sigsgaard T, Mikkelsen S. Do indoor molds in nonindustrial environments threaten workers' health? A review of the epidemiologic evidence. *Epidemiol Rev* 2002;**24**(2):203–17.
8. Valentin N. Microbial contamination in museum collections: organic materials. In: Saiz-Jimenez C, editor. *Molecular biology and cultural heritage*. Netherlands: A.A. Balkema Publishers; 2003. p. 85–91.
9. Salkinoja- Salonen MS, Peltola J, Andersson MA, Saiz-Jimenez C. Microbial toxins in moisture damaged indoor environmental and cultural assets. In: Saiz-Jimenez C, editor. *Molecular biology and cultural heritage*. Netherlands: A.A. Balkema Publishers; 2013. p. 93–105.
10. de Nuntiis P, Maggi O, Mandrioli P, Ranalli G, Sorlini C. Monitoring the biological aerossol. In: Mandrioli P, Caneva G, Sabbioni C, editors. *Cultural heritage and aerobiology—methods and measurement techniques for biodeterioration monitoring*. Kluwer Academic Publishers; 1998. p. 107–44.
11. Favalli MA, Gallo F, Maggi O, Mandrioli P, Pacini E, Pasquariello G, et al. Analysis of the biological aerosol. In: Mandrioli P, Caneva G, Sabbioni C, editors. *Cultural heritage and aerobiology—methods and measurement techniques for biodeterioration monitoring*. Kluwer Academic Publishers; 2003. p. 145–72.
12. Ren P, Jankun TM, Leaderer BP. Comparisons of seasonal fungal prevalence in indoor and outdoor air and in house dusts of dwellings in one northeast American county. *J Expo Anal Environ Epidemiol* 1999;**9**(6):560–8.
13. Zyska B. Fungi isolated from library materials: a review of the literature. *Int Biodeterior Biodegradation* 1997;**40**:43–51.

14. Gallo F, Pasquariello G, Valenti P. Libraries and archives. In: Mandrioli P, Caneva G, Sabbioni C, editors. *Cultural heritage and aerobiology—methods and measurement techniques for biodeterioration monitoring*. Kluwer Academic Publishers; 2003. p. 175–93.
15. Aira MJ, Jato V, Stchigel AM, Rodnguez-Rajo FJ, Piontelli E. Aeromycological study in the Cathedral of Santiago de Compostela (Spain). *Int Biodeterior Biodegrad* 2007;**60**:231–7.
16. Borrego S, Lavin P, Perdomo I, Saraiva SG, Guiamet P. Determination of indoor air quality in archives and biodeterioration of the documentary heritage. *Int Sch Res Netw ISRN Microbiol* 2012. http://dx.doi.org/10.5402/2012/680598.
17. Borrego S, Guiamet P, Saravia SG, Batistini P, Garcia M, Lavin P, et al. The quality of air at archives and the biodeterioration of photographs. *Int Biodeterior Biodegrad* 2010;**64**: 139–45.
18. Bueno DJ, Silva JO, Oliver G. Hongos ambientales en una biblioteca: un año de estúdio. *An Doc* 2003;**6**:27–34.
19. Gambale CJ, Costa-Manso E, Croce M, Sales M. Library fungi at the University of São Paulo and their relationship with respiratory allergy. *J Investig Allergol Clin Immunol* 1993;**3**(1):45–50.
20. Harkawy A, Górny RL, Ogierman L, Wlazło A, Ławniczek-Wałczyk A, Niesler A. Bioaerosol assessment in naturally ventilated historical library building with restricted personnel access. *Ann Agric Environ Med* 2011;**18**(2):323–9.
21. Jain AK. Survey of bioaerossol in different indoor working environments in Central India. *Aerobiologia* 2000;**16**(2):221–5.
22. Lugauskas A, Krikštaponis A. Microscopic fungi found in the libraries of Vilnius and factors affecting their development. *Indoor Built Environ* 2004;**13**:169–82.
23. Pinheiro AC, Viegas C, Jurado V, Sabino R, Veríssimo C, Rosado L, et al. *Fungal presence in selected Portuguese archives: species and fungal load*, in prep.
24. Pinheiro AC. *Fungal communities in archives: assessment strategies and impact on paper conservation and human health* [Ph.D. dissertation]. Lisbon; 2014.
25. Pinheiro AC, Viegas C, Jurado V, Sabino R, Veríssimo C, Rosado L, et al. Estudo de Comunidades Fúngicas em Arquivos: Implicações na Conservação e na Saúde. In: *Anuário do Património Ed. Canto Redondo*. 2012.
26. Pinheiro AC, Viegas C, Jurado V, Brandão J, Rosado L, Macedo MF. Fungal presence in archival settings: assessment using classic culturing and molecular biology methods. In: *Preprints of the ICOM-CC 16th Triennial conference: cultural heritage/cultural identity: the role of conservation*. September 19–23, 2011. Preprints CD: ISBN: 978-989-97522-0-7; Book of Abstracts: ISBN: 978-989-97522-2.
27. Pinheiro AC, Macedo MF, Jurado V, Saiz-Jimenez C, Viegas C, Brandão J, et al. Mould and yeasts identification in archival settings: preliminary results on the use of traditional methods and molecular biology options in portuguese archives. *Int Biodeterior Biodegradation* 2011;**65**(4):619–27.
28. Rakotonirainy M, Fohnrer F, Flieder F. Research on fungicides for aerial disinfection by thermal fogging in libraries and archives. *Int Biodeterior Biodegradation* 1999;**44**:133–9.
29. Ruga L, Bonofiglio T, Orlandi F, Romano B, Fornaciari M. Analysis of the potential fungal biodeteriogen effects in the "Doctorate Library" of the University of Perugia, Italy. *Grana* 2008;**47**(1):60–9.
30. Wlazlo A, Lawniczek A, Ludzen-Izbinska B, Harkawy A, Anczyk E, Gorny RL. *Exposure to bioaerossols at libraries*. Copenhagen, Denmark: Indoor Air; 17–22 August 2008.
31. Cappitelli F, Fermo P, Vecchi R, Piazzalunga A, Valli G, Zanardini E, et al. Chemical–physical and microbiological measurements for indoor air quality assessment at the Ca' GrandaHistorical archive, Milan (Italy). *Water Air Soil Pollut* 2009;**201**:109–20.

32. Brokerhof AW, van Zanen B, den Teuling A. *Fluffy stuff: integrated control of mould in archives*. Netherlands Institute for Cultural Heritage (ICN); 1999.
33. Goyer N, Lavoie J, Lazure L. *Bioaerosols in the workplace: evaluation, control and prevention guide*. Québec: Institut de Recherche en Santé et en Sécurité du Travail du Québec; 2001. Available at: http://www.irsst.qc.ca/files/documents/pubirsst/t-24.pdf.
34. Holmberg K. Mould growth inside buildings. In: Berglund B, Lindvall T, Sundell J, editors. *Indoor air. Sensory and hyperreactivity reactions to sick buildings*, vol. 3. Stockholm: Swedish Council for Building Research; 1984. p. 253–6.
35. Ohgke H, Geers A, Beckert J. Fungal load of indoor air in historical and newly constructed buildings used by public services. In: *Proceedings of the 4th international conference in indoor air quality and climate*, vol. 1; 1987. p. 681–4.
36. Reynolds SJ, Streifel AJ, Mcjilton CE. Elevated airborne concentrations of fungi in residential and office environment. *Am Ind Hyg Assoc J* 1990;**51**(11):601–4.
37. Klánová K. The concentrations of mixed populations of fungi in indoor air: rooms with and without mould problems, rooms with and without health complaints. *Cent Eur J Public Health* 2000;**8**(1):59–61.
38. Ordinance, 353-A/2013, December 4th. *Ministério do Ambiente, Território e Energia, da Saúde e Solidariedade*, Emprego e Segurança Social. Diário da República n° 235-I Série-A. Available at: www.dre.pt.
39. Micalli O, Montacutelli R, Tarsitani G. Pathogenic microorganisms and situations of risk to man. In: Mandrioli P, Caneva G, Sabbioni C, editors. *Cultural heritage and aerobiology—methods and measurement techniques for biodeterioration monitoring*. Kluwer Academic Publishers; 2003. p. 31–43.
40. Maggi O, Persiani AM, Gallo F, Valenti P, Pasquariello G, Sclocchi MC, et al. Airborne fungal spores in dust present in archives: proposal for a detection method, new for archival materials. *Aerobiologia* 2000;**16**:429–34.
41. Buttner MP, Stetzenbach LD. Monitoring airborne fungal spores in an experimental indoor environment to evaluate sampling methods and the effects of human activity on air sampling. *Appl Environ Microbiol* 1993;**59**:219–26.
42. Duchaine C, Mériaux A. The importance of combining air sampling and surface analysis when studying problematic houses for mold biodiversity determination. *Aerobiologia* 2001;**17**: 121–5.
43. Zielinska-Jankiewicz K, Kozajda A, Piotrowska M, Szadkowska-Stanczyk I. Microbiological contamination with moulds in work environment in libraries and archive storage facilities. *Ann Agric Environ Med* 2008;**15**(1):71–8.
44. Viegas C. *Exposição profissional a fungos: o caso dos trabalhadores dos ginásios com piscina*. Caminhos do Conhecimento; 2012.
45. Mesquita N, Portugal A, Videira S, Rodríguez-Echeverría S, Bandeira AML, Santos MJA, et al. Fungal diversity in ancient documents. A case study on the archive of the University of Coimbra. *Int Biodeterior Biodegradation* 2009;**63**:626–9.
46. Harkawy A, Górny RL, Ogierman L, Wlazło A, Ławniczek-Wałczyk A, Niesler A. Bioaerosol assessment in naturally ventilated historical library building with restricted personnel access. *Ann Agric Environ Med* 2011;**18**(2):323–9.
47. Nielsen KF. Mycotoxin production by indoor molds. *Fungal Genet Biol* 2003;**39**:103–17.

Chapter 13

Recreational Environment

Pathogenic Fungi in Public Places, Information Gaps in Assessing Public Health Risk

Helena Solo-Gabriele[1], João Brandão[2], Bari Gordon[1], Alesia Ferguson[3]
[1]Department of Civil, Architectural, and Environmental Engineering, University of Miami, Coral Gables, FL, USA; [2]Reference Unit for Parasitic and Fungal Infections, Department of Infectious Diseases, National Institute of Health Doutor Ricardo Jorge, Lisboa, Portugal; [3]Department of Environmental and Occupational Health, University of Arkansas for Medical Sciences, Little Rock, AR, United States

INTRODUCTION

As reported in previous chapters, fungi cause a wide range of human illness. They can cause allergies and respiratory diseases, including pneumonia and asthma. In North America, asthma rates have been increasing and part of this increase can be attributed to molds that trigger and exacerbate asthma through the inhalation of their spores, fragments, and proteins and the volatile organics that they release. Fungi are also notoriously associated with cutaneous infections of the skin, nails, and hair, estimated to affect 25% of the world's population at any given time. Fungi also infect the mucosal tissues; the respiratory, urinary, and intestinal tracts; the central nervous system; and even the blood. Not including the high incidence of cutaneous and subcutaneous infections, fungi are responsible for over 200 million additional infections yearly.[1] Given the wide range of illness with variable modes of entry, all routes of exposure, including the most common of ingestion, skin, and inhalation, should be considered when evaluating health risks from fungi.

When it comes to disease transmission in the environment, one distinguishing feature of fungi is the association with sediment and soil environments. Fungi are a natural part of the environment, playing a critical role in decomposing organic matter. They grow on land and in aquatic environments and their main source of energy comes from organic carbon sources. As such, fungi are significant from a human health perspective. Fungi can multiply within the human body, which is typically an environment protected from highly energetic light frequencies. This environmental context allows the existence of many pathogenic forms of fungi, with organic matter serving as a growth medium—whether it comes from cells or organic matter found in soil and sediment. It is

Environmental Mycology in Public Health. http://dx.doi.org/10.1016/B978-0-12-411471-5.00013-2

this ability to enter and grow within human tissue that makes it important to understand the distribution of fungi in areas where human contact is likely.

Intimate contact between humans and soil is common in farming and planting. Such activities are usually conducted by adults, who may utilize personal protection (such as gloves) and who should undertake personal hygiene practices to minimize exposures. Our focus in this chapter is on children, a vulnerable population, known for intimate contact with sediments in public play places, including sandpits, sandy areas at playgrounds, and also recreational beaches. In such areas, children are commonly observed to sit, lie, and dig in the sand. Contact with the sand in these areas is not only accepted but expected, given the various toys sold that consist of play buckets and shovels to encourage contact and play with sand. Play should continue to be encouraged, but there may be reasons to protect or modify designated play areas to retard excessive fungal growth or encourage the practice of better hygiene after these play activities. Also, there may be a need to emphasize discouraging incidental hand-to-mouth ingestion and to protect food from contamination in these environments in situations in which the risk of illness may be high.

Although playing with sand is common in public places, rarely is sand monitored for its ability to transmit disease.[2] Only a handful of studies have evaluated the risk of disease transmission through sand, and these studies have typically focused on evaluating the transmission of bacterial and protozoan illnesses and, to a lesser extent, viral illnesses.[3] The information required to understand the risk of disease transmission by fungi through sand play is essentially nonexistent, potentially owing to the lack of fundamental or mechanistic data necessary to conduct risk assessments. The objective of this chapter is to identify information gaps in assessing public health risks from fungi found in public places (beyond air and living organisms), to provide recommendations toward monitoring sands in public play places, and to develop potential mitigation strategies. One common method to assess public health risk from microorganisms is the quantitative microbial risk assessment (QMRA), which will serve as the framework from which we will identify information gaps for assessing public health risks.

QUANTITATIVE MICROBIAL RISK ASSESSMENT

The QMRA is a defined process for evaluating risk. It consists of four steps (Figure 1), including hazard identification, exposure assessment, dose–response, and risk characterization.[4] Hazard identification within the QMRA framework of this chapter requires the identification of fungal pathogens, including their transmission routes and the types of diseases caused by them. For fungi to cause disease, humans must be exposed. Exposure can occur through various routes including inhalation, ingestion through hand-to-mouth activities, and skin contact, in which the skin can be intact or injured (e.g., wounds and abrasions). Sand and sediment, being abrasive, can be of particular concern given their propensity to promote wounds. Additional exposure routes relevant to child play behavior include contact with eyes. The term dose is generally defined in terms of the number of fungal organisms ingested or inhaled during the exposure period, the number that enter the body through

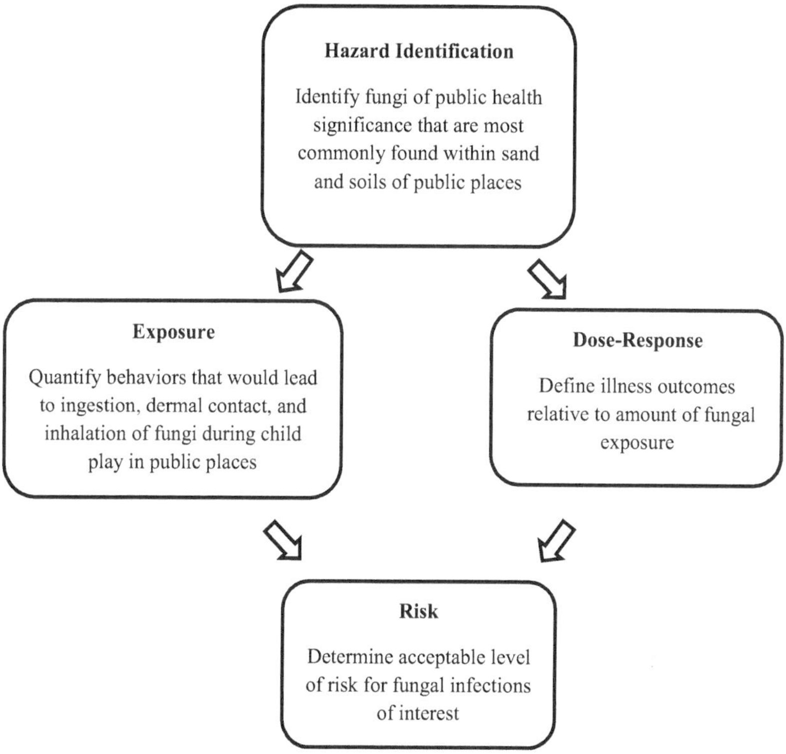

FIGURE 1 Four primary steps of quantitative microbial risk assessment. *Modified from Masters and Ela.*[55]

injured skin, and, potentially, the number of organisms that come in contact with skin or eyes for local effects. Response corresponds to the illness that ensues from the exposure and resulting dose. Different people respond differently to infectious agents, and so, typically, response is defined in terms of a continuum that expresses the probability of infection as a function of the number of pathogens per exposure. Risk is computed from the probability of infection coupled with information about the expected dose. Because of the uncertainty in the parameters used, risk is usually expressed as a range.

Additional details of each of the four steps in the risk assessment process in the context of fungi and children's exposure in public places are described below.

Hazard Identification

A literature review was conducted to identify the fungi that are most commonly identified within sands and soils of public places with an emphasis on summarizing studies that evaluated sandpits, beach sands, school yards, and public parks. The fungi that were detected at frequencies greater than 10% are summarized in Table 1. Fungi appearing consistently among the 20 studies included as part of the review are listed as follows in alphabetical order.

TABLE 1 Summary of Fungi Found in Parks, Beach Sands, and Sandpits

Fungus Name	Relevant Route of Exposure	Human Health Effects	Detection Frequency (%)	Medium	Location	References
Acremonium sp.	Contact	Rare in humans; inflammation of joints or bone	12.6	Public parks soil	Shiraz, Iran	56
			53.4	Beach sand in volcanic environment	Madeira archipelago, Portugal	57
			19.0	5 Ligurian beaches	Northwestern Mediterranean	58
Alternaria alternata	Inhalation	Allergen	67.8	Sandpits	Turin, Italy	59
Aphanoascus fulvescens	Contact	Infections of hair	28	Soil at parks	Pavia Province, Italy	60
			78.5	Sandpits	Turin, Italy	59
Arthroderma multifidum	Unknown	Unknown	25	Soil of national parks	Slovakia	61
Aspergillus sp.	Inhalation, food consumption	Allergic and lung disease, food toxicity	30.4	Ipanema Beach	Rio de Janeiro, Brazil	62
			100	Beaches	South Florida	63
			2.9	Soil in public gardens and parks	Rome, Italy	64
			22.1	Beach sand in volcanic environment	Madeira archipelago, Portugal	57
			14.8	Beach sands	Portuguese coast	65

Chrysosporium anam. A. cuniculi	Contact	Dermatophyte	24.4	30 primary Schools and 15 public parks	Madras City, India	66
Candida albicans	Contact	Oral and genital infections	Not specified	Beach sand	Oahu, Hawaii	67
			1.4	Beach sand	Oahu, Hawaii	68
Candida guilliermondii	Contact	Cutaneous infections with rare systemic infections	11.1	Beach sand	South Florida	47
Candida glabrata		Rare infections of urogenital tract and bloodstream	11.1	Beach sand	South Florida	47
Candida parapsilosis		Wound and tissue infections and sepsis of immunocompromised patients	11.1	Beach sand	South Florida	47
Candida sp.	Contact	Infections of the mouth and genital areas	17.1	Beach sand	Portuguese coast	65
			22.7	Beach sand in volcanic environment: 2010	Madeira archipelago, Portugal	57
Candida tropicalis	Unknown	Rare; blood infections	Frequent	Beach sand	South Florida	63
			66.7	Beach sand	South Florida	47
Chaetomium murorum	Unknown	Unknown	38	Beaches	Terragona city and outskirts, Catalunya, Spain	69

Continued

TABLE 1 Summary of Fungi Found in Parks, Beach Sands, and Sandpits—cont'd

Fungus Name	Relevant Route of Exposure	Human Health Effects	Detection Frequency (%)	Medium	Location	References
Chrysosporium sp.	Contact	Some cause infections of skin, hair, nails	13.1	Public parks soil	Shiraz, Iran	56
			50	Soil of national parks	Slovakia	61
			80.0	Soil in public gardens and parks	Rome, Italy	64
Chrysosporium asperatum (*Myceliophthora vellerea*)	Unknown	Unknown	12.5	Beaches	Terragona city and outskirts, Catalunya, Spain	69
			23.3	Soil at parks	Pavia Province, Italy	60
Chrysosporium evoluceanui	Contact	Infection of skin, hair, nails	6.9	Sandpits at 21 kindergarten schools and 8 public parks	West Bank of Jordan	70
Chrysosporium indicum (*Trichophyton indicum*)	Contact	Infection of skin, hair, nails	25	Sandpits	Turin, Italy	59
			50.1	Soils from gardens, parks, and animal yards	Western Australia	71
			45.5	Soil at parks	Pavia Province, Italy	60

Chrysosporium keratinophilum (*Trichophyton evoluceanui*)	Contact	Dermatophyte, rarely pathogenic	20.7	Sandpits at 21 kindergarten schools and 8 public parks	West Bank of Jordan	70
			42.8	Sandpits	Turin, Italy	59
			25	Soil of national parks	Slovakia	61
			48.8	30 primary schools and 15 public parks	Madras City, India	66
			54.2	13 elementary schools and 7 public parks	Isfahan, Iran	72
			31.1	Soil at parks	Pavia Province, Italy	60
Chrysosporium pannicola	Contact	Infection of skin, hair, nails	26.6	Soil at parks	Pavia Province, Italy	60
Chrysosporium pannorum	Contact	Dermatophyte	40	30 primary schools and 15 public parks	Madras City, India	66
Chrysosporium tropicum (*Sporotrichum exile*)	Contact	Infection of skin, hair, nails	50	Sandpits	Turin, Italy	59
			87	Public parks and gardens	Terragona city and outskirts, Catalunya, Spain	69
			63	Beaches	Terragona city and outskirts, Catalunya, Spain	69

Continued

TABLE 1 Summary of Fungi Found in Parks, Beach Sands, and Sandpits—cont'd

Fungus Name	Relevant Route of Exposure	Human Health Effects	Detection Frequency (%)	Medium	Location	References
			62.2	30 primary schools and 15 public parks	Madras City, India	66
			17.7	Soil at parks	Pavia Province, Italy	60
Cladosporium sp.	Inhalation	Rarely pathogenic, allergen	14.8	Beach sand in volcanic environment: 2011	Madeira archipelago, Portugal	57
			17.9	5 Ligurian beaches	Northwestern Mediterranean	58
Cladosporium cladosporiodes	Inhalation	Rarely pathogenic, allergen	32.1	Sandpits	Turin, Italy	59
Clonostachys rosea	Unknown	Unknown	25	Soil of national parks	Slovakia	61
Coniothyrium fuckelii	Unknown	Rare, one case of liver infection	10.7	Sandpits	Turin, Italy	59,73
Cryptococcus neoformans	Inhalation	Lung infection, meningitis	1.1	Avian-contaminated soils	Oahu, Hawaii	68
			19.0	Children's recreational sites	Lodz, Poland	74

Diheterrespora sp.	Unknown	Unknown	60.0	Soil in public gardens and parks	Rome, Italy	64
Engyodontium album	Unknown	Rarely pathogenic, one case of blood infection.	25	Soil of national parks	Slovakia	61
Fusarium sp.	Food, contact	Some produce toxins; some produce infection in nails and cornea	23.8	Public parks soil	Shiraz, Iran	56
			12.6	Ipanema Beach	Rio de Janeiro, Brazil	62
			100	Beaches	South Florida	63
			10.2	Beach sand in volcanic environment: 2010	Madeira archipelago, Portugal	57
			11.1		Portuguese coast	65
Geomyces pannorum	Contact	Infection of skin and nails	32.1	Sandpits	Turin, Italy	59,75
Geotrichum candidum	Inhalation	Geotrichosis lung illness	15	13 elementary schools and 7 public parks	Isfahan, Iran	72
Gliocladium roseum	Unknown	Unknown	35.7	Sandpits	Turin, Italy	59
Malbranchea sp.			25	Soil of national parks	Slovakia	61
			14.2	Sandpits	Turin, Italy	59

Continued

TABLE 1 Summary of Fungi Found in Parks, Beach Sands, and Sandpits—cont'd

Fungus Name	Relevant Route of Exposure	Human Health Effects	Detection Frequency (%)	Medium	Location	References
Microsporum sp.			25	Soil of national parks	Slovakia	61
Microsporum cookei		Nonpathogenic	21.7	Soils from gardens, parks, and animal yards	Western Australia	71
			34.3	Soil in public gardens and parks	Rome, Italy	64
Microsporum gypseum (*Arthroderma gypseum*)	Contact	Infection of skin, hair, nails	17.2	Sandpits at 21 kindergarten schools and 8 public parks	West Bank of Jordan	70
			Not specified	Beach sand	Oahu, Hawaii	67
			51.1	Soil at parks	Pavia Province, Italy	60
			10.7	Sandpits	Turin, Italy	59
			27	Public parks and gardens	Terragona city and outskirts, Catalunya, Spain	69
			50	Beaches	Terragona city and outskirts, Catalunya, Spain	69

			30.7	Soils from gardens, parks and animal yards	Western Australia	71
			88.6	Soil in public gardens and parks	Rome, Italy	64
			48.8	30 primary schools and 15 public parks	Madras City, India	66
			11.7	13 elementary schools and 7 public parks	Isfahan, Iran	72
			20	Soil at parks	Pavia Province, Italy	60
Microsporum vallerea			25	Soil of national parks	Slovakia	61
Myceliophthora costantin (Ctenomyces serratus)	Contact	Infection of skin, hair, nails	14.4 detect	Soil at parks	Pavia Province, Italy	60
Mycelium sterile *dematiaceum*			21.4	Sandpits	Turin, Italy	59
Paecilomyces sp.			22.9	Soil in public gardens and parks	Rome, Italy	64
Paecilomyces carneus			25	Soil of national parks	Slovakia	61
Paecilomyces farinosus			10.7	Sandpits	Turin, Italy	59

Continued

TABLE 1 Summary of Fungi Found in Parks, Beach Sands, and Sandpits—cont'd

Fungus Name	Relevant Route of Exposure	Human Health Effects	Detection Frequency (%)	Medium	Location	References
Paecilomyces lilacinus	Cutaneous	Rare	46.4	Sandpits	Turin, Italy	59
			100	Soil of national parks	Slovakia	61,76
Penicillium sp.	Inhalation	Allergen	12.4	Public parks soil	Shiraz, Iran	56
			16.2	Ipanema Beach	Rio de Janeiro, Brazil	62
			100	Beaches	South Florida	63
			41.9	Beach sand in volcanic environment: 2010	Madeira archipelago, Portugal	57
			17.3	5 Ligurian beaches	Northwestern Mediterranean	58
Penicillium citrinum	Inhalation	Produces respiratory toxin	10.1	5 Ligurian beaches	Northwestern Mediterranean	58
Penicillium funiculosum			14.2	Sandpits	Turin, Italy	59
Phialophora sp.			10.7	Sandpits	Turin, Italy	59

Pochonia chlamydosporia			25	Soil of national parks	Slovakia	61
Rhinocladiella pedrosoi			21.4	Sandpits	Turin, Italy	59
Rhizophlyctis rosea			44	Soil from national parks, urban reserves and gardens, and agricultural lands	New South Wales, Australia	77
Rhodotorula glutinis			17.9	Children's recreational sites	Lodz, Poland	74
Rhodotorula mucilaginosa			Frequent	Beaches	South Florida	63
Scopulariopsis brevicaulis	Contact	Nail infection	1.1	Beach sand	Oahu, Hawaii	68
Sterile mycelia			40.3	Beach sand in volcanic environment: 2010	Madeira archipelago, Portugal	57
Trichophyton ajelloi (*Arthroderma uncinatum*)	Not applicable	Nonpathogenic	52.2	Soil at parks	Pavia Province, Italy	60
			60.7	Sandpits	Turin, Italy	59
			100	Soil of national parks	Slovakia	61
			8.0	Soils from gardens, parks, and animal yards	Western Australia	71

Continued

TABLE 1 Summary of Fungi Found in Parks, Beach Sands, and Sandpits—cont'd

Fungus Name	Relevant Route of Exposure	Human Health Effects	Detection Frequency (%)	Medium	Location	References
			34.4	Soil at parks	Pavia Province, Italy	60
			50	Soil of national parks	Slovakia	61
			80.0	Soil in public gardens and parks	Rome, Italy	64
Trichophyton cutaneum	Contact	Infections of the skin, hair, nails	0.8	Beach sand	Oahu, Hawaii	68
Trichophyton mentagrophytes	Contact	Infection of skin, hair, nails	6.9	Sandpits at 21 kindergarten schools and 8 public parks	West Bank of Jordan	70
			Not specified	Beach sand	Oahu, Hawaii	67
			37.7	30 primary schools and 15 public parks	Madras City, India	66
Trichophyton terrestre		Dermatophyte?	25	Soil of national parks	Slovakia	61
			17.1	Soil in public gardens and parks	Rome, Italy	64
			31.1	30 primary schools and 15 public parks	Madras City, India	66

Trichosporon sp.	Contact	Some forms cause disease	10.7	Sandpits	Turin, Italy	59
Trichosporon cutaneum	Contact	Hair infection; systemic infections in immunocompromised individuals	16.7	Children's recreational sites	Lodz, Poland	74
			Not specified	Beach sand	Oahu, Hawaii	67
Ulocladium chartarum			14.2	Sandpits	Turin, Italy	59
Verticillium chlamyaosporium			14.2	Sandpits	Turin, Italy	59
Verticuillum sp.			75	Soil of national parks	Slovakia	61

Aspergillus sp. are known to cause allergic responses and lung infection in immune-compromised populations. Specifically, *Aspergillus fumigatus* and *Aspergillus flavus* are known to produce aflatoxin, which is both a toxin and a carcinogen when consumed in food. Of the 20 studies evaluated, five detected *Aspergillus* sp. For the four beach studies that reported detection, the frequency of detection varied from 15% to 100%. The detection rate (2.9%) was lower for the one study that detected *Aspergillus* within the soil of public gardens and parks.

Candida sp. were detected in beach sands in five studies. The specific species detected included *C. albicans*, *C. guilliermondii*, *C. glabrata*, *C. parapsilosis*, and *C. tropicalis*. *C. albicans* is mainly associated with infections of the mouth (thrush) and genital areas. *C. guilliermondii*, *C. glabrata*, *C. parapsilosis*, and *C. tropicalis* are considered to be opportunistic, infecting immunocompromised individuals and surgical patients. Illnesses can be systemic with *C. guilliermondii*, also associated with cutaneous infections; *C. glabrata*, associated with infections of the urogenital tract and bloodstream; *C. parapsilosis*, associated with tissue infections and sepsis; and *C. tropicalis*, associated with blood infections.

Some species of *Chrysosporium* are associated with infections of the skin, hair, and nails. *Chrysosporium* sp. were detected in 10 of the 20 studies that included samples from beaches, park soil, soil from public gardens, soil from schools, and sandpits utilized by young children. The species detected included *C. asperatum*, *C. evoluceanui*, *C. indicum*, *C. keratinophilum*, *C. pannicola*, *C. pannorum*, and *C. tropicum*. The vast majority of these species are classified as dermatophytic fungi, which are known for causing skin diseases.

Cryptococcus neoformans was detected at two sites corresponding to children's recreational sites and another corresponding to avian-contaminated soil. *C. neoformans* can cause serious illness of the lung and can cause meningitis.

By far, the fungi detected most frequently was *Microsporum* sp., with the *Microsporum gypseum* species dominating. *Microsporum* sp. are dermatophytes that can cause skin ailments such as tinea and ringworm. Within the 11 studies documenting the detection of *Microsporum*, four reported detection within beach sites, whereas all 11 studies reported detection within soils.

Penicillium sp. were detected in six studies, five focused on beach sands and one focused on public park soil. The frequency of detection within the beach sites ranged from 10% to 100%. The frequency of detection at the public park was 12%. *Penicillium* sp. produce known allergens. One of the species identified, *Penicillium citrinum*, even produces citrinin, a cell-respiratory mycotoxin responsible for yellow rice fever in humans.

Trichophyton sp. include species that are nonpathogenic (*Trichophyton ajelloi*) and those that are pathogens (*Trichophyton cutaneum*, *Trichophyton mentagrophytes*, and *Trichophyton terrestre*). The pathogenic forms are known to cause infection of the skin, hair, and nails. Of the 20 studies evaluated, six detected pathogenic forms. *T. cutaneum* was documented in one beach sand study, *T. mentagrophytes* was documented in three studies, and *T. terrestre* was documented in three studies.

Trichosporon sp., in particular *Trichosporon cutaneum*, are known to cause a harmless hair infection. Systemic infections have been documented in immunocompromised individuals. *Trichosporon* sp. were documented in three studies that evaluated sandpits, children's recreational beach sites, and beach sand.

Exposure Assessment

Exposure is typically assessed through the evaluation of various exposure scenarios and by quantifying the factors that contribute to the corresponding exposure. Factors considered to compute the magnitude of exposure include the time, frequency, and duration of exposure corresponding to that scenario as well as human behavior parameters and human characteristics. An excellent resource for exposure factors information is the U.S. Environmental Protection Agency's *Exposure Factors Handbook*.[5] This handbook has extensive information concerning ingestion of drinking water, ingestion of water while swimming, food intake, mouthing frequency and duration, soil and dust ingestion, inhalation rates, skin surface areas, and soil adherence to skin. The handbook also includes information about typical activities such as time indoors versus outdoors, time playing on dirt, and time playing on playgrounds. Another excellent source of information is the EPA's Consolidated Human Activity Database (CHAD), which includes detailed diaries of daily behavior for over 30,000 individuals (http://www.epa.gov/heasd/documents/chad_fact_sheet.pdf). CHAD documents the amount of time individuals spend in the various activities. Although there is a wealth of information available in the *Exposure Factors Handbook* and in CHAD, the exposure parameters do not address the amount of time in contact with soil, sand, and sediments during play in public places. Also, data are limited on the amount of time children spend in public places such as beach sites, parks, and school yards.

The factors needed to estimate exposure through ingestion, dermal contact, and inhalation can be expressed through a series of formulas.[6] Exposure is defined as what comes into contact with the human boundaries and dose is what makes it into the body. Readily available algorithms often combine both exposure and dose by using human parameters such as inhalation rates and uptake rate. To illustrate the computation methods, below are classic equations used to estimate the dose of a microbial contaminant (number of fungi per exposure) through ingestion (D_s), dermal contact (D_d), and inhalation (D_i):

$$D_s = C_s \times IR_s \times CF \tag{1}$$

$$D_d = C_s \times SA \times AF \times CF \tag{2}$$

$$D_i = C_s \times \left(\frac{1}{PEF}\right) \times IR_a \times ET \tag{3}$$

where:

D_s, D_d, and D_i are the exposure dose from soil or sand ingestion, dermal contact, and inhalation, respectively (number of fungi per exposure);
C_s is the contaminant concentration (number of fungi per kilogram of sand or n/kg);
IR_s is the soil or sand intake rate (mg/event);
IR_a is the inhalation rate (m^3/h);
SA is the skin surface area available for exposure (cm^2/event);
AF is the adherence factor of sediment to skin (mg/cm^2 per event);
PEF is the soil-to-air particulate emission factor (m^3/kg);
ET is the exposure time (h/event); and
CF is the conversion factor (10^{-6} kg/mg)

Soil ingestion rates, IR_s, have been identified as the most variable parameter when computing risk from contaminated soils at playgrounds, parks, and picnic areas.[7] Many factors can influence the value of IR_s. For example, the amount of sand that transfers to the child's skin is dependent on the condition of the sand and the condition of the child's hand (where moisture plays a significant role in soil/sand adherence) and the time and pressure of contact.[8–12] Soil/sand can also be ingested through mouthing of objects.[13] Additional factors to consider include the number of times a child puts his or her hand in his or her mouth, licks his or her fingers, or touches the outer boundaries of his or her mouth. Xue et al.[14] compiled and evaluated hand-to-mouth activity from nine different studies. They found 6.7–28 contacts/h for hand-to-mouth behavior during indoor settings and 2.9–14.5 contacts/h for outdoor settings. The ranges were observed to be age related. Risk computations should also consider a subgroup of the child population that is known for ingesting abnormally large amounts of soil. This subgroup, characterized as pica, crave and eat nonfood items, such as sand and dirt. For a child to be considered pica the amount of nonfood items ingested must be higher than what is considered normal for a particular age. Pica is most common in children between the ages of 2 and 3 years and in people with developmental disabilities.[15] Average soil ingestion rates for children can vary by a factor of 50, with pica children located at the upper end of the spectrum. Given the role in driving risk, efforts are needed to better define IR_s in the context of child play behavior in public places. Simplified algorithms are needed that consider the ranges caused by differences in children's age and pica behavior and differences in sand/soil conditions.

Dermal exposure can result in infection by fungi through surface contact on the skin given that some fungi, in particular dermatophytes, can colonize skin. Others may need to enter through abrasions or wounds to cause infection.[16] The equation above corresponds to skin surface infections. For skin surface infections, the number of fungi per skin surface area depends upon the surface area of skin exposed, SA, and the adherence of the soil/sediment to skin, AF. Skin can be exposed through direct contact with the sediment or through deposition from air containing suspended sediment particles. The amount of direct contact and possible deposition from air should be evaluated to better estimate exposures during child play scenarios in public places.

Fungi can also potentially be inhaled during play activities. The amount inhaled would be dependent upon the child's inhalation rate, IR_a, which would depend upon the level of physical activity and the amount of sediment suspended in the air (1/PEF). The significant exposure or dose would correspond to the amount of sediment that actually makes it into the nasal contact boundary and, ultimately, into the lungs. Depending upon the mode of infection, the fungi can potentially colonize within the nose or lungs.

Other potential exposure routes relevant to children's play behavior include inhalation of dust from sediments and potential contact with eyes, when the dust contains fungi. For eye exposure, fungi would need to transfer to the hand prior to when the child rubs his or her eye. There is also the possibility of exposure by having sediments blown directly into the eye from contaminated air. These factors should also be evaluated when considering the potential risk of eye infections during child play scenarios in public places.

Dose–Response

The recommended dose–response relationship that relates illness outcomes relative to the amount of microbes in contact with skin or eyes, ingested, or inhaled is available for some microbes. The relationship typically used is an exponential relationship: $P_{inf} = 1 - \exp(-kN)$, where P_{inf} is the probability of infection, N is the number of pathogens per exposure, and k is the probability of a microbe infecting,[4] although other more complex relationships can be used. The units of N are in absolute numbers for oral and inhalation exposures and in numbers per unit surface area for dermal and eye exposures. In reviewing the consolidated dose–response literature established specifically for QMRA (http://qmrawiki.msu.edu/), dose–response relationships have been identified for many bacteria, for several viruses, for a few protozoa, and for prions. No dose–response data are available for fungi within the consolidated QMRA database.

In a review of the literature, Eduard[17] summarized many studies conducted on animals plus the epidemiological literature focusing on human occupational exposures. In his review, he defined lowest observed effect levels for *Aspergillus* and *Penicillium* species for allergic and other respiratory responses and listed allowable concentrations for *Candida* and *Cryptococcus* species. There are additional primary papers that focus on a particular fungal pathogen. For example, Ward et al. (2010)[18] identified threshold values for allergic asthma responses to *Penicillium chrysogenum* within a murine model. Rehman et al.[19] induced infection of *C. albicans* in mice through a dose of 5×10^6 viable blastospores. The majority of the literature focuses on providing dose–response data of fungal inhibition (response) by antifungal agents (dose). In some cases, dose (fungi)–response (illness) information can be gleaned from these drug-focused studies as long as such evaluations require the initiation of illness among the study population. For example, in a study focused on testing the efficacy of the antifungal agent fluconazole, Larsen et al.[20] induced meningitis through the injection of approximately 700 colony-forming units of *C. neoformans* directly into the cranial vault of their test mice, causing infection.

Risk Characterization

Risk characterization combines information from the exposure assessment and dose–response to estimate a probability of illness. Risk can be computed deterministically whereby one value is provided for each step in the exposure assessment and for dose–response, or it can be computed stochastically. Stochastic simulations take into account the distribution of parameters for each step in the process. For example, in the case of hand-to-mouth ingestion, it would consider the number of times the hand is put in the mouth per exposure period, the amount of fungi adhered to the hand either directly or through sand, the amount of the hand put in the mouth, the amount of fungi transferred from the hand into the mouth, and ultimately the amount swallowed. Each parameter in this process has uncertainty. This uncertainty can be defined by a probability distribution. Stochastic methods incorporate the probability distribution for each parameter in the risk computation, thereby providing an overall estimate of risk in terms of a range of values within a set confidence interval. A very common stochastic method of analysis is based upon a Monte Carlo method. The Monte Carlo method is a process that computes risk many times (on the order of 10^5 to 10^6 iterations). For each iteration, the simulation randomly chooses the value of a parameter from its distribution. Confidence limits can then be placed on the estimated risk from all of the iterations. Typically, the range is given in terms of 95% confidence limits, which implies that there is a 95% probability that the true value lies within this range. Acceptable levels of risk depend upon the severity of the illness. For terminal illnesses, acceptable risks are traditionally on the order of 10^{-5} or 10^{-6}. However, for less severe health outcomes, such as self-limiting gastrointestinal ailments, acceptable risk can be as high as 1.9×10^{-2}, the risk value used to establish swimming advisories in the United States.[21]

DISCUSSION/CONCLUSIONS

The QMRA is a useful tool for evaluating health risks from microbiological agents. The steps of the process require an identification of the hazard, exposure, and dose–response and the subsequent computation of risk. Microbial risk assessment has been widely used to evaluate several different means of disease transmission including airborne routes[22–25] and waterborne routes. Waterborne routes include studies focused on drinking water sources,[26–28] recreational water,[29–33] storm water,[34] and multiple-use waters.[35] These studies generally focus on surrogates of pathogenic diseases, the fecal indicator bacteria,[29] or on direct measures of pathogens, including bacterial pathogens,[29,35] viral pathogens,[26,28,30,33,34] and protozoan pathogens. As of this writing, no study is available that evaluates the transmission of fungal pathogens through waterborne pathways.

Another common application of QMRA is to evaluate risks from food intake, especially foods that have been undercooked[36] or contaminated with wastewater,[37] animal waste,[38] or unknown sources.[39] Again, these studies generally focus on evaluating bacterial,[37,38] viral, and protozoan pathogens[36] and helminth eggs.[37] A few studies have evaluated fungal risks associated with food contaminated with *Aspergillus* sp.[40] and *Fusarium* sp.[41]

The bulk of microbial risk assessment from solid media is focused on land application of animal- or sewage-derived waste[42,43] or fomites.[44] For sediment, sand, and soil ingestion, the vast majority of studies focus on risks from chemical constituents.[7,45] Very few[3,46] consider microbial illness from contact with soils and sediments. Doorduyn et al.[46] evaluated risks from bacterial pathogens (*Salmonella* sp.). Shibata and Solo-Gabriele[3] evaluated risks from the protozoan pathogen *Cryptosporidium,* the viral pathogen enterovirus, and the bacterial pathogen *Staphylococcus aureus*. No studies evaluated risks from fungal infections through nonfood solid media.

Although fungal risk assessments have been conducted in the context of food, none have been conducted in the context of sediment/sand exposure. Thus, there is a clear void in evaluating fungal risks from environmental exposures, including water and sands/sediments in public places. There is a need to assess and compile dose–response data available for the various fungi to facilitate fungal risk assessments. Dose–response should be consolidated for pathogenic fungi frequently detected in these environments (e.g., *Aspergillus, Candida, Chrysosporium, Cryptococcus, Microsporum, Penicillium, Trichophyton, Trichosporon*). There is a void in defining exposure factors specific to child play behaviors in these settings. Studies should be conducted to evaluate the levels of exposure of children who play in sandpits, parks, and beach sands to confirm the estimates obtained through scenario assessments as recommended through the QMRA. Children in beach environments are particularly vulnerable because of their play activities with sediments/sands in these settings. Given this vulnerability, more effort is needed to quantify levels of fungi, especially within the intertidal zone area and the supratidal zone immediately above the high-tide line, which is where microbe levels are typically the highest,[47,48] and whether these fungi are capable of transferring to children who play in these areas.

Because many environmental assessment studies have found many different types of fungi, future risk assessment should also consider the impacts of cumulative exposures. If different fungi within the sediment/sand are characterized by the same disease endpoint, the cumulative effects of the exposure to these fungi should be considered in an effort to evaluate overall impacts from contact with sediment and sand during child play activities.

Other chapters address specific situations (endemic species, propagation, and resistance to antimicrobial agents) in an environmental context that may ultimately be extrapolated to sand, although is yet to be documented in scientific references.

Recommendations

To provide guidance to the public health community, we further recommend the computation of guidance fungi levels in sand/sediment media using a QMRA approach. This can be conducted in a fashion similar to what has been done for other microbe groups in sand.[3] In brief, the procedure requires the definition of acceptable risk levels, which would depend upon the severity of the disease (10^{-2} to 10^{-6}). The procedure also requires documenting the various exposure factors involved in the scenario (children playing with sand) and, also, knowledge of dose–response for the various fungal pathogens considered to be of greatest risk. To improve the estimates, we recommend studies that document children's microactivity behaviors during play with sand and sediment. Digital video-recording/video-translation techniques have been previously used to estimate exposures for children in residential and farmworker settings.[49,50] Work is recommended to apply this same technology to evaluating children's activity patterns (e.g., frequency and duration of contact and, importantly, the sequence of soil to skin, hand to mouth, and object to mouth contact) and, therefore, the realistic potential for exposure to microbes, including fungi, during play with sediments and sands. We recommend a consolidation of fungal dose–response data, similar to what has been prepared for viruses, bacteria, and protozoans, for inclusion within databases used to compute risk (e.g., http://qmrawiki.msu.edu/). We recommend additional studies to evaluate the influence of abrasions and wounds on the adherence and uptake of fungi and ultimate impacts on contracting fungal infections. From this information, allowable levels of fungi can then be computed to establish guidelines for sands in public play areas. Potential mitigation strategies can then be developed to address areas characterized by excess risk. Mitigation can be in the form of improving hygiene activities after exposure[49,51–54] and protecting the sediment and sands from conditions that would promote fungal contamination and fungal growth. Also the microbial flora of sands used during beach renourishment should be considered when sand is replaced at eroded beaches. Ultimately, a risk assessment should be conducted as part of all mitigation strategies to confirm whether they reduce risk to acceptable levels.

REFERENCES

1. The Fungal Research Trust. How common are fungal diseases? Fungal Research Trust 20th Anniversary Meeting; June 18, 2011; London, UK.
2. Sabino R, Rodrigues R, Costa I, Carneiro C, Cunha M, Duarte A, et al. Routine screening of harmful microorganisms in beach sands: implications to public health. *Sci Total Environ* 2014;**472**:1062–9.

3. Shibata T, Solo-Gabriele HM. Quantitative microbial risk assessment of human illness from exposure to marine beach sand. *Environ Sci Technol* 2012;**46**:2799–805.
4. Haas CN, Rose JB, Gerba CP. *Quantitative microbial risk assessment.* 1st ed. New York: John Wiley & Sons, Inc.; 1999.
5. U.S. EPA. *Exposure factors handbook.* Final Report. Washington, DC: U.S. EPA; 2011. Report No. EPA/600/R-09/052F.
6. U.S. EPA. *Risk assessment guidance for superfund: Volume I. Human health evaluation manual (Part D, standardized planning, reporting, and review of superfund risk assessments).* Washington, DC: Office of Emergency and Remedial Response; 2001. Report No. OSWER Directive No. 9285.7–47. Available from: http://www.epa.gov/swerrims/riskassessment/ragsd/index.htm.
7. Guney M, Zagury GJ, Dogan N, Onay TT. Exposure assessment and risk characterization from trace elements following soil ingestion by children exposed to playgrounds, parks and picnic areas. *J Hazard Mater* 2010;**182**:656–64.
8. Ferguson A, Bursac Z, Johnson W, Davis J. Combined analysis of results on single and multiple contacts: potting soil, sand and clay transfers to cloth for the estimate of soil adherence to human skin. *Environ Nat Resour Res* 2013;**3**(3):33–44. http://dx.doi.org/10.5539/enrr.v3n3p33.
9. Ferguson A, Bursac Z, Johnson W, Davis J. Computer-controlled chamber measurements for clay adherence relevant for potential exposure through skin. *J Environ Sci Health A* 2012;**47**(3):382–8.
10. Ferguson A, Coleman S, Bursac Z, Johnson W. Computer controlled chamber measurements for multiple contacts for soil-skin adherence from aluminum and carpet surfaces. *Hum Ecol Risk Assess* 2009a;**15**(4):811–30.
11. Ferguson A, Bursac Z, Coleman S, Biddle D, Johnson W. Comparisons of computer controlled chamber measurements for soil-skin adherence from aluminum and carpet surfaces. *Environ Res* 2009b;**109**(3):207–14.
12. Ferguson A, Bursac Z, Coleman S, Biddle D, Johnson W. Soil-skin adherence from carpet: use of a mechanical chamber to control contact parameters. *J Environ Sci Health A* 2008;**43**(12):1451–8.
13. Xue J, Zartarian V, Tulve N, Moya J, Freeman N, Auyeung W, et al. A meta-analysis of children's object-to-mouth frequency data for estimating non-dietary ingestion exposure. *J Exp Sci Environ Epidemiol* September 2010;**20**(6):536–45.
14. Xue J, Zartarian V, Moya J, Freeman N, Beamer P, Black L, et al. A meta-analysis of children's hand-to-mouth frequency data for estimating nondietary ingestion exposure. *Risk Anal* April 2007;**27**(2):411–20.
15. Rose EA, Porcerelli JH, Neale AV. Pica: common but commonly missed. *Am Board Fam Pract* 2000;**13**(5):353–8.
16. Hannigan GD, Grice EA. Microbial ecology of the skin in the era of metagenomics and molecular microbiology. *Cold Spring Harb Perspect Med* 2013;**3**:a015362.
17. Eduard W. Fungal spores: a critical review of the toxicological and epidemiological evidence as a basis for occupational exposure limit setting. *Crit Rev Toxicol* 2009;**39**(10):799–864.
18. Ward MW, Chung YJ, Copeland LB, Doerfler DL. A comparison of the allergic responses induced by *Penicillium chrysogenum* and house dust mite extracts in a mouse model. *Indoor Air* 2010;**20**(5):380–91. http://dx.doi.org/10.1111/j.1600-0668.2010.00660.x.
19. Rehman HH, Mohan AA, Tabassum HH, Ahmad FF, Rahman SS, Parvez SS, et al. Deltamethrin increases *Candida albicans* infection susceptibility in mice. *Scand J Immunol* 2011;**73**(5):459–64.
20. Larsen RA, Bauer M, Thomas AM, Sanchez A, Citron D, Rathbun M, et al. Correspondence of in vitro and in vivo fluconazole dose-response curves for *Cryptococcus neoformans. Antimicrob Agents Chemother* 2005;**49**(8):3297–301.
21. U.S. Environmental Protection Agency (EPA). *Ambient water quality criteria for bacteria 1986.* Washington, DC: U.S. EPA; 1986. Report No. A440/5–84–002.

22. Coleman ME, Thran B, Morse SS, Hugh-Jones M, Massulik S. Inhalation anthrax: dose response and risk analysis. *Biosecu Bioterror: Biodefense Strategy, Pract Sci* 2008;**6**(2):147–59.
23. Jones RM, Masago Y, Bartrand T, Haas CN, Nicas M, Rose JB. Characterizing the risk of infection from *Mycobacterium tuberculosis* in commercial passenger aircraft using quantitative microbial risk assessment. *Risk Anal* 2009;**29**(3):355–65.
24. Sales-Ortells H, Medema G. Screening-level risk assessment of *Coxiella burnetii* (Q fever) transmission via aeration of drinking water. *Environ Sci Technol* 2012;**46**:4125–33.
25. Balderrama-Carmona AP, Gortáres-Moroyoqui P, Álvarez-Valencia LH, Castro-Espinoza L, Mondaca-Fernández I, Balderas-Cortés JdJ, et al. Occurrence and quantitative microbial risk assessment of *Cryptosporidium* and *Giardia* in soil and air samples. *Int J Infect Dis* 2014;**26**:e123–7.
26. Chigor VN, Sibanda T, Okoh AI. Assessment of the risk for human health of Adenoviruses, hepatitis A virus, rotaviruses and enteroviruses in the Buffalo River and three source water dams in the Eastern Cape. *Food Environ Virol* 2014;**6**(2):87–98.
27. Schijven JF, Teunis PFM, Rutjes SA, Bouwknegt M, Husman AMDR. QMRAspot: a tool for quantitative microbial risk assessment from surface water to potable water. *Water Res* 2011;**45**:5564–76.
28. Teunis PFM, Rutjes SA, Westrell T, Husman AMDR. Characterization of drinking water treatment for virus risk assessment. *Water Res* 2009;**43**:395–404.
29. Craig DL, Fallowfield HJ, Cromar NJ. Effectiveness of guideline faecal indicator organism values in estimation of exposure risk at recreational coastal sites. *Water Sci Technol* 2003;**47**(3):191–8.
30. Wong M, Kumar L, Jenkins TM, Xagoraraki I, Phanikumar MS, Rose JB. Evaluation of public health risks at recreational beaches in Lake Michigan via detection of enteric viruses and a human-specific bacteriological marker. *Water Res* 2009;**43**:1137–49.
31. Ashbolt NJ, Schoen ME, Soller JA, Roser DJ. Predicting pathogen risks to aid beach management: the real value of QMRA. *Water Res* 2010;**44**:4692–703.
32. Soller JA, Bartrand T, ashbolt NJ, Ravenscroft J, Wade TJ. Estimating the primary etiologic agents in recreational freshwaters impacted by human sources of faecal contamination. *Water Res* 2010;**44**:4736–47.
33. Viau EJ, Lee D, Boehm AB. Swimmer risk of gastrointestinal illness from exposure to tropical coastal waters impacted by terrestrial dry-weather runoff. *Environ Sci Technol* 2011a;**45**:7158–65.
34. McBride GB, Stott R, Miller W, Bambic D, Wuertz S. Discharge-based QMRA for estimation of public health risks from exposure to storm waterborne pathogens in recreational waters in the United States. *Water Res* 2013;**47**:5282–97.
35. Ishii S, Nakamura T, Ozawa S, Kobayashi A, Sano D, Okabe S. Water quality monitoring and risk assessment by simultaneous multi pathogen quantification. *Environ Sci Technol* 2014;**48**:4744–9.
36. Opsteegh M, Prickaerts S, Frankena K, Evers EG. A quantitative microbial risk assessment for meatborne *Toxoplasma gondii* infection in the Netherlands. *Int J Food Microbiol* 2011;**150**:103–14.
37. Forslund A, Ensink JHJ, Markussen B, Battilani A, Psarras G, Gola S, et al. *Escherichia coli* contamination and health aspects of soil and tomatoes (*Solanum lycopersicum* L.) subsurface drip irrigated with on-site treated domestic wastewater. *Water Res* 2012;**46**:5917–34.
38. Kennedy MC, Clough HE, Turner J. Case studies in Bayesian microbial risk assessments. *Environ Health* 2009;**8**(Suppl I):SI9.
39. Pielaat A, Van Leusden FM, Wijnands LM. Microbiological risk from minimally processed packaged salads in the Dutch food chain. *J Food Prot* 2014;**77**(3):395–403.

40. Bui-Klimke TR, Wu F. Ochratoxin A and human health risk: a review of the evidence. *Crit Rev Food Sci Nutr* May 29, 2014. [Epub ahead of print]. http://dx.doi.org/10.1080/10408398.2012.724480.
41. Pieters MN, Freijer J, Baars B-J, Fiolet DCM, van Klaveren J, Slob W. Risk assessment of deoxynivalenol in food: concentration limits, exposure and effects. In: Devries J, Trucksees M, Jackson LS, editors. *Mycotoxins and food safety*. New York: Kluwer Academic/Plenum Publishers; 2012. p. 235–48.
42. Brooks JP, McLaughlin MR, Gerba CP, Pepper IL. Land application of manure and class B biosolids: an occupational and public quantitative microbial risk assessment. *J Environ Qual* 2012;**41**:2009–23.
43. Viau E, Bibby K, Paez-Rubio T, Peccia J. Toward a consensus view on the infectious risks associated with land application of sewage sludge. *Environ Sci Technol* 2011b;**45**:5459–69.
44. Lopez GU, Gerba CP, Tamimi AH, Kitajima M, Maxwell SL, Rose JB. Transfer efficiency of bacteria and viruses from porous and nonporous fomites to fingers under different relative humidity conditions. *Appl Environ Microb* 2013;**79**(18):5728–34.
45. Hemond HF, Solo-Gabriele HM. Children's exposure to arsenic from CCA-treated wooden decks and playground structures. *Risk Anal* 2014;**24**(1):51–64.
46. Doorduyn Y, Van Den Brandhof WE, Van Duynhoven YTHOP, Wannet WJB, Van Pelt W. Risk factors for *Salmonella enteritidis* and Typhimurium (DT104 and Non-DT104) infections in the Netherlands: predominant roles for raw eggs in Enteritidis and sandboxes in Typhimurium infections. *Epidemol Infect* 2006;**134**(3):617–26.
47. Shah AH, Abdelzaher AM, Phillips M, Hernandez R, Solo-Gabriele HM, Kish J, et al. Indicator microbes correlate with pathogenic bacteria, yeasts and helminthes in sand at a subtropical recreational beach site. *J Appl Microbiol* 2011;**110**:1571–83.
48. Whitman RL, Harwood VJ, Edge T, Nevers M, Byappanahalli M, Vijayavel K, et al. Microbes in beach sands: integrating environment, ecology and public health. *Rev Environ Sci Biotechnol* 2014;**13**:329–68.
49. Freeman NCG, Jimenez M, Reed KJ, Gurunathan S, Edwards RD, Roy A, et al. Quantitative analysis of children's microactivity patterns: The Minnesota Children's Pesticide Exposure Study. *J Expo Anal Environ Epidemiol* 2001;**11**:501–9.
50. AuYeung W, Canales RA, Beamer P, Ferguson AC, Leckie JO. Young children's hand contact activities: an observational study via videotaping in primarily outdoor residential settings. *J Expo Sci Environ Epidemiol* 2006;**16**:434–46.
51. Chance GW, Harmsen E. Children are different: environmental contaminants and children's health. *Can J Public Health* 1998;**89**. S9–13, S10–15.
52. Freeman NCG, Ettinger A, Berry M, Rhoads G. Hygiene- and food-related behaviors associated with blood lead levels of young children from lead-contaminated homes. *J Expo Anal Environ Epidemiol* 1997;**7**:103–18.
53. Freeman NC, Sheldon L, Jimenez M, Melnyk L, Pellizzari E, Berry M. Contribution of children's activities to lead contamination of food. *J Expo Anal Environ Epidemiol* 2001b;**11**:407–13.
54. Freeman NC, Hore P, Black K, Jimenez M, Sheldon L, Tulve N, et al. Contributions of children's activities to pesticide hand loadings following residential pesticide application. *J Expo Anal Environ Epidemiol* 2005;**15**:81–8.
55. Masters GM, Ela W. *Introduction to environmental engineering and science*. 3rd ed. Englewood Cliffs, NJ: Prentice Hall; 2008.
56. Pakshir K, Ghiasi MR, Zomorodian K, Gharavi AR. Isolation and molecular identification of keratinophilic fungi from public parks soil in Shiraz, Iran. *Biomed Res Int* 2013;**2013**:1–5.

57. Pereira E, Figueira C, Aguiar N, Vasconcelos R, Vasconcelos S, Calado G, et al. Microbiological and mycological beach sand quality in a volcanic environment: Madeira archipelago. *Port Sci Total Environ* 2013. 461–462, 469–479.
58. Salvo VS, Fabiano M. Mycological assessment of sediments in Ligurian beaches in the Northwestern Mediterranean: pathogens and opportunistic pathogens. *J Environ Manage* 2007;**83**:365–9.
59. Fillipello Marchisio V. Keratinolytic and keratinophilic fungi of children's sandpits in the city of Turin. *Mycopathologia* 1986;**94**:163–72.
60. Caretta G, Mangiarotti AM, Pintelli E. Keratinophilic fungi isolated from soil of Italian parks in the Province of Pavia. *Eur J Epidemiol* 1992;**8**(3):330–9.
61. Javoreková S, Labuda R, Maková J, Novák J, Medo J, Majerčiková K. Keratinophilic fungi isolated from soils of long-term fold-grazed, degraded pastures in national parks of Slovakia. *Mycopathologia* 2012;**174**:239–45.
62. Sarquis MID, DeOliveira PC. Diversity of microfungi in the sandy soil of Ipanema Beach, Rio de Janeiro, Brazil. *J Basic Microbol* 1996;**36**(1):51–8.
63. Vogel C, Rogerson A, Schatz S, Laubach H, Tallman A, Fell J. Prevalence of yeasts in beach sand at three bathing beaches in South Florida. *Water Res* 2007;**41**:1915–20.
64. Mercantini R, Marsella R, Caprilli F, Dovciallo G. Isolation of dermatophytes and correlated species from the soil of public gardens and parks in Rome. *Sabourandia* 1980;**18**:123–8.
65. Sabino R, Verissimo C, Cunha MA, Wergikoski B, Ferreira FC, Rodriques R, et al. Pathogenic fungi: an unacknowledged risk at coastal resorts? New insights on microbiological sand quality in Portugal. *Mar Pollut Bull* 2011;**62**:1506–11.
66. Ramesh VM, Hilda A. Incidence of keratinophilic fungi in the soil of primary schools and public parks of Madras city, India. *Mycopathologia* 1999;**143**:139–45.
67. Anderson JH. In vitro survival of human pathogenic fungi in Hawaiian beach sand. *Sabouraudia* 1979;**17**:13–22.
68. Kishimoto RA, Baker GE. Pathogenic and potentially pathogenic fungi isolated from beach sands and selected soils of Oahu, Hawaii. *Mycologia* 1969;**61**(3):537–48.
69. Guarro J, Punsola L, Calvo MA. Keratinophilic fungi from soil of Tarragona, Catalunya. *Mycopathologia* 1981;**76**:69–71.
70. Ali-Shtayeh MS. Keratinophilic fungi isolated from children's sandpits in the Nablus area, West Bank of Jordan. *Mycopathologia* 1988;**103**:141–6.
71. McAleer R. Investigation of keratinophilic fungi from soils in Western Australia a preliminary survey. *Mycopathologia* 1980;**72**:155–65.
72. Shadzi S, Chadeganipour M, Alimoradi M. Isolation of keratinophilic fungi from elementary schools and public parks in Isfahan, Iran. *Mycoses* 2002;**45**:496–9.
73. Kiehn TE, Polsky B, Punithalingam E, Edwards F, Brown AE, Armstrong D. Liver infection caused by *Coniothyrium fuckelii* in a patient with acute myelogenous leukemia. *J Clin Microbiol* 1987;**25**(12):2410–2.
74. Wojcik A, Kurnatowski P, Blaszkowska J. Potentially pathogenic yeasts from soil of children's recreational areas in the city of Lodz (Poland). *Int J Occup Med Environ* 2013;**26**(3):477–87.
75. Gianni C, Caretta G, Romano C. Skin infection due to *Geomyces pannorum* var.pannorum. *Mycoses* 2003;**46**(9–10):430–2.
76. Safdar A. Progressive cutaneous hyalohyphomycosis due to *Paecilomyces lilacinus*: rapid response to treatment with caspofungin and itraconazole. *Clin Infect Dis* 2002;**34**(10):1415–7.
77. Gleason FH, Letcher PM, McGee PA. Some Chytridiomycoto in soil recover from drying and high temperatures. *Mycol Res* 2004;**108**(5):583–9.

Chapter 14

Hospital Environment

Raquel Sabino
National Institute of Health Dr. Ricardo Jorge, Infectious Diseases Department, Lisbon, Portugal

The hospital environment has been suggested to play a crucial role in the epidemiology of nosocomial fungal infections. As molds reproduce by releasing spores into the air, the probable sources of fungal spores are ventilation, air conditioning systems, occupants and visitors, dust, ornamental plants, flowers, fresh fruit, food, water, and building works in and around hospitals.[1] The ubiquitous presence of fungi is a potential health threat because transmission occurs via contact, ingestion, aspiration of contaminating particles, aerosolization of potable water, or the hands of health-care workers. Fungi are important causes of life-threatening infections in immunocompromised patients and may lead to severe or fatal outcomes. In many cases there are multiple factors that predispose these patients to infections, such as neutropenia induced by therapy or bone marrow involvement, hypogammaglobulinemia, T-cell dysfunction, and mucosal damage. In general, patients with hematologic malignancies such as leukemia or lymphoma are more severely immunocompromised and at higher risk of developing invasive mycoses than those with solid tumors. Among patients with hematological malignancies, 20–50% have evidence of invasive fungal infections at autopsy.[2] As fungi become increasingly more resistant to the limited antifungal agents available, an alternative or additional means to control fungal infections is to decrease the exposure of the patient to these microorganisms, thereby preventing an infection from occurring.[3]

Aspergillus and *Candida* species account for most fungal infections in immunocompromised patients (Table 1). However, life-threatening infections caused by emerging pathogens such as *Trichosporon beigelii*, *Fusarium* species, *Scedosporium apiospermum*, species belonging to class Mucormycotina (former Zygomycetes and agents of mucormycosis), and *Malassezia furfur* are increasing in frequency (Table 1). Although *Candida* infections are more common than infections caused by filamentous fungi, the number of deaths caused by these last is higher.[4–6] The mortality rate attributed to candidemia varies between 8% and 53% in halogenous blood cell transplant recipients, whereas in disseminated infections caused by filamentous fungi such as *Aspergillus*, *Fusarium*, and Mucormycotina species, this rate varies between 56% and 95%.[4]

Environmental Mycology in Public Health. http://dx.doi.org/10.1016/B978-0-12-411471-5.00014-4

TABLE 1 Most Frequent Fungal Species Associated with Nosocomial Infections[72]

Candida spp.	*C. albicans*	Other yeasts	*Blastoschizomyces* spp.
	C. glabrata		*Cryptococcus neoformans*
	C. guilliermondii		*Malassezia* spp.
	C. kefyr		*Rhodotorula* spp.
	C. krusei		*Saccharomyces* spp.
	C. lusitaniae		*Trichosporon* spp.
	C. parapsilosis	Class Mucormycotina	*Lichteimia* spp.
	C. tropicalis		*Cunninghamella* spp.
Aspergillus spp.	*A. fumigatus*		*Mucor* spp.
	A. niger		*Rhizopus* spp.
	A. terreus		*Rhizomucor* spp.
	A. flavus	Dematiaceous molds	*Alternaria* spp.
Other hyaline molds	*Acremonium* spp.		*Bipolaris* spp.
	Fusarium spp.		*Curvularia* spp.
	Paecilomyces spp.		*Cladophialophora* spp.
	Scedosporium spp.		*Exophiala* spp.
	Scopulariopsis spp.		*Phialophora* spp.
	Trichoderma spp.		

Because nosocomial fungal infections are rapidly progressive, there is a critical need for more efforts toward prevention, early diagnosis, and effective treatment of these infections. Of these, preventive measures are of major importance in the control of nosocomial fungal infections, including environmental surveillance and strict application of cleaning procedures. Training health-care workers and improving their knowledge about fungal infections also has special importance in prevention and controlling nosocomial fungal infections.[1]

Few studies have attempted to correlate the level of fungal pollution with the occurrence of specific diseases among patients or hospital staff. The degree of contamination by fungi in the hospital environment may increase dramatically in combination with various factors, such as geoclimatic factors, the presence of contamination sources, and a favorable microclimate. Because exposure to fungi can cause serious health problems, it is clearly essential to evaluate the degree of contamination in the various environments and to use those evaluations to determine the risk of infection for patients and staff alike.[7]

CONTAMINATION SOURCES

There are two main sources for fungal contamination in hospital environments: external (outdoors) and internal (indoors). Most fungi present indoors originate in external environments[8,9] but some do not. Indoor sources include building materials, foodstuffs, flower pots, pipes or other water distribution systems, heating, building ventilation, air conditioning systems, and dust.[10–12]

The number and species of fungi present in the hospital environment depends on the number of people using the facilities, including patients, medical staff, and visitors; the types of activities performed; the presence of organic matter; the levels of humidity and temperature indoors; and the season of the year.[13] Overall, a high inoculum of a nosocomial pathogen in a cold room with high relative humidity represents the best situation for its survival.

All activities performed inside buildings may vary the fungal particles present. Cleaning, construction work, and any other major dust-raising activities have a particular impact.[8] For example, fungal pollution in operating theaters was found to be influenced by the possible contamination of the sterilization rooms to which they are normally directly connected. Indeed, the activities performed in these rooms, such as the washing of surgical instruments, are conducive to the creation of a microclimate (high relative humidity and air temperature, as well as the release of droplets) that favors fungal growth.[7]

Numerous studies have demonstrated that the hands and gloves of healthcare workers readily acquire pathogens after coming into contact with contaminated hospital surfaces and they can subsequently transfer these organisms to the patients and inanimate surfaces they touch (Figure 1). Inanimate surfaces have thus often been described as the source of outbreaks of

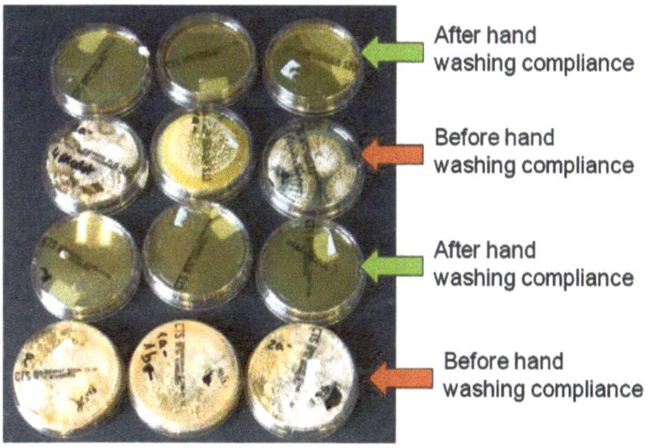

FIGURE 1 Contact plates from the hands of healthcare workers before and after hand washing compliance.

FIGURE 2 Surface sampling by swabbing the surfaces using a 10 × 10-cm-square stencil disinfected with 70% alcohol solution between samples according to the International Standard ISO 18,593 (2004).

FIGURE 3 Water tap sampling using a swab.

nosocomial infections and should be analyzed when studying environmental contamination in a hospital ward. Doorknobs, water taps, and other surfaces with frequent manual handling/contact should be screened (Figures 2–4).

The longer a nosocomial pathogen persists on a surface, the more likely it comes to be the cause of a nosocomial infection. A good example of this is *Candida parapsilosis*, yeast with increasing prevalence in hospital settings because it survives on surfaces for long periods of time. This species has often been reported as responsible for nosocomial outbreaks.[14–16] This may be due to its ability to form biofims[17] and to survive much better than *Candida albicans* on various surfaces.[2] *C. parapsilosis* infects the patient usually by exogenous transmission. Hospital health-care workers have been associated with both sporadic cases and outbreaks of invasive fungal infections in immunocompromised

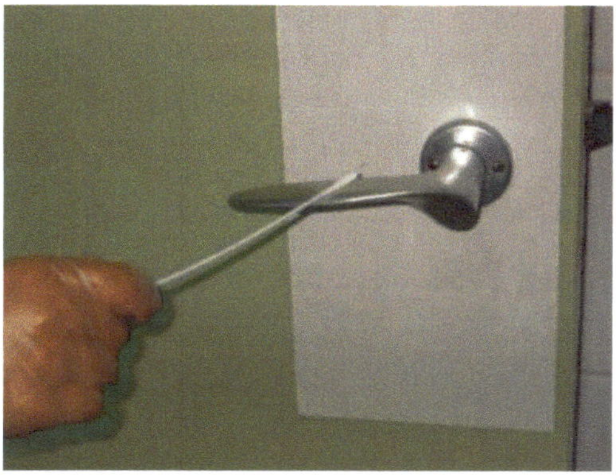

FIGURE 4 Doorknob sampling.

patients.[18,19] In a study performed by Sabino et al.,[20] environmental *C. parapsilosis* strains showed to be more resistant to phagocytic host defenses than bloodstream isolates, being potentially more deleterious in the course of infection than strains from a clinical source. Environmental surveillance and application of strict cleaning procedures should hence be implemented and sustained to prevent cross infections and hospital outbreaks, not only for *C. parapsilosis* but also for all other fungi.

Construction Work

Construction may produce dust and debris that can carry microorganisms into patient care areas. External construction works increase the potential of contamination with dust/debris on air-intake filters and may decrease filtration capacity resulting in airborne spread of microorganisms via ventilation systems. All efforts should be made to reduce patient and personnel exposure to resulting dust and debris.

The most significant source of health-care-associated infections related to construction is the dust that is raised during construction and demolition. Attached to that dust are tiny fungal spores of many different species, but of most significance are the spores of the *Aspergillus* species, particularly *Aspergillus fumigatus*. Large numbers of *Aspergillus* spores may be generated during construction projects and conveyed by air currents to distant sites, posing a serious health hazard.[21]

Specific protective measures against environmental contamination must be taken when there is construction activity on a hospital site,[22] such as keeping doors and windows closed or even sealing the windows and putting barriers between patient care and renovation or construction areas. In addition, biocleaning should be reinforced and surfaces should be regularly and systematically cleaned. Moreover, the air conditioning systems should be checked routinely for spores and an air treatment system should be used. Hospital planners should ensure that rooms for immunocompromised patients have an adequate capacity to minimize fungal spore counts through the use of high-efficiency (>90%) particulate air (HEPA) filtration and high rates of room air exchange (i.e., >12 air changes per hour).

More recently, mobile air-treatment decontamination units have been developed as an alternative to laminar air filtration. To reduce dust, contractors and workmen are instructed to wet the construction site.[21–23]

Heating, Ventilating, and Air Conditioning Systems

In buildings with central HVAC (heating, ventilation, and air conditioning) systems that are properly maintained, filters should remove many of the spores present. Many instances are known in which the HVAC system itself served as an amplification and dissemination site for fungal spores. In these cases fungi have been found growing on air filters as well as in the ducts.

Therefore, although HVAC systems can help remove and/or dilute more than 80% of aerosols from the outdoors, they can also provide favorable conditions for bioaerosols to colonize.[24] Microbiological growth may occur in an HVAC system equipped with low-efficiency filters, humidifiers that use water recycling, or in areas in which water condensation remains stagnant and large recirculation of the air is present.[25,26] Condensation on ductwork or other components is another likely source of moisture. Cooling coils, drains pans, and water pans for humidifiers are likely locations for fungal growth, especially when there is standing water. These deficiencies can occur in the design, installation, maintenance, and operation of the HVAC system. Microorganisms can thus spread in the indoor air via the HVAC system and be inhaled by the people working or especially by the hospitalized patients. *Aspergillus* and *Fusarium* are some of the most frequent genera found in these systems.[27,28]

To prevent infections originating in HVAC systems, these systems should be visually inspected for dirt, dust, and mold before cleaning as well as after cleaning. The ductwork should also be inspected for fungal growth and cleaned with a disinfectant and filters replaced with appropriate periodicity. In the case of water condensation, its cause should be identified. All the system processes should be evaluated, including all ventilation system components, the operability of the air-intake system, the effectiveness of the air-delivery system in occupied spaces, and the reentrainment and infiltration potential of contamination that can enter the building from sources such as cooling towers, exhaust fan outlets, plumbing vents, flues, and motor vehicle exhaust.[24–26] Another factor to bear in mind is windows without correct sealing. Fungal spores from outside can accumulate in the windows' slots and then disperse into the inside air, as observed in the example shown in Figure 5.

Water Reservoirs, Pipes, and Moist Environments Inside the Hospital

Water distribution systems may also be a source of fungal contamination in hospitals. Fungi in water are aerosolized when water is running (taps and

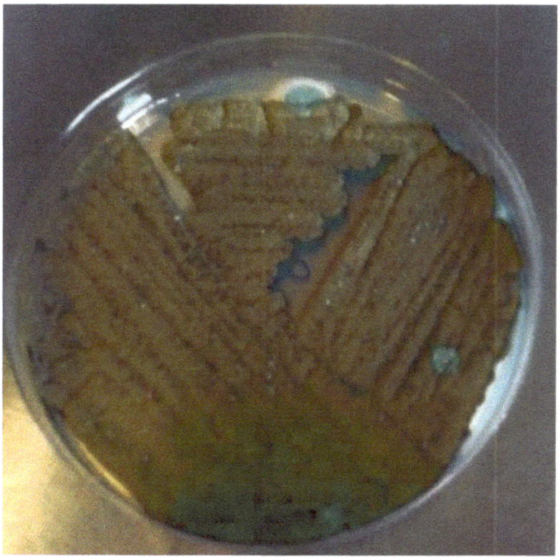

FIGURE 5 Culture (on malt extract agar) of a swab collected from a hospital ward window's slot. Uncountable colony-forming units (cfu) of *Cladosporium* sp. were detected.

showers[28,29]). Potential patient exposure happens with inhalation (i.e., while showering); ingestion; transmission through abraded mucous membrane in the oral cavity, skin, or cornea; or directly into the blood and internal organs in special circumstances such as catheter insertion, peritoneal dialysis, or surgical procedures.[30]

Environmental risk of waterborne fungal infections should be interpreted according to the local situation, and geographical differences may exist, influenced also by different procedures for water collection and treatment among countries. In particular, the origin of the water supply (i.e., underground vs surface water) has a significant impact on the recovery of molds from water.[31–34] Moreover, if water is stored in a manner that allows it to have contact with ambient air (storage in a surface water reservoir) fungal contamination will occur and a diversity of molds can be recovered from outlets.[29] Several studies report that filamentous fungi and yeast are common on water pipe surfaces, even in the presence of free chlorine residuals.[35,36] Elevated open storage tanks and fire hydrants may be a significant source of fungi.[37,38] Low flow rates are frequently maintained in these structures, which can enhance biofilm development.[39]

As water is distributed into downstream pipes of smaller and smaller diameter, biofilms can develop to a considerably greater extent because of the greater degree of water flow variability to which these pipes are exposed as a consequence of varying patterns of water use. When formed at or near the point of use, a biofilm community can act as microbial repository that constantly disperses viable microbes into the adjacent water streams. Independent of regular flora, these microbes can also colonize patients, caregivers, environmental surfaces, medical devices, and instruments that subsequently come into contact with water.[40]

FIGURE 6 Culture (on malt extract agar) of a swab collected from a water tap. Cultures revealed the presence of *Fusarium solani*, *C. parapsilosis*, and *Exophiala* sp.

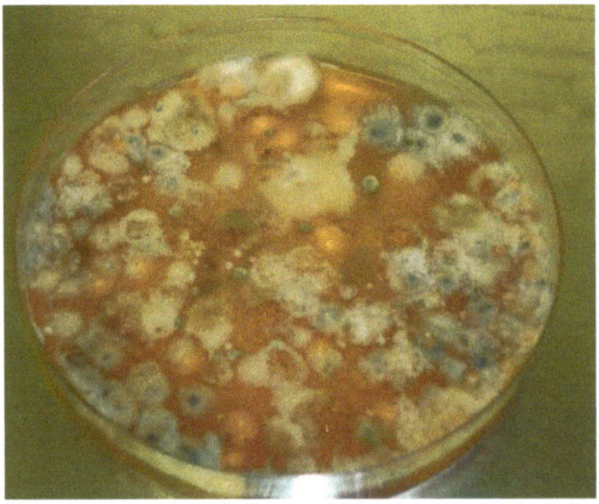

FIGURE 7 Culture (on malt extract agar) of a swab collected from the shower wall of a patients' restroom. Various *Fusarium* species were obtained, totaling 128 colony-forming units.

In addition, because humidity favors fungal growth, high fungal densities near water sources may simply reflect the presence of conidia in the air or on surfaces and indicate the need for new cleaning procedures.[32] Air levels of *Fusarium* and *Aspergillus* were found to increase in hospital environments after running showers multiple times.[28,29] Species from these two genera are described as the most frequently found owing to their conidial dispersion mode or ability to form biofilms, but variation in ecological niches and different methods of collection and times of sampling may account for potential differences among institutions with regard to the rate and type of fungal colonization in water.[32]

Figures 6–8 show cultural results obtained from surfaces and air analyses of a patients' restroom in a hematological ward.

According to Anaissie et al.,[31] in the case of fungal detection, hospitals should formulate policies to avoid or minimize exposure of immunosuppressed

FIGURE 8 Culture (on malt extract agar) of an air sample collected from a patients' restroom. Various *Aspergillus* species were obtained, as well as *Cladosporium* spp., *Alternaria* sp., and *Penicillium* spp.

patients to tap water from any source. The most effective and least expensive approach that can be applied worldwide is the prevention of exposure of immunosuppressed patients to hospital tap water by providing sterile (boiled) water for drinking. In addition, patients should avoid showering during severe immunosuppression because of the risk of acquiring the organisms through aerosolization of contaminated hospital water. We strongly recommend that bed baths provided with sterile disposable sponges be used instead of showering.

Therefore, testing hospitals' water supply for the presence of opportunistic molds is important to avoid fungal infections in severely immunocompromised patients.

Food

The existence of isolated gastrointestinal filamentous fungal infection, without pulmonary or disseminated infection, supports the hypothesis that, in addition to the risk of spore inhalation, water or food can be a vehicle for spores.[32]

In hospital kitchens there are high levels of fungal contamination.[41] Indeed, the preparation of food, the warm and humid atmosphere, and the presence of various materials (perhaps including foodstuffs already contaminated by fungal spores) provide an ideal environment for the growth of fungi and, consequently, the contamination of patients' meals. If airborne fungal spores are adsorbed to larger particles, they can settle on surfaces. Therefore, work surfaces, kitchen utensils, and foodstuffs may also become contaminated.[7]

Several fruits (apples, bananas, peaches, and oranges) brought to the patient by visitors may contain a high number of fungal spores in their skin. In the absence of control measures, one unique ward could be contaminated with thousands of spores brought from external sources through fruit carrying spores of saprophytic or phytopathogenic fungi or even of fungi used as biopesticides.[42] In a study performed by Gangneux et al.[43] in a hematological unit, pepper and herbal teas were also very contaminated by viable *Aspergillus* and non-*Aspergillus* molds, as well as freeze-dried soups, corn, coconuts, cashew nuts, coffee, beans, soy, cheeses, and smoked meats.

It has been previously recommended that immunocompromised patients should avoid such contaminated foods, and granulocytopenic patients should receive sterile or low-microbial-content diets.[44] Several physical (oven heating at 210 °C, microwave irradiation, freezing at 20 °C) and chemical (washing with water, soap, and disinfecting solutions, alone and combined) procedures are applicable to foods and wrappings for fungal eradication. Heating, however, cannot be applied to some types of foods, such as fruits, fruit juices, bread, and cheese, and microwave treatment is not 100% efficient.[45] When foods cannot be exposed to high temperature or microwaves, 70% ethanol only partially reduces the level of surface contamination.[46] In these cases, the most advisable measure is to avoid those foods. Similarly, soft cheeses should not be given to patients, as they are highly contaminated by yeasts and cannot be sterilized.[45]

To achieve low contamination of food, Bouakline et al.[45] proposed that the catering process in hematology wards should include sterilization of foods and also dishes, listing of banned foods, compliance with procedures for disinfection of individual packaging, and regular mycological testing of storage sites and kitchens.

Intravenous Solutions Administered to the Patients

Intravenous fluids, drugs, and nutrition have become an indispensable part of modern therapy, and large quantities of commercially and locally manufactured fluids are used in treatment daily.[47,48]

In modern medical practice, up to 80% of hospitalized patients received intravenous (IV) therapy at some point during their admission.[48] The parenteral route of administration is generally adopted for medicaments that cannot be given orally, either because of patient intolerance, drug instability, or poor absorption via the enteral route. In the unconscious patient, parenteral administration is the only safe and most effective means of administering medicaments through the IV route. Contamination of IV products is a recurrent problem and can have fatal consequences.

Microbial contamination of injections, infusions, and other fluids often results from poor sterilization management, inadequate analytical facilities, lack of properly trained personnel, obsolete equipment, inappropriate production environment, poor-quality packaging, or indeterminate errors during the quality control process.[48–50]

It has been reported that microorganisms can gain access to IV infusions during administration, by external sources of contamination such the influx of unfiltered air, the addition of drugs, and the migration of microorganisms through the cannulae of the administration set.[48] Infusion fluids requiring compounding or the addition of medications to the fluid container were found to produce 7% of primary bloodstream infections when those fluids were prepared.[51] The hands of health-care workers could be the major transmission vehicle and contamination is probably attributable to induction of needle puncture in the body of IV fluid bags by nursing staff.[50] Parenteral nutrition and IV fat emulsions can become contaminated during preparation and infusion, with fungal pathogens,[52] especially *Candida* species, which account for 20–30% of systemic infections associated with central venous catheters.[53] *C. albicans* has the ability to either grow very well or sustain prolonged viability in all nutritional IV products and *C. parapsilosis* grows very well in parenteral solutions rich in glucose. Intraluminal spread of infection may also result from intrinsic contamination of the infusion fluids. Pathogenic fungi such as *Aspergillus*, *Candida*, *Fusarium*, and *Paecilomyces* have been isolated from "commercially sterile" fluids, with no visible fungal growth, suggesting that these products can be potential health hazards.[50] Fungi may also contaminate intact bottles of fluid. The colony of fungus is usually found growing on the underside of the rubber stopper and may be very difficult to see on cursory inspection.[47]

Biofilms as Causes of Exogenous Contamination

Environmental surfaces in health-care settings are often contaminated by microorganisms, and biofilms can also develop on the surfaces in these settings. Fungi are especially adapted to growth on surfaces, as evidenced by their absorptive nutrition mode, their secretion of extracellular enzymes to digest complex molecules, and their apical hyphal growth.[54]

Filamentous fungal biofilms are expected to be common in environments exposed to high moisture with a major air interface (i.e., unsaturated environments).[55] In fact, biofilms generally form on any surface that is exposed to nonsterile water or other liquids and are consequently found in many environmental and medical systems.

Biofilms are therefore defined as highly structured communities of microorganisms that are either surface associated or attached to one another and are enclosed within a self-produced protective extracellular matrix.[56] The advantages to an organism of forming a biofilm include protection from the environment, resistance to physical and chemical stress, metabolic cooperation, and a community-based regulation of gene expression.[53] Several studies have demonstrated that conventional methods to inactivate free-floating microorganisms with antimicrobial agents or disinfection solutions are often less effective against pathogens within a biofilm.[57–59]

The properties of a given biofilm will depend on: (i) the physical and chemical properties of the substratum; (ii) the number and type of cells present within the biofilm; and (iii) the external physical environment, such as the water in a stream or blood in a patient.[53]

Numerous nosocomial fungal infections are associated with indwelling medical devices (e.g., dental implants, catheters, heart valves, vascular bypass grafts, ocular lenses, artificial joints, and central nervous system shunts), which can act as substrates for biofilm growth. Forty percent of patients with microbial colonization of IV catheters develop occult fungemia, with consequences ranging from focal disease to severe sepsis and death.[60] Contamination of those devices with biofilm-producer fungi could arise from external/exogenous sources such as the host's own flora, a health-care professional's transmission, or pathogenic species that exist in the hospital environment, responsible for secondary colonization. The ability to disseminate indwelling biofilm cells to new locations, and thus cause secondary disease or encourage persistence of the organism in a new location, is an important point to keep in mind in the study of hospital environmental fungi and their potentialities in causing disease.

THRESHOLD VALUES USED TO EVALUATE MICROBIOLOGICAL CONTAMINATION IN THE HOSPITAL ENVIRONMENT

The medical literature provides no limits or guidelines for fungal concentrations in hospital environments that are appropriate for assessing whether the contamination in a particular location is acceptable. Because there are no generally accepted threshold limit values for the air of hospitals, the results obtained in studies of indoor air quality of this setting could be compared only with the values recommended by various authors or institutions. Augustowska and Dutkiewicz[61] proposed, for hospital wards, maximum levels of airborne fungi equal to 200 cfu/m^3. Also according to these authors, there are three classes of hospital ward cleanliness: the intensive care units are classified as class 2, which means a presence of microorganisms not exceeding 300 cfu/m^3 of indoor air. The number of cfu/m^3 allowed in operating theaters is 0, whereas in treatment rooms it is 50 cfu/m^3.[62] The World Health Organization, on the other hand, considers 150 cfu/m^3 a reason for concern, especially when "potentially pathogenic species are found."[63]

According to Morris et al.[64] and for *Aspergillus*, the airborne concentration in protective isolation suites should be 15 cfu/m^3.

The National Disease Surveillance Centre of Dublin[65] also published some guidelines regarding the maximal number of *Aspergillus* that should be detected in the hospital environment. Therefore, they suggested that in HEPA-filtered air (>95% efficiency and >10 air changes per hour) the maximal count should be <0.1 cfu/m^3; in wards with no air filtration, 5.0 conidia/m^3; and in places under construction or with defective ventilation, 2.3–5.9 conidia/m^3. Regardless, if total fungal count exceeds 1.0 cfu/m^3 on several occasions, the air systems or procedural practices require intensive evaluation.

The degree of exposure assessed by colony-forming units per unit of air is not predictive of disease; however, every strain present in the environment is a potential pathogen if it encounters the appropriate host.[66] In fact, a significant relationship between the degree of fungal contamination of air and surfaces in hematology wards and the incidence of invasive nosocomial aspergillosis has been demonstrated in nonepidemic situations.[67]

Not only the quantification is important to evaluate the air quality but the confirmed presence of certain species is important as well. This leads to the comparison of fungal species detected indoors and outdoors to detect possible sources of contamination inside the analyzed units.

Moreover, and according to the American Industrial Hygiene Association *Biosafety Guide*,[68] the confirmed presence (in any type of indoor environment) of *Stachybotrys chartarum*, *Aspergillus versicolor*, *Aspergillus flavus*, *A. fumigatus*, and *Fusarium moniliforme* requires appropriate risk management decisions to be made. "Confirmed presence" means the detection of colonies in several samples, many colonies in one sample, or, when a single colony was found in a single sample, there is evidence of growth of these fungi on building materials.

Several authors also advise that further investigation of sources of contamination should be performed in the following circumstances:

- Total indoor counts are greater than outdoor counts;
- Comparison of indoor and outdoor levels of fungal organisms show one of the following:
    - Organisms are present in the indoor sample and not in the outdoor sample;
    - The predominant organism found in the indoor sample is different from the predominant organism in the outdoor sample;
    - A monoculture of an organism is found in the indoor sample; it may be absent from samples taken in other areas of the building;
    - Persistently high counts.

In certain situations, air sampling without concomitant surface sampling may not adequately reflect the level of microbial contamination in indoor environments[69] because spores from different fungal species have different features, which leads to different dispersion and aerosolization times. Furthermore, contamination with one of them can directly influence the others. Therefore, the analysis should always be performed both for the air and for the surfaces to complement the results and have a complete picture of the fungi present. Routine environmental sampling allows determination of reference air and surface contamination limits to be used in everyday practice.

No defined threshold values are established for fungi on surfaces but, according to Sahay et al.,[70] the value $50\,cfu/cm^2$ should be used. This study, however, does not present specific limits for hospital surfaces. According to the European GMP for Pharmaceutical Products (01/97),[71] on surfaces of very high risk and of high risk, the recommended levels of contamination are <1 and 5 cfu/55-mm plate, respectively.

WHEN SHOULD A HOSPITAL ENVIRONMENTAL ANALYSIS BE PERFORMED?

To end this chapter, in the hospital environment the monitoring of the levels of microbiological contamination is advisable in the following circumstances:

- To identify potential sources of nosocomial aspergillosis or other fungal outbreaks when some cases have been identified;
- Prior to occupancy of special controlled environments, e.g., to determine the efficiency of HEPA filters in laminar flow facilities;
- To predict environmental spore contamination from outside sources;
- To identify defects/breakdowns in hospital ventilation/filtration systems;
- To correlate outbreaks of invasive aspergillosis or other fungal infections with hospital construction or demolition work;
- To monitor the efficiency of certain procedures (such as cleaning processes) within hospital building wards, where at-risk patients are managed; and
- To evaluate the microbiological quality of the water used to supply the hospital and that may become inhaled or ingested.

REFERENCES

1. Kordbacheh P, Zaini F, Kamali P, Ansari K, Safara M. Study on the sources of nosocomial fungal infections at intensive care unit and transplant wards at a teaching hospital in Tehran. *Iran J Public Health* 2005;**34**:1–8.
2. Traoré O, Springthorpe V, Sattar S. A quantitative study of the survival of two species of *Candida* on porous and non-porous environmental surfaces and hands. *J Appl Microbiol* 2002;**92**:549–55.
3. Neely A, Orloff M. Survival of some medically important fungi on hospital fabrics and plastics. *J Clin Microbiol* 2001;**39**:3360–1.
4. Martino R, Subirá M, Rovira M, Solano C, Vázquez L, Sanz GF, et al. Invasive fungal infections after allogenic peripheral blood stem cell and solid organ transplantation: incidence and risk factors in 395 patients. *Br J Heamatol* 2002;**116**:475–82.
5. Fukuda T, Boeckh M, Carter RA, Sandmaier BM, Maris MB, Maloney DG, et al. Risks and outcomes of invasive fungal infections in recipients of allogeneic hematopoietic stem cell transplants after nonmyeloablative conditioning. *Blood* 2003;**102**:827–33.
6. Roden MM, Zaoutis TE, Buchanan WL, Knudsen TA, Sarkisova TA, Schaufele RL, et al. Epidemiology and outcome of zygomycosis: a review of 929 reported cases. *Clin Infect Dis* 2005;**41**:634–53.
7. Perdelli F, Cristina ML, Sartini M, Spagnolo AM, Dallera M, Ottria G, et al. Fungal contamination in hospital environments. *Infect Cont Hosp Ep* 2006;**27**:44–7.
8. Flannigan B. Air sampling for fungi in indoor environments. *J Aerosol Sci* 1997;**28**:381–92.
9. Horner WE, Worthan AG, Morey PR. Air- and dustborne mycoflora in houses free of water damage and fungal growth. *Appl Environ Microb* 2004;**70**:6394–400.
10. Chao HJ, Miltom DK, Schwartz J, Burge HA. Dustborne fungi in large office buildings. *Mycopathologia* 2001;**154**:93–106.
11. Pasanen A-L, Heinonen-Tanski H, Kalliokoski P. Fungal microcolonies on indoor surfaces – an explanation for the base-level fungal spore counts in indoor air. *Atmos Environ* 1992a;**26B**:117–20.

12. Pasanen A-L, Juutinen T, Jantunen MJ, Kalliokoskia P. Occurrence and moisture requirements of microbial growth in building materials. *Int Biodeterior Biodegrad* 1992b;**30**:273–83.
13. Araujo R, Cabral JP. Fungal air quality in medical protected environments. In: Kumar A, editor. *Air quality*. InTech; 2010. ISBN: 978-953-307-131-2. Available from: http://www.intechopen.com/books/airquality/fungal-air-quality-in-medical-protected-environments.
14. Wenzel R, Edmond M. The impact of hospital acquired blood stream infections. *Emerg Infect Dis* 2001;**7**:174–7.
15. Lupetti A, Tavanti A, Davini P, Ghelardi E, Corsini V, Merusi I, et al. Horizontal transmission of *Candida parapsilosis* candidemia in a neonatal intensive care unit. *J Clin Microbiol* 2002;**40**:2363–9.
16. Clark TA, Slavinski SA, Morgan J, Lott T, Arthington-Skaggs BA, Brandt ME, et al. Epidemiologic and molecular characterization of an outbreak of *Candida parapsilosis* bloodstream infections in a community hospital. *J Clin Microbiol* 2004;**42**:4468–72.
17. Laffey S, Butler G. Phenotype switching affects biofilm formation by *Candida parapsilosis*. *Microbiology* 2005;**151**:1073–81.
18. Barchesi F, Caggiano G, Falconi Di Francesco L, Montagna MT, Barbuti S, Scalise G, et al. Outbreak of fungemia due to *Candida parapsilosis* in a pediatric oncologic unit. *Diagn Microbiol Infect Dis* 2004;**49**:269–71.
19. van Asbeck EC, Huang YC, Markham AN, Clemons KV, Stevens DA. *Candida parapsilosis* fungemia in neonates: genotyping results suggest healthcare workers hands as source, and review of published studies. *Mycophatologia* 2007;**164**:287–93.
20. Sabino R, Sampaio P, Carneiro C, Rosado L, Pais C. Response to macrophage infection by *Candida parapsilosis sensu stricto* isolates with different multilocus genotypes and origins. *BMC Microbiol* 2011;**11**:180. http://dx.doi.org/10.1186/1471-2180-11-180.
21. Opal SM, Asp AA, Cannady Jr PB, Morse PL, Burton LJ, Hammer 2nd PG , et al. Efficacy of infection control measures during a nosocomial outbreak of disseminated aspergillosis associated with hospital construction. *J Infect Dis* 1986;**153**:634–7.
22. Fournel I, Sautour M, Lafon I, Sixt N, L'Ollivier C, Dalle F, et al. Airborne Aspergillus contamination during hospital construction works: efficacy of protective measures. *Am J Infect Control* 2010;**38**:189–94.
23. Tabbara KF, Jabarti AA. Hospital construction-associated outbreak of ocular aspergillosis after cataract surgery. *Ophthalmology* 1998;**105**:522–6.
24. Law AKY, Chau CK, Chan GYS. Characteristics of bioaerosol profile in office buildings in Hong Kong. *Build Environ* 2001;**36**:527–41.
25. Wu PC, Li YY, Chiang CM, Huang CY, Lee CC, Li FC, et al. Changing microbial concentrations are associated with ventilation performance in Taiwan's air-conditioned office buildings. *Indoor Air* 2005;**15**:19–26.
26. Huang R, Agranovski I, Pyankov O, Grinshpun S. Removal of viable bioaerosol particles with a low-efficiency HVAC filter enhanced by continuous emission of unipolar air ions. *Indoor Air* 2008;**18**:106–12.
27. Lentino JR, Rosenkranz MA, Michaels JA, Kurup VP, Rose HD, Rytel MW, et al. Nosocomial aspergillosis: a retrospective review of airborne disease secondary to road construction and contaminated air conditioners. *Am J Epidemiol* 1982;**116**:430–7.
28. Anaissie EJ, Stratton SL, Dignani MC, Summerbell RC, Rex JH, Monson TP, et al. Pathogenic *Aspergillus* species recovered from a hospital water system: a 3-year prospective study. *Clin Infect Dis* 2002;**34**:780–9.
29. Warris A, Voss A, Verweij PE. Hospital sources of *Aspergillus* species: new routes of transmission? *Rev Iberoam Micol* 2001;**18**:156–62.

30. Grabińska-Łoniewska A, Koniłłowicz-Kowalska T, Wardzyńska G, Boryn K. Occurrence of fungi in water distribution system. *Pol J Environ Stud* 2007;**16**:539–47.
31. Anaissie EJ, Kuchar RT, Rex JH, Francesconi A, Kasai M, Müller FM, et al. Fusariosis associated with pathogenic *Fusarium* species colonization of a hospital water system: a new paradigm for the epidemiology of opportunistic mold infections. *Clin Infect Dis* 2001;**33**: 1871–8.
32. Gangneux JP, Bretagne S, Cordonnier C, Datry A, Derouin F, Grillot R, et al. Prevention of nosocomial fungal infection: the French approach. *Clin Infect Dis* 2002;**35**:343–6.
33. Warris A, Onken A, Gaustad P, Janssen W, van der Lee H, Verweij PE, et al. Point-of-use filtration method for the prevention of fungal contamination of hospital water. *J Hosp Infect* 2010;**76**:56–9.
34. Warris A, Voss A, Abrahamsen TG, Verweij PE. Contamination of hospital water with *Aspergillus fumigatus* and other molds. *Clin Infect Dis* 2002;**34**:1159–60.
35. Nagy LA, Olson BH. Occurrence and significance of bacteria, fungi, and yeasts associated with distributionpipe surfaces. In: *AWWA water qual. tech. conf.* 1985. p. 213–38. Houston.
36. Doggett MS. Characterization of fungal biofilms within a municipal water distribution system. *Appl Environ Microbiol* 2000;**66**:1249–51.
37. Rosenzweig WD, Pipes WO. Fungi from potable water: interaction with chlorine and engineering effects. *Water Sci Technol* 1988;**20**:153–9.
38. Rosenzweig WD, Pipes WO. Presence of fungi in drinking water, Chapter 7. In: Larson RA, editor. *Biohazards of drinking water treatment*. Ann Arbor, MI: Lewis Publishers; 1989. p. 85–93.
39. Geldreich EE. Microbiological quality control in distribution systems. Chapter 18. In: Pontius FW, editor. *Water Quality and Treatment*. 4th ed. *American water works association*. New York: McGraw-Hill; 1990.
40. Ortolano GA, McAlister MB, Angelbeck JA, Schaffer J, Russell RL, Maynard E, et al. Hospital water point-of-use filtration: a complementary strategy to reduce the risk of nosocomial infection. *Filtr Suppl* 2005;**1**:3–26.
41. Viegas C, Ramos C, Almeida-Silva M, Sabino R, Veríssimo C, Rosado L, et al. Air fungal contamination in ten hospitals' food units from Lisbon. In: Popov V, Brebbia CA, editors. *Food and environment: the quest for a sustainable future*. Wessex: WIT Press; 2011. p. 127–32.
42. Milner RJ, Hunter DM. Recent developments in the use of fungi as biopesticides against locusts and grasshoppers in Australia. *J Orthopt Res* 2001;**10**:271–6.
43. Gangneux JP, Noussair L, Bouakline A, Roux N, Lacroix C, Derouin F, et al. Experimental assessment of disinfection procedures for eradication of *Aspergillus fumigatus* in food. *Blood* 2004;**104**:2000–2.
44. Remington JS, Schimpff SC. Please don't eat the salads. *N Engl J Med* 1981;**304**:433–5.
45. Bouakline A, Lacroix C, Roux N, Gangneux JP, Derouin F. Fungal contamination of food in hematology units. *J Clin Microbiol* 2000;**38**:4272–3.
46. Gangneux JP, Noussair L, Bouakline A, Roux N, Lacroix C, Derouin F. Experimental assessment of disinfection procedures for eradication of *Aspergillus fumigatus* in food. *Blood* 2004;**104**:2000–2.
47. Robertson MH. Fungi in fluids—a hazard of intravenous therapy. *J Med Microbiol* 1970;**3**: 99–102.
48. Ali A, Maysoon AM, Huda JM, Mohammed HS. Microbial contamination due to malpractice during administration of intravenous fluids in Baghdad hospitals. *AJPS* 2011;**9**:145–57.

49. Caudron JM, Ford N, Henkens M, Macé C, Kiddle-Monroe R, Pinel J, et al. Substandard medicines in resource-poor settings: a problem that can no longer be ignored. *Trop Med Int Health* 2008;**13**:1062–72.
50. Babalola MO, Akinyanju JA. Mycological analysis and potential health hazards of dextrose intravenous infusions in Nigeria. *IOSR J Pharm (IOSRPHR)* 2012;**2**:20–33.
51. Macias AE, Huertas M, de Leon SP, Munoz JM, Chavez AR, Sifuentes-Osornio J, et al. Contamination of intravenous fluids: a continuing cause of hospital bacteremia. *Am J Infect Control* 2010;**38**:217–21.
52. Kuwahara T, Shimono K, Kaneda S, Tamura T, Ichihara M, Nakashima Y, et al. Growth of microorganisms in total parenteral nutritions without lipids. *Intern J Med Sci* 2010;**7**:43–7.
53. Cooper IR. Microbial biofilms: case reviews of bacterial and fungal pathogens persisting on biomaterials and environmental substrata. In: Mendez-Vilas A, editor. *Current research, technology and education topics in applied microbiology and microbial biotechnology*. Badajoz, Spain: Formatex Research Centre; 2011. p. 807–17. ISBN: 978-84-614-6195-0.
54. Harding MW, Marques LLR, Howard RJ, Olson ME. Can filamentous fungi form biofilms? *Trends Microbiol* 2009;**17**:475–80.
55. Holden PA. Biofilms in unsaturated environments. *Methods Enzymol* 2001;**337**:125–43.
56. Ramage G, Mowat E, Jones B, Williams C, Lopez-Ribot J. Our current understanding of fungal biofilms. *Crit Rev Microbiol* 2009;**35**:340–55.
57. Donlan RM, Costerton JW. Biofilms: survival mechanisms of clinically relevant microorganisms. *Clin Microbiol Rev* 2002;**15**:167–93.
58. Fux CA, Costerton JW, Stewart PS, Stoodley P. Survival strategies of infectious biofilms. *Trends Microbiol* 2005;**13**:34–40.
59. Hoyle BD, Costerton JW. Bacterial resistance to antibiotics: the role of biofilms. *Prog Drug Res* 1991;**37**:91–105.
60. Chandra J, Kuhn DM, Mukherjee PK, Hoyer LL, McCormick T, Ghannoum MA, et al. Biofilm formation by the fungal pathogen *Candida albicans*: development, architecture, and drug resistance. *J Bacteriol* 2001;**183**:5385–94.
61. Augustowska M, Dutkiewicz J. Variability of airborne microflora in a hospital ward within a period of one year. *Ann Agric Environ Med* 2006;**13**:99–106.
62. Gniadek A, Macura AB. Intensive care unit environment contamination with fungi. *Adv Med Sci* 2007;**52**:283–7.
63. Goyer N, Lavoie J, Lazure L, Marchand G. *Bioaerosols in the workplace: evaluation, control and prevention guide*. Institut de Recherche en Santé en Sécurité du Travail du Québec; 2001. T-24, Montréal, IRSST, 2001, 72 pages, ISBN: 2551214432.
64. Morris G, Kokki MH, Anderson K, Richardson MD. Sampling of *Aspergillus* spores in air. *J Hosp Infect* 2000;**44**:81–92.
65. National Disease Surveillance Centre. *National guidelines for the prevention of nosocomial invasive aspergillosis during construction/renovation activities*; 2002. ISBN: 0-9540177-3-0.
66. Muñoz P, Burillo A, Bouza E. Environmental surveillance and other control measures in the prevention of nosocomial fungal infections. *Clin Microbiol Infect* 2001;**7**:38–45.
67. Brenier-Pinchart MP, Lebeau B, Quesada JL, Mallaret MR, Borel JL, Mollard A, et al. Influence of internal and outdoor factors on filamentous fungal flora in hematology wards. *Am J Infect Control* 2009;**37**:631–7.
68. Dillon H, Heinsohn P, Miller D. *Field guide for the determination of biological contaminants in environmental samples—an AIHA biosafety guide*. Fairfax, Virginia, USA: American Industrial Hygiene Association (AIHA) Press; 1996.

69. Buttner M, Stetzenbach L. Monitoring airborne fungal spores in an experimental indoor environment to evaluate sampling methods and the effects of human activity on air sampling. *Appl Environ Microbiol* 1993;**59**:219–26.
70. Sahay R, Parvataneni S, Barnes R. Dynamics of surficial mold in indoor environment. In: *Proceedings of AIHce*; June 2, 2009. [mold abstracts, 92].
71. GMP for pharmaceutical products/BPF Européennes. *Fabrication des médicaments stériles* (01/97).
72. Richardson M, Lass-Flörl C. Changing epidemiology of systemic fungal infections. *Clin Microbiol Infect* 2008;**14**:5–24.

Section IV

Chapter 15

Fungal Disease Outbreaks and Natural Disasters

Kaitlin Benedict, Mary E. Brandt
Mycotic Diseases Branch, Centers for Disease Control and Prevention, Atlanta, GA, USA

INTRODUCTION

Invasive fungal infections (IFIs) are becoming increasingly recognized as an emerging public health issue. These infections are often difficult to treat and can be associated with substantial morbidity and mortality. The observed increases in IFIs are likely to be multifactorial and may include (1) host factors, such as a larger at-risk immunosuppressed population or travel or relocation of immune-naïve persons to areas to which certain IFIs are endemic; (2) pathogen factors, such as new virulence or resistance mechanisms; and (3) environmental factors, including changing land use patterns, agricultural antifungal use, and climate change.[1] Some pathogenic fungi, such as *Coccidioides*, *Blastomyces*, and *Histoplasma*, exist within specific ecological niches and are typically restricted to certain geographic regions, whereas other fungi such as *Aspergillus* and Mucormycetes are thought to be more widespread. Within a defined habitat, weather and seasonal patterns probably influence the growth and distribution of these fungi.

The term "outbreak" refers to a greater than expected number of cases occurring within a given geographic area or time frame. Many infectious disease outbreaks involve person-to-person transmission; however, outbreaks of fungal disease are somewhat unique in this respect because they usually arise as a result of exposures to a common environmental source. Environmental disruption is often a key factor in the dispersal of fungal elements into air, water, or organic matter; this can occur on a relatively small scale, in relation to construction, excavation, or similar activities, or on a larger scale, such as extreme weather events or natural disasters. This chapter highlights fungal disease outbreaks associated with these types of events, with a special emphasis on outbreaks after natural disasters; describes the related public health response methods in the United States; and discusses potential strategies to reduce death and disability due to IFIs.

Environmental Mycology in Public Health. http://dx.doi.org/10.1016/B978-0-12-411471-5.00015-6
2016 Published by Elsevier Inc.

FUNGAL DISEASES AFTER NATURAL DISASTERS

Disasters, by definition, cause human, material, or environmental loss such that the affected community cannot cope using local resources and requires assistance from outside sources.[2] Depending on the type of disaster and the specific sociocultural and economic setting in which it occurs, a disaster can lead to various public health problems, including population displacement, limited access to food and safe water, injury, and infection.[3] Despite the public perception that disasters lead to infectious disease epidemics, outbreaks directly attributable to natural disasters are uncommon.[4] Postdisaster IFIs are similarly unusual, but they are becoming increasingly reported and represent a unique category of disaster-associated health concern. These infections primarily result from inhalation or cutaneous inoculation of fungal spores due to extensive exposures to the natural environment (e.g., near-drowning, wound contamination); multiple fungal pathogens and routes of infection have been reviewed elsewhere.[5] Notably, postdisaster settings can amplify some of the existing clinical and public health challenges associated with the diagnosis and treatment of IFIs.

OUTBREAKS CAUSED BY DIMORPHIC FUNGI

Coccidioidomycosis

Coccidioidomycosis, also called valley fever, is an infection caused by *Coccidioides*, a soil-dwelling fungus endemic to the southwestern United States and parts of Mexico, Central America, and South America.[6] The growth and distribution of *Coccidioides* in the environment are thought to be closely linked to climate and weather: the organism requires moisture to grow, yet the arthroconidia become more easily aerosolized after periods of low rainfall.[7] Therefore, seasonal and climatic changes may play a role in overall coccidioidomycosis incidence and the occurrence of outbreaks.[8] Although most cases of coccidioidomycosis are sporadic, common-source outbreaks have been observed in association with earth-disturbing activities such as construction and archaeological excavation.[9,10]

Two examples of disaster-associated coccidioidomycosis have been described. In 1977, a severe dust storm that originated in California's southern San Joaquin Valley, an area to which coccidioidomycosis is highly endemic, caused a large outbreak of over 115 cases, many of which occurred in areas not previously considered to be endemic, including Sacramento County.[11,12] Similarly, after an earthquake in Northridge, California, in January 1994, over 200 cases in Ventura County were believed to have resulted from exposure to the dust clouds generated by the earthquake.[13] A case–control study showed that being in a dust cloud and the length of exposure were significantly associated with symptomatic disease.[13] Notably, the study also showed that 93% of case patients initially received at least one antibacterial medication before being diagnosed with coccidioidomycosis, suggesting that the infection may not have been considered during their initial diagnosis.

Histoplasmosis

Histoplasmosis, caused by the fungus *Histoplasma capsulatum*, is associated with a wide range of clinical manifestations and is an important opportunistic infection in immunosuppressed persons. Skin test surveys conducted in the 1940s identified the areas surrounding the Ohio and Mississippi River valleys as the endemic regions in the United States.[14] Parts of Latin America are also endemic, and cases have also been seen in other areas of the world.[15] *Histoplasma* grows especially well in soil that is contaminated with bird or bat droppings, and histoplasmosis outbreaks have often been associated with point-source exposure or disturbance of contaminated soil.[16] Other, often large or prolonged, histoplasmosis outbreaks have occurred as a result of soil-disrupting activities not necessarily involving bird or bat droppings, such as construction, bulldozing, and soil tilling.[16] Unlike other infections caused by dimorphic fungi, guidelines for histoplasmosis prevention have been established, which aim to decrease the risk for occupationally acquired histoplasmosis associated with these types of activities.[17]

Blastomycosis

Similar to coccidioidomycosis and histoplasmosis, blastomycosis typically manifests as a pulmonary infection than can range in severity from asymptomatic to life-threatening. *Blastomyces dermatitidis*, the causative agent, appears to thrive best in moist soils near waterways and is considered endemic to regions of the United States and Canada, primarily areas bordering the Ohio and Mississippi Rivers, the Saint Lawrence River, and the Great Lakes. Most documented cases of blastomycosis are sporadic, but at least 15 common-source outbreaks have been described in the literature, many of which involved shared outdoor exposures (e.g., camping, hunting, or fishing), and approximately one-third of these outbreaks were associated with excavation or construction.[18]

OUTBREAKS CAUSED BY MOLDS

Mucormycosis

Mucormycosis (formerly zygomycosis) is a rare infection caused by molds belonging to the order Mucorales, which are believed to be ubiquitous in nature, particularly in soil and decaying organic matter.[19] Mucormycosis most often manifests as a pulmonary or rhinocerebral infection in immunosuppressed persons.[20] Outbreaks related to outdoor and indoor environments have both been described. For example, a few instances of health-care-associated mucormycosis are believed to have resulted from exposure to hospital construction or water damage.[21] Therefore, it has been suggested that among immunosuppressed hospital inpatients, exposure to hospital construction can act as a risk factor for developing an invasive mold infection.[22]

Cutaneous mucormycosis is less common than sinus or pulmonary mucormycosis but is more likely to occur in persons with no underlying medical conditions, and clusters of cases have been observed after several natural disasters owing to traumatic implantation of fungal spores into open wounds. For example, the first known cluster of postdisaster mucormycosis was described after a volcanic eruption in Armero, Colombia, in 1985, which killed approximately 23,000 persons and injured 4500 more.[23] Among 38 injured patients who were hospitalized with necrotizing lesions, eight had infections caused by the Mucormycete *Rhizopus arrhizus* (*oryzae*).[23] A similar cluster of 13 cases of *Apophysomyces trapeziformis* soft-tissue infections occurred after the May 22, 2011 tornado in Joplin, Missouri.[24] Case patients were located in the portion of the tornado path that sustained the greatest amount of damage and were all severely injured; a case–control study showed that penetrating trauma and increasing number of wounds were independent risk factors for developing mucormycosis.[24]

Several years after Hurricane Katrina, in a hospital in New Orleans, Louisiana, five cases of cutaneous mucormycosis caused by *Rhizopus delemar* were attributed to exposure to contaminated linens, which was the only exposure common to these patients.[25] *Rhizopus* was also recovered from clean linens and from bins used to deliver linens from the off-site laundry facility.[25] The role of the hurricane in this outbreak was not clear, but it may have contributed by resulting in an environment with a higher than usual concentration of *Rhizopus* spores.[25] Finally, several isolated case reports of soft-tissue infection after the 2004 Indian Ocean tsunami represent some of the particular clinical challenges encountered with cutaneous mucormycosis, which can appear similar to bacterial wound infections but require early administration of antifungal treatment and aggressive surgical debridement.[26,27]

Aspergillosis

Aspergillus spp. are ubiquitous molds that can affect human health ranging from allergic reaction to severe disseminated disease.[28] Invasive aspergillosis is a serious cause of morbidity and mortality in immunocompromised hosts, and many outbreaks among hospitalized patients have been described. In approximately half of these outbreaks, hospital construction or renovation was implicated as the likely source.[29] Lessons learned from hospital outbreaks of aspergillosis have contributed to several evidence-based guidelines for the prevention of environmentally acquired infections in health-care facilities, such as those developed by the Healthcare Infection Control Practices Advisory Committee.[30,31] These guidelines contain infection prevention and control recommendations relevant to various potential sources of *Aspergillus*, including dust, inadequate ventilation, contaminated medical equipment, and water damage.[30,31]

In general, damp indoor spaces can support mold growth, and there is sufficient evidence that indoor mold can increase the risk for invasive *Aspergillus* infection in severely immunocompromised persons.[32] However, no association has been shown between indoor mold and increased risk for aspergillosis or other IFIs in healthy persons.[32] Similarly, there is little evidence specifically in the context of disasters to suggest that exposure to water-damaged buildings, either in health-care facilities or community settings, results in increased risk for invasive aspergillosis; however, this phenomenon could represent a lack of recognition of a possible association.[33,34] One report described a cluster of unexplained chronic cough in six persons living in a temporary refuge after the 2011 Great East Japan earthquake and tsunami; although the patients' sputum cultures and culture plates exposed to the interior of the refuge both yielded *Aspergillus*, it is unclear if the patients' symptoms were related to airway colonization with *Aspergillus* or true infection.[35] An outbreak of *Aspergillus* meningitis in six women after the 2004 Indian Ocean tsunami may have been indirectly related to indoor mold, as the source was suspected to be contaminated syringes used to administer spinal anesthesia, which were improperly stored in a humid warehouse.[36]

Public Health Response to Fungal Outbreaks in the United States

Because comprehensive surveillance systems are limited or nonexistent for most IFIs, outbreak detection often depends on an astute clinician's recognition of an increased number of cases and communication of the problem to local public health authorities. As with many other types of public health investigations in the United States, response to a cluster of IFIs often begins at the local level, with state and federal involvement if requested. A successful response often requires collaboration between clinicians and members of various public health agencies to develop recommendations to end the existing outbreak and to prevent similar situations from arising in the future.

CONCLUSION

The potential for outbreaks of IFI in association with either small-scale or widespread environmental disruption is well recognized. Less well understood are the ways in which factors such as climate change and patterns of human activity influence the environmental burden and distribution and opportunities for exposure to pathogenic fungi. Because prevention of human exposures to fungal pathogens in the natural environment is often difficult, strategies to reduce the public health burden of IFIs should focus on timely diagnosis and administration of appropriate treatment, which can lead to better outcomes. These issues are particularly relevant in postdisaster situations, in which the local health infrastructure is often damaged or destroyed. In certain settings, adherence to existing guidelines, such as those for health-care settings and occupational histoplasmosis, may help reduce the likelihood of environmentally acquired IFIs among specific at-risk populations.[17,31] Overall, increased awareness about IFIs among health-care providers and the general public is needed, particularly in the context of activities or events that disturb the environment in which pathogenic fungi may be present.

DISCLAIMER

The findings and conclusions in this chapter are those of the authors and do not necessarily represent the official position of the Centers for Disease Control and Prevention.

REFERENCES

1. Brandt ME, Park BJ. Think fungus-prevention and control of fungal infections. *Emerg Infect Dis* 2013;**19**:1688–9.
2. World Health Organization. *WHO definitions: emergencies*. http://www.who.int/hac/about/definitions/en/index.html [accessed January 2014].
3. Noji EK. The public health consequences of disasters. *Prehosp Disaster Med* 2000;**15**:147–57.
4. Watson JT, Gayer M, Connolly MA. Epidemics after natural disasters. *Emerg Infect Dis* 2007;**13**:1–5.
5. Benedict K, Park BJ. Invasive fungal infections after natural disasters. *Emerg Infect Dis* 2014;**20**:349–55.
6. Chiller TM, Galgiani JN, Stevens DA. Coccidioidomycosis. *Infect Dis Clin North Am* 2003;**17**:41–57. viii.
7. Kolivras KN, Comrie AC. Modeling valley fever (coccidioidomycosis) incidence on the basis of climate conditions. *Int J Biometeorol* 2003;**47**:87–101.
8. Park BJ, Sigel K, Vaz V, Komatsu K, McRill C, Phelan M, et al. An epidemic of coccidioidomycosis in Arizona associated with climatic changes, 1998–2001. *J Infect Dis* 2005;**191**:1981–7.
9. Cummings KC, McDowell A, Wheeler C, McNary J, Das R, Vugia DJ, et al. Point-source outbreak of coccidioidomycosis in construction workers. *Epidemiol Infect* 2010;**138**:507–11.
10. Petersen LR, Marshall SL, Barton-Dickson C, Hajjeh RA, Lindsley MD, Warnock DW, et al. Coccidioidomycosis among workers at an archeological site, northeastern Utah. *Emerg Infect Dis* 2004;**10**:637–42.
11. Flynn NM, Hoeprich PD, Kawachi MM, Lee KK, Lawrence RM, Goldstein E, et al. An unusual outbreak of windborne coccidioidomycosis. *N Engl J Med* 1979;**301**:358–61.
12. Williams PL, Sable DL, Mendez P, Smyth LT. Symptomatic coccidioidomycosis following a severe natural dust storm. An outbreak at the Naval Air Station, Lemoore, Calif. *Chest* 1979;**76**:566–70.
13. Schneider E, Hajjeh RA, Spiegel RA, Jibson RW, Harp EL, Marshall GA, et al. A coccidioidomycosis outbreak following the Northridge, Calif, earthquake. *JAMA* 1997;**277**:904–8.
14. Manos NE, Ferebee SH, Kerschbaum WF. Geographic variation in the prevalence of histoplasmin sensitivity. *Dis Chest* 1956;**29**:649–68.
15. Kauffman CA. Histoplasmosis: a clinical and laboratory update. *Clin Microbiol Rev* 2007;**20**:115–32.
16. Cano MV, Hajjeh RA. The epidemiology of histoplasmosis: a review. *Semin Respir Infect* 2001;**16**:109–18.
17. CDC. *Histoplasmosis: protecting workers at risk*; 2004.
18. Benedict K, Roy M, Chiller T, Davis JP. Epidemiologic and ecologic features of blastomycosis: a review. *Curr Fungal Infect Rep* 2012;**6**:327–35.

19. Richardson M. The ecology of the zygomycetes and its impact on environmental exposure. *Clin Microbiol Infect* 2009;**15**(Suppl. 5):2–9.
20. Roden MM, Zaoutis TE, Buchanan WL, Knudsen TA, Sarkisova TA, Schaufele RL, et al. Epidemiology and outcome of zygomycosis: a review of 929 reported cases. *Clin Infect Dis* 2005;**41**:634–53.
21. Rammaert B, Lanternier F, Zahar JR, Dannaoui E, Bougnoux ME, Lecuit M, et al. Healthcare-associated mucormycosis. *Clin Infect Dis* 2012;**54**(Suppl. 1):S44–54.
22. Weems Jr JJ, Davis BJ, Tablan OC, Kaufman L, Martone WJ. Construction activity: an independent risk factor for invasive aspergillosis and zygomycosis in patients with hematologic malignancy. *Infect Control: IC* 1987;**8**:71–5.
23. Patino JF, Castro D, Valencia A, Morales P. Necrotizing soft tissue lesions after a volcanic cataclysm. *World J Surg* 1991;**15**:240–7.
24. Neblett Fanfair R, Benedict K, Bos J, Bennett SD, Lo YC, Adebanjo T, et al. Necrotizing cutaneous mucormycosis after a tornado in Joplin, Missouri, in 2011. *N Engl J Med* 2012;**367**:2214–25.
25. Duffy J, Harris J, Gade L, Sehulster L, Newhouse E, O'Connell H, et al. Mucormycosis outbreak associated with hospital linens. *Pediatr Infect Dis J* 2014;**33**:472–6.
26. Andresen D, Donaldson A, Choo L, Knox A, Klaassen M, Ursic C, et al. Multifocal cutaneous mucormycosis complicating polymicrobial wound infections in a tsunami survivor from Sri Lanka. *Lancet* 2005;**365**:876–8.
27. Snell BJ, Tavakoli K. Necrotizing fasciitis caused by *Apophysomyces elegans* complicating soft-tissue and pelvic injuries in a tsunami survivor from Thailand. *Plastic Reconstr Surg* 2007;**119**:448–9.
28. Barnes PD, Marr KA. Aspergillosis: spectrum of disease, diagnosis, and treatment. *Infect Dis Clin North Am* 2006;**20**:545–61. vi.
29. Vonberg RP, Gastmeier P. Nosocomial aspergillosis in outbreak settings. *J Hosp Infect* 2006;**63**:246–54.
30. CDC. Guidelines for environmental infection control in health-Care facilities. Recommendations of the CDC and the Healthcare Infection Control Practices Advisory Committee. *MMWR* 2003;**55**:1–42.
31. CDC. *Guideline for isolation precautions: preventing transmission of infectious agents in healthcare settings healthcare infection control practices advisory committee*; 2007.
32. Institute of Medicine, Committee on Damp Indoor Spaces and Health. *Damp indoor spaces and health*. Washington, DC: The National Academies Press; 2004.
33. Barbeau DN, Grimsley LF, White LE, El-Dahr JM, Lichtveld M. Mold exposure and health effects following hurricanes Katrina and Rita. *Annu Rev Public Health* 2010;**31**:165–78. 1 p following 78.
34. Rao CY, Reed D, Kemmerly S, Morgan J, Fridkin SK. Assessing invasive mold infections, mold exposures and personal protective equipment use among immunocompromised New Orleans residents after hurricane Katrina. In: *Society for healthcare epidemiology of America 17th annual scientific meeting*; 2007. [Baltimore, MD.].
35. Ogawa H, Fujimura M, Takeuchi Y, Makimura K. Chronic cough in a tsunami-affected town. *Pulm Pharmacol Ther* 2012;**25**:11.
36. Gunaratne PS, Wijeyaratne CN, Chandrasiri P, Sivakumaran S, Sellahewa K, Perera P, et al. An outbreak of *Aspergillus meningitis* following spinal anaesthesia for caesarean section in Sri Lanka: a post-tsunami effect? *Ceylon Med J* 2006;**51**:137–42.

Part III

Fungi and Metabolites

Chapter 16

Dietary Exposure Assessment of European Population to Mycotoxins: A Review

Carlo Brera, Francesca Debegnach, Emanuela Gregori, Sonia Colicchia, Sabina Soricelli, Brunella Miano, Maria Chiara Magri, Barbara De Santis

Istituto Superiore di Sanità – Dipartimento di Sanità Pubblica Veterinaria e Sicurezza Alimentare – Reparto OGM e Xenobiotici di origine fungina - Viale Regina Elena, Rome, Italy

INTRODUCTION

It is well known that if it exists at contamination levels, the presence of a hazard in the diet may pose an acute or chronic risk that can lead to serious consequences for human health and animal welfare, as well as associated huge economic losses for food and feed business operators.

As a consequence of a succession of food safety scandals occurring across the world, the guarantee of high-level safe production is a demanding issue that has been treated exhaustively by the European Union (EU) and other international organizations for the past 15 years.

In the EU, after the *bovine spongiform encephalopathy* scandal in particular, there was a need for full harmonization of risk assessment principles leading to a safe production chain in food and feed sectors, characterized by a high guarantee of consumer health and animal wellness. A harmonized protocol for risk assessment purposes was effectively necessary to provide risk managers and other stakeholders with the proper information, to avoid duplicating the opinions of different organizations, so as to ensure that the overall risk derived from cumulative sources of exposure to the same stressor could be properly assessed and to assist integration within EU of new member states with their approaches toward performing risk assessment.

Indications of the White Paper in 2000 and then the publication of two reports by the Scientific Steering Committee of Health and Consumer Protection Directorate General of the EU led to the establishment of a system based on the harmonization of risk assessment procedures to guarantee a high level of food and feed safety.[1,2]

Environmental Mycology in Public Health. http://dx.doi.org/10.1016/B978-0-12-411471-5.00016-8

Previously, in 1997, the *Codex Alimentarius Commission Procedural Manual*, 10th Edition,[3] defined the terminology inherent to risk analysis that was subsequently adopted.

In these documents, risk assessment was defined as a science-led process for establishing the likelihood of adverse effects to human health and the environment from exposure to risk sources. Risk assessment was also identified as a process consisting of four different stages: hazard identification, implying toxicological and epidemiological assessment; hazard characterization, involving the assessment of dose–response and extrapolation to humans; exposure assessment; and risk characterization.

In 2002, Regulation EC/178/2002[4] laid down the general principles and requirements of food law and procedures for food safety, and among various key points formally set up the Authority for Food Safety, and assigned to stakeholders the full responsibility for the production of safe foods and feeds along the whole agri-food chain.

After increasing implementation of the legislative prescriptions, an effective decrease in the occurrence of food/feed-borne outbreaks occurred in the past recent years, even though high levels of health risk in specific population groups may still persist.

In the same direction and in an increasingly collaborative way, official control activities performed at the level of each member state have been contributing to the common target of a higher level of safety and hygiene requirements for food and feed products.

International institutions such as the World Health Organization (WHO), the Food and Agriculture Organization of the United Nations (FAO), the *International Agency for Research on Cancer*, the Organization for Economic Cooperation and Development, and the *European Food Safety Authority* (EFSA) made a further effort to increase the level of perception of the risks related to diet by publishing Scientific Opinions, guidelines, and codes of practice aimed at providing basic information for minimizing the occurrence of food-borne diseases as much as possible.

In the risk assessment frame, exposure assessment, which refers to the estimation of probability that a population group is exposed to a contaminant, and to what extent, may represent the most challenging issue. This is because some limitations in providing data can occur, such as a restricted number of observations, use of improper consumption databases, the presence of variations related to the inhomogeneity of the occurrence and/or consumption data for a single territory or (still more relevant) for the European scenario, the presence of aggregated occurrence or consumption data, and the presence of only partial evidence of differences between an entire population and a consumers-only risk assessment.

Other sources of error can derive wrong conclusions for at-risk groups such as infants, toddlers, children and adolescents, the elderly, pregnant women, and immunodeficient subjects.

An important distinction to be taken into account is that related to deterministic and stochastic models.

The term "deterministic" describes an approach in which numerical fixed values are used at each step in the risk assessment; for example, the mean or the 95th percentile value of measured data may be used to generate a single risk estimate as the end result.

In a stochastic approach, mathematical modeling of the variability of the involved phenomena is used and the end point is a probability distribution of the risk assessment.

Main advantages to using probabilistic approaches are related to the possibility of arriving at an overview of different scenarios, analyzing variability and uncertainties, and to obtaining more realistic risk patterns. Conversely, the complexity of the model and problems generating data are major disadvantages.

Currently, in chemical risk assessments the point estimate is still used more frequently than a probabilistic approach in the presence of concomitant sources of uncertainties such as providing a "static" observation in the curve of the exposure as actually occurs over the lifespan of humans. Also, the worst-case approach commonly followed in a point estimate of exposure can represent a completely unrealistic situation with related overestimated scenarios.

To overcome this trouble, a quantitative risk assessment (QRA) approach is increasingly gaining in consensus and practicability.

QRA, which is different from a qualitative risk assessment, can be deterministic or probabilistic and entails the use of models whose output is associated with numerical data and analysis.

QRAs also provide information about uncertainty in numerical terms with a representation of uncertainty distributions calculated by diverse statistical methods. One of the most prominent advantages of a QRA is that it provides risk managers with answers at a higher level of detail than a qualitative risk assessment.

Probability distributions used in stochastic risk models may represent variability as well as uncertainty. In this context, uncertainty represents the lack of perfect knowledge about a parameter value, which can be reduced by further measurements. Typical examples of uncertainty are the poor appropriateness of occurrence and/or consumption data or the lack of information on the metabolic pathway of a compound. Variability, on the other hand, represents the real heterogeneity of the population that is unchangeable by other measurements: for instance, the difference in diet habits from one population group to another or the variable levels of a hazard in different servings.[1,2]

Another key element in the development of a model used in stochastic approaches is sensitivity analysis (SA), which studies the correlation between variation in the output of a model and different sources of variation in the model input.[5] The main objectives of SA can be summed up in an assessment of the influence of factors able to affect the quality and importance of a model, to generate a maximum and the minimum impact on output variation and establish research priorities.

Moreover, there are different methods for SA, such as correlation analysis, spider plots, factorial designs, and gradient estimation; more specifically, four kinds of methods were developed to perform global SA: (1) scatterplots; (2) regression analysis, input–output correlation; (3) analysis if variance and response surface method; and (4) variance-based methods.

As previously mentioned, exposure assessment is one of the four elements that compose risk assessment.

Conventionally, occurrence data indicating the level of a compound in foods, consumption data, and body weight (bw) are the three basic groups of data necessary to calculate the exposure of a population group.

Typical information that should characterize occurrence data as reliable are the type of sampling, the quantitative measurement of the effect of first and second processing on the distribution of the compound of interest within the industrial process, and the level of the compound as consumed. As far as consumption data are concerned, primarily disaggregated values for gender, geographical location, and age should be used.

Typically, the intake assessment may be based on the mean intake derived from mean occurrence level and mean consumption value, or the 95th percentile of an entire population or only of consumers, to better provide a real scenario of the exposure.

Another approach to obtain information about the exposure is by measuring biomarkers in human biological fluids such as human milk, serum, plasma, and urine and feces; the intake can be correlated with pharmacokinetic relationships.[6]

In this chapter, the most updated approaches and results derived from research studies and/or monitoring data are presented for a real concerning group of fungal metabolites such as mycotoxins.

As fully reported in the literature, the toxic effects of mycotoxins are related to their carcinogenic and in some case genotoxic power and other effects not as risky for human and animal health, such as teratogenicity, immunotoxicity, and nephrotoxicity.

Because of the peculiar characteristics of mycotoxins, the exposure assessment for these toxic compounds is challenging since a lot of uncertainties can affect the final output of the assessment. Among the most important, the exact metabolic profile of the parent compound is not yet fully known because in most of the cases no information exists regarding the type, amount, and toxic effect of all the metabolites formed in vivo; in addition, a correlation factor does not exist between the intake of mycotoxins via diet and their overall moiety in living organisms. In addition, uncertainty may be derived from errors associated with the sampling procedure when monitoring program and research studies performed to assess the exposure.

EXPOSURE ASSESSMENT METHODOLOGY

The WHO defines exposure assessment as qualitative and/or quantitative evaluation of the likely intake of biological, chemical, or physical agents via food, as well as exposure from other sources if relevant.[7]

Another definition states that exposure assessment corresponds to the process aimed at estimating or measuring the extent, frequency, and length of exposure to an agent, along with the characteristics of the population exposed. Ideally, it describes the sources, pathways, routes, and uncertainties in the assessment.[8]

Several methods can be used to estimate the intake of a food chemical, but the appropriate tool for dietary assessment strictly depends on the purpose for which it is needed. The fitness for purpose of the selected method could be influenced by the specific evaluation needed and data availability, among others. However, whatever the elected methodology, the algorithm used to calculate exposure is the following:

$$\text{Exposure}\left(\left(\text{ng/kg bw}\right)/\text{day}\right) = \frac{\text{concentartion} \times \text{consumption}}{\text{body weight}}$$

According to the United States Environmental Protection Agency,[8] quantitative exposure can be estimated following three different approaches:

1. At the point of contact, measuring both concentration and time of contact and integrating them (point of contact measurement).
2. Separately evaluating the concentration of exposure and the time of contact, and then combining this information (scenario evaluation).
3. Determining the dose, which in turn can be reconstructed through internal indicators such as biomarkers, after exposure has taken place (reconstruction).

These approaches also apply to food chemicals. Duplicate diet studies (DDS) correspond to point-of-contact measurements, occurrence, and consumption data collection, and their combination provide scenario evaluations and biomarkers to be considered as reconstructions. Probabilistic modeling represents an advanced combination method in scenario evaluation.

Exposure assessment methodologies can also be categorized as follows[9]:

1. Screening methods, which are considered a starting point toward setting future priorities for more detailed collections of data (budget methods and per capita approach).
2. Methods based on specific data, such as the point estimate (deterministic approach), that combine a single data point of food consumption with a single mycotoxin value determined for the relevant foods considered. The consumption and contamination values used for this deterministic exposure calculation can be media or high level; the choice depends on the purpose. When variables (consumption and concentration) are described in terms of distribution, a probabilistic analysis can be performed, allowing evaluation of the variabilities and/or uncertainties of considered variables (probabilistic approach).
3. Confirmatory methods that include the use of biomarkers and DDS, which will be described later in this chapter.

As previously mentioned, the most common approach to estimating mycotoxin exposure is to combine contamination data with consumption data using deterministic methods or probabilistic modeling. So far, comparison of results from deterministic exposure assessment with probabilistic models has not shown large differences or extreme estimations; however, probabilistic analysis provides confidence intervals, uncertainty, and variability estimations.[10]

Consumption and Occurrence Data

The objective of the exposure assessment must be clearly identified before appropriate data sources can be selected: for example, pre-regulation and post-regulation exposure assessments are undertaken for different purposes and may have different data sources and default assumptions. The basic consideration is that the data required to assess the exposure are determined by the objective of the assessment. In dietary exposure assessment, it is important to obtain accurate information on both the concentration of chemicals in food (occurrence) and food consumption. Selection of the sampling, analysis, and reporting procedures are critical issues for obtaining consistent and comparable data on chemical concentrations in food, and the quality of food consumption data depends on the survey design, the method and tools used, and the statistical treatment applied.

Concentration Data/Occurrence Data

The sources of mycotoxin concentration data are studies to proposed Maximum Level, national monitoring and surveillance program, GEMS/Food database, scientific literature or research activities, and total diet study (TDS). So, at level of European Union, EFSA invites National food authorities, research institutions, academia, food business operators, and other stakeholders to submit data on occurrence by continuous calls for data. The member states should submit the data using the Standard Sample Description (SSD) format to improve the validity of the concentration data with important sampling details (http://www.efsa.europa.eu/en/data/call/datex101217.htm).

Whatever the data source, it is important to have detailed information on the survey type or design, sampling procedures, sample treatment, analytical method performances, protocols for non-detects or not-quantified results, and uncertainty.

Sampling and Sample Treatment

The sampling procedure and how it is carried out are critical to achieve valid results. Different strategies exist, depending on the objectives of the studies, also recognizing the cost implication of sampling techniques and balancing them against additional information obtained for use in risk assessment. The first sampling approach, called objective sampling, corresponds to random sampling aiming at obtaining a real representation of the levels of chemicals in food, so sample numbers of different varieties or brands can be stratified according to production, consumption, and market share. This sampling strategy is applied when the concentration data are not foreseeable. The second strategy, direct or target or suspect sampling, is aimed at sampling products expected to contain a higher concentration of chemicals: for example, focusing sampling on suspected areas or seasons. In the last case, attention should be paid to data selection or assumptions regarding completeness of the data used in exposure assessment. Many European measurement programs for mycotoxins are based on a mixture of both types of sampling to obtain a representative scenario for contaminants in foodstuffs. For mycotoxins, another important aspect of sampling and sample treatment is the homogeneity of materials, which has a large influence on analysis results. Special care should also be taken to ensure that the analyzed sample size is representative and sufficient for accurate and reproducible determination of the mycotoxin average concentration of the sample.[11] Two main approaches, analysis of food group composites and analysis of individual foods, can be applied to analyzing foods when generating analytical data from survey, including TDS, and both can be significant for the estimated dietary exposure.

Analytical Method Performances

Obtaining best estimates for exposure is critically dependent on the quality of concentration data. Concentration data should be obtained using validated methods fit to the purpose of the assessment. To improve exposure results, it is important to provide information on the accuracy of the method, laboratory-to-laboratory variation, the precision of the method, the level of detection and quantification, and the uncertainty of measurement.

Left-Censored Data

A critical point in dietary exposure assessment comes from data below the limit of detection (LoD) or quantification (LoQ), also known as left-censored data, which are determined by performing the analytical method. Non-detects have a wide impact in the final exposure assessment when the percentage of left-censored data is high for a deterministic approach as well as a probabilistic one. Whereas the most common methods to manage data below LoD and LoQ are based on their substitution by LoD, LoD/2, LoD/$\sqrt{2}$, this approach can lead to inaccurate and irreproducible estimates, even the simplest ones, such as mean and standard deviation.[12]

To handle left-censored data, different approaches are reported.[13–15]

Food Consumption Data

Food consumption data reflect what individuals or groups consume in terms of food, including beverages and drinking water and dietary supplements. These data can be estimated through food consumption surveys at an individual or household level, or using food supply data or a collection of duplicate diet. A food consumption survey includes records or diaries, food-frequency questionnaires (FFQs), dietary recall, and total diet studies.

Consumption data should include information about factors that may influence assessment exposure. Such factors include demographic characteristics of the population sampled, bw, the geographic region, and the period when data are collected. For exposure studies of a particular contaminant as mycotoxins, it is also important to consider food consumption patterns for sensitive subpopulations (e.g., children, people with celiac disease).

Data reporting is a crucial topic because consumption data should be in a format that allows them to be matched with concentration data. A common coding system to classify foods is also crucial when collecting food consumption data, to exchange food composition data efficiently and facilitate epidemiological surveys.[16,17] Another consideration is that food may be consumed as such or as an ingredient as part of a recipe for mixed foods, so the use of standard recipes and the attribution of ingredients to individual foods may introduce some uncertainty into consumption data.[18]

Food Supply Data

This is a population-based method because its result estimates the average value for the entire population or at the per capita level. Food supply data provide gross annual estimates of the national availability of food commodity rather than food consumption. Despite this limitation, these studies may be useful in the first step of risk assessment regarding mycotoxins that are mainly evaluated in raw or semi-processed commodities.

A known example of a population-based method is GEMS/Food consumption cluster diets, developed by the WHO based on selected FAO balance sheets. They represent average per capita food consumption and are used to assess dietary exposure by the Joint FAO/WHO Meeting on Pesticide Residues and the Joint FAO/WHO Expert Committee on Food Additives (http://www.who.int/foodsafety/chem/gems/en/index1.html).

Household Survey

Food availability or consumption at the household level may be estimated by a budget survey and a consumption survey, by collecting data on foodstuffs purchased by a household and making a follow-up of consumed food or changes in food stocks. The data are collected by record keeping, interviews, or both. Such data are useful for comparing food availability among different geographic areas and socioeconomic groups and for tracking dietary changes in the total population.

Individual Dietary Survey

Data collected by individual-based methods provide detailed information on food consumption patterns. To obtain these data, several methods can be used: with the food record, or food diary, the subject reports all foods consumed during a specified period; recall methods reflect past consumption, varying from intake over the previous day (24-h recall) to usual food intake; the dietary history method or food-frequency method can also be used. Overall, the amount of consumed food should be measured as accurately as possible.

Total Diet Study

The essential principles of a TDS are the selection of food based on food consumption data to represent a large portion of a typical diet, their preparation as food to consumed, and subsequent pooling of related foods before analysis.[19] In practice, all selected food items that are part of the average diet are purchased, prepared according to standard household procedures, and aggregated into food groups. Each food group is analyzed for the number of additives, contaminants, and nutrients. The use of pooled samples of individual food items leads to background contamination levels in the general food supply; therefore, TDS results are best suited for calculating chronic exposure to food chemical substances, such as mycotoxins, whereas monitoring and surveillance activities can capture more highly contaminated individual food items. Moreover, with the TDS approach many of the conservative assumptions contained in other dietary exposure assessment methodologies are absent. A TDS most accurately represents the levels of the compounds in the edible portion of food at the point of consumption, and takes into account loss during processing, food preparation, and storage.

A TDS performed on a limited number of broadly pooled food samples is sometimes used for screening purposes. This might be useful as a starting point toward setting future priorities for more detailed collections of data. On the other hand, TDS for a refined dietary exposure assessment requires the analysis of a large number of less pooled samples; the choice of the foods to be included in the different pools should be also made, taking into account the different diet habits related to different seasons and/or regions.

Probabilistic Approach

The probabilistic approach can be considered a refined method aimed at assessing actual exposure.[20] Conversely to the point estimate approach, the probabilistic approach considers all available data, and to describe the parameters that contribute to the risk, probability distributions are involved. As a result, a distribution of risk is produced that characterizes the range of risk that might be experienced by an individual or a population. With this perspective, probabilistic methods in exposure assessment are essential to represent the complexity of real-life situations. Use of this method increases the complexity of calculation and time consumption compared with the point estimate approach but it allows the components of variance (variability and uncertainty) to be included in the assessment. Variability is the result of a natural random process and expresses diversity among units in a population or group. It was pointed out that each type of food production and processing, as well as mycotoxin arrangement and human host response, is highly variable. On the other hand, uncertainty concerns the lack of knowledge about a phenomenon and parameters.[21] Hence, the reliability of results from a probabilistic analysis depends on the quality and nature of model inputs (i.e., empirical or parametric approach, as described below) and on the model-fitting data when a parametric approach is selected.[22]

Two probabilistic approaches to model exposure assessment can be considered: simple distributions or pure probabilistic models. Simple distributions employ distributions of food intake but use a fixed value for the concentration variable. Pure probabilistic models weight each possible scenario by the probability of its occurrence. The structure of a pure probabilistic model may be similar to a deterministic one that links the variables together, except that each uncertain variable is represented by a distribution function instead of a single value. This approach ensures that any variability (true heterogeneity) and uncertainty (lack of knowledge) in variables are reflected in the model output.[23] Here, we will deal with pure probabilistic models, referring to them as probabilistic models. Moreover, the use of SA is recommended as an essential element for the purpose of improving estimates and identifying model uncertainties.[23,24]

Food Consumption and Mycotoxin Concentration Distributions

Probabilistic distributions of food consumption and analyte concentration can be entered into probabilistic models using an empirical or parametric approach. The empirical approach is based on histograms from collected data and represents a mathematical description of their shape; the parametric approach fits theoretical distribution as normal, gamma, lognormal, and others, which determines the shape and range of the mathematical function of the parametric distribution.[25] Nevertheless, the authors stress complexity in the choice of the distribution that better describes food consumption, because little guidance is available on the proper distribution and no agreement has yet been reached.[22]

Limitations to using empirical distribution are, first, that they require access to the raw food consumption data; in addition, they are strongly restricted to the range of observed data.[22] Nonparametric methods, in fact, produce less reliable estimates, especially for high quantiles of exposure (95th or 99th). This problem arises because histograms of food consumption data usually show several totally empty regions.[26] Conversely, a key advantage derived from using parametric distributions is that an exposure assessment on the tail of a distribution is more reliable, and estimates beyond the range of observed data are provided. This aspect is extremely important because exposure assessment tends to focus on distribution tails. On the other hand, some caution with stochastic distributions is required because theoretical distributions do not always reflect the admissible range of real intake, but are characterized by an extended tail to the right.[22]

How to Select From Distributions: Monte Carlo Approach

Whatever the nature of the probabilistic model (empirical or parametric), to define the exposure assessment for mycotoxins, a combination of the distribution of consumption and contamination data is required. With this perspective, a point from the first distribution is combined with a point from the second. A well-known numerical technique that allows selection and combination is the Monte Carlo approach. Accordingly to the literature, the Monte Carlo approach is based on random selection of a single point estimate value from each distribution assigned for each input parameter (consumption and contamination intake). "The randomly selected single values are used to calculate the expression defined by the exposure assessment model and stored. This sequence is repeated several 1000 times (iteration) taking, at each iteration, a different set of values for the input selected. Values that are more likely to occur, according to the defined probability distribution, are selected more frequently. The result of the analysis is a frequency distribution for the output of interest."[21]

Again, a distinction must be made between empirical and parametric distribution. Indeed, whereas the latter depends on random samplings from that distribution function that are better fitted to consumption and contamination data, empirical methods select and join consumption and contamination only via the raw data.[26]

Critical Points

Probabilistic methods are preferred compared with deterministic methods when feasible, because fitting food consumption and contamination data by modeling probabilistic distributions leads to more reliable estimates and allows for exposure assessment beyond raw data. However, the use of probabilistic methods requires caution. Attention should be paid to using a lognormal distribution to model food consumption data. Results from goodness-of-fit tests are usually accompanied by lognormal probability plots (P–P plots) for food consumption. Comparing modeled food intake estimates with observed intakes, deviation of the latter from the former, in particular at the tails, indicates a poorer fit of lognormal distribution to food consumption data.[22] Moreover, to select the best parametric distributions, commonly goodness-of-fit tests such as Anderson–Darling test (AD) are performed. Nevertheless, AD test results are strongly influenced by sample size. Thus, additional caution must be used as parametric distributions are fitted.[22]

Confirmatory Methods

Duplicate Diet Study

The DDS is a method for estimating dietary intakes that involves collecting and analyzing identical portions of food and beverages consumed by an individual.[27] DDS provides information on individual intakes, which makes this method particularly useful for exposure assessment of a subgroup of population such as vegetarian (different diet habits), children (different bw), and celiac patients (special diet). The DDS provides accurate intake data that are not subject to errors resulting from food processing, including the lack of food composition data. On the other hand, the method is expensive and imposes a high individual burden, which makes the DDS unsuitable for large-scale studies.

Biomarkers in Exposure Assessment

Specific molecular biological markers, biomarkers, may be used to measure the extent of exposure to a toxic substance,[28] provided the prerequisite regarding the relationship between the biomarker and the health effect is met.[29]

The biomarker methodology has begun to be used in occupational exposure and the employment of biomarkers as a tool of molecular epidemiology for exposure assessment for a food contaminant began over a number of years with success whenever its measurement has given evidence of the consumption of contaminated food with a good correlation to intakes.[9]

When data on toxicity and exposure estimation of a particular biomarker are available, biomarkers may help interpret the significance and relationship to adverse health effects. Biomarkers may be measured in body fluids or tissues as the molecule itself or as a metabolite; to be functional, a biomarker should correlate with dietary intake, be available with suitable persistence (have a half-life of days), and be measurable in the specimen as a parent compound with suitable specificity and precision.

The biotransformation pathway may differ depending on the contaminant; the biomarker measured may be found in urine, blood, breast milk, or, more rarely, in other fluids or tissues such as sputum and feces or biopsies tissues of lung, liver, or brain.

Biomarkers are divided into three categories: (1) biomarkers of exposure, which indicate that exposure to a particular contaminant has taken place to a certain extent; (2) biomarkers of effect, which indicate the biological response of the exposure; and (3) biomarkers of susceptibility, which act as indicators of the intrinsic sensitivity of individuals to the toxic agent.[28,30]

The measurement of biomarkers of exposure represents an exceptional method to directly confirm an exposure event or substantiate the relevance or applicability of results derived from probabilistic studies.[9] However, biomarkers represent a measure of overall exposure and do not discriminate among different sources (i.e., food or air).[29] If in some cases this represents a limit, it is not the case for some chemical contaminants such as mycotoxins, whose exposure mainly results from food ingestion. Even if they do not represent an alternative method to classic estimates of exposure based on food consumption and concentration levels, biomarkers have indubitable advantages for measuring exposure over time, estimating exposure directly (not relying on models and uncertainty assumptions), and assessing individual estimates (especially for specific subpopulations such as vegetarians and celiac patients). Moreover, biomarker measurements are not subject to drawbacks regarding the heterogeneity of food contamination or to variability in food processing and cooking,[31] both of which are critical in measuring mycotoxin contamination in food. To be reliable and provide an accurate estimate of the contaminant to which an exposure exists, biomarkers must be validated. The validation process is complex and implies a parallel experimental design of animal studies to confirm the biomarker–exposure (or disease) link and human studies to confirm the connection between the marker and the exposure.[28] The best choice to validate a biomarker carefully is to follow it over time in a long-term prospective epidemiological study, to definitively account for events and levels of exposure. On the other hand, the use and analyses of biomarkers have important limitations; in some cases (e.g., children) invasive collection of blood may be problematic regarding availability and collection (involving medical personnel); the cost of the analysis may be high because of the technique used (high sensitivities are required);and analytical methods may be complex and or standard reference material may be absent.[9,32]

Exposure to mycotoxins through the diet poses a threat to human health and is a concern for public health, especially because it mainly results from cereals that are unavoidably contaminated by *Aspergillus* and or *Fusarium* toxigenic fungi.

The first biomarker of exposure that was validated and used for mycotoxin risk assessment was the aflatoxin (AF) albumin adduct in serum (more specifically, the AF-lysine adduct) and AF-N7-guanine, AFB1-mercapturic acid, and AFM1 in urine.[33,34] During observational studies performed in 1968–1985, it was possible to elucidate the correlation between AFB1 ingestion and the incidence of hepatocellular carcinoma in the human population thanks to identification of the AFB1-DNA adduct and its analytical measurement.[35]

Several studies showed good correlations of the biomarker with food intake. Zhu et al.[34] found a correlation coefficient of 0.82 between AF intake and urinary AFM1; Groopman et al.[36,37] found a correlation coefficient of 0.82 and 0.80 between AF intake and urinary AF-N7-guanine in a study located in Gambia and China, respectively.

Table 1 reports more recent studies on the analysis of biomarkers of exposure in humans. Because of the long half-life of albumin in humans, measurement of AF albumin adduct in sera is strongly preferred with respect to urine, indicating an exposure extent over a period of months.[38,39] Conversely, urinary metabolites and adducts, which have a shorter half-life, reflect the intake of the previous day.[28,38]

Because of the extensive occurrence of ochratoxin A (OTA), which is found in a wide range of foods (i.e., cereals, dried vine fruit, wine, coffee, and licorice), molecular epidemiology studies are particularly important and helpful to biomonitor the extent of exposure. As a biomarker, OTA has been measured in serum, urine, and milk in several studies worldwide (Table 1), evidencing

TABLE 1 Recent Studies on Analyses of Biomarkers of Exposure in Human Specimens

Biomarker	Country	Specimen	Population Group	Levels	References
AFM1	Brazil	Urine		AFM1: 1.8–3.99 pg/ml	56
AFB1-N7-guanine AFM1 AFQ1	China	Urine		AFB1-N7-G: 0.38 (median) ng/ml AFM1: 0.04 ng/ml AFQ1: 10.4 ng/ml	57
AFB1-lysine adduct	Ghana	Serum	Pregnant women	AFB1-L: 0.44–0.269 pg/mg albumin	58
AFM1 AFB1	Guinea	Urine	Children (1–2.5 years old)	AFB1: 189 pg/ml (average) AFM1: 5.0–6.2 pg/ml	59
Aflatoxin-albumin	Gambia	Serum	Children	AF-alb: 24 pg/mg (average)	60
AFB1	Portugal	Serum	18 poultry workers	AFB1:1–4.23 ng/ml	61
AFB1 AFM1 Aflatoxicol	Gambia	Serum		AFB1: 0.2–74 ng/ml AFM1: 0.03–6.8 ng/ml Aflatoxicol: 0.01–3.2 ng/ml	62
AFM1	Turkey	Breast milk	Women	AFM1: 60.10–299.99 ng/l	63
OTA	Czech Republic	Serum	Adults	OTA: 0.28 ng/ml (average)	64
OTA	Sierra Leone Njala	Serum	Children (5–14 years old)	OTA: 2.4 ng/ml (average)	62
OTA	Portugal	Serum	Healthy population	OTA: 0.96–2.49 ng/ml	65
OTA	Tunisia	Serum	Non-nephropathic individuals	OTA: 0.1 ng/ml (LoD) 0.2 ng/ml (LoQ)	66

OTA	Spain	Plasma	Blood donors	OTA: 0.075 ng/ml (LoD) 0.23 ng/ml (LoQ)	67
OTA	Turkey	Serum	Healthy population	OTA: 0.887–3.43 ng/ml	68
OTA	UK	Urine	Volunteers	OTA: 0.01 ng/ml (LoD)	40
OTA	Italy	Urine	Healthy	OTA: 0.012–0.046 ng/ml	69
OTA	Hungary	Urine	Healthy	OTA: 0.006–0.065 ng/ml	70
OTA	Portugal, Coibra	Urine	Healthy	OTA: 0.019 ng/ml (average)	71
OTA	Germany	Urine	Non-specified	OTA: 0.02 ng/ml (LoD)	72
OTA	Sweden	Breast milk	Women	OTA: 0.01 ng/ml (LoD) 0.04 ng/ml (LoQ)	73
OTA	Italy	Breast milk	Women	OTA: 0.1 ng/ml (LoD)	74
OTA	Italy, Lombardy	Breast milk	Women	OTA: 0.0005 ng/ml (LoD)	75
OTA	Brazil, S. Paulo	Breast milk	Women	OTA: 0.01 ng/ml (LoQ)	76
OTA	Slovakia	Breast milk	Women	OTA: 0.0048 ng/ml (LoD) 0.0144 ng/ml (LoQ)	77
OTA	Turkey, Ankara	Breast milk	Women	OTA: 0.62–13 ng/ml	63
DON	China	Urine	Adults	DON: 4–94 ng/ml	78
DON	France	Urine	Male farmers	DON: 7 ng/ml creatinine (average)	45
DON	Swedish	Urine	Adults	DON: 11 ng/ml creatinine (average)	79
DON	UK	Urine	Adults	DON: 1–61 ng/ml	78

Continued

TABLE 1 Recent Studies on Analyses of Biomarkers of Exposure in Human Specimens—cont'd

Biomarker	Country	Specimen	Population Group	Levels	References
FB1	Centane, South Africa	Urine	Adults	FB1: 225 pg/ml (average)	46
FB1	China, Huaian	Urine	Adults	FB1: 13630 pg/ml creatinine (average)	79
FB1	China, Fusui	Urine	Adults	FB1: 720 pg/ml creatinine (average)	79
FB1	Mexico	Urine	Adults	FB1: 147 pg/ml (average)	80
AFM1, FB1, FB2, OTA, DON, DON-15-GlcA, DON-3-GlcA, ZEA, ZEA-14-GlcA, αZOL	Cameroon		Adults	AFM1: 0.17–1.38 ng/ml FB1: 1.7–14.8 ng/ml FB2: <1.7 ng/ml OTA: 0.17–1. 87 ng/ml DON: <1–3 ng/ml DON-15-GlcA: 11.0–96.2 ng/ml DON-3-GlcA: 20.0–22.5 ng/ml ZEA: 1.3–1.4 ng/ml ZEA-14-GlcA: 3.38–31.2 ng/ml αZOL: < 1.7 ng/ml NIV: 10.0–22.0 ng/ml	81

AFM1, DON, DON-15-Ogluc, FB1, FB2, OTA, ZEA, ZEA-14-Ogluc	Nigeria	Urine	Children, adolescent, adults	AFM1: 0.3–1.5 ng/ml DON: 2.0 DON-15-Ogluc: 3.5–8.0 ng/ml FB1: 4.6–12.8 ng/ml FB2: 1.0 ng/ml OTA: 0.2–0.6 ng/ml ZEA: 0.3 ng/ml ZEA-14-Ogluc: 9.5–44.5 ng/ml	82
AFM1, OTA, DON, DOM1, FB1, FB2, αZOL, βZOL	Italy	Urine	Pilot study (n=10)	OTA: 0.02–0.25 ng/ml DON: 1.1–14.0 ng/ml	48
AFM1, OTA, FB1, FB2, OTα	Korea	Urine	Pilot study (n=20)	OTA: 0.012–0.093 ng/ml AFM1: 0.009 ng/ml	83
ZEA, αZOL, βZOL, OTA, DON, FB1	South Africa	Urine	Adults female	ZEA: 0.529 ng/mg creatinine (average); FB1: 1.52 ng/mg creatinine (average); αZOL: 0.614 ng/mg creatinine (average); βZOL: 0.702 ng/mg creatinine (average); OTA: 0.041 ng/mg creatinine (average); DON: 11.3 ng/mg creatinine (average);	84
ZEA, αZOL, βZOL, OTA, DON, FB1, AFM1	Southern Italy	Urine	Adults	ZEA: 0.056 ng/ml (average); αZOL:0.074 ng/ml (average); βZOL:0.088 ng/ml (average); OTA: 0.061 ng/ml (average); DON: 10.32 ng/ml (average); FB1:0.029 ng/ml (average); AFM1:0.10 ng/ml (average)	85

a large geographical variation. Klassen's equation and its modifications[40,41] are currently used to obtain estimated exposure values to OTA starting from plasma concentrations (nanograms per milliliter from serum analyses), plasma clearance (i.e., renal filtration rate), and OTA bioavailability (for most animals, 50%). Comparative studies failed to correlate blood levels and dietary intake,[42] probably because of the lack of information on toxicokinetic mechanisms and the fact that the long half-life and continuous OTA exposure result in a steady-state concentration.[43] Hence, the use of OTA in blood as a validated biomarker has been replaced by measurement of urinary OTA, whose content is lower but constitutes a promising alternative, especially when further developments regarding the relationship between OTA intake and the urinary biomarker will be defined. OTA levels have been extensively measured in breast milk to assess the exposure of a particular subgroup (lactating women) and potential exposure of the offspring. To date, the pattern of OTA distribution in milk is unclear and few studies have attempted to correlate dietary intake with OTA concentration in milk; nevertheless, because newborns exclusively breast feed during the first few weeks, the biomonitoring of OTA in human milk is an assessment of risk for infants.

Deoxynivalenol (DON) contamination is associated with wheat and maize; thus exposure is linked to cereal-based diets. Using biomarker monitoring, a dietary intervention study demonstrated that by avoiding wheat, subjects' urinary levels of DON were significantly reduced.[44] Deoxynivalenol and two metabolites, DON glucuronide and the de-epoxy-deoxynivalenol (DOM1), may be present in the urine of exposed individuals. A number of studies have been formed to monitor the three biomarkers, especially in the United Kingdom (UK), northern France, Sweden, and China, whose climatic conditions favor the proliferation of *Fusarium* species responsible for DON production (Table 1). Turner et al.[45] developed a highly sensitive and robust analytical method to measure free DON and DOM1, including DON-glucuronide after an enzyme treatment. To date, only total DON (the sum of free DON and DON-glucuronide) has been reported in literature and samples positive for DOM1 have been found only in France,[45] and not in the UK or China.

Biomarkers of fumonisins have been sought for a long time to define putative biomarkers. because of the limited metabolism of fumonisins, measurement of free urinary fumonisins (B1 and B2) seems to be the best choice to biomonitor exposure. Van der Westhuizen et al.[46] presented a study in which the relationship between urinary FB1 and FB1 ingestion gave good results and confirmed the potential use of free FB1 as a validated biomarker.

The sphinganine (Sa)/sphingosine (So) ratio has also been proposed as putative biomarker, on the basis that FB1 potently inhibits the enzyme ceramide synthase, which catalyses acylation of Sa and reacylation of So. This inhibition accumulates intracellular Sa, altering the Sa/So ratio. In a large range of animal species, the changed ratio occurs in a dose-dependent[47] manner, but the same ratio in humans was not related to fumonisin exposure.

Simultaneous determination of more than one mycotoxin in human biological fluids by measuring with the multi-biomarker method is a new challenge. To this purpose, urine seems to be the preferred specimen because it may contain biomarkers from different mycotoxins and is easier to collect and handle than serum.[48] Until now, only a small number of scientific works have been published, mainly reporting validation of the analytical method and applying it to a pilot survey with a small number of human urine samples.

In conclusion, biomarkers are a tool with promise for use in monitoring the most important mycotoxins. Multi-mycotoxin analyses will help establish the relevance of occurrences. Nevertheless, to better exploit the biomarker methodology in exposure assessment, more work is needed to validate each biomarker strongly and link it to food intake.

EXPOSURE ASSESSMENT OF EUROPEAN POPULATION

Aflatoxins

Aflatoxins are among the most important genotoxic carcinogens of food for which there is no fixed tolerable daily intake; therefore, it is appropriate to limit the content under the principle of as much as reasonable achievable. Aflatoxins occur in grains and foodstuff (e.g., corn, peanuts) and contamination values may vary from less than 1 μg/kg to more than 1 mg/kg.

Following EFSA opinion,[49] the risk characterization approach is based on assessing the margin of exposure that derives from the benchmark dose and exposure estimates ratio. As shown in Table 2, in Europe, values of exposure for adults are generally under 1 ng/kg bw/day with some exceptions for the worst scenarios (upper bound values or special population). Special attention deserves to be paid to exposure rates of children; in fact, they are always higher compared with adults because despite almost equal consumption values, their bw is considerable lower.

However, during the past 2 to 3 years, unfavorable climate conditions with higher temperatures and humidity caused the proliferation of *Aspergillus* ssp., provoking the persistent presence of AF, especially in the southern EU countries. The combined action of official control plans and prospective epidemiological studies should be put in place to monitor and take the necessary actions for risk assessment.

Aflatoxin M1 is a metabolite of AFB1 that is found in the milk of lactating animals such as dairy cows and derived milk products as a result of carryover from contaminated feed to edible animal origin products.

As for AFs, the exposure values of children merit special attention because they account for both higher consumption values and lower bw compared with adults. Table 3 lists exposure values for some EU countries. The highest values are all around 0.1 μg/kg bw/day, which is to be considered, especially, for lactovegetarian and infants exposures, a worryingly high value compared

TABLE 2 Exposure Assessment of European Population to Aflatoxins

Country	Consumption Data	Survey Methodology	Population Group	Exposure LB–UB (ng/kg bw/day)	References
France	National survey	7-day dietary record	Children	0.323–0.888[a]	50
			Adults	0.117–0.345[a]	
		5-day dietary record	Ovolactovegetarians	0.6–1.6[a]	
			Lactovegetarians	0.4–0.9[a]	
			Vegans/macrobiotics	0.9–2.1	
France	National survey	7-day dietary record	Children	0.001–0.39 (AFB1)	86
			Adults	0.002–0.22 (AFB1)	
Greece		Assumption	Children	0.04–10.75	87
			Adolescents	0.03–6.6	
			Adults	0.02–3.07	
Spain	Catalonian survey	FFQ	Children	0.106–0.337	88
			Adults male	0.135–0.532[a]	
			Adults female	0.078–0.299[a]	
			Infants[b]	0.17–37.47 0.12–29.6 (AFB1)	89
			Infants[c]	0.08–0.94 0.01–0.62 (AFB1)	89
Sweden	National survey	Dietary questionnaire	Adults	0.760–2.100	90
The Netherlands	National survey	Duplicate diet	Children	0.02–0.44	91

[a]*Mean value 95th percentile.*
[b]*Ecological infants' cereals.*
[c]*Conventional infants' cereals.*

TABLE 3 Exposure Assessment of European Population to Aflatoxin M1

Country	Consumption Data	Survey Methodology	Population Group	Exposure LB–UB (μg/kg bw/day)	References
France	National survey	7-day dietary record	Children	0.054[a]	86
			Adults	0.03[a]	
		5-day dietary record	Ovolactovegetarians	0.1–0.2[b]	50
			Lactovegetarians	0.1–0.3[b]	
Ireland	National survey	7-day estimated food records	Adults male	8.62–212.6[b,c]	92
		7-day weighting food records	Adults female	9.3–238[b,c]	
Portugal			Adults	0.08	93
Spain			Infants, starter formula	0.56–2.65[d]	94
			Infants, follow-up formula	0.67–3.35[d]	
			Infants, toddler formula	1.01–2.81[d]	

[a]*Mean value.*
[b]*Mean value 95th percentile.*
[c]*pg/kg bw/day.*
[d]*ng/kg bw/week.*

with 0.14 ng/kg bw/day corresponding to the estimated daily intake of an adult exposed to a contaminated milk at maximum level (0.05 μg/kg) with a consumption of 200 g/day milk.

Ochratoxin A

Because the tolerable OTA intake set by EFSA[43] is expressed as weekly intake, the value was divided by 7 with the aim of comparing the tolerable value with assessed exposures reported in Table 4. Whatever the hypothesis (LB or UB), none of the exposure scenarios exceeded the tolerable intake. However, children always account for the highest exposure because of the strong influence of bw on exposure calculation with respect to consumption value.

Fumonisins

Estimation of dietary intake has been also focused on vulnerable groups. Particular attention was devoted to children, vegetarians, and subgroups of population, with a warning about possible higher exposure owing to an unfavorable bw–intake ratio, dietary habits, or pathologies. Table 5 reports the results of some articles assessing the exposure of vegetarians and celiac patients. According to the First French Total Diet Study,[50] vegetarians have higher exposure values than other adults, but exposure assessed for infants and children are even higher when comparing both with non-vegetarian adults and vegetarian adults. Regarding the exposure assessment calculated for celiac patients, results reported in the Spanish article[15] are the lowest compared with exposure values reported for other groups of population investigated. Conversely, in the report of Dall'Asta et al.,[51] the data showed significantly higher intake in celiac patients (0.39 μg/kg bw/day) compared with the control group (0.03 μg/kg bw/day).

Zearalenone

Consistent with EFSA Scientific Opinion, chronic dietary exposure to zearalenone based on the data in Table 6 is below or in the region of the TDI for all age groups and does not seem to be considered a health concern. An exception is dietary exposure for vegans and macrobiotic eaters in France,[50] where zearalenone dietary exposure was higher in younger consumers than in adults. Dietary exposure levels decreased with increasing age as a consequence of the higher intake of food per kilogram bw in these age groups.

Trichothecenes (DON, T-2 and HT-2, and NIV)

Table 7 sums up results of the most recent exposure assessments on trichothecenes. In general, for all toxins, exposure values for children were highest for all reported countries, because bw has a strong influence on exposure calculations.

TABLE 4 Exposure Assessment of European Population to Ochratoxin A

Country	Consumption Data	Survey Methodology	Population Group	Exposure LB–UB (ng/kg bw/day)	References
France	National survey	7-day dietary record	Children	4.07–7.77[a]	50
			Adults	2.16–3.63[a]	
France	National survey	7-day dietary record	Children	0.23–2.82	86
			Adults	0.28–1.92	
Greece	Not available		Children	0.03–2.17	87
			Adolescents	0.02–1.74	
			Adults	0.01–1.24	
Italy	National survey	3-day dietary record	Children	1.16–2.42[b]	95
			Adolescents	0.56–1.00[b]	
			Adults	0.28–0.37[b]	
Spain	Catalonian survey	FFQ	Infants	0.28–7.23[d]	96
			Children	0.09–0.98[d]	
			Adolescents	0.14–0.68[d]	
			Adults	0.37–1.31[d]	
Sweden	National survey	Dietary questionnaire	Children	1.4-2.6[a]	90
			Adults	1.2-1.9[d]	
The Netherlands	National survey	Duplicate diet	Children	4.1[c]	91

[a]*Mean value 95th percentile.*
[b]*Cocoa and chocolate, weekly intake, consumers only.*
[c]*Mean value*
[d]*Stochastic approach.*

TABLE 5 Exposure Assessment of European Population to Fumonisins

Country	Consumption Data	Survey Methodology	Population Group	Exposure LB–UB (ng/kg bw/day)	References
France	National survey	7-day dietary record	Children	15–44.6 (FB1) 6.48–30.4 (FB2)	86
			Adults	7.45–29.1 (FB1) 2.44–15.89 (FB2)	
France	National survey	7-day dietary record	Children	46–75[b]	50
			Adults	14–64[b]	
			Ovolactovegetarians	40–100[b]	
			Lactovegetarians	50–120[b]	
			Vegans/macrobiotic	30–90[b]	
Italy	National survey	3-day dietary record	Children	348–582[a]	97
			Adolescents	88–119[a]	
			Adults	55–78[a]	
			Elderly	34–46[a]	
Italy		7-day dietary record	Celiac sufferers	0.39[e,f]	51
			Adults	0.03[e,f]	

Nordic countries[d]	National survey	Not available	Children	400[f]	98
Spain	Catalonian survey	FFQ	Infants	195.19–508.20[c]	15
			Children	85.30–293.66[c]	
			Adolescents	79.25–241.05[c]	
			Adults male	100.41–313.33[c]	
			Adults female	103.02–320.76[c]	
			Celiac patients	33.16–132.39[c]	
The Netherlands	National survey	Duplicate diet	Children	28[e]	91

[a]*Consumers only, gluten-free products.*
[b]*Mean value 95th percentile.*
[c]*Stochastic approach.*
[d]*Denmark, Norway, Sweden, Finland, and Iceland.*
[e]*μg/kg bw/day.*
[f]*Mean value.*

TABLE 6 Exposure Assessment of European Population to Zearalenone

Country	Consumption Data	Survey Methodology	Population Group	Exposure LB–UB (ng/kg bw/day)	References
Austria	National survey	24-h recall	Adults	28–116	99
Belgium	National survey	24-h recall	Adults	37–45	100
Finland	National survey	24-h recall	Adults	27–36	99
France	National survey	7-day dietary record	Children	66–132[a]	50
			Adults	33–70[a]	
			Ovolactovegetarians	50–110[a]	
			Lactovegetarians	60–120[a]	
			Vegans/macrobiotic	200–570[a]	
Italy	National survey	3-day dietary record	Children	22–237[b]	97
			Teenagers	16–173[b]	
			Adults	12–117[b]	
			Elderly	12–111[b]	
Norway	National survey	FFQ	Adults male	8–19	99
			Adults female	7–17	

Portugal	National survey	Food balance sheets	Adults	4	99
Spain	Catalonian survey	FFQ	Infants and toddlers	12.2–17.9	101
			Children	2.3–6.2	
			Adolescents	1.5–2.2	
			Adults	0.9–1.5	
			Elders	0.3–0.5	
The Netherlands	National survey	2-day dietary record	Children	46–50	99
		7-day estimated weight consumed	Infants, toddlers young people	21–55[c]	
		2-day dietary record	Adults	20.7	
		7-day weighted record	Adults female	13[c]	
		4-day diary record	Elderly male	14[c]	
		4-day diary record	Elderly female	12[c]	

[a]*Mean value 95th percentile.*
[b]*Gluten-free products, consumers only.*
[c]*Mean value.*

TABLE 7 Exposure Assessment of European Population to Trichothecenes (DON, T-2 and HT-2 toxins, and NIV)

Country	Consumption Data	Survey Methodology	Population Group	Exposure LB–UB (μg/kg bw/day)	References
Deoxynivalenol					
Belgium	National survey	2-day dietary record	Adults	0.035–0.091	100
France	National survey	7-day dietary record	Children	451–929[a]	50
			Adults	281–57[a]	
		5-day dietary record	Ovolactovegetarians	360–720[a]	
			Lactovegetarians	320–830[a]	
			Vegans/macrobiotics	410–960[a]	
France	National survey	7-day dietary record	Children	544–1029[a]	86
			Adults	373–722[a]	
Ireland	IUNA survey		Adults male	681–13,580[b,c]	92
			Adults female	651-13,834 [b,c]	
Spain	Catalonian survey		Infants	0.90–3.57	102
			Children	0.68–2.49	
			Adolescents	0.43–1.46	
			Adults male	0.37–1.18	
			Adults female	0.56–1.84	
			Elderly	0.280–1.08	
			Ethnics	0.96–3.82	
			Celiac patients	0.15–0.64	

T-2 HT-2 toxins					
Belgium	National survey		Adults	0.017–0.030 0.0258[d,e]	100
France	National survey	7-day dietary record	Children	4.00–38 (T-2)[a] 10.5–53.1 (HT-2)[a] 0.11–1.92 (T-2)[a]	86
			Adults	0.07–1.82 (HT-2)[a]	
Spain	Catalonian survey	FFQ	Children	0.041–0.091[d]	103
			Adolescents	0.029–0.055[d]	
			Adults male	0.018–0.036[d]	
			Adults female	0.034–0.041[d]	
Spain	National survey		Infants	79.29[a,e]	104
			Children	76.54[a,e]	
			Adults	8.64[a,e]	
Nivalenol				(ng/kg bw/day)	
Austria		24-h recall	All population	78–274	99
Denmark		Food record	All population	30–72	99
Finland	National survey	Food record	Adults male	24–51	105
			Adults female	22–44	
France	National survey	7-day dietary record	Children	163–300[c]	50
			Adults	88–157[c]	
			Ovolactovegetarians	120–230[c]	
			Lactovegetarians	120–190	
			Vegans/microbiotics	210–420	

Continued

TABLE 7 Exposure Assessment of European Population to Trichothecenes (DON, T-2 and HT-2 toxins, and NIV)—cont'd

Country	Consumption Data	Survey Methodology	Population Group	Exposure LB–UB (μg/kg bw/day)	References
Norway	National survey	Food record	Infants	46–92	106
			Children	12–63	
			Adolescents	8–16	
			Adults	7–14	
Sweden		Food record	All population	70–140	107
United Kingdom	National survey	Food record	Toddler	64[e]	99
			Children	64[e]	
			Adolescents	34[e]	
			Adults male	25[e]	
			Adults female	17[e]	
			Elderly male	27[e]	
			Elderly female	21[e]	

[a] *ng/kg bw/day.*
[b] *pg/kg bw/day.*
[c] *Mean value 95th percentile.*
[d] *Stochastic approach.*
[e] *Mean value.*

Focusing on the exposure assessment for DON, even if the lower bound values are always lower than the TDI (1 μg/kg bw/day), the upper bound values often exceed this value, especially for children. This scenario suggests that the children's exposure to DON should be kept under observation.

Emerging Mycotoxins

In addition to the major mycotoxins, a considerable number of minor toxins (emerging mycotoxins) are known to be naturally produced by molds. For many of them, information is limited regarding outbreaks of disease or experimental data, and evaluation of their possible negative impact on human or animal health, and data on their occurrence in food and feed.[52] However, the EFSA published a Scientific Opinion for some of them, such as ergot alkaloids (EAs),[53] citrinin,[54] and *Alternaria* toxins.[55]

The Scientific Opinion on EAs was published by the EFSA on 2012.[53] The EFSA based its risk assessment on the main *Claviceps purpurea* EAs, namely ergometrine, ergotamine, ergosine, ergocristine, ergocryptine (which is a mixture of α- and β-isomers), ergocornine, and the corresponding –inine epimers. The highest chronic exposure to EAs was estimated in children for all of the reported countries. However, for the highest exposure value reported for children in Denmark (0.075–0.170 μg/kg bw/day), results were about 3.5 times lower than the TDI value, set by EFSA at 0.6 μg/kg bw/day.[53]

The children's age group showed the highest exposure level to citrinin, as well. The EFSA Scientific Opinion[54] reported an LB and UB exposure range of 9.2–23.3 μg/kg bw/day for Swedish children as the highest value. Although the CONTAM Panel concluded that the derivation of a TDI is not appropriate, the exposure-assessed values are to be considered safe.[54]

Regarding *Alternaria* toxins, a limited amount of data are available for the purpose of assessing exposure. Moreover, regarding existing data on occurrence, samples reported to be lower than LoD/LoQ ranged between 87% and 100% depending on the toxin. However, the EFSA reported exposure scenarios for alternariol, alternariol methyl ether, tenuazonic acid (TeA), and tentoxin. The assessment was made only for an adult population and the highest values were reported for TeA, for which the mean chronic exposure ranged from 36 to 141 ng/kg bw/day; the lowest exposure estimates were reported for tentoxin, ranging from 0.01 to 7 ng/kg bw/day (EFSA, 2011).[55]

CONCLUSION AND FUTURE NEEDS

There is a need to obtain reliable results on exposure assessments. Typical uncertainties derive from errors in the performance of sampling procedures or from using an incomplete or improper consumption database or of non-accredited analytical methods, or for misusing deterministic approaches for exposure assessment studies.

This chapter presented the most recent data available in the literature on exposure assessment to mycotoxins from the European population. For all combinations of matrix/mycotoxin, the most challenging situations were for a broad consumer group such as infants, toddlers, and children owing to the unfavorable ratio intake per body weight.

As concerns children, this aspect assumes even more relevance when it is considered that their dietary habits usually are the same as those of the adults in terms of both amount and type. In fact, it is known that existing legal maximum limits were set only for food for adults, with no consideration for children who usually consume the same foods.

Another highly ignored aspect regarding risk related to exposure assessment is the consideration of a toxicological threshold derived from the potential cumulative presence of more mycotoxins in a food. As is known, the concomitant presence of mycotoxins in a food product can lead to additive or even synergistic effects, and in chronic exposures this issue becomes very relevant for the acquisition of information on possible sources of risk to human health.

More attention should be paid to the performance of TIS in which all sources of contamination are contemporary, taking into account an evaluation of exposure of xenobiotics such as mycotoxins to a human population.

In fact, these toxic compounds can be present in houses where poor environmental conditions are present, or in airborne particulates, especially in risky workplaces such as cereal storage centers and mills, or even in specific products such as organic and whole foods that, in providing a certain benefit for human health, can pose a risk as potentially affected by higher contamination of mycotoxins.

A final consideration is that, regarding the toxic nature of mycotoxins and relevant information on toxicity mechanisms derived from studies on animals, the correlation between human pathologies and the etiological role of mycotoxins is still far from known.

For all of the issues, at the highest level of authorities in food safety, further effort should be encouraged to obtain the information necessary to fill in existing gaps, especially to carry out preventive actions from the prenatal period onward.

REFERENCES

1. European Commission (EC) Health & Consumer Protection Directorate-General (DG SANCO). The future of risk assessment in the european union the second report on the harmonisation of risk assessment procedures. In: *First report on the harmonisation of risk assessment procedures. Part 2: appendices 26–27*. October 2000.
2. European Commission (EC) Health & Consumer Protection Directorate-General (DG SANCO). The future of risk assessment in the european union the second report on the harmonisation of risk assessment procedures. In: *The second report on the harmonisation of risk assessment procedures*. April 10-11, 2003.
3. Food and Agriculture Organization of the United Nations World Health Organization. *Codex alimentarius commission procedural manual*. 10th ed. Rome; 1997.

4. European Commission (EC). Regulation (EC)/178/2002 of the European Parliament and of the Council of 28 January 2002. Laying down the general principles and requirements of food law, establishing the European food safety authority and laying down procedures in matters of food safety. *Official J Eur Communities*. L 31/1–24 of 1.2.2002.
5. Saltelli A, Chan K, Scott M. *Sensitivity analysis*. New York: Wiley; 2000.
6. Kuiper-Goodman T. Food safety: mycotoxins and phycotoxins in perspective. In: Miraglia M, van Egmond HP, Brera C, Gilbert J, editors. *Mycotoxins and phycotoxins – developments in chemistry, toxicology and food safety*. 305 West Magnolia Street, Suite 196 Ft. Collins, Colorado, CO 80521: Alaken, Inc; 1999.
7. World Health Organization (WHO). *Report of the joint FAO/WHO consultation on food consumption and exposure assessment of chemicals*. WHO/FSF/FOS/97.5. Geneva: World Health Organization; 1997.
8. US Environmental Protection Agency. Guidelines for exposure assessment; notice. Part VI. *Fed Regist* 1992;**57**(104):22888–938.
9. Kroes R, Müller D, Lambe J, et al. Assessment of intake from the diet. *Food Chem Toxicol* 2002;**40**:327–85.
10. Marin S, Ramos AJ, Cano-Sancho G, et al. Mycotoxins: occurrence, toxicology, and exposure assessment. *Food Chem Toxicol* 2013;**60**:218–37.
11. FAO/WHO. *Food and agriculture organization of the United Nations*. Rome: CAC/GL40-Rev.1-2003 Guidelines on GOOD Laboratory Practice in pesticide residue analysis; 2003. [Codex Alimentarius Commission].
12. Helsel DR. Fabricating data: how substituting values for non-detects can ruin results and what can be done about it. *Chemosphere* 2006;**65**:2434–9.
13. European Food Safety Authority (EFSA). Management of left-censored data in dietary exposure assessment of chemical substances. *EFSA J* 2010;**8**(3):1557.
14. Helsel DR. *Non-detects and data analysis: statistics for censored environmental data*. New York: John Wiley and Sons; 2005.
15. Cano-Sancho G, Ramos AJ, Marìn S, et al. Occurrence of fumonisins in Catalonia (Spain) and an exposure assessment of specific population groups. *Food Addit Contam* 2011b;**29**(5): 799–808.
16. Ireland J, van Erp-Baart AMJ, Charrondière UR, et al. Selection of a food classification system and a food composition database for future food consumption surveys. *Eur J Clin Nutr* 2002;**56**(2):S34–45.
17. Pakkala H, Christensen T, Presser K. Towards harmonized data interchange in food consumption data. *Comp Stand Interf* 2014;**36**:592–7.
18. Reinivuo H, Bell S, Ovaskainen ML. Harmonisation of recipe calculation procedures in European food composition databases. *J Food Compos Anal* 2009;**22**(5):410–3.
19. European Food Safety Authority (EFSA). Towards a harmonized total diet study approach: a guidance document. *EFSA J* 2011;**9**:2450.
20. European Food Safety Authority (EFSA). Overview of the procedures currently used at EFSA for the assessment of dietary exposure to different chemical substances. *EFSA J* 2011;**9**(12):2490.
21. Lammerding AM, Fazil A. Hazard identification and exposure assessment for microbial food safety risk assessment. *Int J Food Microbiol* 2000;**58**:147–57.
22. Gilsenan MB, Lambe J, Gibney MJ. Assessment of food intake input distributions for use in probabilistic exposure assessments of food additives. *Food Addit Contam* 2003;**20**(11):1023–33.
23. Lambe J. The use of food consumption data in assessments of exposure to food chemicals including the application of probabilistic modelling. *P Nutr Soc* 2002;**61**:11–8.

24. Delgado JE, Wolt JD. Fumonisin B_1 and implications in nursery swine productivity: a quantitative exposure assessment. *J Anim Sci* 2010;**88**:3767–77.
25. Vose D. *Risk Analysis: a quantitative guide*. 2nd ed. Chichester, West Sussex: John Wiley and Sons Ltd; 2000.
26. Gauchi JP, Leblanc JC. Quantitative assessment of exposure to the mycotoxin ochratoxin A in food. *Risk Anal* 2002;**22**:219–34.
27. FAO/WHO. *IPCS. Principles and methods for the risk assessment of chemicals in food. Environ health criteria*. Geneva: WHO; 2009. p. 240.
28. Groopman JD, Wogan GN, Roebuck BD, et al. Molecular biomarkers for aflatoxins and their application to human cancer prevention. *Cancer Res* 1994;**54**:1907s–11s.
29. Aitio A, Kallio A. Exposure and effect monitoring: a critical appraisal of their practical application. *Toxicol Lett* 1999;**108**:137–47.
30. Timbrell JA. Biomarkers in toxicology. *Toxicology* 1998;**129**:1–12.
31. Wild CP, Gong YY. Mycotoxins and human disease: a largely ignored global health issue. *Carcinogenesis* 2010;**31**(1):71–82.
32. Van den Brandt P, Voorrips L, Hertz-Picciotto I, et al. The contribution of epidemiology. *Food Chem Toxicol* 2002;**40**:387–424.
33. Gan LS, Skipper PL, Peng XC, et al. Serum albumin adducts in the molecular epidemiology of aflatoxin carcinogenesis: correlation with aflatoxin B1 intake and urinary excretion of aflatoxin M1. *Carcinogenesis* 1988;**9**:1323–5.
34. Zhu JQ, Zhang LS, Hu X, et al. Correlation of dietary aflatoxin B1 levels with excretion of aflatoxin M1 in human urine. *Cancer Res* 1987;**47**:1848–52.
35. Carvajal M. Aflatoxin-DNA adducts as biomarkers cancer: nature, formation, kinds of AF-DNA adducts, methodology, effects, and control. In: Siantar DP, Trucksess MW, Scott PM, Herman EM, editors. *Food contaminants mycotoxins and food allergens*; 2008. p. 13–5. 1001 Chapter 2.
36. Groopman JD, Hall A, Whittle H, et al. Molecular dosimetry of aflatoxin–N7-guanine in human urine obtained in the Gambia, West Africa. *Cancer Epidemiol Biomarkers Prev* 1992;**1**:221–8.
37. Groopman JD, Zhu JQ, Donahue PR, et al. Molecular dosimetry of urinary aflatoxin–DNA adducts in people living in Guangxi autonomous region, people's Republic of China. *Cancer Res* 1992;**52**:45–52.
38. Leong YH, Latiff AA, Ahmad NI, et al. Exposure measurement of aflatoxins and aflatoxin metabolites in human body fluids. A short review. *Mycotoxin Res* 2012;**28**:79–87.
39. Sabbioni G, Skipper P, Buchi G, et al. Isolation and characterization of the major serum albumin adduct formed by aflatoxin B1. *Carcinogenesis* 1987;**8**:819–24.
40. Gilbert J, Brereton P, MacDonald S. Assessment of dietary exposure to ochratoxin A in the UK using a duplicate diet approach and analysis of urine and plasma samples. *Food Addit Contam* 2001;**18**:1088–93.
41. Scott PM. Biomarkers of human exposure to ochratoxin A. *Food Addit Contam* 2005;**21**(Supp 1): 99–107.
42. Duarte SC, Pena A, Lino CM. Human ochratoxin A biomarkers from exposure to effect. *Crit Rev Toxicol* 2011;**41**(3):187–212.
43. European Food Safety Authority (EFSA). Opinion of the scientific panel on contaminants in the food chain on a request from the commission related to ochratoxin A in food. *EFSA J* 2006;**365**:1–56.
44. Turner PC, Burley VJ, Rothwell JA, et al. Dietary wheat reduction decreases the level of urinary deoxynivalenol in UK adults. *J Exp Sci Environ Epidemiol* 2008;**18**:392–9.
45. Turner PC, Hopton RP, Lecluse Y, et al. Determinants of urinary deoxynivalenol and de-epoxy deoxynivalenol in male farmers from Normandy. *Fr J Agric Food Chem* 2010b;**58**:5206–12.

46. Van der Westhuizen L, Shephard GS, Burger HM, et al. Fumonisin B1 as a urinary biomarker of exposure in a maize intervention study among South African subsistence farmers. *Cancer Epidemiol Biomarkers Prev* 2011;**20**:483–9.
47. Shephard GS, Van der Westhuizen L, Sewram V. Biomarkers of exposure to mycotoxins: a review. *Food Addit Contam* 2007;**24**(10):1196–201.
48. Solfrizzo M, Gambacorta L, Warth B, et al. Comparison of single and multi-analyte methods based on LC-MS/MS for mycotoxin biomarker determination in human urine. *World Mycotoxin J* 2013;**6**:355–66.
49. European Food Safety Authority (EFSA). Opinion of the Scientific Committee on a request from EFSA related to a harmonised approach for risk assessment of substances which are both genotoxic and carcinogenic. *EFSA J* 2005;**282**:1–31.
50. Leblanc JC, Tard A, Volatier JL, et al. Estimated dietary exposure to principal food mycotoxins from the first French total diet study. *Food Addit Contam* 2005;**22**(7):652–72.
51. Dall'Asta C, Scarlato AP, Galaverna G, et al. Dietary exposure to fumonisins and evaluation of nutrient intake in a group of adult celiac patients on a gluten-free diet. *Mol Nutr Food Res* 2012;**56**(4):632–40.
52. Miraglia M, Debegnach F, Brera C. Mycotoxins: detection and control. In: Watson DH, editor. *Pesticide, veterinary and other residues in food*. Cambridge (England): Woodhead Publishing in Food Science and Technology; 2004.
53. European Food Safety Authority (EFSA). Scientific opinion on Ergot alkaloids in food and feed. *EFSA J* 2012b;**10**(7):2798.
54. European Food Safety Authority (EFSA). Scientific opinion on the risks for public and animal health related to the presence of citrinin in food and feed. *EFSA J* 2012a;**10**(3):2605.
55. European Food Safety Authority (EFSA). Scientific opinion on the risks for animal and public health related to the presence of *Alternaria* toxins in feed and food. *EFSA J* 2011;**9**(10):2407.
56. Romero AC, Ferreira TRB, Dias CTS, et al. Occurrence of AFM1 in urine samples of a Brazilian population and association with food consumption. *Food Control* 2010;**21**:554–8.
57. Mykkänen H, Zhu H, Salminen E, et al. Fecal and urinary excretion of aflatoxin B1 metabolites (AFQ1, AFM1 and AFB-N7-guanine) in young Chinese males. *Int J Cancer* 2005;**115**:879–84.
58. Shuaib FMB, Jolly PE, Ehiri JE, et al. Association between birth outcomes and aflatoxin B1 biomarker blood levels in pregnant women in Kumasi, Ghana. *Trop Med Int Health* 2010;**15**(2):160–7 (8).
59. Polychronaki N, Wild CP, Mykkänen H, et al. Urinary biomarkers of aflatoxin exposure in young children from Egypt and Guinea. *Food Chem Toxicol* 2008;**46**:519–26.
60. Turner PC, Moore SE, Hall AJ, et al. Modification of immune function through exposure to dietary Aflatoxin in Gambian children. *Environ Health Persp* 2003;**111**(2):217–20.
61. Viegas S, Veiga L, Figueredo P, et al. Occupational exposure to aflatoxin B1 in swine production and possible contamination sources. *J Toxicol Environ Health A* 2013;**76**(15):944–51.
62. Jonsyn-Ellis FE. Aflatoxins and ochratoxins in serum samples of school children. *J Nutr Environ Med* 2007;**16**(1):52–8.
63. Gurbay A, Girgin G, Sabuncuoglu SA, et al. Ochratoxin A: is it present in human breast milk samples obtained mothers from Ankara, Turkey. In: abstracts of the 46th Congress of the European societies of toxicology. *Toxicol Lett* 2009;**189**(Suppl.):S57–273.
64. Ostry V, Malir F, Roubal T, et al. Monitoring of mycotoxin biomarkers in the Czech Republic. *Mycot Res* 2005;**21**:49–52.
65. Lino CM, Baeta ML, Henri M, et al. Levels of ochratoxin A in serum from urban and rural Portuguese populations and estimation of exposure degree. *Food Chem Toxicol* 2008;**46**:879–85.
66. Ghali R, Hmaissia -Khlifa K, Mezigh CH, et al. HPLC determination of ochratoxin A in a low volume of human blood serum. *Anal Lett* 2008;**41**:757–66.

67. Coronel MB, Sanchis V, Ramos AJ, et al. Prevention of nephrotoxicity of ochratoxin A, a food contaminant. *Toxicol Lett* 2009;**82**:2847–52.
68. Karima HK, Ridha G, Zied A, et al. Estimation of Ochratoxin A in human blood of healthy Tunisian population. *Exp Toxicol Pathol* 2010;**62**(5):539–42.
69. Pascale M, Visconti A. Rapid method for the determination of ochratoxin A in urine by immunoaffinity column clean-up and high performance liquid chromatography. *Mycopathologia* 2000;**152**:91–5.
70. Fazekas B, Tar A, Kovacs M. Ochratoxin a content of urine samples of healthy humans in Hungary. *Acta Vet Hung* 2005;**53**:35–44.
71. Manique R, Pena A, Lino CM, et al. Ochratoxin A in themorning and afternoon portions of urine from Coibra and Valencia populations. *Toxicon* 2008;**51**:1281–7.
72. Munoz K, Blaszkewicz M, Degen G. Simultaneous analysis of ochratoxin A and its major metabolite ochratoxin alpha in plasma and urine for an advanced biomonitoring of the mycotoxin. *J Chromatogr B* 2010;**878**:2623–9.
73. Breitholtz-Emanuelsson A, Olsen M, Oskarsson A, et al. Ochratoxin A in cow's milk and in human milk with corresponding human blood samples. *JAOAC Int* 1993;**76**:842–6.
74. Micco C, Miraglia M, Brera C, et al. Evaluation of ochratoxin A level in human milk in italy. *Food Addit Contam Part A* 1995;**12**:351–4.
75. Turconi G, Guarcello M, Livieri C, et al. Evaluation of xenobiotics in human milk and ingestion by the newborn- an epidemiological survey in Lombardy (Northern Italy). *Eur J Nutr* 2004;**43**:191–7.
76. Navas SA, Sabino M, Rodriguez-Amaya DB. Aflatoxin M1 and ochratoxin A in a human milk bank in the city of Sao Paulo, Brazil. *Food Addit Contam* 2005;**22**:457–62.
77. Dostal A, Jakusova L, CajdovaJ, et al. Results of the first studies of occurrence of Ochratoxin A in human milk in Slovakia. *Bratisl Lek Listy* 2008;**109**:276–8.
78. Meky FA, Turner PC, Ashcroft AE, et al. Development of a urinary biomarker of human exposure to deoxynivalenol. *Food Chem Toxicol* 2003;**41**:265–73.
79. Xu L, Cai Q, Tang L, et al. Evaluation of fumonisin biomarkers in a cross-sectional study with two high-risk populations in China. *Food Addit Contam Part A Chem Anal Control Expo Risk Assess* 2010;**27**:1161–9.
80. Gong YY, Torres-Sanchez L, Lopez-Carrillo L, et al. Association between tortilla consumption and human urinary fumonisin B1 levels in a Mexican population. *Cancer Epidemiol Biomarkers Prev* 2008;**17**(3):688–94.
81. Abia WA, Warth B, Sulyok M, et al. Bio-monitoring of mycotoxin exposure in Cameroon using a urinary multi-biomarker approach. *Food Chem Toxicol* 2013;**62**:927–34.
82. Ezekiel CN, Warth B, Ogara IM, et al. Mycotoxin exposure in rural residents in northern Nigeria: a pilot study using multi-urinary biomarkers. *Environ Intl* 2014;**66**:138–45.
83. Ahn J, Kim D, Kim H, et al. Quantitative determination of mycotoxins in urine by LC-MS/MS. *Food Add Contam* 2010;**27**(12):1674–82.
84. Shephard GS, Burger HM, Gambacorta L, et al. Multiple mycotoxin exposure determined by urinary biomarkers in rural subsistence farmers in the former Transkei, South Africa. *Food Chem Toxicol* 2013;**62**:217–25.
85. Solfrizzo M, Gambacorta L, Lattanzio V, et al. Simultaneous LC-MS/MS determination of aflatoxin M1, ochratoxin A, deoxynivalenol, de-epoxydeoxynivaleol, α and β-zearalenols and fumonisin B1 in urine as a multi-biomarker method to assess exposure to mycotoxins. *Anal Bioanal Chem* 2011;**401**:2831–41.
86. Sirot V, Fremy JM, Leblanc JC. Dietary exposure to mycotoxins and health risk assessment in second French total diet study. *Food Chem Toxicol* 2013;**52**:1–11.

87. Villa P, Markaki P. Aflatoxin B1 and ochratoxin A in breakfast cereals from Athens market: occurrence and risk assessment. *Food Control* 2009;**20**:455–61.
88. Cano-Sancho G, Sanchis V, Marín S, et al. Occurrence and exposure assessment of aflatoxins in Catalonia (Spain). *Food Chem Toxicol* 2013;**51**:188–93.
89. Hernandez-Martınez R, Navarro-Blasco I. Aflatoxin levels and exposure assessment of Spanish infant cereals. *Food Addit Contam* 2010;**B3**:275–88.
90. Thuvander A, Möller T, Enghardt Barbieri H, et al. Dietary intake of some important mycotoxins by the Swedish population. *Food Addit Contam* 2001;**18**(8):696–706.
91. Bakker G, Sizoo E, Pereboom-De Fauw DP, et al. Determination of mean daily intakes of aflatoxin B1, aflatoxin M1, ochratoxinA, trichotecenes and fumonisins in 24 hours-diets of children in the Netherlands. *World Mycotoxin J* 2009;**2**:451–9.
92. Coffey R, Cummins E, Ward S. Exposure assessment of mycotoxins in dairy milk. *Food Control* 2009;**20**:239–49.
93. Duarte SC, Almeida AM, Teixeira AS, et al. Aflatoxin M1 in marketed milk in Portugal: assessment of human and animal exposure. *Food Control* 2013;**30**:411–7.
94. Gomez-Arranz E, Navarro-Blasco I. Aflatoxin M1 in Spanish infant formulae: occurrence and dietary intake regarding type, protein-base and phisical state. *Food Addit Contam B* 2010;**3**:193–9.
95. Brera C, Debegnach F, De Santis B, et al. Ochratoxin A in cocoa and chocolate products from the Italian market: occurrence and exposure assessment. *Food Control* 2011;**22**:1663–7.
96. Coronel MB, Marìn S, Cano-Sancho G, et al. Exposure assessment to ochratoxin A in Catalonia (Spain) based on the consumption of cereals, nuts, coffee, wine, and beer. *Food Addit Contam* 2012;**29**(6):979–93.
97. Brera C, Debegnach F, De Santis B, et al. Exposure assessment to mycotoxins in gluten-free diet for celiac patients. *Food Chem Toxicol* 2014;**69**:7–13.
98. Petersen A, Throup I. Preliminary evaluation of fumonisins by the Nordic countries and occurrence of fumonisins (FB1 and FB2) in corn-based foods on the Danish market. *Food Addit Contam* 2001;**18**(3):221–6.
99. Gareis M, Schothorst RC, Vidnes A, et al. SCOOP task 3.2.10. *Collection of occurrence data of Fusarium toxins in food and assessment of dietary intake by the population of eu member states*; 2003.
100. De Boevre M, Jacxens L, Lachat C, et al. Human exposure to mycotoxins and their masked forms through cereal-based foods in Belgium. *Toxicoloy Lett* 2013;**218**:271–92.
101. Cano-Sancho G, Marin S, Ramos AJ. Occurrence of zearalenone, an estrogenic mycotoxin, in Catalonia (Spain) and exposure assessment. *Food Chem Toxicol* 2012;**50**:835–9.
102. Cano-Sancho G, Gauchi JP, Sanchis V, et al. Quantitative dietary exposure assessment of the Catalonian population (Spain) to the mycotoxin deoxynivalenol. *Food Addit Contam* 2011;**8**:1098–109.
103. Cano-Sancho G, Ramos AJ, Sanchis V. Exposure assessment of T2 and HT2 toxins in Catalonia (Spain). *Food Chem Toxicol* 2012;**50**:511–7.
104. Rodríguez-Carrasco Y, Ruiz MJ, Font G. Exposure estimates to *Fusarium* mycotoxins through cereals intake. *Chemosphere* 2013;**93**:2297–303.
105. Rautala T, Hietaniemi V, Rämö S, et al. *Fusarium* toxins: adult intake from cereals and cereal-based products in Finland. *Evira Res Reports* 2008;**5**.
106. Norwegian Scientific Committee for Food Safety (VKM). *Risk assessment of mycotoxins in cereal grain in Norway*; 2013.
107. Nordic Council of Ministers (NCM). *Fusarium toxins in cereals – risk assessment.* Copenhagen. Tema Nord Food: Nordic Council of Ministers; 1998.

Chapter 17

Mycotoxins as Food Carcinogens

Ariane Vettorazzi, Adela López de Cerain
Department of Pharmacology and Toxicology, School of Pharmacy, University of Navarra, Spain

MYCOTOXINS CONTAMINATING FOOD

Food is an important source of exposure to different intentionally added food ingredients or additives and also to unintentionally added substances such as abiotic and biotic contaminants. Among substances that can be found in different foodstuffs and processed foods, chemicals are a source of health concern. In general, chemicals that appear in food can be naturally occurring substances (mycotoxins, marine biotoxins or some metals) or substances from anthropogenic sources such as industrial activities (polychlorinated biphenyls and dioxins), food processing methods (polycyclic aromatic hydrocarbons, acrylamide, or additives), or agricultural and animal production practices (pesticides or veterinary drug residues).

Mycotoxins are an important group of naturally occurring substances that contaminate a huge variety of raw materials, feed, and food commodities. These toxins are secondary metabolites produced mainly by different species of *Aspergillus*, *Penicillium*, *Fusarium*, *Claviceps*, and *Alternaria* genus.[1] The most relevant groups of mycotoxins found in food are produced by the following five fungal genera: aflatoxins (AFs), produced by *Aspergillus* species; ochratoxin A (OTA), produced by both *Aspergillus* and *Penicillium*; trichothecenes (type A: HT-2 and T-2 toxin; and type B: deoxynivalenol), zearalenone, fumonisins B1 and B2, and the emerging mycotoxins (fusaproliferin, moniliformin, beauvericin, and enniatins) produced mainly by *Fusarium* species; and ergot alkaloids produced by *Claviceps*.[2] These fungi can contaminate raw materials or processed foods throughout the entire food chain. The main contaminated products are cereals and cereal-based foods, feed, dried fruits, fruit and vegetables, juices and wine, herbs and spices, nuts and seeds, coffee, and cocoa. Moreover, some mycotoxins or their metabolites have been found in animal products such as meat and milk. In the annual report of the Rapid Alert System for Food and Feed,[3] aflatoxins were the primary mycotoxins associated with the notifications, followed by OTA, deoxynivalenol, fumonisins, and zearalenone. Nuts, nut products and seeds, fruits and vegetables, feed, herbs and spices, cereals, and bakery products were the most affected categories.

Environmental Mycology in Public Health. http://dx.doi.org/10.1016/B978-0-12-411471-5.00017-X

Mycotoxins and the diseases caused by them were relatively obscure in the scientific literature until the discovery of aflatoxin in the early 1960s, when outbreaks of mass mortality in turkeys in the United Kingdom were linked to the ingestion of imported Brazil nut meal containing mycelium of a toxic mold, *Aspergillus flavus*, which was later shown to produce the aflatoxins.[4] Since then, there has been growing interest in research related to them, which has resulted in an increase in the number of publications regarding mycological, chemical, toxicological, and epidemiological aspects of mycotoxins. Thus, the possibility of foods being contaminated by mycotoxins was known early on; as a consequence, countries across the world have implemented control measures directed to specific mycotoxins in foodstuffs, particularly grains and oil seeds.

Acute toxic effects are easier to detect and therefore to prevent than are chronic effects because in general, symptoms appear in a short period after consumption of the contaminated food commodities. This, together with the fact that acute intoxications are usually produced by a relatively higher concentration of the chemical than chronic intoxication, facilitates tracing to the source of exposure. However, a current health concern in our society is determination of the etiology of degenerative or chronic diseases such as cancer. Because food is consumed daily over our lifespan, there is a need to identify and control all chemicals that appear at high or low doses in food, intentionally or unintentionally. Indeed, carcinogenic or potential carcinogenic effects of chemicals present in food are one of the main toxicological priorities worldwide.[5] Public health authorities and other governmental/international agencies such as the World Health Organization (WHO), Food Agriculture Organization (FAO), Food and Drug Administration (FDA), the European Food Safety Authority (EFSA), the International Agency for Research on Cancer (IARC), and others are paying deep attention to this.

In this respect, for many years toxicology has proved scientific background to decision makers who regulate food safety. The field of food toxicology has always focused on an analysis of the presence of genotoxic/mutagenic or carcinogenic substances in raw materials and food commodities. Moreover, past decades, there has been increased interest in assessing chemicals based on their mode of action[6] rather than using the traditional approach of adverse end points. In the case of carcinogens this is especially relevant, because different approaches are recommended for different modes of action. Non-genotoxic carcinogens are compounds that produce cancer but without reacting directly with DNA; they may affect other biological processes involved in spindle function and organization, inhibition of topoisomerases, stimulation of cell proliferation, inhibition of apoptosis, immune suppression, endocrine disruption, and many more.[7] These processes are considered to have a biological threshold in their dose–response relationships, and thus health-based guidance values (such as tolerable/acceptable daily/weekly/monthly intakes) are derived for these non-genotoxic carcinogens.[8] For genotoxic carcinogens (DNA-reactive compounds), a non-threshold mechanism has traditionally been assumed, and thus precautionary principles are used for risk management.[5] In the specific case of chemicals causing DNA damage but through indirect mechanisms (such as radical oxygen species production), the consensus is that a practical threshold may also exist.[9]

Mycotoxins have a huge range of miscellaneous acute and chronic toxic effects in farm animals and humans. A wide range of toxic effects from gastrointestinal, immune, estrogenic, nervous, liver, or kidney disorders, and genotoxic, teratogenic, and cancerogenic properties have been described (for a general review, see Marín et al.[2]). Mycotoxins (or groups of mycotoxins) appearing in food, whose toxicity has been considered relevant to animal or human health, have been regulated through legislation limiting their levels in food. Many countries have adopted regulations to limit mycotoxin exposure. Their presence is related to the effect they might have on consumer health, but they may also have an impact on world trade.[2] Ideally, regulation should be implemented after the application of a risk assessment procedure that needs both toxicity information and exposure data. In past years, analytical techniques have improved considerably, reaching lower detection and quantification levels in food matrices. In consequence, there is increasing information regarding the presence of some mycotoxins in food.

In this chapter, information is compiled regarding safe exposure levels in humans to different mycotoxins and maximum permitted concentrations in food matrices. The review is limited to the European scenario, which is under the scope of the EFSA, which provides independent scientific advice and clear communication about existing and emerging risks. With respect to regulations limiting the level of mycotoxins in food, only European legislation has been taken into account. Regarding toxic effects, the focus is on carcinogenic aspects and the different modes of action. Special attention is given to evaluations performed by the IARC and the National Toxicology Program (NTP), two of the main organizations evaluating the carcinogenic effects of chemicals (not only food contaminants), which provide relevant data for food risk assessment performed by other international bodies such as the WHO, the EFSA, the FDA, and Health Canada. Finally, two well-studied mycotoxins (aflatoxin B1 (AFB1) and OTA) are reviewed from the point of view of carcinogenicity. AFB1 is a well-known human carcinogen with a clear genotoxic mechanism, whereas OTA is one of the most potent rat nephrocarcinogens, but with an unknown mode of action and insufficient human epidemiological data.

TOLERABLE DAILY INTAKES AND MAXIMUM LEVELS IN FOODSTUFFS

For most kinds of toxicity, it is generally believed that there is a dose below which no adverse effect will occur. Threshold doses (no observed adverse effect level (NOAEL), lowest observed effect level (LOEL), etc.) can be experimentally determined under controlled conditions; these values are reference doses used

TABLE 1 Tolerable Daily Intakes

Mycotoxins	Tolerable Daily Intakes	References/year
Ochratoxin A	TWI: 120 ng/kg bw	2006[11]
Patulin	PMTDI: 0.4 μg/kg bw	2000[13]
Deoxynivalenol	TDI: 1 μg/kg bw	1999[15]
Fumonisins	TDI: 2 μg/kg bw	2000[16], 2003[17]
Zearalenone	TDI: 0.25 μg/kg bw	2011[18]
T2 and HT-2	TDI: 0.1 μg/kg bw	2011[20]
Nivalenol	TDI: 1.2 μg/kg bw	2013[19]
Ergot alkaloids	TDI: 0.6 μg/kg bw ARf dose: 1 μg/kg bw	2012[22]

TWI: tolerable weekly intake; PMTDI: provisional maximum tolerable daily intake; TDI: tolerable daily intake; ARf: acute reference dose; bw: body weight.

to estimate tolerable daily intakes (TDIs). The TDI is an estimate of the amount of a substance in air, food, or drinking water that can be taken daily over a lifetime without appreciable health risk. TDIs are calculated on the basis of laboratory toxicity data to which uncertainty factors are applied. They are normally used for food contaminants, substances that do not have a reason to be found in food (as opposed to substances that do, such as additives, pesticide residues, or veterinary drugs in foods). A summary of TDIs for different mycotoxins is presented in Table 1. On the contrary, for genotoxic carcinogens such as AFB1, the principle of "as low as reasonably achievable" (ALARA) is applied because it is assumed that genotoxicity is a non-threshold effect. For this reason, a TDI has not been estimated for aflatoxins.

The EU Scientific Committee for Food clearly expressed in its opinion on September 23, 1994, that aflatoxins are genotoxic carcinogens,[10] and consequently exposure to aflatoxins from all sources should be as low as reasonably anticipated. Therefore, to achieve this aim, regulation has been implemented at the European level that limits the total aflatoxin content (sum of aflatoxins B1, B2, G1, and G2) as well as the AFB1 content alone, because AFB1 is by far the most toxic compound. A summary of the maximum levels of aflatoxins and the metabolite found in milk, AFM1, permitted in different foodstuffs is presented in Table 2.

Regarding OTA, the last updated EFSA Scientific Opinion derived a tolerable weekly intake (TWI) of 120 ng/kg body weight (bw)[11] (Table 1). Based on the former and on information about human exposure to OTA by the European population of European Union (EU) member states,[12] maximum levels of OTA were set in cereals, cereal products, dried vine fruit,

TABLE 2 Maximum Levels of Mycotoxins Permitted in Foodstuffs (References are Shown in Parentheses)

Foodstuffs	Maximum Levels (μg/kg)		
Aflatoxins[152–155]	**B1**	**B1+B2+G1+G2**	**M1**
Groundnuts; almonds, pistachios, and apricot kernels; hazelnuts and Brazil nuts; and other tree nuts	2–12	4–15	–
Dried fruit other than dried figs	2–5	4–10	–
Dried figs	6	10	
Raw milk, heat-treated milk, and milk for the manufacture of milk-based products	–	–	0.050
Maize and rice	5	10	–
Cereals and all products derived from cereals, including processed cereal products	2	4	–
Processed cereal-based foods and baby foods for infants and young children	0.1	–	–
Infant formulas and follow-on formulas, including infant milk and follow-on milk	–	–	0.025
Dietary foods for special medical purposes intended specifically for infants	0.1	–	0.025
Spices	5	10	
Ochratoxin A[153,156,157]			
Cereals and all products derived from unprocessed cereals, including processed cereal products and cereals intended for direct human consumption	3		
Processed cereal-based foods and baby foods for infants and young children Dietary foods for special medical purposes intended specifically for infants	0.50		
Wine and grape juice and other related products	2		
Dried vine fruit (currants, raisins, and sultanas)	10		
Roasted coffee beans and ground roasted coffee, excluding soluble coffee)	5		
Soluble coffee (instant coffee)	10		

Continued

TABLE 2 Maximum Levels of Mycotoxins Permitted in Foodstuffs (References are Shown in Parentheses)—cont'd

Ochratoxin A[153,156,157]	
Spices, including dried spices	15–30
Wheat gluten not sold directly to the consumer	8
Licorice root and extract	20–80
Patulin[153]	
Fruit juices, concentrated fruit juices as reconstituted, and fruit nectars Spirit drinks, cider, and other fermented drinks derived from apples or containing apple juice	50
Solid apple products, including apple compote and apple puree intended for direct consumption, with the exception of foodstuffs listed below	25
Apple juice and solid apple products, including apple compote and apple puree, for infants and young children and labeled and sold as such Baby foods other than processed cereal-based foods for infants and young children	10
Deoxynivalenol[153,158]	
Cereals	750–1750
Processed cereal-based foods and baby foods for infants and young children	200
Pasta (dry)	750
Bread (including small bakery wares), pastries, biscuits, cereal snacks, and breakfast cereals	500
Zearalenone[153,158]	
Cereals	75–350
Refined maize oil	400
Processed cereal-based foods and baby foods for infants and young children	20

TABLE 2 **Maximum Levels of Mycotoxins Permitted in Foodstuffs (References are Shown in Parentheses)—cont'd**

Fumonisind(153,158)	**Sums of B1 + B2**
Maize	1000–4000
Maize-based breakfast cereals and maize-base	800
Processed maize-based foods and baby foods for infants and young children	200

roasted coffee, wine, grape juice, spices, licorice, wheat gluten, and foods for infants and young children, all of which contribute significantly to general human exposure to OTA or to the exposure of vulnerable groups of consumers such as children (Table 2).

With respect to patulin, the Scientific Committee for Food (SCF) endorsed the provisional maximum tolerable daily intake (PMTDI) of 0.4 μg/kg bw[13] (Table 1). In 2001, a SCOOP task, "Assessment of the dietary intake of patulin by the population of EU Member States" in the framework of Directive 93/5/EEC was performed.[14] Based on this assessment and taking into account the PMTDI, maximum levels for patulin were set for fruit juices, spirit drinks, cider and other fermented drinks derived from apples, solid apple products, and some foods derived from apple for infants and young children (Table 2).

Regarding *Fusarium* toxins, the SCF and the EFSA CONTAM Panel adopted several opinions evaluating deoxynivalenol[15] establishing a TDI of 1 μg/kg bw, fumonisins[16,17] (TDI of 2 μg/kg bw), zearalenone[18] (TDI of 0.25 μg/kg bw), nivalenol[19] (TDI of 1.2 μg/kg bw) and T-2 and HT-2 toxins[20] (TDI of 100 ng/kg bw for the sum of T-2 and HT-2 toxins) (Table 1). In the framework of Directive 93/5/EEC, SCOOP task "Collection of occurrence data on *Fusarium* toxins in food and assessment of dietary intake by the population of EU member states" was performed.[21] Based on that assessment and taking into account the TDIs, maximum levels for *Fusarium* toxins were set in 2006 and reviewed in 2007 (Table 2).

In Europe, *Claviceps purpurea* is the most widespread species producing ergot alkaloids including ergometrine, ergotamine, ergosine, ergocristine, ergocryptine (which is a mixture of α- and β-isomers), ergocornine, and the corresponding -inine epimers.[22] The most characteristic effect is vasoconstriction. The EFSA CONTAM Panel performed estimates of both chronic and acute exposure for various age groups across European countries and established a group acute reference dose (ARf) of 1 μg/kg bw and a group TDI of 0.6 μg/kg bw/day (Table 1). Available data do not indicate concern for any population subgroup, but the dietary exposure estimates relate to a limited number of food groups, and a possible unknown contribution from other foods cannot be discounted.[22]

CARCINOGENIC RISK TO HUMANS: IARC AND NTP CLASSIFICATIONS

From the point of view of public health, mycotoxins with demonstrated evidence of being causal agents of human cancer have the highest priority so as to protect the human population. Several mycotoxins have been evaluated by the IARC and the NTP, two main international organisms that prepare scientific documents identifying substances or circumstances of exposure that pose or may pose a cancer risk to humans.

Since the 1970s, the WHO IARC has identified environmental factors that can increase the risk of human cancer. These include chemicals, complex mixtures, occupational exposure, physical agents, biological agents, and lifestyle factors. The information is obtained from studies of cancer in humans, studies of cancer in experimental animals, and mechanistic and other relevant data. The results (critical reviews and evaluations of evidence on carcinogenicity) are published in the *IARC Monographs on the Evaluation of the Carcinogenic Risks to Humans*. Agents are selected for review on the basis of two main criteria: (1) there is evidence of human exposure and (2) there is some evidence or suspicion of carcinogenicity.[23] National and international health agencies can use this information as scientific support for actions to prevent exposure to potential carcinogens.

The IARC classifies five different groups, depending on the strength of evidence for carcinogenicity in humans and experimental animals, using standard terms. Mechanistic and other relevant data are also taken into account. The classification system and number of agents classified in each group up to 2013 are[24]:

Group 1: Carcinogenic to humans (113 agents)
Group 2A: Probably carcinogenic to humans (66 agents)
Group 2B: Possibly carcinogenic to humans (285 agents)
Group 3: Not classifiable as to its carcinogenicity to humans (505 agents)
Group 4: Probably not carcinogenic to humans (1 agent)

According to the IARC, Group 1 classification is used when there is sufficient evidence of carcinogenicity in humans. As an exception, an agent may be placed in this category when evidence of carcinogenicity in humans is less than sufficient but there is sufficient evidence of carcinogenicity in experimental animals and strong evidence in exposed humans that the agent acts through a relevant mechanism of carcinogenicity.[23]

Group 2 includes agents for which, at one extreme, the degree of evidence of carcinogenicity in humans is almost sufficient, as well as those for which, at the other extreme, there are no human data but for which there is evidence of carcinogenicity in experimental animals. Agents are assigned to either Group 2A (probably carcinogenic to humans) or Group 2B (possibly carcinogenic to humans) on the basis of epidemiological and experimental evidence of carcinogenicity and mechanistic and other relevant data. The terms "probably carcinogenic" and "possibly carcinogenic" have no quantitative significance and are used simply as descriptors of different levels of evidence of human carcinogenicity, with "probably carcinogenic" signifying a higher level of evidence than "possibly carcinogenic."[23]

Group 3 is used most commonly for agents for which evidence of carcinogenicity is inadequate in humans and inadequate or limited in experimental animals. Category 4 is used for agents for which there is evidence suggesting a lack of carcinogenicity in humans and in experimental animals.[23]

Every 2 years the NTP prepares a Report on Carcinogens (RoC), a congressionally mandated, science-based document prepared for the Secretary of the United States (US) Department of Health and Human Services. It should be taken into account that the RoC lists only substances to which a significant number of people living in the US are exposed. The RoC classifies the substances into two categories:

1. Known to be a human carcinogen: When there is sufficient evidence of carcinogenicity from human studies that indicates a causal relationship between exposure to the agent (or mixture) and human cancer. In the 13th RoC, the latest edition, 56 substances have been listed in this category.[25]
2. Reasonably anticipated to be a human carcinogen: This category includes substances for which there is limited evidence of cancer in humans or sufficient evidence of cancer in experimental animals showing a cause-and-effect relationship between exposure to the substance and cancer. Alternatively, a substance can be listed in this category if there is evidence that it is a member of a class of substances already listed in the RoC or causes biological effects known to lead to the development of cancer. In the 13th edition of the RoC, 187 substances were listed in this category.[25]

Neither the RoC nor the IARC monographs program performs cancer risk assessment of the substances evaluated. They indicate only the potential hazard and do not establish the exposure condition that would pose cancer risks to individuals in their daily lives. Such formal risks, as well the establishment of different measures to reduce human exposure to carcinogens, are performed by the appropriate international and national health and research agencies.[25] Moreover, evaluation of a compound by any of these organizations does not necessarily mean that the substance is carcinogenic for humans; and vice versa, because a substance has not yet been considered for evaluation does not imply that it is noncarcinogenic.[23]

Mycotoxins Evaluated by IARC and NTP

Table 3 classifies several mycotoxins appearing in food commodities that have been evaluated to date by the IARC and the NTP. Only two mycotoxins are included in the 13th RoC: aflatoxins, as a "known human carcinogen," and OTA, considered "reasonably anticipated to be a human carcinogen." With respect to

TABLE 3 Mycotoxins Evaluated by the IARC or NTP (for Each Evaluation, Volume and Year of Publication are Given)

Mycotoxins	CAS Number	IARC Classification	IARC Monograph	NTP Classification (NTP, 2014)[25]
Aflatoxins[a]	001402-68-2	1	100F, 2012[33]	Known to be human carcinogen
Aflatoxin M1	001402-68-2	2B	56, 1993[31]	
Ochratoxin A[b]	000303-47-9	2B	56, 1993[31]	Reasonably anticipated to be human carcinogen
Toxins derived from *Fusarium graminearum*, *Fusarium culmorum*, and *Fusarium crookwellense*: zearalenone, deoxynivalenol, nivalenol, and fusarenone X[c]		3	56, 1993[31]	
Toxins derived from *Fusarium sporotrichioides*: T-2 toxin[d]		3	56, 1993[31]	
Toxins derived from *Fusarium moniliforme*: fumonisin B1, fumonisin B2, and fusarin C		2B	56, 1993[31]	
Fumonisin B1	116355-83-0	2B	82, 2002[32]	
Sterigmatocystin[e]	010048-13-2	2B	10, 1976; Suppl. 7, 1987[29]	
Penicillic acid	000090-65-3	3	10, 1976[27]; Suppl. 7, 1987[29]	
Citrinin	000518-75-2	3	40, 1986[28]; Suppl. 7, 1987[29]	
Patulin[f]	000149-29-1	3	40, 1986[28]; Suppl. 7, 1987[29]	

[a]*Aflatoxins previous evaluations: Vol. 1 (IARC, 1972),[26] Vol. 10 (IARC, 1976),[27] Suppl. 7 (IARC, 1987)[29] superseded by Vol. 56 (IARC, 1993) and Vol. 82 (IARC, 2002).*
[b]*Ochratoxin A previous evaluations: Vol. 10 (IARC, 1976)[27]; Vol. 31 (IARC, 1983)[30]; Suppl. 7 (IARC, 1987)[29] superseded by Vol. 56 (IARC, 1993).*
[c]*Toxins derived from F. graminearum, F. culmorum, and F. crookwellense previous evaluations. Vol. 11 (IARC, 1976)[40]; Vol, 31 (IARC, 1983)[30]; Suppl. 7 (IARC, 1987).[29]*
[d]*Toxins derived from* Fusarium sporotrichioides *previous evaluations Vol. 31 (IARC, 1983)[30]; Suppl. 7 (IARC, 1987).[29]*
[e]*Sterigmatocystin previous evaluations: Vol. 1 (IARC, 1972).[26]*
[f]*Patulin previous evaluations: Vol. 10 (IARC, 1976)[27].*

the IARC, aflatoxins are also considered human carcinogens (Class 1) and OTA is "probably carcinogenic to humans" (Class 2B). In addition, a long list of mycotoxins has been evaluated since 1972, although for a great proportion of them the last revision was carried out before 1987. In the case of several revisions and monographs, only the last IARC classification is indicated, together with a reference of the last monograph that has evaluated the mycotoxin or group of toxins (Table 3). Previous evaluations are indicated at the bottom of the table. From the point of view of carcinogenic classification, the most dangerous mycotoxin known to date is aflatoxin, in particular AFB1.

The first volume of IARC monographs on the evaluation of carcinogens,[26] which reviewed evidence from 19 known or suspected human carcinogens, included two mycotoxins or groups of toxins: aflatoxins and sterigmatocystin. At that time, chemicals with experimental evidence of carcinogenicity and/or evidence of human exposure had priority in the evaluation. There was evidence of carcinogenicity in animals for both mycotoxins. In the case of aflatoxins, there was considerable evidence of food contamination and some records of increased frequency of liver cancer in some human populations that consumed diets contaminated with aflatoxins. However, no causal relationship had yet been established. In the case of sterigmatocystin, surveys of human foods did not produce evidence of human exposure. Moreover, at that time, AFM1, an in vivo AFB1 metabolite, was considered carcinogenic for rainbow trout and, although its carcinogenicity in rats had not yet been evaluated, it was shown to be equally toxic as AFB1 in rats in other toxicological studies.

In an IARC Volume 10,[27] aflatoxins and sterigmatocystin were reevaluated and OTA, patulin, and penicillic acid were reviewed for the first time; cyclochlorothine and luteoskyrin were included in this monograph. All of them were fungal metabolites not produced for commercial purposes and susceptible to contaminating foodstuffs. In IARC Volume 40,[28] patulin was reevaluated and citrinin and rugulosin were evaluated for the first time. Evaluations of these mycotoxins and other compounds were revised in 1987[29] and aflatoxins were considered human carcinogens (Group 1) whereas sterigmatocystin was included in Group 2B of possibly carcinogenicity to humans, and citrinin, OTA, patulin, penicillic acid, cyclochlorothine, luteoskyrin, and rugulosin were considered not classifiable as to their carcinogenicity to humans (Group 3). Other fungal metabolites used mainly as pharmaccutical or vcterinary drugs, such as actinomycins, adriamycin, azaserine, chloramphenicol, and griseofulvin, although reviewed by IARC monographs (mainly in Volume 10[27] and Supplement 7[29]) and/or the RoC 13th edition, will not be discussed. Substances isolated from edible mushrooms, such as agaritine and gyromitrin (Class 3) (IARC Volume 31,[30] Supplement 7[29]), will not be commented on in this chapter.

In IARC Volume 56,[31] aflatoxins (B1, B2, G1, G2, and M1), OTA, and several toxins derived from different species of *Fusarium* were reviewed. For OTA this is the last IARC revision, but for aflatoxins there are updates in Volumes 82[32] and 100F.[33] The mechanism of carcinogenicity of AFB1 is well known and will be described more in depth; there is also strong evidence of food contamination and human exposure.[34] Regarding OTA, there is evidence of carcinogenicity in rodents but not in humans. Although there has been a great amount of research in past years, its mechanism of carcinogenicity is debated. There is evidence of food contamination and human exposure.[11]

Regarding *Fusarium* toxins, the most widely distributed toxigenic *Fusarium* species is *Fusarium graminearum*, which causes disease in wheat and maize all over the world, except in dry land wheat and subtropical maize. This fungus produces type B triochothecenes deoxynivalenol and nivalenol and zearalenone, depending on the strain. The closely related species, *Fusarium culmorum* and *Fusarium crookwellense*, produce the same toxins and occur in cooler and slightly warmer areas, respectively. *Fusarium crookwellense* and some strains of *F. graminearum* also produce type B trichothecene fusarenone X. Toxins produced by these fungi were not classifiable as to their carcinogenicity to humans (Group 3) because evidence from human studies was inadequate or did not exist and evidence in experimental animals was inadequate.[31]

Fusarium moniliforme and a number of related species are ubiquitous in maize. These fungi produce fumonisins and fusarins. Fumonisin B1, B2, and B3 are the major ones produced in nature. The most prevalent in contaminated maize is fumonisin B1. Fusarin C is a member of a family of unstable compounds that include fusarins A, B, C, D, E, and F. Toxins derived from *F. moniliforme* were classified as possibly carcinogenic to humans (Group 2B) because there was inadequate evidence in humans for the carcinogenicity of toxins derived from *F. moniliforme* but sufficient evidence in experimental animals for the carcinogenicity of cultures of *F. moniliforme* that contain significant amount of mycotoxins. At that time there was limited evidence in experimental animals of the carcinogenicity of fumonisin B1 and fusarin C, and inadequate evidence for fumonisin B2.[31]

Fusarium sporotrichioides may occur in cereals, particularly in northern temperate climates. This species produces T-2 toxin, a type A trichothecene, and other metabolites. Toxins derived from *F. sporotrichioides* were considered not classifiable as to their carcinogenicity to humans (Group 3) because no data were available on the carcinogenicity to humans and there was limited evidence in experimental animals for the carcinogenicity of T-2 toxin.[31]

In Volume 82,[32] aflatoxins and fumonisin B1 were evaluated. Fumonisin B1 is the most prevalent member of a family of toxins produced by several species of *Fusarium* molds that occur mainly in maize. Fumonisin B1 contamination of maize has been reported worldwide; human exposure has also been demonstrated and is greatest in regions where maize products are the dietary staple. Fumonisin B1 was considered possibly carcinogenic to humans (Group 2B) because there was inadequate evidence in humans for the carcinogenicity of fumonisins but sufficient evidence in experimental animals. Fumonisin B1 was tested for carcinogenicity by oral administration in one study in mice, one study in male rats, and one study in male and female rats. In female mice, it caused an increase in hepatocellular adenomas and carcinomas. In one study in male rats, it caused an increase in cholangiocarcinomas and hepatocellular carcinomas. In the other rat study, it induced renal tubule carcinomas in male rats, over half of which were classified as a rare highly malignant variant. Fumonisin B1 has also been shown to promote tumors in mouse skin and trout livers when 7,12-dimethylbenz[a]anthracene and AFB1, respectively, were used as tumor initiators.[32]

AFLATOXIN B1: GENOTOXIC CARCINOGEN

In general, the word "aflatoxins" involves a mixture of aflatoxins but usually refers to AFB1. This terminology will be used as well in this section, but where necessary the specific aflatoxin designation may be used.

Aflatoxins were discovered in the 1960s owing to an outbreak of a disease of unknown etiology in turkeys in England.[35] Since then, many scientific evaluations and publications have focused on this group of mycotoxins. Currently, aflatoxins are considered well-known human liver carcinogens: the only group of mycotoxins classified as Group 1 by IARC[33] and as "known to be human carcinogens" by the NTP.[25] Indeed, aflatoxins have been considered to be among the most potent mutagenic and carcinogenic substances known.[36] AFB1 is the most potent carcinogen of the aflatoxins and the mechanism involved in its carcinogenicity is now well understood: genotoxicity from direct interaction of a reactive metabolite (epoxide) with DNA. Indeed, nowadays AFB1 is used in many publications with toxicological mechanistic/predictive purposes as a reference compound for genotoxic and carcinogenic mechanisms.[37–39]

Carcinogenicity in Experimental Animals

Several long-term toxicological (carcinogenicity) studies have been performed with aflatoxins. The main carcinogenicity studies in experimental animals, with administration of aflatoxin mixtures and AFB1, AFB2, AFG1, and AFG2 or the metabolite M1 to rats, mice, hamsters, salmon, trout, ducks, tree shrews, woodchucks, and monkeys, by several routes of exposure, have been extensively reviewed and evaluated by the IARC (1993,[31] 2002,[32] 2012[33]).

The liver is the main target organ, inducing a high incidence of hepatocellular carcinoma (HCC) in almost all animal species tested to date. Depending on the route of administration, aflatoxins also caused tumors at several different tissues sites in several species of experimental animals.[25] Some differences in sensitivity, mainly attributed to differences in aflatoxin-metabolizing enzymes, between species have been also described.[33]

Studies in Mice

Adult mice are considered to be almost completely refractory to AFB1-induced carcinogenesis after oral administration (1000 μg/kg of diet).[32,33] This resistance has been attributed to the constitutive hepatic expression of a detoxifying enzyme (α-class GST: mGSTA3-3) with high affinity for AFB1 8,9-epoxide, the DNA reactive metabolite of AFB1.[41,42] However, intraperitoneal administration of AFB1 to mice increased the incidence of lung adenomas.[43,44] Some lung adenocarcinomas have also been detected after oral administration of AFG1 (3 and 30 μg/kg bw for 24 weeks) to NIH mice.[45] Sarcomas at the injection site after subcutaneous injection (10 μg of a mixture of aflatoxins) have also been described in mice.[46]

In recent years, studies using different transgenic mouse models and different dosing regimens of intraperitoneal AFB1 administration have been performed. Treatment of $XPA^{-/-}$ mice (unable to repair DNA damage by the nucleotide excision repair process),[47] Hupki mice (human TP53 knock-in mice)[48] and $p53^{+/-}$ mice[49] with AFB1 resulted in an increased incidence of liver carcinomas compared with respective wild-type mice.

Studies in Rats

Since the first study that reported the induction of hepatocellular tumors in rats with certain samples of groundnuts,[50] many carcinogenicity studies by various routes of administration have been performed with mixtures of aflatoxins or with AFB1, AFB2, AFG1, AFG2, or AFM1.

A huge number of these studies were performed before 1993 and demonstrated the carcinogenic potency of aflatoxins. These studies have been extensively reviewed by the IARC.[31] After oral administration, mixtures of aflatoxins and AFB1 caused hepatocellular and/or cholangiocellular liver tumors, including carcinomas, in rats.[51–61] Moreover, renal-cell tumors and a low incidence of tumors at other sites were found.[62–65] Intraperitoneal administration of AFB1 also caused a high incidence of liver-cell tumors in adult rats[66], and when administered during pregnancy or lactation, induced tumors (malign and benign) in mothers and progeny in liver and other organs (from digestive tract and urogenital and nervous systems).[67] As occurred in mice, subcutaneous injection of AFB1 resulted in local sarcomas at the administration site.[46]

Regarding other aflatoxins, AFB2 induced hepatocellular adenomas after oral administration and a low incidence of hepatocellular carcinomas after intraperitoneal administration.[66] AFG1 induced hepatocellular adenomas and carcinomas, and renal-cell tumors in rats after oral administration[64,66] and local sarcomas after subcutaneous injection.[46] However, hepatocarcinogenic effects of AFG1 were considered to be weaker than those of AFB1.[31] AFB1 metabolites such as AFM1 and aflatoxicol[55,68,69] also produced hepatocellular carcinomas. However, the incidence of tumors was lower than in animals treated with AFB1 under the same conditions.

Hao et al.[70] carried out carcinogenicity studies in Wistar rats, in which AFB1 was administered intraperitoneally (100–200 μg/kg bw, one to three times a week for 64 weeks). This study further confirmed that AFB1 is a liver carcinogen in the rat.

Studies in Fish

Aflatoxins have also been extensively tested in fish (mainly rainbow trout and salmon). As was the case for rat studies, a high number of studies were performed before 1993. Exposure of different strains of rainbow trout to mixtures of aflatoxins or AFB1 caused hepatocellular and/or cholangiocellular liver tumors in all studies performed.[31] The incidence of liver lesions after AFB1 (from 0 to 5000 µg/kg diet) in salmon (basophilic focus and hepatic adenomas) is much lower than in trout (mainly hepatocellular carcinomas).[71] A more recent study also confirmed that AFB1 is a liver carcinogen in trout.[72]

AFB1 (6000 µg/kg of diet) also induced hepatic tumors in guppies (*Lebistes reticularis*) after 9 and 11 months.[73] Moreover, exposure of fish (trout and salmon) embryos to AFB1 induced a high incidence of hepatocellular carcinomas.[74,75]

Regarding the rest of the aflatoxins, as in rats, AFG1, AFM1, and aflatoxicol had carcinogenic effects in trout but with a lower effect than AFB1.[76–80] AFB2 had little effect and AFG2 did not show hepatocarcinogenic effects at the dose levels tested.[78] The metabolite AFQ1 also induced hepatocellular carcinoma in trout, but with a lower incidence than AFB1.[81]

Studies in Other Animal Species

AFB1 (or a mixture of aflatoxins) also induced hepatocellular carcinoma in ducks, Syrian golden hamsters, and tree shrew after oral administration and in toad after AFB1 administration in the dorsal lymph sac. Experiments with rhesus, cynomolgus, and African green monkeys also revealed liver carcinoma after intraperitoneal, intramuscular, or oral administration.[31,33]

Factors Affecting Carcinogenicity in Animal Studies

Modulating effects of different agents such as diets viruses, parasites, and other chemicals on aflatoxin-induced carcinogenicity have been widely studied. Some of the most relevant modulating agents and their principal effects are cited below (non-exhaustive list):

- Ammoniation. It has been demonstrated that decontamination of feed containing aflatoxins by ammoniation significantly reduced or completely eliminated the induction of hepatic tumors in trout[82] and rats,[61] respectively. Moreover, in trout fed nonfat dried milk from cows fed ammoniated or non-ammoniated aflatoxin-contaminated whole cottonseed, ammoniation almost eliminated the liver tumor response.[83]
- Interaction with other chemicals. Several studies evaluating the effects or interaction of aflatoxins with different substances have been carried out either for mechanistic purposes or to evaluate carcinogenic end points (detailed information in IARC[31]). In general, *N*-nitrosodimethylamine, nafenopin (a peroxisome proliferator), and ethanol demonstrated an increase in the carcinogenic effects of AFB1 in rats. In contrast,

β-naphthoflavone, butylated hydroxyanisole or butylated hydroxytoluene, α-benzene hexachloride, ethoxyquin, oltipraz, lindane, 1-methyl-2-mercaptoimidazole, sodium selenite, and extracts of *Rhizopus delemar* (edible yeast) diminished AFB1 carcinogenic potential when administered before, simultaneously, or after AFB1 administration in rats. Different studies in trout have also demonstrated that Aroclor 1254 (polychlorinated biphenyl), β-naphthoflavone, and indole-3-carbinol (present in cruciferous vegetables) reduced the incidence of AFB1-hepatocellular carcinoma. However, in some studies in which doses of indole-3-carbinol were given after (not before) AFB1 exposure, a significant increase in tumor formation was observed compared with that in fish treated with AFB1 alone.[31] Taken into account the ability of indole-3-carbinol (depending on the exposure protocol) to both inhibit and promote AFB1-induced carcinogenesis,[84] other studies evaluated the influence of dietary indole-3-carbinol (0.2% w/w) on relative levels of CYP isozymes known to metabolize AFB1, the AFB1 glutathione detoxification pathway, and AFB1–DNA adduct formation.[85,86] Seven days of feeding the indole-3-carbinol diet increased microsomal concentrations of CYP1A1, 1A2, and 3A1/2, with a smaller effect on 2B1/2 and no effect on CYP2C11. Moreover, the liver glutathione *S*-transferase subunit (Yc2) appeared to be substantially elevated by a diet containing indole-3-carbinol. This effect was also observed, but to a lower extent, in a diet containing β-naphthoflavone. Indeed, the induction of this enzyme has been considered to have a major role in the resistance of rats to AFB1-induced hepatocarcinogenicity after treatment with enzyme inducers including oltipraz, ethoxyquin, and butylated hydroxyanisole, as well as in known mouse resistance to AFB1 carcinogenicity.[33]

- Diet. Several studies have evaluated the effects of diet composition on AFB1 carcinogenicity.[31] In general, malnourishment, marginal lipotrope diets, or high-protein (casein) diets have been shown to increase the carcinogenic effects of AFB1 orally or intraperitoneally administered to rats. Fatty acids, such as cyclopropenoid fatty acids,[87] or different concentrations of vitamin A[88] had little or no modifying effect on the response to AFB1 in rats.
- Viruses. One of the most important agents that modulates AFB1 carcinogenicity is the hepatitis B virus (HBV). Several studies carried out in woodchucks,[89] tree shrews,[90,91] and transgenic mice ($p53^{+/-}$)[92,93] demonstrated that HBV-infected animals were more sensitive than uninfected ones. In general, combined HBV–AFB1 treatment not only reduced the time of appearance, but also resulted in a higher incidence of liver tumors. This agent is especially relevant for human studies, because HBV has been considering a confounding factor in many epidemiological studies performed for aflatoxins (see the section on cancer in humans).
- Hepatectomy. Certain liver insults such as partial hepatectomies may contribute to tumor formation. Indeed, mice are considered refractory to AFB1 tumor formation except under conditions of partial hepatectomy or HBV infection.[32]
- *Conclusions for Animal Studies*

Taking into account all of the information regarding aflatoxins in experimental animals, the IARC in its last evaluation[33] concluded that there was sufficient evidence in experimental animals for the carcinogenicity of naturally occurring mixtures of aflatoxins and of AFB1, AFG1, and AFM1. However, it was considered that there was limited evidence for AFB2 and inadequate evidence for AFG2.

Aflatoxin B1 Mechanism of Action

The mechanism involved in AFB1 carcinogenicity is well known and involves several key steps: activation to reactive metabolites, DNA and protein adduct formation.

- Metabolism: activation to reactive metabolites

The liver is the major site of aflatoxin metabolism; the CYPs that are implicated are CYP1A2, 2B6, 3A4, 3A5, and 3A7 (Figure 1).[33] CYP3A4 is predominant in human liver and mediates the formation of AFB1-exo-8,9-epoxide, the highly reactive metabolite that binds to DNA, and AFQ1. CYP1A2 can also lead to the formation of exo-epoxide but mainly generates a high proportion of endo-epoxide, which does not bind to DNA, and AFM1. However, CYP1A2 has been reported to be more efficient in producing AFB1-exo-8,9-epoxide at low AFB1 concentrations that may be found after dietary exposures.[34] Finally, CYP3A5 has been described to metabolize AFB1, mainly to the exo-8,9 epoxide, but it is

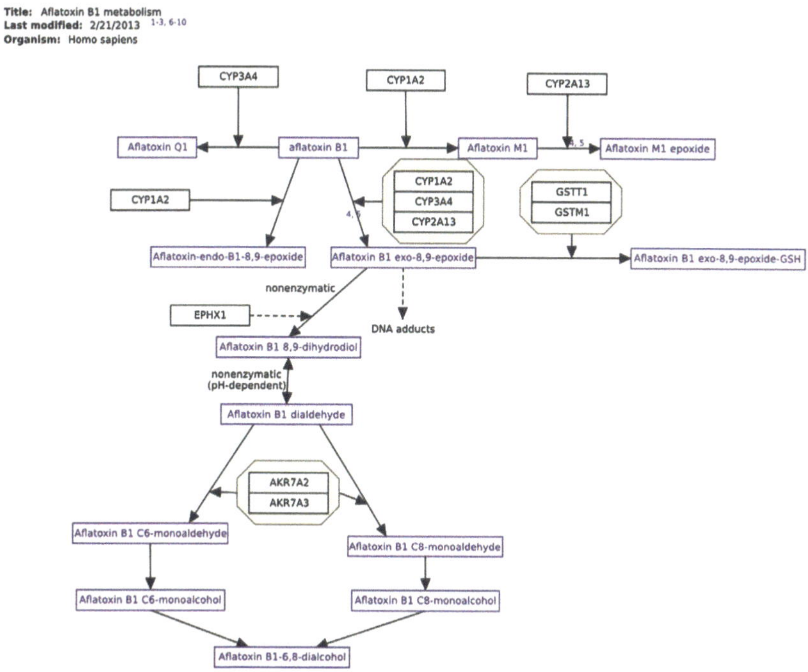

FIGURE 1 Aflatoxin B1 metabolism. *From Wikipathways.*[171]

much less efficient at forming the detoxification product, AFQ1. Some studies in human liver samples have reported the importance of CYP3A5 in liver with low CYP3A4 expression.[34]

The DNA reactive metabolite, AFB1-exo-8,9-epoxide, can hydrolyze to AFB1-8,9-dihydrodiol, an unstable intermediate that undergoes base-catalyzed rearrangement to a dialdehyde. This AFB1 dialdehyde can react with proteins, such as albumin, but not with DNA.[34]

Some species differences as well as interindividual variations among humans in response to AFB1 may be explained by different expression of different CYPs involved in AFB1 metabolism yielding to different proportions of 8,9-exo- and endo-epoxide or to less toxic metabolites.[33]

Some chemopreventive effects of certain modulating agents, previously commented upon, as well as the increased risk for liver cancer in HBV patients exposed to AFB1, may also be partly explained by the interaction of these agents with different CYPs. For example, the chemopreventive agent oltipraz has demonstrated to inhibit CYP1A2,[94] whereas in HBV-transgenic mice, liver injury was associated with increased expression of CYP enzymes.[95]

DNA Adduct Formation

As explained, AFB1 is activated to the DNA reactive metabolite AFB1-exo-8,9-epoxide. This epoxide can covalently bind to the N7 position of the guanine base of the DNA and form the 8,9-dihydro-8-(N7-guanosinyl)-9-hydroxy aflatoxin B1 adduct (AFB1-N7-Gua adduct).[96,97] This adduct represents more than 98% of total adducts formed by the AFB1-exo-8,9-epoxide[98] and can break down to form two secondary lesions: the apurinic (AP) site and the ring-opened AFB1–formamidopyrimidine (FAPY) adduct.[99] Some authors consider the FAPY adduct to be the prime candidate for both the genotoxicity and mutagenicity of aflatoxin that ultimately may lead to liver cancer. Indeed, AFB1–FAPY was found to cause a much higher frequency of G to T mutations than the original AFB1–N7–Gua adduct and was able to block the replication.[99] This is consistent with many other studies that found G to T transversion mutations to be predominant in cell and animal model systems, and that AFB1 is genotoxic in prokaryotic and eukaryotic systems in vitro, including human cells, and in vivo in humans and in a variety of animal species model systems.[31,32] Indeed, in human HCC in geographical areas where aflatoxin exposure is high, up to 50% of tumors have been shown to harbor a specific AGG to AGT point mutation in codon 249 of the TP53 tumor suppressor gene.[33]

However, the repair mechanisms for AFB1-induced DNA damage are not well understood. AFB1–N7–guanine and AFB1–FAPY are thought to be predominantly repaired by nucleotide excision repair in bacteria, yeast, and mammals. Although AFB1–FAPY is removed less efficiently than AFB1–N7–guanine in mammals, both lesions are repaired with equal efficiency in bacteria, which reflects differences in damage recognition between bacterial and mammalian repair systems. Some authors reported that DNA repair activity and modulation of repair by AFB1 seem to be major determinants of susceptibility to AFB1-induced carcinogenesis.[100] Moreover, polymorphisms in some DNA repair genes have been considered to have a role in HCC risk of AFB1-exposed populations.[34]

The AFB1-exo-8,9-epoxide reactive metabolite can be detoxified through conjugation with glutathione, mediated by the enzyme glutathione *S*-transferase (GST). Differences in AFB1-carcinogenicity response among animal species have been attributed to different activities of GST, because the enzymatic activity is much higher in resistant species. Indeed, humans are considered to have a lower GST activity than mice or rats, which suggests that humans are less capable of detoxifying AFB1-exo-8,9-epoxide.[25] Moreover, the chemopreventive agent oltipraz has been demonstrated to induce GST.[101] This enzyme has also been considered a cause of the increased risk for liver cancer in HBV patients exposed to AFB1, because HBV may decrease the activity of GST.[102]

- Protein adduct formation

As mentioned, the AFB1 dialdehyde is able to form adducts with proteins such as albumin by a Schiff base mechanism, and these adducts have been theorized to be at least one cause of the acute toxicity of AFB1.[35] Apart from the acute effects, cytotoxic effects have also been considered to contribute to AFB1 carcinogenicity, by leading to compensatory liver hyperplasia that might promote the incorporation of mutations into the DNA of dividing cells already initiated by AFB1-exo-epoxide.[103]

The formation of such protein adducts may be influenced by a further detoxification step involving reduction of the cytotoxic AFB1 dialdehyde to AFB1-dialcohol catalyzed by aldo-keto reductase (AKR) enzymes such as rat AKR7A1[103] and human AKR7A2 or AKR7A3[104,105] (Figure 1). Indeed, overexpression of AKR7A1 in a cell line leads to increased protection against the cytotoxicity of the dialdehyde metabolite of AFB1 as well as a reduction in protein-adduct formation, which indicates that AKR7A1 may be responsible for protecting against toxicity in vivo.[105] Surprisingly, this increase in activity did not protect the liver against AFB1 toxicity, as measured by bile duct proliferation, nor did it protect against carcinogenesis, as judged by the formation of GST-P–positive foci in a study performed in transgenic rats.[106]

The role of these AKR enzymes in AFB1 carcinogenesis and chemoprotection needs further investigation.

Conclusions for AFB1 Mechanism of Action

Because of the mechanistic information of aflatoxins, the IARC concluded in its last evaluation[33] that there is strong evidence that the carcinogenicity of aflatoxins operates by a genotoxic mechanism of action involving metabolic activation to a genotoxic epoxide metabolite, formation of DNA adducts, and modification of the TP53 gene.

Cancer in Humans

Because HCC is considered the sixth most common cancer worldwide and mortality for this disease is almost synonymous with incidence, the epidemiology of HCC in relation to exposure to chronic infection with HBV and aflatoxins has been reviewed in depth on a number of occasions.[34]

Some records of increased frequency of liver cancer in some human populations (Uganda, Thailand, and Kenya) that consumed diets contaminated with aflatoxins were included in the first volume of the IARC monographs.[26] However, no causal relationship had been established. In the next IARC evaluation,[27] two more epidemiological studies were reported (one in Mozambique and one related to occupational exposure via inhalation in a mill). The IARC considered at that time that these studies of liver cancer incidence in relation to aflatoxin intake provided circumstantial evidence of a causal relationship. In 1987, IARC considered there was sufficient evidence for carcinogenicity to humans.[29]

Since then, many studies have confirmed this evaluation; several cohort, case series, and case–control studies have shown clear associations between aflatoxin exposure and the incidence of HCC (for a detailed review, check IARC, 1993,[31] IARC, 2002,[32] and IARC, 2012[33]). In general, these studies were performed in areas with high aflatoxin exposure and a high prevalence of chronic hepatitis B, which is a recognized risk factor for liver cancer.[34] The association between biomarkers of exposure to AFB1 and primary liver-cell cancer also remained when analyses controlled for hepatitis B infection.[25] However, in its last evaluation, IARC[33] considered that in the presence of HBV exposure, there is a greater multiplicative interaction between aflatoxin and HBV, increasing the risk for HCC.

Aflatoxin Biomarkers

Aflatoxin molecular biomarkers have had an important role in establishing the etiologic role of AFB1 in HCC. The development, validation, and application of molecular biomarkers of AFB1 were reviewed by Kensler et al.[35]

Several biomarkers have been used for epidemiological studies and intervention trials to reduce aflatoxin exposure and dose. The most widely used were aflatoxin–albumin adducts in blood, aflatoxin–mercapturic acid (product of the conjugation of the reactive epoxide by GST enzymes), the metabolite AFM1, and the aflatoxin-N7-guanine adducts measured in urine.[35] A high correlation between the presence of aflatoxin–DNA adducts in the liver, their urinary excretion, and the formation of the serum albumin adduct has been found. Owing to their different half-lives, urinary and serum aflatoxin adduct levels reflected recent (1- to 2-day) and chronic (2- to 3-month) exposure, respectively.[34]

Moreover, several studies have linked exposure to aflatoxin with formation of the specific mutation in codon 249 in the TP53 tumor-suppressor gene, providing an important biological target for risk assessment. Finally, because of the important interaction between aflatoxin and HBV, the hepatitis-B surface antigen is an important tool for detecting individuals with high risk for HCC.[33] As noted previously, enhance detoxification or activation pathways may directly affect AFB1 toxicity and carcinogenicity. A few studies were undertaken with the purpose of analyzing a variety of genetic polymorphisms as probable modifiers of risk from aflatoxins. Even though IARC[33] considered these studies to be limited, an increased risk for HCC was found among GSTM1-null genotype individuals.

Vulnerable Population Groups

As discussed regarding the mechanism of action, several factors may explain interindividual differences, such as genetic or acquired variability in metabolism of aflatoxins or variability in DNA repair mechanisms. Moreover, different exposure scenarios may lead to different risks. For risk assessment purposes, the EFSA[34] considered the following groups as possibly most susceptible to aflatoxin toxicity:

- High-level consumers of nuts
- Children
- Vegetarians and vegans
- Subgroups with chronic hepatitis infection

Conclusions for Human Studies

Taking into account all information regarding aflatoxins in humans, in its last evaluation the IARC[33] concluded that there was sufficient evidence in humans for the carcinogenicity of aflatoxins. The overall evaluation clearly stated that aflatoxins cause cancer of the liver (HCC).

OCHRATOXIN A: LONG GENOTOXIC–EPIGENETIC DILEMMA

For many decades, the main concern regarding OTA contamination in food has been its potential carcinogenicity. This is still valid.[107] There is strong scientific evidence of the ability of OTA to cause tumors in rodent bioassays (rats and mice). However, there is still a lack of human epidemiological data, and its mode of action as a carcinogen is still under continuous debate.

Carcinogenicity in Experimental Animals

As is the case for short-term toxicity, the kidney is the main target organ of OTA carcinogenicity. The mycotoxin produces renal tumors in mice and rats, but with some sex and species differences. In general, male animals are more sensitive than females, and rats are considered to be more sensitive than mice.[108] In an article in which the incidence of renal tubule carcinogenesis in NTP carcinogenesis bioassays was reviewed, OTA was considered to be the most potent chemical tested to date by the NTP.[109]

Studies in Mice

One of the first studies carried out for OTA was one performed by Kanisawa and Suzuki.[110] In that study, 19 male ddY mice were offered diets with 0 or 40 mg/kg diet of OTA (alone or in combination with dimethyl sulfoxide) for 44 weeks. Of the 19 animals, three developed nodules in the liver and eight developed liver tumors. In the kidney, 10 animals had atypical hyperplasia, 18 cystadenomas, and five renal cell tumors (data from Huff et al.).[111] A few years later, and in the same mouse strain, the same authors performed a study (50-mg/kg diet of OTA) with different exposure durations (0, 5, 10, 15, 20, 25, and 30 weeks, followed by basal diet through 70 weeks) and found hepatomas in the 20-, 25- and 30-week treated groups (two of 14, five of 15, and six of 17, respectively) and renal cell tumors in the 15-, 20-, 25- and 30-week treated groups (three of 15, one of 14, two of 15, and four of 17, respectively). Lung tumors were also observed but were not taken into account because no dose–response relationship was found (data from Huff et al.).[111] No metastases attributable to the kidney or liver tumors were found.[112]

The most relevant carcinogenicity study carried out on mice was one performed by the FDA.[113] A diet containing OTA at a concentration of 0, 1, or 40 mg/kg was fed to groups of mice B6C3F1 of each sex for 24 months. Renal neoplasms, both carcinomas and adenomas, were found only in male mice in the 40-mg/kg dose group. All had nephropathy characterized by renal tubular dilatation, attenuation, and hyperplasia of lining epithelium, and proliferation of regenerative tubules. Females in the 40-mg/kg dose group had similar but less severe renal changes but no carcinomas or adenomas. The incidence of hepatocellular carcinomas slightly increased in both male and females, but statistical significance was evident only for females. The authors noted that OTB and benzene were present in high concentrations in the sample of OTA and that the possibility of a synergistic effect should be considered.

Studies in Rats

The most comprehensive studies on OTA carcinogenicity performed to date in rats are those carried out by the NTP.[114] Toxicology and carcinogenicity studies were conducted by administering OTA in corn oil by gavage (0, 21, 70, and 210 μg/kg, respectively, 5 days/week) to groups of Fisher 344/N rats of each sex for 16 days, 13 weeks, 9 months, 15 months, and 2 years. Tables 4 and 5 show the LOELs and NOELs for karyomegaly (considered to be an early pathological indicator of renal toxicity) and carcinogenicity of OTA in male and female rats.[108] Although LOELs and NOELs were similar between females and males (Table 4), males had a higher incidence of renal tumors than did females (Table 5).

The significance of OTA-induced renal carcinomas in rats was increased by the high frequency of metastases attributed to renal-cell carcinomas (males: none of 50, none of 51, four of 51, and 13 of 50; females: none of 50, none of 51,

TABLE 4 Lowest Observed Effect Levels (LOELs) and No Observed Effect Levels (NOELs) for Karyomegaly and Carcinogenicity for Male and Female F344/N Rats (Boorman et al.[159]; NTP[114])

Effect	Study Duration	LOEL (ug/kg bw)		NOEL (ug/kg bw)	
		Male	Female	Male	Female
Karyomegaly of proximal tubule cells	90 days[a]	62.5	62.5	ne	ne
	9 and 15 months[b]	70	70	21	21
Kidney tumors	2 years[b]	70	70	21	21

ne: not established.
[a]*Dose: 0, 62.5, 125, 250, 500, or 1000 μg/kg bw OTA in corn oil administered by gavage 5 days/week for 90 days.*
[b]*Dose: 0, 21, 70, or 210 μg/kg bw OTA in corn oil administered by gavage 5 days/week for 9 months, 15 months, or 2 years.*
Adapted from WHO.[108]

one of 50, and none of 50 for doses of 0, 21, 70, and 210 μg/kg, respectively), mainly in the lung and lymph nodes. Also, at the high dose females also had a greater multiplicity of fibroadenomas in the mammary gland (14 of 50) than did controls and rats at lower doses (four to five of 50).[112,114]

Since the NTP study, more carcinogenicity studies have been performed. However, the NTP study is still considered the reference standard carcinogenicity bioassay performed with OTA. Indeed, the occurrence of combined adenomas and carcinomas in the kidneys of male rats, as the most sensitive sex and species for kidney carcinogenicity of OTA, was considered the most appropriate data for quantitative risk assessment (benchmark dose modeling) by the WHO.[108] However, this calculation did not provide a lower point of the departure than the LOEL of 8 μg/kg bw/day for deterioration of renal functions in pigs used to establish the current TWI for OTA by the EFSA (120 ng/kg bw)[11] or the WHO (100 ng/kg bw).[108]

Subsequent carcinogenicity studies conducted after the NTP study also confirmed OTA-induced kidney tumorigenesis, as well as sex and strain specificity of the mycotoxin at a dose of 0.4 mg/kg bw (3 days/week, in sodium bicarbonate).[115,116] In general, those studies found that Dark Agouti male were more susceptible to OTA-induced kidney tumorigenesis, whereas Dark Agouti females were resistant and Lewis rats from both sexes had an intermediate response. However, increased mammary proliferative lesions were observed in female Lewis rats but not in female Dark Agouti rats.

TABLE 5 Renal Tumors and Karyomegaly in Male and Female F344/N Rats Exposed to OTA[a] (Boorman et al.[159]; NTP[114])

OTA Dose (ug/kg bw)	Adenomas		Carcinomas		Adenomas or Carcinomas		Karyomegaly	
	Male	Female	Male	Female	Male	Female	Male	Female
0	1/50	0/50	0/50	0/50	1/50	0/50	0/50	0/50
21	1/51	0/51	0/51	0/51	1/51	0/51	1/51	8/51
70	6/51	1/50	16/51	1/50	20/51	2/50	51/51	50/50
210	10/50	5/50	30/50	3/50	36/50	8/50	50/50	50/50

[a]*Administered by oral gavage 5 days/week for 2 years.*
Adapted from WHO.[108]

Other studies in which F344, F344x Sprague–Dawley hybrids, or Dark Agouti rats were given OTA daily via feed instead of the oral gavage administration used in previous studies also showed tumor formation in kidneys with different dosage regimens.[117–120]

Because the main source of exposure to OTA is via the consumption of contaminated food, all of these carcinogenesis studies were performed using oral or diet administrations. However, because human OTA exposure can also take place via dermal contact,[121] for example, during manual agricultural labor, some authors also analyzed the potential of OTA to cause dermal tumors in a two-stage mouse skin model of multistage carcinogenesis. OTA showed the ability to induce skin tumors when tested as a tumor initiator at a single dose of 200 nmol/mouse[122] and as a tumor promoter when applied at a dose of 50 nmol/mouse (twice weekly for 24 weeks) to initiated mouse skin.[123]

Conclusions for Animal Studies

Taking into account all of the information regarding OTA in experimental animals, the IARC concluded that there was sufficient evidence in experimental animals for the carcinogenicity of OTA.[31] More concretely, it was concluded that OTA increased the incidence of hepatocellular tumors in mice of each sex and produced renal-cell adenomas and carcinomas in male mice and in rats of each sex. The EFSA[11] also concluded that long-term exposure to OTA induces kidney and liver tumors in rodents, but only at nephrotoxic doses.

More recently, the NTP RoC[25] also considered that OTA is reasonably anticipated to be a human carcinogen based on sufficient evidence of carcinogenicity from studies in experimental animals.

Ochratoxin A Mechanism of Action: The Dilemma

Unfortunately, the mechanism or mechanisms of action involved in OTA-mediated tumorigenicity and toxicity are still not well understood. Unraveling the mode of action (MoA) of OTA has been an important scientific objective for many years and has been a matter of discussion and debate. Nothing other than direct genotoxic mechanisms (direct DNA binding of OTA), indirect oxidative DNA damage, and several epigenetic mechanisms (such as disruption of mitosis, cell proliferation, activation of cell signaling pathways, and protein synthesis inhibition) have been proposed for OTA. Differentiation among these different mechanisms, as discussed in the introduction of this chapter, is as a key aspect for risk assessment and management, because non-genotoxic carcinogens are considered to be substances that produce threshold effects[5] whereas direct genotoxic carcinogens are considered to cause non-threshold effects. In the case of indirect genotoxins, such as substances that damage DNA through radical oxygen species, a practical threshold can be considered.[9] The controversy regarding OTA is that evidence supporting each of these mechanisms has been provided. The most recent reviews supporting the different mechanisms of action are cited in Table 6.

TABLE 6 Most Recent Scientific Articles (since 2007) Reviewing Ochratoxin A Mode of Action (MoA)

References	Author's Final Comments Regarding OTA MoA
Pfohl-Leskowicz and Manderville[161]	Available evidence suggests that OTA is a genotoxic carcinogen by induction of oxidative DNA lesions coupled with direct DNA adducts via quinone formation.
Marin Kuan et al.[162]	A network of interacting mechanisms such as protein synthesis inhibition, oxidative stress, modulation of transcription factors (Nrf2, HFN4α, and NF-κB), and activation of specific cell signaling pathways (MAPKS and calcium homeostasis) is proposed as responsible for OTA carcinogenicity.
Mally and Dekant[163]	Disruption of mitosis and subsequent stimulation of cell proliferation proposed as a primary MoA.
Rumora and Zanic-Grubisic[164]	Relation between OTA and MAPK is extensively reviewed. The role of OTA-produced oxidative stress in these pathways is also commented on. However, MAPK was not considered the most important player in OTA MoA because many other molecules could contribute to the complex network of OTA action.
Marin Kuan et al.[165]	Plausible role of oxidative stress in OTA carcinogenicity based on strong evidence for the induction of an oxidative stress response observed.
Mally[166]	Sequence of key events: uptake into proximal tubule epithelium, inhibition of histone acetyltransferase, disruption of mitosis, cell proliferation, and genetic instability, proposed as a MoA of OTA.
Pfohl-Leskowicz and Manderville[167]	Direct mechanism involving OTA bioactivation and DNA adduct formation is proposed as most plausible MoA.
Haighton et al.[168]	Mechanistic data reviewed supported a threshold-based mechanism as most plausible.
Sorrenti et al.[169]	Review of different studies performed to counteract the adverse effects of oxygen radicals generated after OTA exposure demonstrates that antioxidants are able to protect OTA-induced DNA damage, lipid peroxidation, and cytotoxicity.
Limonciel and Jennings[170]	There is compelling evidence that OTA's inhibition of Nrf2 is the mechanism for both nephrotoxicity and carcinogenicity effects.

Updated from Vettorazzi et al.[160]

This controversy at a scientific level can also be observed in the different evaluations performed by various international agencies. Some agencies have used a threshold-based approach and have derived provisional tolerable weekly intakes (PTWI) of 100[108] or 120 ng/kg bw[11] whereas others such as Health Canada recommended that OTA be regulated as a non-threshold carcinogen and derived a provisional TDI of 3 ng/kg bw.[124]

The latest evaluation performed by the WHO[108] takes into account the different scientific evidence published to date and summarizes the main hypotheses that may contribute, totally or partially, to the possible MoA of OTA as follows:

- genotoxicity from direct interaction of OTA or a reactive metabolite with DNA
- generation of tumors resulting from chronic renal toxicity and compensatory cell proliferation
- generation of tumors resulting from inhibition of phenylalanine–tRNAPhe synthetase and protein synthesis
- disruption of cell–cell signaling pathways and the process of cell division
- alteration of intracellular calcium homeostasis
- mitochondrial dysfunction leading to oxidative stress and indirect induction of DNA damage

Some of these hypotheses were considered by the committee to account completely for tumor formation, whereas others were considered only as possible contributors.[108] Overall, the committee concluded that the evidence pointed to a number of non-genotoxic modes of action and thus supporting a PTWI of 100 ng/kg bw.

Conclusions for OTA Mechanism of Action

Taking into account the mechanistic information of OTA, the EFSA concluded in its most recent evaluation[11] that studies on the genotoxicity of OTA remain controversial, whereas a few years later, the WHO considered in its last evaluation[108] that non-genotoxic modes of action could be involved in the generation of renal tumors.

Cancer in Humans

It has been suggested that OTA may be the main etiological factor of Balkan Endemic Nephropathy (BEN), a kidney disease occurring in geographically limited areas of Balkan countries such as Bulgaria, Bosnia, Serbia, Croatia, and Romania.[112,125,126] Balkan Endemic Nephropathy has been defined as a chronic tubulointerstitial kidney disease characterized by chronic interstitial fibrosis with slow progression to end-stage kidney disease and urothelial malignant disease.[127]

However, there is still a lack of convincing epidemiological evidence associated with OTA exposure.[11,31,108,112] OTA has been found more frequently and/or in higher concentration in food and blood of inhabitants in regions with BEN than in other regions (for a review, see Fuchs and Peraica[128]). Some regional differences in exposure to OTA have also been found[129–131] and higher OTA blood concentrations have been measured in patients with some nephropathies compared with control individuals.[132–134] However, a review noted that although some studies indicated that contamination of food by OTA may be more widespread in rural endemic areas than in nonendemic areas, this is not reflected by markedly elevated blood concentrations of OTA. Moreover, blood concentrations of OTA in the same range as those observed in endemic areas have been found in countries with no history of endemic nephropathy.[135]

Furthermore, OTA exposure is only one of several hypotheses concerning an environmental etiology for BEN. The aristolochic acid hypothesis considers the disease to be produced by chronic intoxication with seeds of *Aristolochia clematitis*, which contaminate fields in endemic regions.[136] Aristolochic acids are considered to be carcinogenic to humans by the IARC[137] and the NTP.[25] Another hypothesis is the Pliocene lignite hypothesis, which proposes that the disease is caused by long-term exposure to polycyclic aromatic hydrocarbons and other toxic organic compounds leaching into well drinking water from low-rank coal in the vicinity of the endemic settlements.[136] The last is the multi-mycotoxic hypothesis, in which the synergism between OTA and various other mycotoxins such as penicillic acid, citrinin, fumonisin, and a not yet chemically identified *Penicillium polonicum* nephrotoxin causes enhanced OTA toxicity.[138] Finally, it was suggested that BEN risk is also influenced by inherited susceptibility.[139,136]

Ochratoxin Biomarkers

Several biomarkers of exposure have been used for OTA biomonitoring, such as OTA concentration in blood, urine, or breast milk.

Because of the long serum half-life of OTA, blood concentration has been used extensively in epidemiological studies. The OTA blood concentration is considered to be a convenient biomarker of exposure because similar estimates of exposure have been derived from dietary surveys and blood analyses.[108] An extended summary of previous findings on the occurrence of OTA in blood samples of healthy persons was presented in a report prepared for the 56th meeting of the Joint FAO/WHO Expert Committee on Food Additives[112] and more recently by the WHO.[108]

Advances in analytical methods have enabled researchers to monitor the generally low levels of OTA in urine.[140] Gilbert et al.[141] found that urine was a better indicator of OTA intake than plasma, thereby confirming urine as a suitable biomarker to estimate exposure to the mycotoxin. Some studies measuring OTA levels in urine of healthy people have been performed by several authors.[142–147]

In humans (as well as in monogastric animals), OTA is excreted in breast milk. OTA has been found in human milk at a wide range of concentrations.[108,148–151]

Regarding the use of biomarkers of effect, OTA nephrotoxicity can be detected by urinary analysis (markers of proximal tubule damage), but this is a relatively nonspecific effect and late in onset. Anemia is an early manifestation but is also nonspecific, and early diagnosis is difficult.[11,112] Micronuclei in human lymphocytes or DNA damage measured by means of Comet assays have been found to be markers of OTA effects but are considered nonspecific.[11]

A study performed in Tunisia[132] found an association between high levels of OTA, high levels of β_2-microglobulin, and chronic interstitial nephropathy (CIN), a disease of unknown etiology similar to BEN. However, patients with CIN of known etiology also had elevated levels of β_2-microglobulin. The authors concluded that β_2-microglobulin cannot be considered a diagnostic parameter of unexplained CIN, but it may be used as inclusion criteria in the presence of a high blood OTA concentration.

Conclusions for Human Studies

Taking into account all information regarding OTA in humans, the IARC[31] concluded that there was inadequate evidence in humans for the carcinogenicity of OTA. This is still valid nowadays. Indeed, in its last evaluation, the EFSA[11] concluded that various studies in humans have associated OTA with an endemic kidney disease observed in the Balkans (BEN and related urinary tract tumors), but convincing epidemiological evidence associated with OTA exposure is lacking.

REFERENCES

1. Bräse S, Encinas A, Keck J, Nising CF. Chemistry and biology of mycotoxins and related fungal metabolites. *Chem Rev* 2009;**109**:3903–90.
2. Marin S, Ramos AJ, Cano-Sancho G, Sanchis V. Mycotoxins: occurrence, toxicology, and exposure assessment. *Food Chem Toxicol* 2013;**60**:218–37.
3. *RASFF (The Rapid Alert System for Food and Feed)* Annual Report. 2012. http://dx.doi.org/10.2772/48887.
4. Krogh P. Mycotoxins in food. Food Science and Technology. *A series of monographs*. Academic Press; 1987.
5. WHO (World Health Organization Principles and Methods for the Risk Assessment of Chemicals in Food. *International Programme on Chemical Safety*; 2009. Environmental health criteria 240.
6. WHO (World Health Organization). *IPCS Harmonization Project Document*; 2007. No. 4.
7. Hernández LG, van Steeg H, Luijten M, van Benthem J. Mechanisms of non-genotoxic carcinogens and importance of a weight of evidence approach. *Mutat Res* 2009;**682**:94–109.
8. Barlow S, Schlatter J. Risk assessment of carcinogens in food. *Toxicol Appl Pharmacol* 2010;**243**:180–90.
9. Bolt HM, Foth H, Hengstler JG, Degen GH. Carcinogenicity categorization of chemicals-new aspects to be considered in a European perspective. *Toxicol Lett* 2004;**151**:29–41.
10. SCF (Scientific Committee for Food). *Reports of the scientific committee for food, 35th series, opinion of the scientific committee for food on aflatoxins, ochratoxin A and patulin*. 1996. p. 45 http://ec.europa.eu/food/fs/sc/scf/reports/scf_reports_35.pdf.
11. EFSA (European Food Safety Authority). Opinion of the scientific panel on contaminants in the food chain on a request from the commission related to ochratoxin A in food. *EFSA J* 2006;**365**:1–56.

12. Reports on tasks for scientific cooperation. *Task 3.2.7 assessment of dietary intake of ochratoxin A by the population of EU Member States*. 2002. http://ec.europa.eu/food/food/chemicalsafety/contaminants/task_3-2-7_en.pdf.
13. SCF (Scientific Committee for Food). Minutes of the 120th meeting of the scientific committee on food held on 8 and 9 March 2000 in Brussels, minute statement on patulin 2000. http://ec.europa.eu/food/fs/sc/scf/out55_en.pdf.
14. Reports on tasks for scientific cooperation. *Task 3.2.8, assessment of dietary intake of patulin by the population of EU Member States*. 2002. http://ec.europa.eu/food/food/chemicalsafety/contaminants/3.2.8.en.pdf.
15. SCF (Scientific Committee for Food). *Opinion of the scientific committee on food on Fusarium-toxins part 1: deoxynivalenol (DON)*. 1999. http://ec.europa.eu/food/fs/sc/scf/out44_en.pdf.
16. SCF (Scientific Committee for Food). *Opinion of the scientific committee on food on Fusarium-toxins part 3: fumonisin B1 (FB1)*. 2000. http://ec.europa.eu/food/fs/sc/scf/out73_en.pdf.
17. SCF (Scientific Committee for Food). *Updated opinion of the scientific committee on food on fumonisin B1, B2 and B3*. 2003. http://ec.europa.eu/food/fs/sc/scf/out185_en.pdf.
18. EFSA (European Food Safety Authority). Scientific opinion on the risks for public health related to the presence of zearalenone in food. *EFSA J* 2011;**9**:2197.
19. EFSA (European Food Safety Authority). Scientific opinion on risks for animal and public health related to the presence of nivalenol in food and feed. *EFSA J* 2013;**11**:3262.
20. EFSA (European Food Safety Authority). Scientific opinion on the risks for animal and public health related to the presence of T-2 and HT-2 toxin in food and feed. *EFSA J* 2011;**9**:2481.
21. Reports on tasks for scientific cooperation, Task 3.2.10, collection of occurrence data of Fusarium toxins in food and assessment of dietary intake by the population of EU Member States. http://ec.europa.eu/food/fs/scoop/task3210.pdf.
22. EFSA (European Food Safety Authority). Scientific opinion on ergot alkaloids in food and feed. *EFSA J* 2012;**10**:2798.
23. IARC monographs on the evaluation of carcinogenic risks to humans. *Preamble IARC Monogr* January 2006. Amended.
24. IARC (International Agency for Research on Cancer) monographs website: http://monographs.iarc.fr/index.php [classification: last update 30.10.13].
25. NTP (National Toxicology Program). *Report on carcinogens*. 13th ed. Research Triangle Park, NC: US: Department of Health and Human Services, Public Health Service. 2014.
26. IARC Monographs on the evaluation of carcinogenic risk of chemicals to man, vol. 1. 1972.
27. IARC monographs on the evaluation of carcinogenic risk of chemicals to man. *Some naturally occurring substances*, vol. 10. 1976.
28. IARC monographs on the evaluation of carcinogenic risk of chemicals to humans. *Some naturally occurring and synthetic food components, furocoumarins and ultraviolet radiation*, vol. 40. 1986.
29. IARC monographs on the evaluation of carcinogenic risks to humans. *Overall evaluations of carcinogenicity: an updating of IARC monographs*, vols 1–42. 1987. Supplement 7.
30. IARC monographs on the evaluation of carcinogenic risk of chemicals to humans. *Some food additives, feed additives and naturally occurring substances*, vol. 31. 1983.
31. IARC monographs on the evaluation of carcinogenic risks to humans. *Some naturally occurring substances: food items and constituents, heterocyclic aromatic amines and mycotoxins*, vol. 56. 1993.

32. IARC monographs on the evaluation of carcinogenic risks to humans. *Some traditional herbal medicines, some mycotoxins, naphthalene and styrene*, vol. 82. 2002.
33. IARC monographs on the evaluation of carcinogenic risks to humans. *A review of human carcinogens: chemical agents and related occupations*, vol. 100F. 2012.
34. EFSA (European Food Safety Authority). Opinion of the scientific panel on contaminants in the food chain related to the potential increase of consumer health risk by a possible increase of the existing maximum levels for aflatoxins in almonds, hazelnuts and pistachios and derived products. *EFSA J* 2007;**446**:1–127.
35. Kensler TW, Roebuck BD, Wogan GN, Groopman JD. Aflatoxin: a 50-year odyssey of mechanistic and translational toxicology. *Toxicol Sci* 2011;**120**(Suppl. 1):S28–48.
36. WHO (World Health Organization). Safety evaluation of certain food additives and contaminants. *WHO food additives series*, vol. 40. 1998.
37. Doktorova TY, Ellinger-Ziegelbauer H, Vinken M, Vanhaecke T, van Delft J, Kleinjans J, et al. Comparison of hepatocarcinogen-induced gene expression profiles in conventional primary rat hepatocytes with in vivo rat liver. *Arch Toxicol* 2012;**86**:1399–411.
38. Jennen DG, Magkoufopoulou C, Ketelslegers HB, van Herwijnen MH, Kleinjans JC, van Delft JH, et al. Comparison of HepG2 and HepaRG by whole-genome gene expression analysis for the purpose of chemical hazard identification. *Toxicol Sci* 2010;**115**:66–79.
39. Magkoufopoulou C, Claessen SM, Jennen DG, Kleinjans JC, van Delft JH. Comparison of phenotypic and transcriptomic effects of false-positive genotoxins, true genotoxins and non-genotoxins using HepG2 cells. *Mutagenesis* 2011;**26**:593–604.
40. IARC monographs on the evaluation of carcinogenic risk of chemicals to man. *Cadmium, nickel, some epoxides, miscellaneous industrial chemicals and general considerations on volatile anaesthetics*, vol. 11. 1976.
41. Buetler TM, Eaton DL. Complementary DNA cloning, messenger RNA expression, and induction of alpha-class glutathione *S*-transferases in mouse tissues. *Cancer Res* 1992;**52**:314–8.
42. Hayes JD, Judah DJ, Neal GE, Nguyen T. Molecular cloning and heterologous expression of a cDNA encoding a mouse glutathione *S*-transferase Yc subunit possessing high catalytic activity for aflatoxin B1-8,9-epoxide. *Biochem J* 1992;**285**:173–80.
43. Wieder R, Wogan GN, Shimkin MB. Pulmonary tumors in strain A mice given injections of aflatoxin B1. *J Natl Cancer Inst* 1968;**40**:1195–7.
44. Stoner GD, Coman PB, Greisiger EA, Stober J, Morgan M, Pereira MA, et al. Comparison of two routes of chemical administration on the lung adenoma response in strain NJ mice. *Toxicol Appl Pharmacol* 1986;**82**:19–31.
45. Huang XH, Zhang XH, Li YH, Wang JL, Yan X, Xing LX, et al. Experimental lung carcinogenic in vivo study of aflatoxin G1 in NIH mice. *Zhonghua Bing Li Xue Za Zhi* 2004;**33**:260–3.
46. Dickens F, Jones HE. Further studies on the carcinogenic action of certain lactones and related substances in the rat and mouse. *Br J Cancer* 1965;**19**:392–403.
47. Takahashi Y, Nakatsuru Y, Zhang S, Shimizu Y, Kume H, Tanaka K, et al. Enhanced spontaneous and aflatoxin-induced liver tumorigenesis in xeroderma pigmentosum group A gene-deficient mice. *Carcinogenesis* 2002;**23**:627–33.
48. Tong W-M, Lee M-K, Galendo D, Wang ZQ, Sabapathy K. Aflatoxin-B exposure does not lead to p53 mutations but results in enhanced liver cancer of Hupki (human p53 knock-in) mice. *Int J Cancer* 2006;**119**:745–9.
49. Cullen JM, Brown DL, Kissling GE, Foley JF, Rizzo J, Marion PL, et al. Aflatoxin B1 and/or hepatitis B virus induced tumor spectrum in a genetically engineered hepatitis B virus expression and Trp53 haploinsufficient mouse model system for hepatocarcinogenesis. *Toxicol Pathol* 2009;**37**:333–42.

50. Lancaster MC, Jenkins FP, Philp JMcL. Toxicity associated with certain samples of groundnuts. *Nature* 1961;**192**:1095–6.
51. Butler WH, Barnes JM. Carcinogenic action of groundnut meal containing aflatoxin in rats. *Food Cosmet Toxicol* 1968;**6**:135–41.
52. Dickens F, Jones H, Waynforth HB. Oral, subcutaneous and intratracheal administration of carcinogenic lactones and related substances: the intratracheal administration of cigarrete tar in the rat. *Br J Cancer* 1966;**20**:134–44.
53. Carnaghan RBA. Hepatic tumours and other chronic liver changes in rats following a single oral administration of aflatoxin. *Br J Cancer* 1967;**21**:811–4.
54. Wogan GN, Newberne PM. Dose-response characteristics of aflatoxin B1 carcinogenesis in the rat. *Cancer Res* 1967;**27**:2370–6.
55. Wogan GN, Paglialunga S. Carcinogenicity of synthetic aflatoxin M1 in rats. *Food Cosmet Toxicol* 1974;**12**:381–4.
56. Kalengayi MMR, Ronchi G, Desmet VJ. Histochemistry of gamma-glutamyl transpeptidase in rat liver during aflatoxin B1-induced carcinogenesis. *J Natl Cancer Inst* 1975;**55**:579–88.
57. Nishizumi M, Albert RE, Burns FJ, Bilger L. Hepatic cell loss and proliferation induced by *N*-2-fluorenylacetamide, diethylnitrosamine, and aflatoxin B1 in relation to hepatoma induction. *Br J Cancer* 1977;**36**:192–7.
58. Zawirska B, Bednarz W. The particular traits of carcinogenesis induced in Wistar rats by aflatoxin B1. III. Porphyrins and the activity of gamma-glutamyltranspeptidase in primary hepatomas and in their tissue of origin. *Neoplasma* 1981;**28**:35–49.
59. Gil R, Callaghan R, Boix J, Pellin A, Llombart-Bosch A. Morphometric and cytophotometric nuclear analysis of altered hepatocyte foci induced by *N*-nitrosomorpholine (NNM) and atlatoxin B1 (AFB1) in liver of Wistar rats. *Virchows Arch B Cell Pathol* 1988;**54**:341–9.
60. Soffritti M, McConnell EE. Liver foci formation during aflatoxin B1 carcinogenesis in the rat. *Ann N Y Acad Sci* 1988;**534**:531–40.
61. Frayssinet C, Lafarge-Frayssinet C. Effect of ammoniation on the carcinogenicity of aflatoxin- contaminated groundnut oil cakes: long-term feeding study in the rat. *Food Addit Contam* 1990;**7**:63–8.
62. Butler WH, Barnes JM. Carcinoma of the glandular stomach in rats given diets containing aflatoxin. *Nature* 1966;**209**:90.
63. Fong LYY, Chan WC. Long-term effects of feeding aflatoxin-contaminated marketpeanut oil to Sprague-Dawley rats. *Food Cosmet Toxicol* 1981;**19**:179–83.
64. Butler WH, Greenblatt M, Lijinsky W. Carcinogenesis in rats by aflatoxins B1, G1, and B2. *Cancer Res* 1969;**29**:2206–11.
65. Epstein S, Bartus B, Farber E. Renal epithelial neoplasms induced in male Wistar rats by oral aflatoxin B1. *Cancer Res* 1969;**29**:1045–50.
66. Wogan GN, Edwards GS, Newberne PM. Structure-activity relationships in toxicity and carcinogenicity of aflatoxins and analogs. *Cancer Res* 1971;**31**:1936–42.
67. Goerttler K, Löhrke H, Schweizer HJ. Effects of atlatoxin B1 on pregnant inbred Sprague-Dawley rats and their F1 generation. A contribution to transplacental carcinogenesis. *J Natl Cancer Inst* 1980;**64**:1349–54.
68. Hsieh DPH, Cullen JM, Ruebner RH. Comparative hepatocarcinogenicity of aflatoxins B2 and M1 in the rat. *Food Chem Toxicol* 1984;**22**:1027–8.
69. Nixon JE, Hendricks JD, Pawlowski NE, Loveland PM, Sinnhuber RO. Carcinogenicity of aflatoxicol in Fischer 34 rats. *J Natl Cancer Inst* 1981;**66**:1159–63.
70. Hao YR, Yang F, Cao J, Ou C, Zhang JJ, Yang C, et al. Ginkgo biloba extracts (EGb761) inhibits aflatoxin B1-induced hepatocarcinogenesis in Wistar rats. *Zhong Yao Cai* 2009;**32**:92–6.

71. Bailey GS, Willams DE, Wilcox JS, Loveland PM, Coulombe RA, Hendricks JD, et al. Aflatoxin B1 carcinogenesis and its relation to DNA adduct formation and adduct persistence in sensitive and resistant salmonid fish. *Carcinogenesis* 1988;**9**:1919–26.
72. Tilton SC, Gerwick LG, Hendricks JD, Rosato CS, Corley-Smith G, Givan SA, et al. Use of a rainbow trout oligonucleotide microarray to determine transcriptional patterns in aflatoxin B1-induced hepatocellular carcinoma compared to adjacent liver. *Toxicol Sci* 2005;**88**:319–30.
73. Sato S, Matsushima T, Tanaka N, Sugimura T, Takashima F. Hepatic tumors in the guppy (*Lebistes reticulatus*) induced by aflatoxin B1, dimethylnitrosamine, and 2-acetylaminofluorene. *J Natl Cancer Inst* 1973;**50**:767–78.
74. Wales JH, Sinnhuber RO, Hendricks JD, Nixon JE, Eisele TA. Aflatoxin B1 induction of hepatocellular carcinoma in the embryos of rainbow trout (*Salmo gairdneri*). *J Natl Cancer Inst* 1978;**60**:1133–9.
75. Black JJ, Maccubbin AE, Myers, Zeigel RF. Aflatoxin B1 induced hepatic neoplasia in Great Lakes coho salmon. *Bull Environ Contam Toxicol* 1988;**41**:742–5.
76. Canton JH, Kroes R, van Logten MJ, van Schothorst M, Stavenuiter JF, Verhülsdonk CA, et al. The carcinogenicity of aflatoxin M1 in rainbow trout. *Food Cosmet Toxicol* 1975;**13**:441–3.
77. Schoenhard GL, Hendricks JD, Nixon JE, Lee DJ, Wales JH, Sinnhuber RO, et al. Aflatoxicol-induced hepatocellular carcinoma in rainbow trout (*Salmo gairdneri*) and the synergistic effects of cyclopropenoid fatty acids. *Cancer Res* 1981;**41**:1011–4.
78. Ayres JL, Lee DJ, Wales JH. Sinnhuber RO. Aflatoxin structure and hepatocarcinogenicity in rainbow trout (*Salmo gairdneri*). *J Natl Cancer Inst* 1971;**46**:561–4.
79. Bailey GS, Loveland PM, Pereira C, Pierce D, Hendricks JD, Groopman JD, et al. Quantitative carcinogenesis and dosimetry in rainbow trout for aflatoxin B1 and aflatoxicol, two aflatoxins that form the same DNA adduct. *Mutat Res* 1994;**313**:25–38.
80. Bailey GS, Dashwood R, Loveland PM, Pereira C, Hendricks JD. Molecular dosimetry in fish: quantitative target organ DNA adduction and hepatocarcinogenicity for four aflatoxins by two exposure routes in rainbow trout. *Mutat Res* 1998;**399**:233–44.
81. Hendricks JD, Sinnhuber RO, Nixon JE, Wales JH, Masri MS, Hsieh DP, et al. Carcinogenic response of rainbow trout (Salmo gairdneri) to aflatoxin Q1 and synergistic effect of cyclopropenoid fatty acids. *J Nat Cancer lnst* 1980;**64**:523–7.
82. Brekke OL, Sinnhuber RO, Peplinski AJ, Wales JH, Putnam GB, Lee DJ, et al. Aflatoxin in corn: ammonia inactivation and bioassay with rainbow trout. *Appl Environ Microbiol* 1977;**34**:34–7.
83. Bailey GS, Price RL, Park DL, Hendricks JD. Effect of ammoniation of aflatoxin B1-contaminated cottonseed feedstock on the aflatoxin M1 content of cows' milk and hepatocarcinogenicity in the trout bioassay. *Food Chem Toxicol* 1994;**32**:707–15.
84. Dashwood RH, Fong AT, Williams DE, Hendricks JD, Bailey GS. Promotion of aflatoxin B1 carcinogenesis by the natural tumor modulator indole-3-carbinol: influence of dose, duration, and intermittent exposure on indole-3-carbinol promotional potency. *Cancer Res* 1991;**51**:2362–5.
85. Stresser DM, Bailey GS, Williams DE. Indole-3-carbinol and beta-naphthoflavone induction of aflatoxin B1 metabolism and cytochromes P-450 associated with bioactivation and detoxication of aflatoxin B1 in the rat. *Drug Metab Dispos* 1994;**22**:383–91.
86. Stresser DM, Williams DE, McLellan LI, Harris TM, Bailey GS. Indole-3-carbinol induces a rat liver glutathione transferase subunit (Yc2) with high activity toward aflatoxin B1 exo-epoxide. Association with reduced levels of hepatic aflatoxin-DNA adducts in vivo. *Drug Dispos* 1994;**22**:392–9.

87. Nixon JE, Sinnhuber RO, Lee DJ, Landers MK, Harr JR. Effect of cyclopropenoid compounds on the carcinogenic activity of diethylnitrosamine and aflatoxin B1 in rats. *J Natl Cancer Inst* 1974;**53**:453–8.
88. Newberne PM, Rogers AE. Rat colon carcinomas associated with aflatoxin and marginal vitamin A. *J Natl Cancer Inst* 1973;**50**:439–48.
89. Bannasch P, Khoshkhou NI, Hacker HJ, Radaeva S, Mrozek M, Zillmann U, et al. Synergistic hepatocarcinogenic effect on hepadnaviral infection and dietary aflatoxin B1 in woodchucks. *Cancer Res* 1995;**55**:3318–30.
90. Li Y, Su JJ, Qin LL, Yang C, Ban KC, Yan RQ, et al. Synergistic effect of hepatitis B virus and aflatoxin B1 in hepatocarcinogenesis in tree shrews. *Ann Acad Med Singap* 1999;**28**:67–71.
91. Su JJ, Ban KC, Li Y, Qin LL, Wang HY, Yang C, et al. Alteration of p53 and p21 during hepatocarcinogenesis in tree shrews. *World J Gastroenterol* 2004;**10**:3559–63.
92. Ghebranious N, Sell S. Hepatitis B injury, male gender, aflatoxin, and p53 expression each contribute to hepatocarcinogenesis in transgenic mice. *Hepatology* 1998;**27**:383–91.
93. Ghebranious N, Sell S. The mouse equivalent of the human p53ser246 mutation p53ser246 enhances aflatoxin hepatocarcinogenesis in hepatitis B surface antigen transgenic and p53 heterozygous null mice. *Hepatology* 1998;**27**:967–73.
94. Sofowora GG, Choo EF, Mayo G, Shyr Y, Wilkinson GR. In vivo inhibition of human CYP1A2 activity by oltipraz. *Cancer Chemother Pharmacol* 2001;**47**:505–10.
95. Kirby GM, Chemin I, Montesano R, Chisari FV, Lang MA, Wild CP, et al. Induction of specific cytochrome P450s involved in aflatoxin B1 metabolism in hepatitis B virus transgenic mice. *Mol Carcinog* 1994;**11**:74–80.
96. McLean M, Dutton MF. Cellular interactions and metabolism of aflatoxin: an update. *Pharmacol Ther* 1995;**65**:163–92. [Review].
97. Wild CP, Turner PC. The toxicology of aflatoxins as a basis for public health decisions. *Mutagenesis* 2002;**17**:471–81.
98. Guengerich FP, Johnson WW, Shimada T, Ueng YF, Yamazaki H, Langouët S, et al. Activation and detoxication of aflatoxin B1. *Mutat Res* 1998;**402**:121–8.
99. Smela ME, Hamm ML, Henderson PT, Harris CM, Harris TM, Essigmann JM, et al. The aflatoxin B(1) formamidopyrimidine adduct plays a major role in causing the types of mutations observed in human hepatocellular carcinoma. *Proc Natl Acad Sci USA* 2002;**99**:6655–60.
100. Bedard LL, Massey TE. Aflatoxin B1-induced DNA damage and its repair. *Cancer Lett* 2006;**241**:174–83.
101. Gross-Steinmeyer K, Eaton DL. Dietary modulation of the biotransformation and genotoxicity of aflatoxin B(1). *Toxicology* 2012;**299**:69–79.
102. Zhou T, Evans AA, London WT, Xia X, Zou H, Shen F, et al. Glutathione *S*-transferase expression in hepatitis B virus-associated human hepatocellular carcinogenesis. *Cancer Res* 1997;**57**:2749–53.
103. Ellis EM. Protection against aflatoxin B1 in rat—a new look at the link between toxicity, carcinogenicity, and metabolism. *Toxicol Sci* 2009;**109**:1–3.
104. Li D, Ferrari M, Ellis EM. Human aldo-keto reductase AKR7A2 protects against the cytotoxicity and mutagenicity of reactive aldehydes and lowers intracellular reactive oxygen species in hamster V79-4 cells. *Chem Biol Interact* 2012;**195**:25–34.
105. Bodreddigari S, Jones LK, Egner PA, Groopman JD, Sutter CH, Roebuck BD, et al. Protection against aflatoxin B1-induced cytotoxicity by expression of the cloned aflatoxin B1-aldehyde reductases rat AKR7A1 and human AKR7A3. *Chem Res Toxicol* 2008;**21**:1134–42.

106. Roebuck BD, Johnson DN, Sutter CH, Egner PA, Scholl PF, Friesen MD, et al. Transgenic expression of aflatoxin aldehyde reductase (AKR7A1) modulates aflatoxin B1 metabolism but not hepatic carcinogenesis in the rat. *Toxicol Sci* 2009;**109**:41–9.
107. Clark HA, Snedeker SM. Ochratoxin a: its cancer risk and potential for exposure. *J Toxicol Environ Health B Crit Rev* 2006;**9**:265–96. [Review].
108. WHO (World Health Organization). Safety evaluation of certain food additives and contaminants. *WHO food additives series*, vol. 59. 2008.
109. Lock EA, Hard GC. Chemically induced renal tubule tumors in the laboratory rat and mouse: review of the NCI/NTP database and categorization of renal carcinogens based on mechanistic information. *Crit Rev Toxicol* 2004;**34**:211–99. [Review].
110. Kanisawa M, Suzuki S. Induction of renal and hepatic tumors in mice by ochratoxin A, a mycotoxin. *Gann* 1978;**69**:599–600.
111. Huff JE. Carcinogenicity of ochratoxin A in experimental animals. *IARC Sci Publ* 1991;**115**: 229–44. [Review].
112. WHO (World Health Organisation). Safety evaluation of certain mycotoxins in food. *WHO food additives series*, vol. 47. 2001.
113. Bendele AM, Carlton WW, Krogh P, Lillehoj EB. Ochratoxin A carcinogenesis in the (C57BL/6J×C3H)F1 mouse. *J Natl Cancer Inst* 1985;**75**:733–42.
114. NTP (National Toxicology Program). Toxicology and carcinogenesis studies of ochratoxin A (CAS No 303-47-9) in F344/N rats (gavage studies). *Nat I Toxicol Program Tech Rep Ser*, vol. 358. 1989. pp. 1–142.
115. Castegnaro M, Mohr U, Pfohl-Leszkowicz A, Estève J, Steinmann J, Tillmann T, et al. Sex- and strain-specific induction of renal tumors by ochratoxin A in rats correlates with DNA adduction. *Int J Cancer* 1998;**77**:70–5.
116. Son WC, Kamino K, Lee YS, Kang KS. Strain-specific mammary proliferative lesion development following lifetime oral administration of ochratoxin A in DA and Lewis rats. *Int J Cancer* 2003;**105**:305–11.
117. Mantle P, Kulinskaya E. Lifetime, low-dose ochratoxin A dietary study on renal carcinogenesis in male Fischer rats. *Food Addit Contam Part A Chem Anal Control Expo Risk Assess* 2010;**27**:1566–73.
118. Mantle PG, Dobrota M, Gillett CE, Odell EW, Pinder SE. Oncological outcomes in rats given nephrocarcinogenic exposure to dietary ochratoxin a, followed by the tumour promoter sodium barbital for life: a pilot study. *Toxins* 2010;**2**:552–71.
119. Mantle PG. Minimum tolerable exposure period and maximum threshold dietary intake of ochratoxin A for causing renal cancer in male Dark Agouti rats. *Food Chem Toxicol* 2009;**47**:2419–24.
120. Mantle P, Kulinskaya E, Nestler S. Renal tumourigenesis in male rats in response to chronic dietary ochratoxin A. *Food Addit Contam* 2005;**22**(Suppl. 1):58–64.
121. Boonen J, Malysheva SV, Taevernier L, Diana Di Mavungu J, De Saeger S, De Spiegeleer B, et al. Human skin penetration of selected model mycotoxins. *Toxicology* 2012;**301**:21–32.
122. Kumar R, Ansari KM, Chaudhari BP, Dhawan A, Dwivedi PD, Jain SK, et al. Topical application of ochratoxin A causes DNA damage and tumor initiation in mouse skin. *PLoS One* 2012:7.
123. Kumar R, Alam S, Chaudhari BP, Dwivedi PD, Jain SK, Ansari KM, et al. Ochratoxin A-induced cell proliferation and tumor promotion in mouse skin by activating the expression of cyclin-D1 and cyclooxygenase-2 through nuclear factor-kappa B and activator protein-1. *Carcinogenesis* 2013;**34**:647–57.

124. Kuiper-Goodman T, Hilts C, Billiard SM, Kiparissis Y, Richard ID, Hayward S, et al. Health risk assessment of ochratoxin A for all age-sex strata in a market economy. *Food Addit Contam Part A Chem Anal Control Expo Risk Assess* 2010;**27**:212–40.
125. Castegnaro M, Canadas D, Vrabcheva T, Petkova-Bocharova T, Chernozemsky IN, Pfohl-Leszkowicz A, et al. Balkan endemic nephropathy: role of ochratoxins A through biomarkers. *Mol Nutr Food Res* 2006;**50**:519–29.
126. Petkova-Bocharova T, Chernozemsky IN, Castegnaro M. Ochratoxin A in human blood in relation to Balkan endemic nephropathy and urinary system tumours in Bulgaria. *Food Addit Contam* 1988;**5**:299–301.
127. Jha V, Garcia-Garcia G, Iseki K, Li Z, Naicker S, Plattner B, et al. Chronic kidney disease: global dimension and perspectives. *Lancet* 2013;**382**:260–72.
128. Fuchs R, Peraica M. Ochratoxin A in human kidney diseases. *Food Addit Contam* 2005;**22**(Suppl. 1):53–7.
129. Breitholtz-Emanuelsson A, Olsen M, Dahlback A, Hult K. Plasma ochratoxin A levels in three Swedish populations surveyed using an ion-pair HPLC technique. *Food Addit Contam* 1991;**8**:183–92.
130. Peraica M, Domijan AM, Fuchs R, Lucić A, Radić B. The occurrence of ochratoxin A in blood in general population of Croatia. *Toxicol Lett* 1999;**110**:105–12.
131. Scott PM. Biomarkers of human exposure to ochratoxin A. *Food Addit Contam* 2005;**22** (Suppl. 1):99–107.
132. Hassen W, Abid S, Achour A, Creppy E, Bacha H. Ochratoxin A and beta2-microglobulinuria in healthy individuals and in chronic interstitial nephropathy patients in the centre of Tunisia: a hot spot of ochratoxin A exposure. *Toxicology* 2004;**199**:185–93.
133. Jimenez AM, Lopez de Cerain A, Gonzalez-Peñas E, Bétbéder AM, Créppy E, Bello J, et al. Ochratoxin A in human plasma from healthy donors and patients in haemodialysis from the navarra community (Spain). *Toxicol Lett* 1998;**95**:156.
134. Maaroufi K, Achour A, Betbeder AM, Hammami M, Ellouz F, Creppy EE, et al. Foodstuffs and human blood contamination by the mycotoxin ochratoxin A: correlation with chronic interstitial nephropathy in Tunisia. *Arch Toxicol* 1995;**69**:552–8.
135. Mally A, Hard GC, Dekant W. Ochratoxin A as a potential etiologic factor in endemic nephropathy: lessons from toxicity studies in rats. *Food Chem Toxicol* 2007;**45**:2254–60.
136. Stefanovic V, Toncheva D, Atanasova S, Polenakovic M. Etiology of Balkan endemic nephropathy and associated urothelial cancer. *Am J Nephrol* 2006;**26**:1–11.
137. IARC monographs on the evaluation of carcinogenic risks to humans. *A Review of human carcinogens: pharmaceuticals*, vol. 100A. 2012.
138. Stoev SD, Dutton MF, Njobeh PB, Mosonik JS, Steenkamp PA. Mycotoxic nephropathy in Bulgarian pigs and chickens: complex aetiology and similarity to Balkan endemic nephropathy. *Food Addit Contam Part A: Chem Anal Control, Expo Risk Assess* 2010;**27**:72–88.
139. Creppy EE, Moukha S, Bacha H, Carratu MR. How much should we involve genetic and environmental factors in the risk assessment of mycotoxins in humans? *Int J Environ Res Public Health* 2005;**2**:186–93.
140. Warth B, Sulyok M, Fruhmann P, Mikula H, Berthiller F, Schuhmacher R, et al. Development and validation of a rapid multi-biomarker liquid chromatography/tandem mass spectrometry method to assess human exposure to mycotoxins. *Rapid Commun Mass Spectrom* 2012;**26**:1533–40.
141. Gilbert J, Brereton P, MacDonald S. Assessment of dietary exposure to ochratoxin A in the UK using a duplicate diet approach and analysis of urine and plasma samples. *Food Addit Contam* 2001;**18**:1088–93.

142. Fazekas B, Tar A, Kovacs M. Ochratoxin a content of urine samples of healthy humans in Hungary. *Acta Vet Hung* 2005;**53**:35–44.
143. Pascale M, Visconti A. Rapid method for the determination of ochratoxin A in urine by immunoaffinity column clean-up and high-performance liquid chromatography. *Mycopathologia* 2001;**152**:91–5.
144. Pena A, Seifrtova M, Lino C, Silveira I, Solich P. Estimation of ochratoxin A in Portuguese population: new data on the occurrence in human urine by high performance liquid chromatography with fluorescence detection. *Food Chem Toxicol* 2006;**44**:1449–54.
145. Ezekiel CN, Warth B, Ogara IM, Abia WA, Ezekiel VC, Atehnkeng J, et al. Mycotoxin exposure in rural residents in Northern Nigeria: a pilot study using multi-urinary biomarkers. *Environ Int* 2014;**66**:138–45.
146. Solfrizzo M, Gambacorta L, Visconti A. Assessment of multi-mycotoxin exposure in Southern Italy by urinary multi-biomarker determination. *Toxins* 2014;**6**:523–38.
147. Abia WA, Warth B, Sulyok M, Krska R, Tchana A, Njobeh PB, et al. Bio-monitoring of mycotoxin exposure in Cameroon using a urinary multi-biomarker approach. *Food Chem Toxicol* 2013;**62**:927–34.
148. Breitholtz-Emanuelsson A, Olsen M, Oskarsson A, Palminger I, Hult K. Ochratoxin A in cow's milk and in human milk with corresponding human blood samples. *J AOAC Int* 1993;**76**:842–6.
149. Postupolski J, Karłowski K, Kubik P. Ochratoxin A in maternal and foetal blood and in maternal milk. *Rocz Panstw Zakl Hig* 2006;**57**:23–30.
150. Skaug MA, Helland I, Solvoll K, Saugstad OD. Presence of ochratoxin A in human milk in relation to dietary intake. *Food Addit Contam* 2001;**18**:321–7.
151. Turconi G, Guarcello M, Livieri C, Comizzoli S, Maccarini L, Castellazzi AM, et al. Evaluation of xenobiotics in human milk and ingestion by the newborn-an epidemiological survey in Lombardy (Northern Italy). *Eur J Nutr* 2004;**43**:191–7.
152. Commission Regulation (EC) No 466/2001 of 8 March 2001 setting maximum levels for certain contaminants in food stuffs. OJ L 77, 16.3.2001, p. 1.
153. Commission Regulation (EC) No 1881/2006 of 19 December 2006 setting maximum levels for certain contaminants in foodstuffs. OJ L 364, 20.12.2006, p. 16.
154. Commission Regulation (EC) No 165/2010 of 26 February 2010 amending Regulation (EC) No 1881/2006 setting maximum levels for certain contaminants in foodstuffs as regards aflatoxins. OJ L 50, 27.2.2010, p. 8.
155. Commission Regulation (EC) No 1058/2012 of 12 November 2012 amending Regulation (EC) No 1881/2006 as regards maximum levels for aflatoxins in dried figs. OJ L 313, 13.11.2012, p. 14.
156. Commission Regulation (EC) No 594/2012 of 5 July 2012 amending Regulation (EC) No 1881/2006 as regards the maximum levels of the contaminants ochratoxin A, non dioxin-like PCBs and melamine on food stuffs OJ L 176, 6.7.2012, p. 43.
157. Commission Regulation (EC) No 105/2010 of 5 February 2010 amending Regulation (EC) No 1881/2006 setting maximum levels for certain contaminants in food stuffs as regards as regards ochratoxin A. OJ L 35, 6.2.2010, p. 7.
158. Commission Regulation (EC) No 1126/2007 of 28 September 2007 amending Regulation (EC) No 1881/2006 setting maximum levels for certain contaminants in food stuffs as regards as regards Fusarium toxins in maize and maize products. OJ L 255, 29.9.2007, p. 14.
159. Boorman GA, McDonald MR, Imoto S, Persing R. Renal lesions induced by ochratoxin A exposure in the F344 rat. *Toxicol Pathol* 1992;**20**:236–45.
160. Vettorazzi A, van Delft J, López de Cerain A. A review on ochratoxin A transcriptomic studies. *Food Chem Toxicol* 2013;**59**:766–83.

161. Pfohl-Leszkowicz A, Manderville RA, Ochratoxin A. an overview on toxicityand carcinogenicity in animals and humans. *Mol Nutr Food Res* 2007;**51**:61–99.
162. Marin-Kuan M, Cavin C, Delatour T, Schilter B. Ochratoxin A carcinogenicity involves a complex network of epigenetic mechanisms. *Toxicon* 2008;**52**:195–202.
163. Mally A, Dekant W. Mycotoxins and the kidney: modes of action for renal tumor formation by ochratoxin A in rodents. *Mol Nutr Food Res* 2009;**53**:467–78.
164. Rumora L, Zanic-Grubisic´ T. A journey through mitogen-activated protein kinase and ochratoxin A interactions. *Arh Hig Rada Toksikol* 2009;**60**:449–56.
165. Marin-Kuan M, Ehrlich V, Delatour T, Cavin C, Schilter B. Evidence for a role of oxidative stress in the carcinogenicity of ochratoxin a. *J Toxicol* 2011;**2011**:1–15.
166. Mally A. Ochratoxin A and mitotic disruption: mode of action analysis of renal tumor formation by ochratoxin A. *Toxicol Sci* 2012;**127**:315–30.
167. Pfohl-Leszkowicz A, Manderville RA. An update on direct genotoxicity as amolecular mechanism of ochratoxin a carcinogenicity. *Chem Res Toxicol* 2012;**25**:252–62.
168. Haighton LA, Lynch BS, Magnuson BA, Nestmann ER. A reassessment of risk associated with dietary intake of ochratoxin A based on a lifetime exposure model. *Crit Rev Toxicol* 2012;**42**:147–68. [Review].
169. Sorrenti V, Di Giacomo C, Acquaviva R, Barbagallo I, Bognanno M, Galvano F, et al. Toxicity of ochratoxin A and its modulation by antioxidants: a review. *Toxins* 2013;**5**:1742–66.
170. Limonciel A, Jennings P. A review of the evidence that ochratoxin A is an Nrf2 inhibitor: implications for nephrotoxicity and renal carcinogenicity. *Toxins* 2014;**6**:371–9.
171. http://wikipathways.org/index.php/Pathway:WP699.

Chapter 18

Occurrence of Mycotoxins in Indoor Environments

Martin Täubel, Anne Hyvärinen
National Institute for Health and Welfare, Department of Health Protection, Living Environment and Health Unit, Kuopio, Finland

INTRODUCTION AND SCOPE

Moisture damage, dampness, and visible mold in buildings are consistently associated with adverse health outcomes. A 2011 review of the epidemiological evidence has concluded that conditions of dampness and mold are associated with multiple allergic and respiratory effects, including asthma development and exacerbation, current asthma, upper and lower respiratory tract symptoms, respiratory infections, allergic rhinitis, and eczema.[1] Although the link between building dampness and adverse health is well established the causal agents and mechanisms underlying the observed health effects are not well understood. Microbial proliferation on indoor surfaces and in dust following the increased availability of water in damp buildings is one of the main mechanisms in generating dampness-related indoor pollutants and links to observations of visible mold in such buildings. The World Health Organization considers, in their *WHO guidelines for indoor air quality: dampness and mould*, the following dampness-related indoor pollutants as most relevant: allergens (from house dust mites and fungi), bacteria, bacterial and fungal cell wall components (such as endotoxin and fungal β-D-glucans), mycotoxins, and microbial and other volatile organic compounds.[2] This WHO document concludes that even though causative agents of adverse health effects in damp buildings have not been clearly identified, excess levels of various microbial agents, including mycotoxins, in the indoor environment need to be considered as a potential health hazard; microbial growth in response to moisture problems in buildings needs to be avoided or removed.

The WHO report and other reviews[1–4] provide the background in front of which essentially all of the research on mycotoxins in indoor environments is placed: dampness and visible mold are consistently associated with ill health, but our knowledge as concerns the causal agents and disease mechanisms is poor (Figure 1). Mycotoxins (and also the bacterial equivalents, i.e., toxic bacterial secondary metabolites) are among the candidates that have been suggested to be involved in health problems observed in damp buildings. In this chapter,

Environmental Mycology in Public Health. http://dx.doi.org/10.1016/B978-0-12-411471-5.00018-1

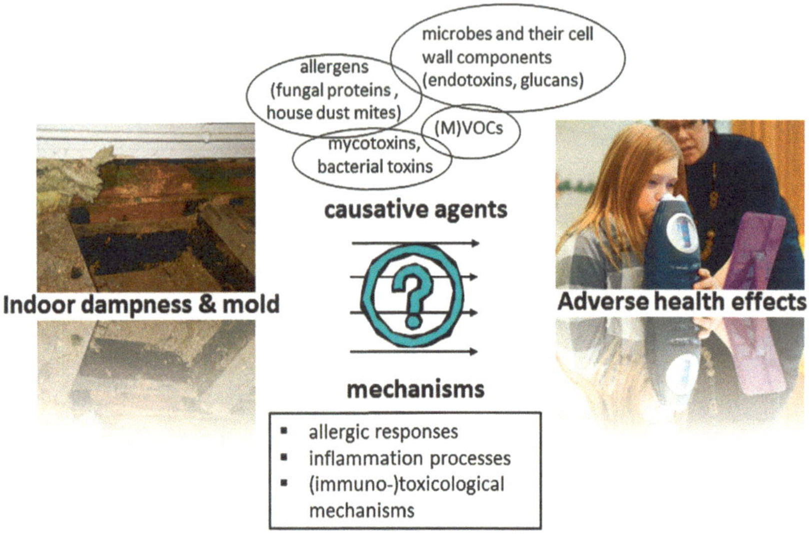

FIGURE 1 Moisture damage, dampness, and visible mold in indoor spaces and the related exposures and adverse health.

we attempt to summarize what is known today from scientific studies that have reported mycotoxins specifically in indoor environmental samples.

The potential involvement of mycotoxins in building dampness-related illness was probably first clearly postulated in 1986, when Croft et al. reported an outbreak of trichothecene mycotoxicoses in a family living in a Chicago home.[5] The family members suffered from symptoms resembling those of stachybotrytoxicosis in livestock, with central nervous symptoms that included neuropsychiatric manifestations. Exposure to *Stachybotrys* spores and to macrocyclic trichothecene mycotoxins produced by these molds was linked to the health effects. In the late 1990s, the "toxic black mold" *Stachybotrys chartarum* and its partly highly toxic secondary metabolites, that is, mycotoxins such as satratoxins, roridins, verrucarins, and spirocyclic drimanes, hit the newspaper headlines. This came after reports of a cluster of cases of pulmonary hemorrhage in infants in Cleveland, Ohio, and several other cases that linked severe health outcomes to exposure to *S. chartarum* and its mycotoxins.[6–10] Subsequently, mycotoxins produced by *Stachybotrys* strains became *the* major target of research on indoor mycotoxins for almost two decades. Only more recently, analytical methods targeting mycotoxins in indoor environmental samples have been developed with the aim of detecting multiple different mycotoxins rather than only a few specific target compounds, for example, Refs 11–13.

In the context of this book, more detailed discussion on what mycotoxins are and why these compounds are produced by fungi is provided elsewhere (Chapter 1.1). Here, mycotoxins are defined as fungal secondary metabolites that pose a potential health risk to humans and/or animals when introduced by a natural route.[14] Generally, secondary metabolites either have an intrinsic function within the producing species, as for example initiation of growth and differentiation, or act on targets external to the producing species. These compounds are proposed to improve survival fitness of the producing species.[15] In this concept, mycotoxins may in many cases be relevant in the mediation of competitive interactions between microorganisms, for example, in competition for nutrients on a building material surface.

To date, several hundreds of different mycotoxins have been identified and characterized, the majority of which in regard to food and feed contaminants in agricultural settings.[16] Estimates, however, reach from 20,000 up to 300,000 unique mycotoxins being present in the environment.[17] Characteristic for mycotoxins is that they are nonvolatile, low-molecular-weight natural products that are typically very stable. In indoor environmental settings—even though not volatile—these compounds do get airborne attached to spores, fragments, and particulate matter. Mycotoxins comprise a wide variety of chemical structures and subsequently different biological activities and act on various organ systems in the human or animal body. Most of our knowledge on the modes of action of mycotoxins relates to ingestion exposure; for example, inflammatory, immune-suppressive, cytotoxic, and carcinogenic effects in various organs have been described (detailed in Chapter 3.6).

Oral ingestion of contaminated food stuffs is the most studied, most recognized, and main route for human exposure to mycotoxins. Whereas the health threat posed by food-borne mycotoxins is a worldwide phenomenon, it is a severe problem in countries with lower agricultural standards, high pressure in terms of nutritional needs, and environmental conditions favoring fungal growth. The topic of food-borne exposure to mycotoxins is addressed in Chapters 3.2 and 3.6 of this book. Dermal contact to mycotoxins as an exposure route is primarily a problem in occupational settings, for example, through handling of mycotoxin-contaminated grains in farms, dealing with mycotoxin-containing materials in laboratories, or in the process of remediating fungal- and mycotoxin-contaminated building structures. Symptoms of irritation upon dermal exposure have been reported for several mycotoxins, but the health relevance of such exposure is little explored. Interestingly, skin irritation is also very commonly reported in residents of "moldy buildings." It is, however, unclear whether such symptoms are actually caused by the mycotoxins present in these buildings or by other exposures. Workers in farms, sawmills, or the feed processing industry can also be exposed to high levels of very potent mycotoxins, such as aflatoxins and ochratoxin A, via inhalation of heavily contaminated dusts.[18–20] Such exposure situations are considered occupational health risks, and research on the potential health implications is ongoing (Chapter 3.4).

In this chapter, we consider non-occupation-related exposure to mycotoxins in indoor environments—such as residential homes, schools, and offices—through inhalation. The focus is on mycotoxins but also other fungal secondary metabolites, for which human health risks are not yet well established, that is, compounds for which the definition "mycotoxin" is not applicable following strict definitions. Whereas most of the general discussion with respect to human exposure is restricted to fungal toxins—and so is the content of this chapter—it should be mentioned that also (toxic) bacterial secondary metabolites may be relevant in the context of indoor exposure. Bacteria produce an enormous variety of bioactive secondary metabolites; many thousands of such compounds have been characterized so far.[21] Such metabolites are widely used in pharmacological products, exploiting their antibiotic, immune-suppressive, enzyme-inhibiting, antitumor, or antiparasitic potential, to give a few examples. Some of the most potent bacterial producers of pharmacologically active compounds are commonly encountered indoors and are particularly linked to conditions of indoor dampness, for example, species of the bacterial genus *Streptomyces*. The finding that toxic bacterial secondary metabolites co-occur alongside mycotoxins in indoor sample materials in damp buildings[22] implies that these compounds should also be considered as part of a complex and diverse microbial exposure situation in mold-contaminated buildings.

MYCOTOXINS IN BUILDING MATERIALS, DUST, AND AIR FROM INDOOR ENVIRONMENTS

This chapter attempts to provide an overview of studies that report on the *actual indoor occurrence* of mycotoxins and fungal secondary metabolites, that is, studies that are supported by the detection of these compounds in naturally infested indoor sample materials. Mycotoxin production of a given fungal strain under laboratory conditions does not imply that any of the same mycotoxins are necessarily produced by this fungus under "real-life conditions", for example in a damp building. The availability of nutrients and growth substrate on the indoor material and water, temperature, light, and other environmental factors,[23–25] as well as species succession and interaction with other microbes present in this particular ecological niche, all are factors that might have an impact on the production of secondary metabolites. It is a well-established fact that toxigenic fungi, that is, fungi that have the potential to produce toxic secondary metabolites, occur in indoor environments and can proliferate on various building materials.[26,27] However, the occurrence of a toxigenic fungus, for example, in an indoor air sample or on a building material does not necessarily mean that mycotoxins of this particular mold are present in the indoor environment as well. In fact, the opposite is also true, that is, detecting mycotoxins in an indoor sample does not necessarily predict the presence of the mycotoxin-producing molds.

There is a good body of literature that has documented mycotoxin production under laboratory conditions by fungal strains that have been isolated from moisture-damaged indoor environments or from species that are known to occur indoors for example, Refs 9,24,28–37. Some authors have taken another step forward to imitate real-life situations and investigated the secondary metabolite production on artificially infested building materials, such as gypsum board or wood for example, Refs 24,25,37–41. In particular the work conducted by K.F. Nielsen and colleagues should be highlighted here. They have laid the foundations for indoor mycotoxin research in describing the vast variety of mycotoxins and other fungal secondary metabolites produced by indoor fungi on building materials under experimental conditions. These studies have not only shown that mycotoxin production can occur on building materials, but have also established that it is species or even strain specific, dependent on growth conditions and substrate (building material) and the competition/coculture with other microbes. It has also been established that one given mycotoxin may be produced by different fungal strains, and one fungal strain may produce different mycotoxins.

Table 1 provides an overview of studies that have reported mycotoxin occurrence in indoor sample materials. We have sorted the studies in chronological order and provided a summary description of the aspects of the study design or locations, sample materials, mycotoxins targeted, and analytical methodology used and the main findings as concerns the occurrence of mycotoxins in indoor environments. We list a total of 30 studies published between 1986 and 2013; although we have done our best to be complete, we cannot exclude the possibility that one or another study published may be missing from this table.

There are a few general observations that can be made from almost 30 years of studies on the indoor occurrence of mycotoxins. To start with it is obvious that mycotoxins do occur indoors, as they have been readily detected in various indoor sample materials in multiple studies, using different analytical methodologies. Most commonly, building materials (in 15 studies) and dust samples (in 17 samples) have been collected for analyses of mycotoxins. Where building materials are concerned, typically mold-infested or water-damaged materials have been sampled. House dust—be it from swabs of ventilation ducts, bulk, or vacuumed samples of floor dust, vacuum cleaner dust bag dust, or settled airborne dust collected from elevated surfaces—has been utilized extensively, in particular also in more recent studies. The rationale behind using house dust as a sample material is that indoor dust links to airborne exposure through mechanisms of deposition and resuspension. Such samples may be more representative in terms of human exposure indoors as they act as a sink for airborne particles from multiple sources (e.g., mold-affected areas in a home), whereas a material sample typically reflects the situation from a single spot in the building. House dust is moreover less affected by the known large temporal variation of microbial concentrations in indoor air.[42] In particular vacuumed floor dust is commonly used in population health studies for determination of contaminants people might be exposed to in indoor environments. Concentrations reported for various mycotoxins in building material samples (or fungal matter scraped off mold-infested materials) are typically in the order of nanograms to micrograms per gram of building material (range pg–mg/g) or nanograms to micrograms per square centimeter of sampled surface. For ventilation duct dust, concentrations of picograms to nanograms per gram of dust (up to μg/g) and picograms to

TABLE 1 Overview of Studies that Report on the Occurrence of Mycotoxins (and Other Fungal and Bacterial Secondary Metabolites) in Indoor Sample Materials

No.	Study Sites and Design	Sample Materials (Number of Samples)	Target Compounds	Analytical Methodology	Main Findings Concerning Indoor Mycotoxin Occurrence	References
1	Residential house with moisture problems in Chicago, IL, USA	Ceiling fiberboard infested with *Stachybotrys chartarum* (N=1)	Trichothecenes	TLC, HPLC, GC-MS	Verrucarol, verrucarins B+J, satratoxin H, trichoverrins A+B detected	Croft et al.[5]
2	3 office spaces with "sick building syndrome"; Montreal, QC, Canada	Dust from ventilation system, floor, surfaces (N=3)	Trichothecenes	TLC, HPLC	T-2 toxin, diacetoxyscirpenol, roridine A, T-2 tetraol detected in the dust samples	Smoragiewicz et al.[57]
3	Office building with water and mold damage in New York, NY, USA	Bulk samples of water-damaged, *Stachybotrys*-contaminated paper material (N=2)	Trichothecenes	HPLC	Satratoxin H (c. 1 μg in 60 mg of scrape-off sample), stachybotrylactone, stachybotrylactone acetate	Johannig et al.[10]
4	Child day care center with moisture damage, Finland	Gypsum board liner from a site with and without water damage (N=2)	Satratoxins G and H, verrucarol	HPLC	Satratoxins G and H (17 μg/g) and verrucarol were found in water-damaged but not in undamaged gypsum board	Andersson et al.[75]

5	Courthouse and office buildings with moisture problems; Florida, USA	Moldy ceiling tiles and vinyl wall covering (N = 3)	Satratoxins G + H, sterigmatocystin, deoxynivalenol	HPLC	Satratoxins G and H at 2 and 5 ppm in ceiling tiles; deoxynivalenol in vinyl wall covering	Hodgson et al.[58]
6	One school, one domestic residence with water/mold damage; Denmark	Moldy wall scrapings, insulation material, outer cardboard layer (N = 3)	Verrucarol and trichodermol	GC-MS	Verrucarol (indicative of macrocyclic trichothecenes) in 3/3, trichodermol in 1/3 samples	Nielsen et al.[76]
7	Residential, moisture-damaged home of an infant with pulmonary hemorrhage; USA	10 cm^2 of a *Stachybotrys*-contaminated closet ceiling (N = 1)	Trichothecenes	HPLC	Roridin L-2 (0.5 ng/cm^2), roridin E (0.7 ng/cm^2), satratoxin H (3.2 ng/cm^2) on building material surface	Flappan et al.[8]
8	23 water-damaged buildings; Denmark	*Stachybotrys*-infested building materials (N = 4)	Trichothecenes, sterigmatocystins	GC-MS, HPLC, TLC	Macrocyclic trichothecenes (verrucarol-type) in 4/4 materials (2–15 ng/cm^2); trichodermol in 2/4 materials	Gravesen et al.[77]
9	Mold-infested domestic residence, several rooms/locations	Vacuumed samples or scrapings from wall paper (N = 22)	Multiple mycotoxins	HPLC-DAD (diode array detection), TLC	Meleagrin, sterigmatocystin, 5-methoxysterigmatocystin, chaetoglobosins A + C detected in multiple samples of the home	Nielsen et al.[40]

Continued

TABLE 1 Overview of Studies that Report on the Occurrence of Mycotoxins (and Other Fungal and Bacterial Secondary Metabolites) in Indoor Sample Materials—cont'd

No.	Study Sites and Design	Sample Materials (Number of Samples)	Target Compounds	Analytical Methodology	Main Findings Concerning Indoor Mycotoxin Occurrence	References
10	Residential building with symptomatic residents and pets; USA	Composite and individual samples of dust (wipes) in the heating system (N = 7)	Ochratoxin A	HPLC, LC-MS, TLC	All samples from the heating ducts contained at least traces of ochratoxin A, up to a maximum conc. of >1500 ppb	Richard et al.[59]
11	Finnish buildings with moisture problems and symptomatic individuals	Bulk samples of interior finishing (N = 79; e.g., wallpaper, cardboard, wood, plasterboard)	17 mycotoxins	HPLC-MS/MS	43% of samples with at least one of the target compounds; most common sterigmatocystin (24%; 0.2 ng/g–31 μg/g), satratoxins G + H, verrucarol, diacetoxyscirpenol, deoxynivalenol (DON), 3-acetyl-DON, T2-tetraol, citrinin	Tuomi et al.[49]
12	Residential homes with visible mold/dampness problems; Germany	Carpet dust samples (N = 11)	Sterigmatocystin	HPLC-MS/MS	Sterigmatocystin detected in 2/11 carpet dust samples from mold-infested homes (2–4 ng/g)	Engelhart et al.[78]

13	Schools, residential homes, commercial building	Dust samples	Stachylysin (N = 8)	ELISA	Stachylysin concentrations in dust samples ranging from 2.2 to 162 ng/mg	Van Emon et al.[60]
14	8 buildings with water damage and *Stachybotrys* contamination, 4 control buildings, outdoor air (USA)	High-volume liquid impaction bioaerosols samples; case buildings N = 40; control buildings N = 30; outdoor air N = 4	Macrocyclic trichothecenes	ELISA	Trichothecenes in contaminated buildings <10 to >1300 pg/m³ of sampled air, significantly lower levels in control homes (<10–120 pg/m³); not detected in outdoor air	Brasel et al.[43]
15	15 buildings with and 9 buildings without mold/ moisture damage	Surface swab, floor dust, and air samples	Macrocyclic trichothecenes	ELISA	Higher concentration of macrocyclic trichothecenes in floor dust, surface swabs, and air from moldy dwellings vs controls, but significant only for floor dust	Charpin-Kadouch et al.[44]
16	Mold-contaminated indoor environments	Moldy material samples (N = 15)	Satratoxins G + H	LC-MS/MS	Satratoxins G and H in 4 samples of wallpaper (max. 9.7 and 12 μg/cm²); roridin E and L-2, satratoxin F, verrucarin J	Gottschalk et al.[79]
17	Buildings with history of water damage	62 building materials (N = 62), settled dust samples (N = 8)	Verrucarol, trichodermol, sterigmatocystin, satratoxins G + H	GC-MS/MS, LC-MS/MS	Verrucarol (19–43 pg/mg), trichodermol (2.4–3.4 pg/mg), sterigmatocystin (17 pg/mg) in 3/8 settled dust samples; toxins in 45 of 62 building material samples	Bloom et al.[80]

Continued

TABLE 1 Overview of Studies that Report on the Occurrence of Mycotoxins (and Other Fungal and Bacterial Secondary Metabolites) in Indoor Sample Materials—cont'd

No.	Study Sites and Design	Sample Materials (Number of Samples)	Target Compounds	Analytical Methodology	Main Findings Concerning Indoor Mycotoxin Occurrence	References
18	Dwelling with water damage and *Stachybotrys* contamination	Air sample (5 m^3/h) for 15 h onto 0.8 µm polycarbonate filter	Satratoxins G and H	LC-MS/MS	Satratoxins G (0.25 ng/m^3) and H (0.43 ng/m^3) detected in air sample from a water-damaged building	Gottschalk et al.[45]
19	5 severely mold-contaminated buildings in New Orleans, LA, USA	Bulk samples from floors from selected rooms (N = 7)	Verrucarol, trichodermol, satratoxins G + H, sterigmatocystin	GC-MS/MS, LC-MS/MS	Verrucarol in 3 samples (0.6–18 pg/mg), sterigmatocystin in 2 homes (18–28 pg/mg)	Bloom et al.[81]
20	57 water-damaged buildings, Sweden	Building materials (N = 100), settled dust samples (N = 18)	Verrucarol, trichodermol, sterigmatocystin, gliotoxin, aflatoxin B1, satratoxins G + H	GC-MS/MS, LC-MS/MS	66% of material and 11% of settled dust positive for at least one mycotoxin; building materials: gliotoxin (0.43–1.12 pg/mg), sterigmatocystin (4.9–150,000 pg/mg), trichodermol (0.9–8700 pg/mg), verrucarol (8.8–17,000 pg/mg), satratoxins G + H; settled dust: verrucarol (0.6–1.7 pg/mg)	Bloom et al.[82]

21	3 schools with nonspecific indoor air problems, 1 control school; Finland	Dust swabs and filter materials from ventilation systems (N = 26)	40 mycotoxins	LC-MS/MS	Mycotoxins detected in all intake and exhaust systems (pg–ng/g or pg–ng/cm^2), in total 10 compounds: beauvericin, enniatins, verrucarol most common; penicillic acid, sterigmatocystin, satratoxins, trichodermol, chaetoglobosin A, gliotoxin, aflatoxin B_1	Hintikka et al.[11]
22	7 water damaged buildings in Belgium	Samples of air, dust, wall paper, mycelium, or silicone (N = 99)	20 mycotoxins	LC-MS/MS, LC-Q-TOF (Time-Of-Flight)-MS	62/99 samples positive for at least one of the mycotoxins, mainly roquefortine C, chaetoglobosin A, and sterigmatocystin; also roridine E, ochratoxin A, aflatoxin B_1 and B_2; concentrations in air: 0.3 pg–3.4 ng/m^3; material samples: 8 pg–13 μg/cm^2	Polizzi et al.[12]
23	Damp buildings in Slovakia and Austria	Mold-infested material samples (N = 14; gypsum board, wall scrapings, paper, soil/wood, etc.)	186 fungal and bacterial secondary metabolites	HPLC-MS/MS	20 different compounds detected, all samples positive for min. 3 metabolites, concentration range ng–mg/kg; meleagrin, roquefortine C, sterigmatocystin, enniatins most common; alamethicin, alternariol, alternariol methyl ether, beauvericin, chaetoglobosin A, chaetomin, citrinin, cytochalasins (B, D), emodin, equisetin, meleagrin, stachybotrylactam, viridicatin	Vishwanath et al.[13]

Continued

TABLE 1 Overview of Studies that Report on the Occurrence of Mycotoxins (and Other Fungal and Bacterial Secondary Metabolites) in Indoor Sample Materials—cont'd

No.	Study Sites and Design	Sample Materials (Number of Samples)	Target Compounds	Analytical Methodology	Main Findings Concerning Indoor Mycotoxin Occurrence	References
24	8 secondary schools in Malaysia	Cotton swabs of airborne settled dust (N=32)	Aflatoxin B_1, satratoxins G+H, sterigmatocystin, verrucarol, trichodermol	LS-MS/MS, GC-MS/MS	Aflatoxin B_1 in 1/32 classrooms (67 pg/m^2), sterigmatocystin in 2/32 samples (max. 50.5 ng/m^2), verrucarol in 4/32 samples (max. 467 pg/m^2)	Cai et al.[63]
25	9 residential homes and 2 public buildings with moisture problems; Finland, Sweden	Building materials (N=42), dust bag dust (7), floor dust (13), and settled airborne dust (7)	186 fungal and bacterial secondary metabolites	LC-MS/MS	All samples positive for min. 1 of the target compounds; 33 different fungal and bacterial secondary metabolites; concentrations ranging from pg/g to μg/g; emodin, enniatin B, beauvericin, equisetin, physcion, sterigmatocystin, and meleagrin most common	Täubel et al.[22]
26	5 residential homes with moisture damage (Finland), 2 control house dusts (USA, India)	Vacuumed floor dust (N=7)	186 fungal and bacterial secondary metabolites	LC-MS/MS	15 different metabolites in moisture-damaged homes, 10 in control dust; sterigmatocystin, equisetin, enniatins, cytochalasin D detected in floor dust of Finnish homes with water damage only	Vishwanath et al.[50]

27	97 school buildings with and without moisture damage in Spain, the Netherlands, Finland	Settled dust and surface swab samples (N=741)	186 fungal and bacterial secondary metabolites; verrucarol, trichodermol	LC-MS/MS, GC-MS/MS	42–58% of dust samples positive for min. 1 of the metabolites; 30 different fungal and bacterial metabolites detected; emodin, enniatins, physcion most common (up to 37% prevalence); higher number of mycotoxins at elevated concentration in moisture-damaged school buildings	Peitzsch et al.[51]
28	A residential home with moisture/mold problems	Bulk samples from bathroom, bedroom, crawl space (N=6)	Trichothecenes, aflatoxins, ochratoxin A	ELISA	Trichothecenes (0.47–11.7 ppb), aflatoxins (3.5 ppb), and ochratoxin A (2.1–7.7 ppb)	Trasher et al.[61]
29	5 residential homes and one office building	Dust specimens	Trichothecenes, aflatoxins, ochratoxin A	ELISA	Macrocyclic trichothecenes in all dust samples, small amounts of ochratoxin A detected in 4/6 samples	Brewer et al.[62]
30	95 residential homes in eastern Finland, 14 with moisture damage	Living room floor dust samples (N=95)	330 fungal and bacterial secondary metabolites	LC-MS/MS	42 different mycotoxins and bacterial metabolites (in the order of magnitude of ng/g dust), up to 29 in a single home; only suggestive associations with moisture damage for individual compounds	Pekkanen et al.[52] Kirjavainen et al.[53]

ELISA, enzyme-linked immunosorbent assay; GC, gas chromatography; HPLC, high-performance liquid chromatography; LC, liquid chromatography; MS, mass spectrometry; TLC, thin-layer chromatography; DAD, diode array detection; TOF, Time-Of-Flight.

nanograms per square centimeter of sampling area have been reported. Typical concentrations in floor dust or settled dust are in the range of nanograms per gram (can reach up to μg/g) or picograms per square centimeter of sampled surface, for settled dust.

There are only few reports of mycotoxin findings from actively sampled indoor air[12,43–45] (see Table 1). Reasons for this are (1) active air sampling is far more cumbersome compared to, for example, obtaining a house dust or building material sample and (2) air concentrations of mycotoxins in indoor environments such as residential homes, schools, or offices are usually very low—disregarding here occupational environments with expected high exposure levels, such as grain-handling facilities. The expected low concentrations set the requirements for using highly sensitive analytical methodology and/or on obtaining high-volume air samples, which implies either extended sampling periods or more expensive equipment capable of high-volume sample collection.

All four studies that report mycotoxins from indoor air are rather consistent in that the concentrations reported are in the range of subnanograms to low nanograms per cubic meter of actively collected air. Based on these studies and also on some more experimental work[46] it can be concluded that even though mycotoxins are not volatile as such, they do get airborne on fungal spores, fragments, and other particles of inhalable size, so that exposure of occupants of mold-infested buildings takes place. Brasel et al.[43] and Charpin-Kadouch et al.[44] used a macrocyclic trichothecene-specific enzyme-linked immunosorbent assay (ELISA) as analytical methodology for determination of mycotoxins from indoor air samples. Whereas the authors of the first study reported significantly higher levels of airborne trichothecenes in water-damaged buildings with known *Stachybotrys* contamination compared to control buildings and outdoor air, the French study failed to show such significant difference in air samples, but reported the presence of macrocyclic trichothecenes (MCTs) also in buildings without mold problems. Gottschalk et al.[45] provided the first report of specific mycotoxins in indoor air determined with specific methodology, using liquid chromatography coupled with tandem mass spectrometry (LC-MS/MS). These authors used a method fulfilling the prerequisites for being accepted as a "confirmatory method for organic residues and contaminants" according to guidelines provided by the European Commission.[47] The use of ELISA or other bio/immunoassays, on the other hand, is somewhat problematic in terms of specificity and assay interferences. Air sampling was performed in a moisture-damaged building with known *Stachybotrys* contamination, in which earlier satratoxins G and H had been found in a moldy wall paper. The same mycotoxins were detected also in indoor air, at concentrations of 0.25 and 0.43 ng/m^3 of air, respectively. Polizzi et al.[12] sampled air in seven moisture-damaged homes in Belgium and detected mycotoxins in 6 out of total 20 air samples, using LC-MS methodology. These authors report the detection of roquefortine C, roridin E, sterigmatocystin, chaetoglobosin A as well as ochratoxin A, and aflatoxins B_1 and B_2 from indoor air. These mycotoxins are likely to be produced by certain *Penicillium* spp., *Stachybotrys* spp., *Chaetomium* spp., *Aspergillus versicolor*, and other *Aspergillus* spp. In particular the detection of aflatoxins (max. 0.15 ng/m^3) in indoor air in the study of Polizzi et al.[12] is surprising as these mycotoxins are generally very rarely detected as indoor environmental contaminants.

When looking at the sampling sites listed for the individual studies in Table 1 it becomes obvious that the mycotoxin occurrence indoors is almost exclusively considered in the context of dampness and mold contamination in indoor spaces. The focus of the sample collection is on buildings (and materials) with severe moisture damage and/or known mold contamination. It is comprehensible to assume that moisture damage and mold contamination indoors relate to excess microbial proliferation, which ultimately increases the occurrence of mycotoxins, potentially to a level that might affect human health. It is, however, surprising that testing such a hypothesis has not been recognized in the design of almost any of the studies that have been conducted. Only a handful of studies have considered also sampling undamaged control environments in addition to damaged premises, and even fewer have done so with sufficient sample numbers. This implies that in fact very little is known about what is "normal" background with respect to the indoor occurrence of mycotoxins and which compounds are strongly related to conditions of dampness and mold. This issue is further discussed below.

Development of analytical techniques and instrumentation during the past decades with respect to their specificity and sensitivity has naturally reflected on the methods used in studies on indoor mycotoxins (Table 1). A good number of studies have attempted to follow recommendations to use chromatographic separation combined with mass spectrometry to imply specificity in the detection of the target compounds.[47,48] For a more detailed discussion on analytical methodology in mycotoxin research and monitoring we refer here to Chapter 3.6 in this book. Mycotoxin determination from indoor samples is a complex and challenging task, given the multitude of potentially relevant secondary metabolites being produced by indoor molds and considering that house dust or building materials are very complex and sample matrices are difficult to deal with.[48] Some of the mycotoxin findings listed in Table 1 have been doubted as being false positive reports. For this more analytical discussion, however, we refer to the papers in which this criticism has been formulated.[26,48]

Initial studies on indoor mycotoxin occurrence focused on a few, specific target analytes that were mostly selected based on their toxicological relevance and suggested adverse effects on human health. Until the late 1990s, MCTs were almost exclusively the mycotoxins of interest, owing to being partly highly toxic and being produced by *S. chartarum*, which had been linked to severe disease outcomes.[6,7,9,10] Only since Nielsen and colleagues reported on the immense potential of indoor molds to produce numerous fungal secondary metabolites on indoor surfaces have multimetabolite methods been developed and applied more frequently in indoor studies.[11–13,22,43,49–52] Nevertheless, the number of studies that have considered a large variety of mycotoxins and sufficient sample numbers is still small. Target compounds and sample matrices only partly overlap between studies, and differences in detection limits and recovery rate for the various mycotoxins complicate an objective assessment of prevalence based on different studies. Therefore it is not possible yet at this point to present a final conclusion on the most important mycotoxins or fungal secondary metabolites regarding their indoor occurrence. Considering those studies that have used multimetabolite methods to analyze larger numbers of samples from several indoor locations so far, the following compounds have been listed repeatedly as being more prevalent in indoor samples: enniatins and beauvericin (depsipeptides produced by *Fusarium* spp.), meleagrin and roquefortine C (produced by *Penicillium* spp.), emodin and physcion (anthraquinone derivatives produced by *Eurotium* and *Aspergillus* spp.), sterigmatocystin and its precursors, as well as 3-nitropropionic acid (produced mainly by *A. versicolor* and *Aspergillus* spp., respectively), and chaetoglobosins (mostly produced by *Chaetomium* spp.). The detection of MCTs, most prominently satratoxins G and H, either directly or through their hydrolysis product verrucarol, is reported also in some of these later studies. Given the earlier, multiple literature reports on findings of MCTs in severely mold-affected indoor environments, this group of mycotoxins should be added to the list of mycotoxins most commonly reported indoors.

Concerning the compounds listed here, there are two points to be made: one, not all of them can be considered real mycotoxins in the strict sense, as their effects on human or animal health are little explored or not well established. Two, a source attribution for the occurrence of these metabolites indoors is not always straightforward. Although the production of most of these compounds by indoor molds on building materials or on cultivation medium has been shown, the high prevalence of compounds such as enniatins, emodin, or physcion in indoor dust samples suggests that also an influx from outdoor sources is relevant.[22,51,53] Some plants produce secondary metabolites identical to fungal compounds, for example, the anthraquinone derivatives emodin and physcion. Indoor contamination by outdoor dust and soil particles and also by plant material may be sources of low levels of indoor mycotoxins.

MYCOTOXINS IN THE CONTEXT OF MOISTURE DAMAGE

As pointed out earlier in this chapter, the interest in mycotoxin occurrence indoors is closely linked to moisture damage and dampness in buildings. Obviously, indoor environments with moisture problems generally provide good growth conditions for microbes because of the higher availability of water, which explains general observations of higher microbial levels. However, molds are present and sustain metabolic activity also in "normal" buildings without moisture problems and may find microenvironments in which to proliferate. Thus, the occurrence of mycotoxins in buildings per se cannot be assumed to be a phenomenon of damp indoor environments only. Among the building-associated fungi with mycotoxin-producing potential are various *Aspergillus* and *Penicillium* species, *Chaetomium globosum*, *Wallemia sebi*, *Eurotium* spp., *Trichoderma* spp., and *S. chartarum*, to give a few examples. Certainly, not all of those fungi are "moisture-damage molds"; several are recognized as primary colonizers in buildings based on their lower water activity requirements[26] and may be present also in buildings without moisture problems.

Against this background it is striking to notice how few of the studies conducted so far have used an index/reference design, that is, have collected sample materials from buildings with and without dampness and mold problems. Only studies of such design will ultimately allow the pinning down of mycotoxins that are actually associated with dampness observations in buildings. The vast majority of papers deal with the detection of a few selected mycotoxins in heavily moisture-damaged and/or mold-infested indoor spaces. As of this writing, we have very limited knowledge on which fungal secondary metabolites occur in *undamaged* indoor environments and at what levels; we have insufficient understanding of what can be considered a normal baseline in terms of the presence of mycotoxins indoors.

The studies that used active air sampling in buildings with severe moisture damage and mold contamination and in control buildings without such observations were briefly presented earlier.[43,44] We wish to add here the work published by Yike et al.,[54] who collected a total of 15 air samples from seven mold-contaminated residential homes and six air samples from uncontaminated control rooms. In their paper they describe the development of a sensitive protein translation assay for the detection of trichothecene mycotoxin activity in airborne particles. The assay used measures one of the biological effects of trichothecenes, which is the inhibition of protein translation in target cells, but does not specifically measure the exact amount of a given mycotoxin in a sample. This is the reason we did not include this work in the list of studies in Table 1, as we restricted that review to reports on actual mycotoxin detection in indoor matrices. The authors found a strong inhibition in the assay by air particulate extracts from mold-contaminated homes, but not from control homes, and allocated this inhibition to the presence of trichothecene mycotoxins. *Stachybotrys* spp. were cultivated from the air in almost all of the case buildings, which makes the presence of (macrocyclic) trichothecenes in indoor air in these homes plausible.

Brasel et al.[43] applied a macrocyclic trichothecene-specific ELISA on 40 air samples collected in eight mold-contaminated buildings (from 16 rooms), 30 air samples from four buildings with no visible contamination and history of water damage (14 rooms), and four outdoor air samples; all samples were collected in the state of Texas in the United States. Whereas MCTs were also detected in the air of control environments (<10–120 pg of trichothecene equivalents per cubic meter of air), the authors reported significantly higher levels in air samples from *S. chartarum*-infested rooms (<10 to >1300 pg/m^3). MCTs were not detected in samples of outdoor air.

Charpin-Kadouch et al.[44] also used the ELISA for determination of MCTs in samples of surface swabs from walls, floor dust, and indoor air from 15 French buildings with and nine buildings without moisture and mold problems. MCTs were detected in all samples, both from index and from reference houses. Mean values for MCTs were higher in moldy buildings for all three sample types; however, a significant difference between index and reference homes was observed only for floor dust samples, not in air or surface swab samples.

The study in schools presented by Hintikka et al.[11] is not a study with a strict index and reference design, but is nevertheless briefly mentioned here. The "index" schools were schools that reported some rather nonspecific indoor air problems and were referred to as typical schools in Finland, whereas the "reference" school was chosen based on not having a history of moisture damage. Dust samples were collected in the ventilation ducts of these schools. The authors reported a variety of mycotoxins at concentrations of picograms to nanograms per square centimeter of swabbed surface in ventilation ducts from all schools, with the control school building actually showing the highest number of different mycotoxins in both the supply and the exhaust air. Compounds detected in the ventilation system of this school included MCTs (only in exhaust); chaetoglobosins, beauvericin, and enniatins in both exhaust and intake ducts; and penicillic acid, sterigmatocystin, gliotoxin, and aflatoxin B_1 in supply air only. The authors concluded—based on the presence of some of the mycotoxins in intake and exhaust air ducts—that a good part of the mycotoxins in school buildings might not originate from sources within the building, but either are introduced through outdoor air or may be produced in the ventilation systems, which would indicate insufficient maintenance.

A more recent study in schools compared the mycotoxin occurrence in moisture-damaged versus non-moisture-damaged school buildings following a robust, epidemiological design.[51] The authors targeted more than 180 fungal and bacterial secondary metabolites and presented data from 675 settled dust samples collected in 66 index and reference schools. Index and reference status of the schools located in the Netherlands, Spain, and Finland were based on standardized school building inspections recording observations of moisture damage and dampness. The authors did not find clear statistically significant differences in the occurrence of individual mycotoxins in moisture-damaged versus non-moisture-damaged school buildings. However, they showed a tendency for the occurrence of a larger number of mycotoxins at elevated levels in moisture-damaged schools, reaching significance when considering all samples from three countries. The study concluded that microbial toxins are also present in undamaged buildings as part of the "normal" microbial flora indoors. The authors suggested that not only does moisture damage—by triggering microbial proliferation and metabolite production—act as a source of mycotoxins in indoor settled dust, but also outdoor air and particulate matter seem to have an impact on the indoor occurrence of microbial secondary metabolites. Indeed, the presence of mycotoxins in outdoor air particulate matter has been shown in a follow-up study conducted by some of the same authors.[55]

Vishwanath and colleagues[50] report mycotoxin findings from floor dust from residential homes with moisture damage in Finland as well as from floor dust from two reference homes in the United States and in India. The geographical difference in the origins of the samples forbids a direct comparison of mycotoxin occurrence in index and reference house dusts, as differences may be related to geographical/climatic differences. However, it can be mentioned here that in the reference house dust fungal secondary metabolites were also detected.

A Finnish birth cohort study presented multimetabolite analysis data from floor dust collected from 95 residential homes.[52,53] Fourteen of these homes had inspection-assessed major moisture damage in the living rooms and kitchen. The authors found a nonsignificant tendency toward elevated total number and load of fungal and bacterial secondary metabolites in moisture-damaged homes compared to undamaged homes. Differences in the number of metabolites in index versus reference homes reached statistical significance when compounds occurring at elevated concentrations were considered (similar to the finding reported by Peitzsch et al. in schools[51]). There were suggestive associations of a number of mycotoxins with moisture damage, but these associations were weak and did not survive correction for multiple testing.

SUMMARY, CONCLUDING REMARKS, FUTURE CHALLENGES

This chapter identifies 30 reports in the literature that have attempted to determine mycotoxins from indoor sample materials, including building material, surface swab, dust, and active air samples. The majority of reports refer to residential homes; schools, offices, and other public buildings are also considered in a few studies. The research on the indoor occurrence of mycotoxins is essentially exclusively conducted in the context of dampness and mold contamination in buildings. Typically, samples are collected from severely moisture-damaged and/or mold-infested premises; where building materials are concerned, these are usually from mold-affected areas. The results of an analysis of samples collected in undamaged, control environments are rarely presented.

In particular earlier studies, but also some of the more recent reports, used analytical methodology that targeted only a few specific mycotoxins. Those target compounds have been selected primarily based on the toxicological properties of some of the fungi known to occur in damp indoor environments. Examples here are the MCTs produced by *S. chartarum* or also sterigmatocystin of *A. versicolor*—compounds that are partly highly toxic and are known to pose a threat to human health upon exposure. Reports of MCTs or also sterigmatocystin indoors are therefore frequently found in the literature. However, there is an enormous variety of mycotoxins produced by fungi in our environment, with estimates starting at 20,000 unique mycotoxins. Considering this and realizing the limited availability of occurrence data for a wider range of mycotoxins, it is unclear whether compounds that have been the focus of early indoor mycotoxin studies are also the most relevant targets with respect to their prevalence indoors. Not all mycotoxins that are detected in indoor dust or indoor air originate from indoor sources. The high prevalence of some fungal secondary metabolites—such as enniatins, beauvericin, emodin, and physcion—observed in indoor samples in a number of studies suggests influx from outdoor sources.

Conclusions about which would be the most relevant mycotoxins in indoor spaces that are *associated with observations of indoor dampness and mold* are impossible at this stage. There is a striking lack of studies that would have considered a large variety of microbial secondary metabolites from a sufficient number of samples collected in a standardized manner from indoor environments with and without moisture damage. The few studies that provide such index/reference data have not shown very clear differences in the prevalence or levels of individual mycotoxins. Such studies rather indicate more subtle differences, for example, a higher number of mycotoxins occurring at elevated concentrations in damp compared to reference buildings. The respective studies were, however, limited to some extent as concerns the severity of the moisture damage conditions[51] and/or the sample numbers.[52,53] There is a need for future studies recruiting severely damaged indoor environments into a sound, epidemiological index/reference study design.

Ultimately, the aim of our research must be to answer the question "... if mycotoxins at concentrations found in mould damaged indoor environments make us sick," as was formulated by Bloom in her Ph.D. thesis.[56] Despite some 30 years of research dealing with the indoor occurrence of mycotoxins, sound information on actual health effects is limited. Several of the studies on the occurrence of mycotoxins indoors listed in Table 1 present some information on health complaints or in a few cases clinical data for the occupants of these buildings.[5,8,10,57–62] However, no formal statistical analyses relating to the measured mycotoxin exposure of the individuals to their health outcomes were performed in any of these studies, mostly for reasons of too few samples or insufficient patient or mycotoxin exposure data. Support from epidemiological studies in clarifying potential health effects upon indoor mycotoxin exposure is almost completely absent. The fact that there are no commonly accepted biomarkers of airborne exposure to multiple mycotoxins in non-occupational settings is certainly contributing to this situation. Cai et al.[63] analyzed associations of verrucarol (the hydrolysis product of MCTs) in surface dust collected in 32 classrooms on self-reported respiratory symptoms of 462 pupils in eight schools in Malaysia. The authors reported an inverse association of verrucarol with daytime breathlessness. The study was, however, limited by the fact that verrucarol was detected in only 4 of 32 classrooms. Two studies that used multimetabolite analysis methods in an epidemiological study setting are being prepared for publication as of this writing, with part of these data being published in conference proceedings. Kirjavainen et al.[53] (manuscript under review; see also Ref. 52) explored associations of mycotoxins in floor dust at early life with development of asthma in 95 pupils in a Finnish birth cohort study. The authors reported that neither sum load nor number of individual toxins were associated with a risk of doctor-diagnosed asthma; positive and negative indicative associations were observed for individual mycotoxins with the development of asthma. Zock et al.[64] (full manuscript under preparation see also Ref. 55) studied respiratory symptoms in 645 teachers in relation to dampness and levels of microbial secondary metabolites in schools in Finland, the Netherlands, and Spain. Preliminary analyses indicated a significant dose–response association of mycotoxin load in dust with the asthma symtpom score and nasal symptoms in Finnish teachers.

There is good knowledge of the toxicological mechanisms that mycotoxins may exert—those are discussed in another section of this book (Chapter 3.6). However, most of what is known refers to food-borne mycotoxins and ingestion as an exposure route. Apart from a few exceptions, much less information is available from the more indoor-relevant metabolites and their toxicology upon inhalation exposure. It has been suggested in the literature that inhalation of mycotoxins may be many times more toxic than ingestion,[65,66] but this is in fact very little understood. Mycotoxins are present in indoor environments and inhalation exposure takes place. Mechanistic work[46,67,68] has shown that not only fungal spores carrying mycotoxins but also fragments may be crucial in terms of exposure. Based on the few reports of mycotoxins in the air of mold-infested indoor spaces, "usual" exposure levels are very low, in the concentration range of picograms to nanograms per cubic meter of air. Thus—unless high-exposure occupational settings are concerned, or exceptionally severe cases of indoor contamination—acute toxic effects of mycotoxin exposure may be rare. Nevertheless, chronic low-level exposure to mycotoxins through inhalation could also represent a health hazard. For example, Miller et al.[69,70] have shown alterations in the expression of inflammation-related genes in mouse lungs upon exposure to "real-world" levels of various indoor mycotoxins. There are in vitro studies that indicate that synergistic effects in evoking cellular responses may be highly relevant when considering multiple mycotoxins and mycotoxins and other microbial compounds.[71–74] These findings also link to the initial scheme presented (Figure 1), illustrating the multitude of biological and chemical exposures present in damp buildings. Measurement of single agents is probably not sufficient if we want to attempt to elucidate the adverse health effects observed in exposed occupants, but rather we will have to take into consideration the complexity of the exposure situation in damp buildings in future assessments.

REFERENCES

1. Mendell MJ, Mirer AG, Cheung K, Tong M, Douwes J. Respiratory and allergic health effects of dampness, mold, and dampness-related agents: a review of the epidemiological evidence. *Environ Health* 2011;**119**:748–56.
2. WHO. *WHO Guidelines for indoor air quality: dampness and mould.* Copenhagen (Denmark): World Health Organization; 2009.
3. IOM (Institute of Medicine). *Damp indoor spaces and health.* Washington DC: National Academic Press; 2004.
4. Tischer C, Chen C-M, Heinrich J. Association between domestic mould and mould components, and asthma and allergy in children: a systematic review. *Eur Resp J* 2011;**38**:812–24.
5. Croft WA, Jarvis BB, Yatatwara CS. Airborne outbreak of trichothecene toxicosis. *Atmos Environ* 1986;**20**:549–52.

6. Dearborn DG, Yike I, Sorenson WG, Miller MJ, Etzel RA. Overview of investigations into pulmonary hemorrhage among infants in Cleveland, Ohio. *Environ Health Perspect* 1999;**107**:495–9.
7. Etzel RA, Montaña E, Sorenson WG, Kullman GJ, Allan TM, Dearborn DG, et al. Acute pulmonary hemorrhage in infants associated with exposure to *Stachybotrys atra* and other fungi. *Arch Pediatr Adolesc Med* 1998;**152**:757–62.
8. Flappan SM, Portnoy J, Jones P, Barnes C. Infant pulmonary hemorrhage in a suburban home with water damage and mold (*Stachybotrys atra*). *Environ Health Perspect* 1999;**107**:927–30.
9. Jarvis BB, Sorenson WG, Hintikka EL, Nikulin M, Zhou Y, Jiang J, et al. Study of toxin production by isolates of *Stachybotrys chartarum* and *Memnoniella echinata* isolated during a study of pulmonary hemosiderosis in infants. *Appl Environ Microbiol* 1998;**64**:3620–5.
10. Johanning E, Biagini R, Hull D, Morey P, Jarvis B, Landsbergis P. Health and immunology study following exposure to toxigenic fungi (*Stachybotrys chartarum*) in a water-damaged office environment. *Int Arch Occup Environ Health* 1996;**68**:207–18.
11. Hintikka E-L, Holopainen R, Asola A, Jestoi M, Peitzsch M, Kalso S. Mycotoxins in the ventilation systems of four schools in Finland. *World Mycotox J* 2009;**2**:369–79.
12. Polizzi V, Delmulle B, Adams A, Moretti A, Susca A, Picco AM, et al. JEM Spotlight: fungi, mycotoxins and microbial volatile organic compounds in mouldy interiors from water-damaged buildings. *J Environ Monit* 2009;**11**:1849–58.
13. Vishwanath V, Sulyok M, Labuda R, Bicker W, Krska R. Simultaneous determination of 186 fungal and bacterial metabolites in indoor matrices by liquid chromatography/tandem mass spectrometry. *Anal Bioanal Chem* 2009;**395**:1355–72.
14. Smith JE, Solomons GL. *Mycotoxins in human nutrition and health*. EUR16048 EN. European Commission, Directorate-General XII; 1994. [Report].
15. Williams DH, Stone MJ, Hauck PR, Rahman SK. Why are secondary metabolites (natural products) biosynthesized? *J Nat Prod* 1989;**52**:1189–208.
16. Bennet JW, Klich M. Mycotoxins. *Clin Microbiol Rev* 2003;**16**:497–516.
17. Council for Agricultural Science and Technology (CAST). *Mycotoxins: risk in plant, animal, and human systems*. Task force report 139. Ames (Iowa): Council for Agricultural Science and Technology; 2003.
18. Brera C, Caputi R, Miraglia M, Iavicoli I, Salerno A, Carelli G. Exposure assessment to mycotoxins in workplaces: aflatoxins and ochratoxin A occurrence in airborne dusts and human sera. *Microchem J* 2002;**73**:167–73.
19. Selim MI, Juchems AM, Popendorf W. Assessing airborne aflatoxin B1 during on-farm grain handling activities. *Am Ind Hyg Assoc J* 1998;**59**:252–6.
20. Viegas S, Veiga L, Figueredo P, Almeida A, Carolinao E, Sabino R, et al. Occupational exposure to aflatoxin B1: the case of poultry and swine production. *World Mycotoxin J* 2013;**6**:309–15.
21. Demain AL. Pharmaceutically active secondary metabolites of microorganisms. *Appl Microbiol Biotechnol* 1999;**52**:455–63.
22. Täubel M, Sulyok M, Vishwanath V, Bloom E, Turunen M, Järvi K, et al. Co-occurrence of toxic bacterial and fungal secondary metabolites in moisture-damaged indoor environments. *Indoor Air* 2011;**21**:368–75.
23. Frisvad JC, Gravesen S. *Penicillium* and *Aspergillus* from Danish homes and working places with indoor air problems: identification and mycotoxin determination. In: Samson RA, Flannigan B, Flannigan ME, Verhoeff AP, editors. *Health implications of fungi in indoor air environment*. Amsterdam (The Netherlands): Elsevier; 1994. p. 281–90.
24. Ren P, Ahearn DG, Crow SA. Comparative study of *Aspergillus* mycotoxin production on enriched media and construction material. *J Ind Microbiol Biotechnol* 1999;**23**:209–13.

25. Nielsen KF. *Mold growth on building materials – secondary metabolites, mycotoxins and biomarkers* [Ph.D. thesis]. Technical University of Denmark; 2002.
26. Nielsen KF. Mycotoxin production by indoor molds. *Fung Gen Biol* 2003;**39**:103–17.
27. Jarvis BB, Miller JD. Mycotoxins as harmful indoor air contaminants. *Appl Microbiol Biotechnol* 2005;**66**:367–72.
28. Andersen B, Nielsen KF, Jarvis BB. Characterization of *Stachybotrys* from water-damaged buildings based on morphology, growth, and metabolite production. *Mycologia* 2002;**94**:392–403.
29. Andersen B, Nielsen KF, Thrane U, Szaro T, Taylor JW, Jarvis BB. Molecular and phenotypic descriptions of *Stachybotrys chlorohalonata* sp. nov. and two chemotypes of *Stachybotrys chartarum* found in water damaged buildings. *Mycologia* 2003;**95**:1227–38.
30. Andersen B, Hollensted M. Metabolite production by different *Ulocladium* species. *Int J Food Microbiol* 2008;**126**:172–9.
31. Andersson MA, Mikkola R, Raulio M, Kredics L, Maijala P, Salkinoja-Salonen MS. Acrebol, a novel toxic peptaibol produced by an *Acremonium exuvarium* indoor isolate. *J Appl Microbiol* 2009;**106**:909–23.
32. De La Campa R, Seifert K, Miller JD. Toxins from strains of *Penicillium chrysogenum* isolated from buildings and other sources. *Mycopathologia* 2007;**163**:161–8.
33. Fogle MR, Douglas DR, Jumper CA, Straus DC. Growth and mycotoxin production by *Chaetomium globosum. Mycopathologia* 2007;**164**:49–56.
34. McMullin DR, Sumarah MW, Miller JD. Chaetoglobosins and azaphilones produced by Canadian strains of *Chaetomium globosum* isolated from the indoor environment. *Mycotoxin Res* 2013;**29**:47–54.
35. Mikkola R, Andersson MA, Kredics L, Grigoriev PA, Sundell N, Salkinoja-Salonen MS. 20-Residue and 11-residue peptaibols from the fungus *Trichoderma longibrachiatum* are synergistic in forming NA^+/K^+-permeable channels and averse action towards mammalian cells. *FEBS J* 2012;**279**:4172–90.
36. Nielsen KF, Huttunen K, Hyvärinen A, Andersen B, Jarvis BB, Hirvonen MR. Metabolite profiles of *Stachybotrys* isolates from water-damaged buildings and their induction of inflammatory mediators and cytotoxicity in macrophages. *Mycopathologia* 2001;**154**:201–5.
37. Nieminen SM, Kärki R, Auriola S, Toivola M, Laatsch H, Laatikainen R, et al. Isolation and identification of *Aspergillus fumigatus* mycotoxins on growth medium and some building materials. *Appl Environ Microbiol* 2002;**68**:4871–5.
38. Gutarowska B, Sulyok M, Krska R. A study of the toxicity of moulds isolated from dwellings. *Indoor Built Environ* 2010;**19**:668–75.
39. Nielsen KF, Thrane U, Larsen TO, Nielsen PA, Gravesen S. Production of mycotoxins on artificially inoculated building materials. *Int Biodeter Biodegr* 1998;**42**:9–16.
40. Nielsen KF, Gravesen S, Nielsen PA, Andersen B, Thrane U, Frisvad JC. Production of mycotoxins on artificially and naturally infested building materials. *Mycopathologia* 1999;**145**:43–56.
41. Nielsen KF, Holm G, Uttrup LP, Nielsen PA. Mould growth on building materials under low water activities. Influence of humidity and temperature on fungal growth and secondary metabolism. *Int Biodeter Biodegr* 2004;**54**:325–36.
42. Hyvärinen A, Vahteristo M, Meklin T, Jantunen M, Nevalainen A, Moschandreas D. Temporal and spatial variation of fungal concentrations in indoor air. *Aerosol Sci Tech* 2001;**35**:688–95.
43. Brasel TL, Martin JM, Carriker CG, Wilson SC, Straus DC. Detection of airborne *Stachybotrys chartarum* macrocyclic trichothecene mycotoxins in the indoor environment. *Appl Environ Microbiol* 2005;**71**:7376–88.
44. Charpin-Kadouch C, Maurel G, Felipo R, Queralt J, Ramadour M, Dumon H, et al. Mycotoxin identification in mold dwellings. *J Appl Toxicol* 2006;**26**:475–9.

45. Gottschalk C, Bauer J, Meyer K. Detection of satratoxin G and H in indoor air from a water-damaged building. *Mycopathologia* 2008;**166**:103–7.
46. Brasel TL, Douglas DR, Wilson SC, Straus DC. Detection of airborne *Stachybotrys chartarum* macrocyclic trichothecene mycotoxin on particulates smaller than conidia. *Appl Environ Microbiol* 2005;**71**:114–22.
47. Commission Decision implementing Council Directive 96/23/EC concerning the performance of analytical methods and the interpretation of results. *Official J Eur Communities* 2002. L221/8–L221/36.
48. Nielsen KF, Frisvad JC. Mycotoxins in building materials. In: Adan OCG, Samson RA, editors. *Fundamentals of mould growth in indoor environments and strategies for healthy living*. Wageningen Academic Publishers; 2011. p. 245–75.
49. Tuomi T, Reijula K, Johnsson T, Hemminki K, Hintikka EL, Lindroos O, et al. Mycotoxins in crude building materials form water-damaged buildings. *Appl Environ Microbiol* 2000;**66**:1899–904.
50. Vishwanath V, Sulyok M, Weingart G, Kluger B, Täubel M, Mayer S, et al. Evaluation of settled floor dust for the presence of microbial metabolites and volatile anthropogenic chemicals in indoor environments by LC-MS/MS and GC-MS methods. *Talanta* 2011;**85**:2027–38.
51. Peitzsch M, Sulyok M, Täubel M, Vishwanath V, Krop E, Borràs-Santos A, et al. Microbial secondary metabolites in school buildings inspected for moisture damage in Finland, The Netherlands and Spain. *J Environ Monit* 2012;**14**:2044–53.
52. Pekkanen J, Kirjavainen P, Sulyok M, Täubel M, Karvonen A, Krska R, et al. Microbial toxins in residential indoor environment. In: *Abstract # 4560 of the 2013 conference of the International Society of Environmental Epidemiology (ISEE), the International Society of Exposure Science (ISES), and the International Society of Indoor Air Quality and Climate (ISIAQ), August 19–23, 2013, Basel, Switzerland*. Environ Health Perspect; 2013. http://dx.doi.org/10.1289/ehp.ehbasel13.
53. Kirjavainen PV, Täubel M, Karvonen AM, Sulyok M, Tiittanen P, Krska R, et al. Microbial secondary metabolites in homes in association with moisture damage and asthma. *Indoor Air* 2015, accepted for publication.
54. Yike I, Allan T, Sorenson WG, Dearborn DG. Highly sensitive protein translation assay for trichothecene toxicity in airborne particulates: comparison with cytotoxicity assays. *Appl Environ Microbiol* 1999;**65**:88–94.
55. Täubel M, Sulyok M, Krop E, Borrás-Santos A, Pekkanen J, Heederik D, et al. Microbial secondary metabolites in indoor environments. In: *Abstract # 5929 of the 2013 conference of the International Society of Environmental Epidemiology (ISEE), the International Society of Exposure Science (ISES), and the International Society of Indoor Air Quality and Climate (ISIAQ), August 19–23, 2013, Basel, Switzerland*. Environ Health Perspect; 2013. http://dx.doi.org/10.1289/ehp.ehbasel13.
56. Bloom E. *Mycotoxins in indoor environments – determination using mass spectrometry* [Ph.D. thesis]. Lund University; 2008.
57. Smoragiewicz W, Cossette B, Boutard A, Krystyniak K. Trichothecene mycotoxins in the dust of ventilation systems in office buildings. *Int Arch Occup Environ Health* 1993;**65**:113–7.
58. Hodgson MJ, Morey P, Leung WY, Morrow L, Miller D, Jarvis BB, et al. Building-associated pulmonary disease from exposure to *Stachybotrys chartarum* and *Aspergillus versicolor*. *JOEM* 1998;**40**:241–9.
59. Richard JL, Plattner RD, May J, Liska SL. The occurrence of ochratoxin A in dust collected from a problem household. *Mycopathologia* 1999;**146**:99–103.
60. Van Emon JM, Reed AW, Yike I, Vesper SJ. ELISA measurement of stachylysin^{TM} in serum to quantify human exposures to the indoor mold *Stachybotrys chartarum*. *JOEM* 2003;**45**:582–91.
61. Thrasher JD, Gray MR, Kilburn KH, Dennis DP, Yu A. A water-damaged home and health of occupants: a case study. *J Environ Public Health* 2012:1–10.

62. Brewer JH, Thrasher JD, Straus DC, Madison RA, Hooper D. Detection of mycotoxins in patients with chronic fatigue syndrome. *Toxins* 2013;**5**:605–17.
63. Cai GH, Hashim JH, Hashim Z, Ali F, Bloom E, Larsson L, et al. Fungal DNA, allergens, mycotoxins and associations with asthmatic symptoms among pupils in schools from Johor Bahru, Malaysia. *Ped Allerg Immunol* 2011;**22**:290–7.
64. Zock J-P, Borras-Santós A, Jacobs J, Espinosa A, Haverinen-Shaughnessy U, Krop E, et al. Dampness and microbial secondary metabolites in schools and respiratory symptoms in teachers. European Respiratory Society Annual Conference, Munich, Abstract 2161.
65. Creasia DA, Thurman JD, Jones 3rd LJ, Nealley ML, York CG, Wannemacher Jr RW, et al. Acute inhalation toxicity of T-2 mycotoxin in mice. *Fund Appl Toxicol* 1987;**8**:230–5.
66. Creasia DA, Thurman JD, Wannemacher Jr RW, Bunner DL. Acute inhalation toxicity of T-2 mycotoxin in the rat and guinea pig. *Fund Appl Toxicol* 1990;**14**:54–9.
67. Kildesø J, Würtz H, Nielsen KF, Kruse P, Wilkins K, Thrane U, et al. Determination of fungal spore release from wet building materials. *Indoor Air* 2003;**13**:48–55.
68. Górny RL, Reponen T, Willeke K, Schmechel D, Robine E, Boissier M, et al. Fungal fragments as indoor air biocontaminants. *Appl Environ Microbiol* 2002;**68**:3522–31.
69. Miller JD, Sun M, Gilyan A, Roy J, Rand TG. Inflammation-associated gene transcription and expression in mouse lungs induced by low molecular weight compounds from fungi from the built environment. *Chem Biol Interact* 2010;**183**:113–24.
70. Rand TG, Dipenta J, Robbins C, Miller JD. Effects of low molecular weight fungal compounds on inflammatory gene transcription and expression in mouse alveolar macrophages. *Chem Biol Interact* 2011;**190**:139–47.
71. Huttunen K, Pelkonen J, Nielsen KF, Nuutinen U, Jussila J, Hirvonen MR. Synergistic interaction in simultaneous exposure to *Streptomyces californicus* and *Stachybotrys chartarum*. *Environ Health Perspect* 2004;**112**:659–65.
72. Kankkunen P, Rintahaka J, Aalto A, Leino M, Majuri ML, Alenius H, et al. Trichothecene mycotoxins activate inflammatory response in human macrophages. *J Immunol* 2009;**182**:6418–25.
73. Islam Z, Amuzie CJ, Harkema JR, Pestka JJ. Neurotoxicity and inflammation in the nasal airways of mice exposed to the macrocyclic trichothecene mycotoxin roridin a: kinetics and potentiation by bacterial lipopolysaccharide coexposure. *Toxicol Sci* 2007;**98**:526–41.
74. Mueller A, Schlink U, Wichmann G, Bauer M, Graebsch C, Schüürmann G, et al. Individual and combined effects of mycotoxins from typical indoor moulds. *Toxicol Vitro* 2013;**27**:1970–8.
75. Andersson MA, Nikulin M, Köljalg U, Andersson MC, Rainey F, Reijula K, et al. Bacteria, molds, and toxins in water-damaged building materials. *Appl Environ Microbiol* 1997;**63**:387–93.
76. Nielsen KF, Hansen MO, Larsen TO, Thrane U. Production of trichothecene mycotoxins on water damaged gypsum boards in Danish buildings. *Int Biodeter Biodegr* 1998;**42**:1–7.
77. Gravesen S, Nielsen PA, Iversen R, Nielsen KF. Microfungal contamination of damp buildings – examples of risk constructions and risk materials. *Environ Health Perspect* 1999;**107**:505–8.
78. Engelhart S, Loock A, Skutlarek D, Sagunski H, Lommel A, Färber H, et al. Occurrence of toxigenic *Aspergillus versicolor* isolates and sterigmatocystin in carpet dust from damp indoor environments. *Appl Environ Microbiol* 2002;**68**:3886–90.
79. Gottschalk C, Bauer J, Meyer K. Determination of macrocyclic trichothecenes in mouldy indoor materials by LC-MS/MS. *Mycotoxin Res* 2006;**22**:189–92.
80. Bloom E, Bal K, Nyman E, Must A, Larsson L. Mass spectrometry-based strategy for direct detection and quantification of some mycotoxins produced by *Stachybotrys* and *Aspergillus* spp. in indoor environments. *Appl Environ Microbiol* 2007;**73**:4211–7.
81. Bloom E, Grimsley LF, Pehrson C, Lewis J, Larsson L. Molds and mycotoxins in dust from water-damaged homes in New Orleans after Hurricane Katrina. *Indoor Air* 2009;**19**:153–8.
82. Bloom E, Nyman E, Must A, Pehrson C, Larsson L. Molds and mycotoxins in indoor environments – a survey in water-damaged buildings. *J Occup Environ Hyg* 2009;**6**:671–8.

Chapter 19

Occupational Exposure to Mycotoxins and Preventive Measures

Stefan Mayer
Berufsgenossenschaft Handel und Warendistribution, Department of Prevention, Mannheim, Germany

CHARACTERISTICS OF OCCUPATIONAL MYCOTOXIN EXPOSURE

An important characteristic of occupational mycotoxin exposure is the route of exposure. Whereas the most important exposure for the general population is oral intake of contaminated food, at workplaces inhalation and dermal contact are typical routes. However, the state of knowledge on occupational mycotoxin exposure lags behind knowledge about dietary exposure. This is caused by major uncertainties with regard to transfer from contaminated material into air (inhalable mycotoxin concentration) and/or the toxin fraction absorbed upon dermal contact or after respiratory intake.[1,2] Airborne exposure is also discussed for indoor environments. However, the evidence that mycotoxins have a role in health problems related to indoor air is weak and it is highly unlikely that the exposure is high enough to cause adverse health effects.[3–5] As opposed to indoor environments, workplace-related exposure to airborne molds and mycotoxins can be much higher.

Another characteristic of occupational exposure is the handling of products and materials on an industrial scale. For instance, when a moldy orange has to be disposed in a private kitchen, exposure to mold and mycotoxins will be limited whereas in workplaces for wholesale, tons of oranges have to deposited in case of mold infestation, which can result in much higher exposure. Because of the high amounts of products and materials that are handled in occupational environments, often high amounts of dust are generated. In the dust an enrichment of mycotoxins can occur. This is probably because mycotoxin concentration is usually higher in the outer shell layer than in the product itself. This is well demonstrated for grain kernels.[6–8] Jargot and Melin observed up to 15-fold higher mycotoxin concentration in airborne dust compared with bulk material.[9]

Environmental Mycology in Public Health. http://dx.doi.org/10.1016/B978-0-12-411471-5.00019-3

INDICATIONS OF OCCURRENCE OF MYCOTOXINS IN OCCUPATIONAL SETTINGS

Prerequisites for mycotoxin exposure are the presence of mold and favorable conditions for mycotoxin production. The production of mycotoxins depends on several factors such as nutrient supply, temperature, pH in the commodities, and even light wavelength.[10,11] The most important factor determining mold growth is the extent of available water and humidity. However, mycotoxin production usually needs higher available water content as precondition.[12] In some cases, a slight increase in humidity is sufficient to promote mold growth and mycotoxin production. Nuts have only a limited ability to bind water. A small increase in absolute water content may therefore result in a marked increase in available water, which can make nuts susceptible to mold growth.[13]

Mycotoxin production is often species specific.[14] Anderson and Thrane published a list of fungal species found on apples, cherries, barley, and wheat from the northern temperate zone, together with a list of mycotoxins known to be produced by these fungi.[15] This offers the opportunity to predict and curtail the list of potentially occurring mycotoxins. However, this approach requires correct identification of the mold as a key feature.[16]

In workplaces, the natural background concentration of mold in the air is often exceeded, especially when organic products are handled. In the case of waste handling or building restoration after water damage, the presence of mold and exposure to it are obvious.[17–20] However, visible mold is not necessarily a precondition for mold and mycotoxin exposure. Mold and mycotoxin exposure can occur even if products do not show visible mold growth and even if they meet the strict requirements of food and feed safety regulations. This can be illustrated by grain kernels. Although grain kernels usually are optically free of mold growth, there is a thin, invisible layer of microorganisms on each kernel. Grain usually contains 10^3–10^4 colony-forming units (CFU) of mold per gram (personal communication, Dr Münzing, Department of Safety and Quality of Cereals, Detmold, Germany). During harvest and transport, the kernels rub against each other and microorganisms are released from the surface to the ambient air. This process is also the reason for elevated endotoxin concentrations observed during grain handling.[21,22] Various dry products such as onions, hay, spices, and herbs tend to release mold during handling.[23–25] Iamanaka et al. found that up to 90% of black sultanas were infected by toxigenic species *Aspergillus carbonarius*, *Aspergillus niger*, and *Aspergillus ochraceus* without being visible.[26] Mayer et al. analyzed settled dust samples from grain elevators.[27] Although there was no evidence of mold growth on the grains, each dust sample contained mycotoxins. Thus, mold and mycotoxin exposure can occur even if mold is not clearly visible.

More obvious is the presence of molds on waste and wet organic materials, such as plant pots made of recycled paper. After handling plants grown in such decomposable pots, three horticulture workers developed a painful inflamed efflorescence on their fingertips followed by scaling of the skin.[28] *Stachybotrys chartarum* detected on the surface of the plant pots was suspected to be responsible for the skin lesions.

Indications for the occurrence of mycotoxins can often be derived from food and feed monitoring data and scientific literature and can provide useful information about possible occupational exposure to mycotoxins, at least at a qualitative level. For instance, mycotoxins have been frequently identified in grain products, coffee beans, spices, and nuts.[27,29–33] In a study by Brera et al., during processing of black pepper with no sign of mold, maximum ochratoxin A (OTA) concentrations of 0.43 ng/m^3 in ambient air and 8.30 ng/m^3 in personal air samples were detected.[34]

A recent example is the detection of aflatoxin M1 in a milk sample, which occurred as a result of feeding highly aflatoxin B1–contaminated maize to cows.[35] When this was discovered, further use of remaining batches of the contaminated maize was forbidden, and appropriate protective measures were ascertained to avoid occupational exposure upon disposal of the contaminated material.[36] A good overview of mycotoxins in food and feed was given, among other things, by Jestoi and Marin et al., and provides a good data basis for the occurrence of mycotoxin in food commodities.[37,38] However, it has to be verified whether the occurrence of mycotoxins in products or materials is also associated with exposure to workers.

Table 1 summarizes information on some workplaces where mycotoxin occurrence has been investigated to date. Such investigations are usually restricted to industrial sectors with a high throughput of organic materials.

Because the conditions for mycotoxin production differ between mold species and even between mold strains, it is difficult to predict mycotoxin production from the occurrence of mold.[16] Anderson et al. identified two different chemotypes of *S. chartarum*, one producing macrocyclic trichothecenes and one producing atranones and dolabellanes.[59] Gareis et al. described the mold species *Stachybotrys chorohalonata*, which is phenotypically identical to *S. chartarum* but produces neither macrocyclic trichothecenes nor atranones or dolabellanes.[60] Thus, it is difficult to make predictions about mycotoxin occurrence for a given condition based only on information about mold occurrence.

A quantitative correlation between exposure to airborne molds and exposure to mycotoxins is not obligatory.[61] Possibly, a small but active fungal biomass may produce the same amount of mycotoxins as a big fungal biomass that produces only small amounts of mycotoxins, what may complicate assessment of mycotoxin exposure. Often mycotoxin production is tested in isolated species from workplaces grown on different media.[57,62,63] Muñoz et al. investigated mycotoxin production by different ochratoxigenic *Aspergillus* and *Penicillium* species on coffee- and wheat-based media.[64] OTA production clearly depended on the time of incubation, fungal species, and medium composition. On a coffee-based medium, moderate OTA levels were produced by *A. ochraceus* BFE635 (9.8 µg/g) and by *A. niger* BFE632 (10.6 µg/g) on day 8 of incubation. In a wheat-based medium, these strains produced much more OTA than

TABLE 1 Occurrence of Mycotoxins in Different Occupational Environments

Occupational Environment	Sampling	Mycotoxin Concentration	References
Peanut mill	Settled dust In air	250–410 ng AfB_1/g dust 0.9–72 ng AfB_1/m^3	39
Peanut shelling plant	In air	0.4–7.6 ng AfB_1/m^3	40
Grain elevator	In settled dust In air (calculated)	4 ng OTA/g (2–128) 40 pg/m^3 (2–600)	30
Grain elevator	In settled dust	104 ng OTA/g dust (17–318) 244 ng Citrinin/g dust (137–344)	41
Grain elevator	In air (calculated)	2 pg OTA/m^3 (0.07–690) 2 ng DON/m^3 (0.2–720) 1 ng ZEA/m^3 (0.1–501)	27
Grain harvest	In settled dust	0.4–2.8 ng OTA/g 0.1–239 ng DON/g 4–339 ng NIV/g	7
Handling of bulk wheat	In air	36 ng OTA/m^3	9
Handling of bulk maize	In air	127 pg fumonisins/m^3	9
Coffee, cocoa, spices	In air personal sampling	0.006–0.087 ng OTA/m^3	29
Blending of spices	In air	1.6–53.4 ng OTA/m^3 3.6–26.5 pg total aflatoxins/m^3	9
Chicory pellets and powder	In air	1.2 pg total aflatoxins/m^3	9
Malt house	In settled dust	0.99 ng OTA/g (0.05–9.90 ng/g) in brewing barley 1.11 ng OTA/g (0.05–3.5) in malt	42

TABLE 1 Occurrence of Mycotoxins in Different Occupational Environments—cont'd

Occupational Environment	Sampling	Mycotoxin Concentration	References
Grain harvest	In settled dust	<20 ng DON/g 54 ng HT-2/g <50 ng T-2/g	43
Grain harvest	In air, personal, and stationary sampling	0.023–6.86 ng aflatoxins/m^3	44
Grain harvest	In air, personal, and stationary sampling	0.3–91 ng AfB_1/m^3	45
Grain handling	In air, personal, and stationary sampling	0.23–54.5 ng aflatoxins/m^3	44
Pigsty	In air, stationary	5–421 ng AfB_1/m^3	45
Pigsty, cleaning	In air, personal, and stationary sampling	124–4849 ng AfB_1/m^3	45
Cowshed	In settled dust	0.2–70 ng OTA/g	46
Cattle feeding	In air, personal sampling	2.60 μg gliotoxins/m^3	47
Poultry houses	In air	0.08 ng AfB_1/m^3 8.53 ng OTA/m^3 2.3 ng ZEA/m^3	48
Indoor	Carpet dust	2–4 ng sterigmatocystins/g	49
Indoor	In air	<10–> 1300 pg/m^3 trichothecene equivalents in room with *Stachybotrys chartarum* 2.2–120 pg/m^3 trichothecene equivalents in room without *S. chartarum*	50
Indoor	In air	Satratoxin G 0.25 ng/m^3 satratoxin H 0.43 ng/m^3	51

Continued

TABLE 1 Occurrence of Mycotoxins in Different Occupational Environments—cont'd

Occupational Environment	Sampling	Mycotoxin Concentration	References
Indoor	In settled dust and building materials	Up to 28 fungal and 5 bacterial metabolites	52
Indoor	In air	0.009–4 ng/m^3 roquefortine C 0.003–1.77 ng/m^3 sterigmatocystin 0.007–3.42 ng/m^3 chaetoglobosin 0.002–0.146 ng/m^3 AfB_1 0.0003–0.0211 ng/m^3 AfB_2 0.003–0.082 ng/m^3 roridin E 0.011–0.288 ng/m^3 OTA	53
Waste treatment	In air	1.1–6.1 pg AfB_1/m^3 3.3–31 pg OTA/m^3	54
Waste treatment	In air	2–830 pg AfB_1 equivalents/m^3	55
Waste recycling	In settled dust, calculated airborne concentrations	Up to 33 different fungal and 5 bacterial metabolites	56
Compost facilities	In air	Qualitative detection of tryptoquivalin and trypacidin	57
Ventilation system	In material samples	Qualitative detection of roridin A, T-2 toxin, diacetoxyscirpenol and T-2 tetraol	58

AfB_1 = aflatoxin B_1; DON = deoxynivalenol; OTA = ochratoxin A; ZEA = zearalenone.

in coffee. The results underline that a correlation does not necessarily exist between occurrence of mycotoxigenic mold species and the presence of mycotoxins. However, from a precautionary viewpoint and for practical reasons, it is prudent to assume that an elevated mold exposure is associated with an elevated mycotoxin exposure unless specific information is available.

AIRBORNE CONCENTRATION, DURATION, AND FREQUENCY AS CRITERIA FOR OCCUPATIONAL MYCOTOXIN EXPOSURE

To assess potential health risks from occupational mycotoxin exposure, it is important to consider the airborne mycotoxin concentration, but also the duration and frequency of exposure. For instance, in industrial processes where products or materials are handled in closed containment systems, workers are usually not exposed. In the case of industrial production of citric acid, strains of *A. niger* are used. Some industrially used strains of *A. niger* can produce fumonisins and OTA, which is clearly undesirable with respect to food safety.[65,66] However, from an occupational safety point of view, this is not critical as long as the products or materials are handled in closed containment systems and the workers are not exposed. For a sound assessment of exposure, it is essential to consider all factors that can influence the concentration of airborne mycotoxins, as well as all factors influencing the duration and frequency of exposure.

Factors Influencing the Concentration of Airborne Mycotoxins

One central determinant of the concentration of airborne mycotoxins is the concentration of mycotoxins in the handled material or product. In cases of handling food commodities or animal feed, it can be assumed that mycotoxin concentrations are usually low because of strict legislation. However, in many other cases, information about mycotoxin concentrations is scarce or completely missing. Thus, information about the extend of mold contamination may provide a rough estimate for the presence of mycotoxin, although one has to be aware of the uncertainties of this approach (see previous discussion). Mold is often present when organic materials or products such as wooden products for home decoration, furniture, cotton, wool clothes, or wood chips are stored, transported, or processed under humid or wet conditions (Mayer, not published).

Another important factor is the amount of handled materials and products, which exceed by far the amounts handled in indoor environments and at home. Even if the mycotoxin concentration in a product or material is low, handling of high amounts can result in an elevated airborne mycotoxin concentration at the workplace.

Because mycotoxins are not volatile and bound to particulate matter, all factors that facilitate the release of contaminated particulate matter and dust into ambient air can increase the chances of airborne mycotoxin exposure. In particular, dry products or materials with a high specific surface, such as hay or plant fibers such as jute or hemp, tend to release dust as a vector for mold and mycotoxins. Also, procedures with a high intensity of handling, for example, mechanical sieving or manual sorting, contribute to an elevated release of contaminated dust. Other tasks often associated with high dust concentrations are cleaning activities such as sweeping or when compressed air is used for cleaning. In particular, the latter activity leads to high dust concentrations and possibly to peak mycotoxin exposures, with a duration of several minutes to several days in the case of basic cleaning in storage warehouses or in large production halls. However, a high concentration of airborne dust is not necessarily associated with a high concentration of airborne mycotoxin.

Duration and Frequency of Occupational Mycotoxin Exposure

It can be assumed that the health risk resulting from mycotoxin exposure increases with the duration and frequency of exposure. Besides the importance of risk assessment, duration and frequency are also important to deduce proportionate protective measures. In the case of a rare and short exposure, it seems adequate to use personal protective devices, whereas in the case of more regular exposure, technical protective measures are more proportionate and are preferred. However, often studies about occupational mycotoxin exposure focus only on the aspect of concentration.

With respect to duration and frequency of airborne mycotoxin exposure, mainly two situations exist: (1) regular exposure at low concentrations and (2) occasional exposure at high (peak) concentrations.

Examples of regular occupational exposure to low airborne mold and probably mycotoxin concentrations can be found in many industrial sectors in which products and materials are handled, which are more or less naturally colonized with molds, particularly in agriculture and the downstream processing industry, but also in the handling of spices or tea, processing of plant raw fibers, and so forth. Available data for grain handling indicate that under usual working conditions, airborne mycotoxin concentrations are normally low and unlikely to add significantly to human exposure (see the section on Assessment Strategies).[27,30,41] Nevertheless, occupational exposure to mycotoxins has to be minimized, in line with the central demand of health and safety regulations (see the section on Prevention).

Although exposure to mycotoxins in the food and feed processing industry can be supposed to be usually low owing to strict regulations for these contaminants. Batches of moldy materials may have to be handled, at least for disposal. Reasons for such situations include wet harvest conditions along with late or insufficient drying, water damage, and improper storage conditions. Although such situations occur only occasionally, handling of highly mycotoxin-contaminated batches can result in peak exposures, depending on the amount of contaminated materials and the working conditions. In such situations, mycotoxin exposure can even result in acute health effects: Di Paolo et al. described a case of acute renal failure and tubular necrosis in a woman who shoveled moldy wheat from an abandoned silo for an entire day.[67] There is evidence that OTA was the causative agent. Stange et al. reported the case of a farmer who had repeatedly ground visibly ergot-infected grain and developed progressive occlusion of the leg arteries over a 6-month period.[68] High blood plasma ergotamine levels indicated inhalational ergot alkaloid intake.

Occasional exposure to mold and possibly to mycotoxins can also occur in other occupational environments, such as the restoration of moldy books after water damage.[69] Another example is moldy products and materials in freight containers. Dunnage, wooden transport crates, and furniture or other organic products or materials show at least sometimes extensive mold growth after overseas transport in freight containers. Exposure to molds and mycotoxins can occur during removal of dunnage or transport crates, but also during manual cleaning of the products. Our own measurements revealed mold concentrations of up to 5×10^5 CFU/m^3 during removal of moldy dunnage with a hammer under open air conditions (data not published). Data on associated mycotoxin exposure are not yet available. Another occupational setting of interest is water-damaged buildings, such as after large floods, when over an extended time moldy dwellings are restored or broken down. Airborne mold concentrations of up to 10^7 CFU/m^3 have been observed, depending on the method of removal of infested plaster.[70] In a water-damaged building without restoration activities Gottschalk et al. found airborne satratoxin G and H at levels of 0.25 and 0.43 ng/m^3, respectively.[51]

Severity of Work

Finally, a factor that can influence the extent of exposure via inhalation is the severity of work. Under demanding work conditions, the rate of breathing and thus the inhalative intake is elevated. Mayer et al. assumed a breathing rate of 600 and 3000 l/h for usual and worst-case conditions, respectively.[27] Halstensen et al. assumed a breathing rate of 1800 l/h for moderate physical activity.[30] Degen estimated a breathing volume of up to 3000 l/h during physical activity.[1] Gottschalk et al. used a respiratory minute volume of 6 l/min to calculate inhalative intake of mycotoxins at rest.[51]

Dermal Exposure

As evident in Table 1, to date, studies on occupational mycotoxin exposure have focused on inhalative exposure. However, dermal contact may also contribute to intake of mycotoxins. According to in vitro studies some lipophilic low-molecular-weight mycotoxins may penetrate human skin. Kemppainen et al. described penetration of T-2 toxin (dissolved in methanol) through excised human skin.[71] Boonen et al. showed skin permeability for aflatoxin B1, OTA, citrinin, zearalenone, and T-2 toxin (dissolved in an ethanol/water mixture (70/30)).[72] However, organic solvents probably facilitate dermal absorption.

Dermal exposure may be of particular relevance in cases of hot and demanding working conditions, when workers wear only light clothes and large areas of skin are exposed. Under such conditions, when workers are sweating, exposure may be increased by absorption of dust on the wet skin surface. However, the relevance of dermal mycotoxin exposure remains unclear.

ASSESSMENT STRATEGIES

According to European directive 2000/54/EC on biological agents, employers have to assess the health risks of biological agents such as mold. This includes an assessment of the potential health effects of mycotoxin exposure. To assess risks from workplace-related mycotoxin exposure requires a case-by-case approach.[73] Adverse effects will depend on the toxin type, the intensity (dose, duration, and route) of exposure, and the amount or fraction of mycotoxin absorbed. Reviews by Mayer et al. and Degen compiled information on occupational mycotoxin exposure; these data are summarized and complemented in Table 1.[1,2] A health risk assessment remains a challenging task, among other things, because of the lack of evaluation standards.[73] Limit values for mycotoxins in air and workplace-related inhalatory intake have not been set by official regulatory bodies. Therefore, alternative assessment strategies are necessary.

One approach compares occupational mycotoxin exposure (calculated from airborne levels) with recommendations for tolerable daily intake (TDI) values that have been set for a small number of mycotoxins in food (Table 2) as a preliminary risk assessment.

For a comparison of airborne occupational mycotoxin exposure with TDI values, the inhalative intake has to be calculated as long as direct measurements are not feasible. For this purpose, mycotoxin concentrations in settled dust can be used in combination with airborne dust concentrations to calculate the inhalative intake. An analysis of settled dust is advantageous because settled dust acts as integrating matrix over a longer period of time. It is assumed that the settled dust has been airborne and respirable. A further (worst-case) assumption is that 100% of inhaled mycotoxins will be absorbed. Mayer et al., Halstensen et al., and Tangni and Pussemier calculated the inhalatory intake of mycotoxins by workers in grain elevators under regular working conditions.[27,30,41]

TABLE 2 **Recommendations for Maximum Tolerable Intake of Certain Mycotoxins through Food Related to 1 kg Body Weight (bw) in Each Case**

Mycotoxin	Recommended Maximum Tolerable Oral Intake
Fumonisin B1, B2, and B3, individually or combined	2 µg/kg bw per day
Ochratoxin A[a]	0.1 µg/kg bw per week
Deoxynivalenol	1 µg/kg bw per day
T-2 and HT-2 toxin	0.06 µg/kg bw per day
Zearalenone	0.2 µg/kg bw per day

[a]*Value has been set provisionally.*

In all studies, calculated intake was at least two orders of magnitude lower than the lowest TDI value for the respective mycotoxins. Mayer et al. analyzed mycotoxin concentrations in settled dust in waste recycling plants and determined airborne dust concentrations.[74] In that study, the concentrations of all detected mycotoxins were summed up and compared with the lowest TDI for T-2/HT-2 toxins. Despite the summation of all mycotoxins, an inhalative intake was more than two-fold lower than the TDI. This indicates that no acute health hazard exists from inhalative intake under regular working conditions.

Although such approaches are feasible to assess occupational airborne exposure, some caveats have to be kept in mind: The TDI values have been derived only for a few mycotoxins, but not for others that may be of interest at particular workplaces (e.g., atranones, satratoxins). Moreover, they do not account for the complexity of simultaneously occurring mycotoxins and/or interactions between mycotoxins and other hazardous compounds. The complexity of compounds present in waste recycling plants is illustrated by the study of Mayer et al., who detected 33 mycotoxins and five bacterial metabolites in settled dust samples.[74] Even in indoor environments a broad spectrum of microbial metabolites can be observed. Täubel et al. detected in total 33 microbial metabolites in different building materials and dust samples.[52] The broad spectrum of mycotoxins and microbial metabolites underlines the problem of how to deal with interactions between mycotoxins or with other hazardous compounds.

Despite these limitations and in the light of missing alternatives, comparison of airborne mycotoxin concentrations with tolerable oral intake values is currently an approach that gives at least an impression of potential health risk.

However, because employers have to assess health risks preferably before work, they cannot wait until sufficient scientific information allows a comprehensive risk assessment to decide which protective measures are appropriate. Thus, a precautionary approach is necessary.

PREVENTION

In most cases insufficient data are available about mycotoxin exposure. Nevertheless the employer has to define appropriate protective measures. Occupational health and safety legislation already considered the lack of assessment criteria and forces therefore the employer to minimize the risk of exposure to biological agents to a level according to the state of the art of protective measures. Occupational health and safety legislation prescribes a strict hierarchy of protective measures. This hierarchy targets first of all at the prevention of exposure. Necessary to that end are measures to avoid the occurrence of mycotoxins.

In the case of food and feed production, this starts at the level of plant breeding and farming practice. Over a course of 5 years, Vogelgsang et al. studied different maize residue treatments conducted on 14 zero-tillage on-farm sites in Switzerland to evaluate their effect on the development of Fusarium head blight and contamination with the mycotoxin deoxynivalenol (DON) in winter wheat grains and wheat straw after grain maize.[74] The authors were able to demonstrate that under a grain maize/winter wheat rotation, the DON content in wheat grains frequently exceeded the European maximum limit, even with thorough treatment of maize residues and less susceptible wheat varieties. Hence, modified crop rotation can contribute to a reduced risk of contamination. Also, the selection of crop cultivars with reduced susceptibility to mold infestation can contribute to a reduction of mycotoxin exposure during subsequent processing.[75]

An approach to reduce crop aflatoxin contamination is the application of atoxigenic strains of *Aspergillus flavus*.[76] This strategy uses components of the endemic diversity to alter structures of *A. flavus* populations.

The next step is to avoid occurrence at workplaces. Appropriate measures start with quality management measures ensuring a high quality of products and materials in terms of a low mycotoxin contamination. Along with quality management measures, appropriate storage conditions are important to avoid or at least limit fungal growth and the (additional) production of mycotoxins. Bucheli et al. demonstrated the influence of storage conditions on fungal growth and OTA production in raw coffee.[77] A key feature of avoiding OTA production was ensuring a low content of 0.75 or less of available water. Denziel et al. demonstrated the influence of the cleaning status of storage rooms on the contamination of pistachios.[78] With respect to storage workers, cleaning of such storage rooms may be associated with elevated mycotoxin exposure as long as sweeping or other dusty cleaning methods are applied. Landers et al. observed no aflatoxin production at a O_2 concentration less than 1%.[79] However, the reduction of O_2 concentration may replace one health hazard with another that may be even more serious.

Because materials such as grains may already be contaminated with mycotoxins in the freshly harvested state, proper storage alone will not always suffice as a protective measure. Accordingly, if mycotoxins have been formed before storage or where dry conditions cannot be ensured for operational reasons, prevention must be initiated at the level of avoiding and minimizing exposure to fungi or fungi-containing dusts. This includes minimization of the airborne concentration as well as of the duration and frequency of exposure.

A high level of protection can be achieved with containment measures such as encapsulation of machines. When avoidance of dust release is not an option, dust should be exhausted at its point of origin. Contamination of adjacent workplaces must be avoided, such as by separating workplaces. High exposure is often observed during cleaning activities owing to the use of brooms or even compressed air. For cleaning activities, the use of vacuum cleaners with sufficient dust retention is indicated. Without high dust retention, small particles will be re-released into the ambient air.

When technical measures are not feasible, organizational measures are necessary. To this kind of measures belongs restriction of the number of exposed employees. An important part of organizational measures is to inform and educate the workers. In many cases, workers are unaware that mycotoxins may be present or of the potential nature of associated risks. Employees must be informed about the potential risks and corresponding protective measures as part of periodic instruction according to occupational safety legislation.

At the end of the hierarchy of protective measures, personal protective devices are indicated. Respiratory protection devices conforming at least to particle filter class 2 according to EN 149 (2009) must be worn where fungal concentrations prevail.[80] Fan-assisted hoods that provide not only respiratory protection but also cover the face, and hence, the mucosae of the eyes are advantageous. Because mycotoxins can also be absorbed via the skin, body-covering apparel must be worn in workplaces subject to dust exposure. Exposed skin parts should be washed once the work has been completed.

Exposure to mold or mold-contaminated dust is also connected with a potential health hazard because of the sensitizing properties of molds and, at least under certain conditions, owing to infections. However, in most cases, protective measures against mycotoxins will form part of the strategies required to combat the sensitizing potential of the fungi. In a laboratory environment where pure mycotoxins are handled, specific protective measures against exposure are conceivable and are part of good laboratory practice.

ACKNOWLEDGMENT

The author gratefully acknowledges the valuable input and constructive criticism of Prof. Dr Gisela H. Degen (IfADo, Dortmund) for an earlier version of the manuscript.

REFERENCES

1. Degen GH. The challenge to assess workplace related risks from mycotoxin exposure (Guest editorial). *Mycotoxin Res* 2008;**24**:i–ii.
2. Mayer S, Engelhart S, Kolk A, Blome H. The significance of mycotoxins in the framework of assessing workplace related risks. *Mycotoxin Res* 2008;**24**:151–64.
3. Kelman BJ, Robbins CA, Swenson LJ, Hardin BD. Risk from inhaled mycotoxins in indoor office and residential environments. *Int J Toxicol* 2004;**23**:3–10.
4. Fung F, Clark RF. Health effects of mycotoxins: a toxicological overview. *J Toxicol Clin Toxicol* 2004;**42**:217–34.
5. WHO. *WHO guidelines for indoor air quality: dampness and mould.* Copenhagen: WHO Regional Office for Europe; 2009.
6. Sorensen WG, Simpson JP, Peach MJ, Thedell TD, Olenchock SA. Aflatoxin in respirable corn dust particles. *J Toxicol Environ Health* 1981;**7**:669–72.
7. Krysinska-Traczyk E, Perkowski J, Dutkiewicz J. Levels of fungi and mycotoxins in the samples of grain and grain dust collected from five various cereal crops in eastern Poland. *Ann Agric Environ Med* 2007;**14**:159–67.
8. Halstensen AS, Nordby KC, Kristensen P, Eduard W. Mycotoxins in grain dust. *Stewart Postharvest Rev* 2008;**4**:1–9.
9. Jargot D, Melin S. Characterization and validation of sampling and analytical methods for mycotoxins in workplace air. *Environ Sci Process Impacts* 2013;**15**:633–44.

10. Magan N, Aldred D. Post-harvest control strategies: minimizing mycotoxins in the food chain. *Int J Food Microbiol* 2007;**119**:131–9.
11. Schmidt-Heydt M, Rüfer C, Raupp F, Bruchmann A, Perrone G, Geisen R. Influence of light on food relevant fungi with emphasis on ochratoxin producing species. *Int J Food Microbiol* 2011;**145**:229–37.
12. Corey JEI. Relationships of water activity to fungal growth. In: Beuchat LR, editor. *Food and beverage mycology*. New York, US: Van Nostrand Reinhold International Company Limited; 1987.
13. Weidenbörner M. *Lebensmittelmykologie*. Behr's Verlag Hamburg; 1999.
14. Frisvad JC, Thrane U, Filtenborg O. Role and use of secondary metabolites in fungal taxonomy. In: Frisvad JC, Bridge PD, Arora DK, editors. *Chemical fungal taxonomy*. New York: Marcel Dekker Inc.; 1998.
15. Andersen B, Thrane U. Food borne fungi in fruit and cereals and their production of mycotoxins. In: Hocking AD, Thrane U, Pitt JI, Samson RA, editors. *Advances in food mycology*. New York: Springer; 2006.
16. Frisvad JC, Thrane U, Samson RA, Pitt JI. Important mycotoxins and the fungi which produce them. In: Hocking AD, Thrane U, Pitt JI, Samson RA, editors. *Advances in food mycology*. New York: Springer; 2006.
17. Malta-Vacas J, Viegas J, Sabino R, Viegas C. Fungal and microbial volatile organic compounds exposure assessment in a waste sorting plant. *J Toxicol Environ Health* 2012;**75**:1410–7.
18. Ławniczek-Wałczyk A, Gołofit-Szymczak M, Cyprowski M, Górny RL. Exposure to harmful microbiological agents during the handling of biomass for power production purposes. *Med Pr* 2012;**63**:395–407.
19. Johanning E, Auger Philip P, Morey R, Yang CS, Olmsted E. Review of health hazards and prevention measures for response and recovery workers and volunteers after natural disasters, flooding, and water damage: mold and dampness. *Environ Health Prev Med* 2014;**19**:93–9.
20. Viegas C, Gomes AQ, Abegão J, Sabinoa R, Graça T, Viegas S. Assessment of fungal contamination in waste sorting and incineration—case study in Portugal. *J Toxicol Environ Health* 2014;**77**:57–68.
21. Spankie S, Cherrie JW. Exposure to grain dust in Great Britain. *Ann Occup Hyg* 2012;**56**:25–36.
22. Halstensen AS, Heldal KK, Wouters IM, Skogstad M, Ellingsen DG, Eduard W. Exposure to grain dust and microbial components in the Norwegian grain and compound feed industry. *Ann Occup Hyg* 2013;**57**:1105–14.
23. Skorska C, Sitkowska J, Krysinska-Traczyk E, Cholewa G, Dutkiewicz J. Exposure to airborne microorganisms, dust and endotoxins during processing of valerian root on farms. *Ann Agric Environ Med* 2005;**12**:119–26.
24. Merget R, Sander I, Rozynek P, Raulf-Heimsoth M, Bruening T. Occupational hypersensitivity pneumonitis due to molds in an onion and potato sorter. *Am J Ind Med* 2008;**51**:117–9.
25. Tournas VH, Katsoudas EJ. Microbiological quality of various medicinal herbal teas and coffee substitutes. *Microbiol Insights* 2008;**1**:47–55.
26. Iamanaka BT, Taniwaki MH, Vicente E, Menezes HC. Fungi producing ochratoxin in dried fruits. In: Hocking AD, Thrane U, Pitt JI, Samson RA, editors. *Advances in food mycology*. New York: Springer; 2006.
27. Mayer S, Curtui V, Usleber E, Gareis M. Airborne mycotoxins in dust of grain elevators. *Mycotoxin Res* 2007;**23**:94–100.
28. Dill I, Trautmann C, Szewzyk R. Massenentwicklung von *Stachybotrys chartarum* auf kompostierbaren Pflanztöpfen aus Altpapier. *Mycoses* 1997;**40**:110–4.
29. Iavicoli I, Brera C, Carelli G, Caputi R, Marinaccio A, Miraglia M. External and internal dose in subjects occupationally exposed to ochratoxin A. *Int Arch Occup Environ Health* 2002;**75**:381–6.

30. Halstensen AS, Nordby KC, Elen O, Eduard W. Ochratoxin A in grain dust—estimated exposure and relations to agricultural practices in grain production. *Ann Agric Environ Med* 2004;**11**:245–54.
31. Tarín A, Rosell MG, Guardino X. Use of high-performance chromatography to assess airborne mycotoxins. Aflatoxins and ochratoxin A. *J Chromatogr A* 2004;**1047**:235–40.
32. Hosseininia AR, Vahabzadeh M, Rashedinia M, Riahi-Zanjani B, Karimi G. A survey of aflatoxins in sesame seeds imported to Khorasan Province, Iran. *Mycotoxin Res* 2014;**30**:43–6.
33. Barberis CL, DAlcero AM, Magnoli CE. Evaluation of aflatoxin B1 and ochratoxin A in interacting mixed cultures of *Aspergillus* sections *Flavi* and *Nigri* on peanut grains. *Mycotoxin Res* 2012;**28**:149–56.
34. Brera C, Caputi R, Miraglia M, Iavicoli I, Salerno A, Carelli G. Exposure assessment to mycotoxins in workplaces: aflatoxins and ochratoxin A occurrence in airborne dusts and human sera. *Microchem J* 2002;**73**:167–73.
35. German Federal Institute for Risk Assessment. *Increased aflatoxin concentrations detected in raw milk*. Information No. 08/2013 of March 01, 2013.
36. Mayer S, Reichel M, Biselli S. *Multi mycotoxin assay as a tool to check airborne exposure to mycotoxins during grain handling*, in preparation.
37. Jestoi M. Emerging fusarium-mycotoxins fusaproliferin, beauvericin, enniatins, and moniliformin—a review. *Crit Rev Food Sci Nutr* 2008;**48**:21–49.
38. Marin S, Ramos AJ, Cano-Sancho G, Sanchis V. Mycotoxins: occurrence, toxicology, and exposure assessment. *Food Chem Toxicol* 2013;**60**:218–37.
39. Niewenhuize van JP, Herber RFM, De Bruin A, Meyer PB, Duba WC. Epidemiologisch onderzoek naar carcinogeniteit bij langdurige low level expositie van een fabriekspopulatie. *T Soc Geneesk* 1973;**51**:754–60.
40. Sorenson WG, Jones W, Simpson J, Davidson JI. Aflatoxin in respirable airborne peanut dust. *J Toxicol Environ Health* 1984;**14**:525–33.
41. Tangni EK, Pussemier L. Ochratoxin A and citrinin loads in stored wheat grains: impact of grain dust and possible prediction using ergosterol measurement. *Food Addit Contam* 2006;**23**:181–9.
42. Gareis M, Meussdoerffer F. Dust of grains and malts as a source of ochratoxin A exposure. *Mycotoxin Res* 2000;**16**:127–30.
43. Nordby KC, Halstensen AS, Elen O, Clasen PE, Langseth W, Kristensen P, et al. Trichothecene mycotoxins and their determinants in settled dust related to grain production. *Ann Agric Environ Med* 2004;**11**:75–83.
44. Burg WA, Shotwell OL, Saltzmann BE. Measurements of airborne aflatoxins during the handling of contaminated corn. *Am Ind Hyg Assoc J* 1981;**42**:1–11.
45. Selim MI, Juchems AM, Popendorf W. Assessing airborne aflatoxin B1 during on-farm grain handling activities. *Am Ind Hyg Assoc J* 1988;**59**:252–6.
46. Skaug MA, Eduard W, Stormer FC. Ochratoxin A in airborne dust and fungal conidia. *Mycopathologica* 2000;**151**:93–8.
47. Lanier C, Richard E, Heutte N, Picquet R, Bouchart V, Garon D. Airborne molds and mycotoxins associated with handling of corn silage and oilseed cakes in agricultural environment. *Atmos Environ* 2010;**44**:1980–6.
48. Wanga Z, Chai T, Lu G, Quan C, Duan H, Yao M, et al. Simultaneous detection of airborne aflatoxin, ochratoxin and zearalenone in a poultry house by immunoaffinity clean-up and high-performance liquid chromatography. *Environ Res* 2008;**107**:139–44.
49. Engelhart S, Loock A, Skutlarek D, Sagunski H, Lommel A, Färber H, et al. Occurrence of toxigenic *Aspergillus versicolor* isolates and sterigmatocystin in carpet dust from damp indoor environments. *Appl Environ Microbiol* 2002;**68**:3886–90.

50. Brasel TL, Martin JM, Carriker CG, Wilson SC, Straus DC. Detection of airborne *Stachybotrys chartarum* macrocyclic trichothecene mycotoxins in the indoor environment. *Appl Environ Microbiol* 2005;**71**:7376–88.
51. Gottschalk C, Bauer J, Meyer K. Detection of satratoxin G and H in indoor air from a water-damaged building. *Mycopathologia* 2008;**166**:103–7.
52. Täubel M, Sulyok M, Vishwanath V, Bloom E, Turunen M, Järvi K, et al. Co-occurrence of toxic bacterial and fungal secondary metabolites in moisture-damaged indoor environments. *Indoor Air* 2011;**21**:368–75.
53. Polizzi V, Delmulle B, Adams A, Moretti A, Susca A, Picco AM, et al. JEM spotlight: fungi, mycotoxins and microbial volatile organic compounds in mouldy interiors from water-damaged buildings. *J Environ Monit* 2009;**11**:1849–58.
54. Gerbl-Rieger S, Hoppenheidt K, Mücke W, Wallnöfer P. Keimemissionen aus Kompostierungs- und Vergärungsanlagen. In: Mücke W, editor. *Research projects E32 and E33 commissioned by the Bavarian state ministry of regional development and environmental affairs*; 1999.
55. Thißen R, Müller T, Braun S, Schwalbe R, Dohms E, Dott W, et al. Mykotoxine von luftgetragenen Fadenpilzen und ihre Relevanz für die Arbeitsplatz- und Umwelthygiene. *Mycotoxin Res* 2000;**16**:99.
56. Mayer S, Vishwanath V, Sulyok M. Airborne workplace exposure to microbial metabolites in waste sorting plants. In: Johanning E, Morrey PR, Auger P, editors. *Bioaerols*. Albany, New York: Fungal Research Group Foundation Inc.; 2012.
57. Fischer G, Müller T, Ostrowski R, Dott W. Mycotoxins of *Aspergillius fumigatus* in pure culture and in native bioaerosols from compost facilities. *Chemosphere* 1999;**38**:1745–55.
58 Smoragiewicz W, Cossette B, Boutard A, Krzystyniak K. Trichothecene mycotoxins in the dust of ventilation systems in office buildings. *Int Arch Occup Environ Health* 1993;**65**:113–7.
59. Andersen B, Nielsen KF, Thrane U, Szaro T, Taylor JW. Molecular and phenotypic descriptions of *Stachybotrys chlorohalonata* sp. nov. and two chemotypes of *Stachybotrys hartarum* found in water-damaged buildings. *Mycologia* 2003;**95**:1227–38.
60. Gareis M, Gottschalk C. Stachybotrys spp. and the guttation phenomenon. *Mycotoxin Res* 2014;**30**:151–159.
61. Halstensen AS, Nordby KC, Klemsdal SS, Elen O, Clasen PE, Eduard W. Toxigenic *Fusarium* spp. as determinants of trichothecene mycotoxins in settled grain dust. *J Occup Environ Hyg* 2006;**3**:651–9.
62. Land CJ, Hult K, Fuchs R, Hagelberg S, Lundström H. Tremorgenic mycotoxins from *Aspergillus fumigatus* as a possible occupational health problem in sawmills. *Appl Environ Microbiol* 1987;**53**:787–90.
63. Bünger J, Westphal G, Mönnich A, Hinnendahl B, Hallier E, Müller M. Cytotoxicity of occupationally and environmentally relevant mycotoxins. *Toxicology* 2004;**202**:199–211.
64. Muñoz K, Vega M, Rios G, Geisen R, Degen GH. Mycotoxin production by different ochratoxigenic *Aspergillus* and *Penicillium* species on coffee- and wheat-based media. *Mycotoxin Res* 2011;**27**:239–47.
65. Frisvad JC, Larsen TO, Thrane U, Meijer M, Varga J, Samson RA. Fumonisin and Ochratoxin A production in industrial Aspergillus niger strains. PLoS ONE 6:e23496. http://dx.doi.org/10.1371/Journal pone 0023496.
66. Jarvis BB, Frisvad JC, Larsen TO, Thrane U, Meijer M, Varga J, et al. Fumonisin and ochratoxin production in industrial *Aspergillus niger* strains. *PLoS One* 2011;**6**(8):e23496. http://dx.doi.org/10.1371/journal.pone.0023496.
67. Di Paolo N, Guarnieri A, Loi A, Sacchi G, Mangiarotti AM, Di Paolo M. Acute renal failure from inhalation of mycotoxins. *Nephron* 1993;**64**:621–5.

68. Stange K, Pohlmeier H, Lübbesmeyer A, Gumbinger G, Schmitz W, Baumgart P. Vaskulärer Ergotismus durch Getreidestaubinhalation. *Dtsch Med Wochenschr* 1998;**123**:1547–50.
69 Zielińska-Jankiewicz K, Kozajda A, Piotrowska M, Szadkowska-Stańczyk I. Microbiological contamination with moulds in work environment in libraries and archive storage facilities. *Ann Agric Environ Med* 2008;**15**:71–8.
70. Dott W, Fischer G, Müller T, Thißen R, Wiesmüller GA. *Belastung der Arbeitnehmer bei Schimmelpilzsanierungsarbeiten in Innenräumen*. Literaturstudie im Auftrag der Tiefbauberufsgenossenschaft; 2004. [Bezugsquelle: BG BAU Berufsgenossenschaft der Bauwirtschaft, Prävention Tiefbau, Landsberger Straße 309, 80687 München] www.bgbau.de.
71. Kemppainen BW, Riley RT, Pace JG, Hoerr FJ. Effects of skin storage conditions and concentration of applied dose on (3H)T-2 toxin penetration through excised human and monkey skin. *Food Chem Toxicol* 1986;**24**:221–7.
72. Boonen J, Malysheva SV, Taevernier L, Di Mavungu JD, De Saeger S, De Spiegeleer B. Human skin penetration of selected model mycotoxins. *Toxicology* 2012;**301**:21–32.
73. Degen GH. Tools for investigating workplace-related risks from mycotoxin exposure. *World Mycotoxin J* 2011;**4**:315–27.
74. Vogelgsang S, Hecker A, Musa T, Dorn B, Forrer HR. On-farm experiments over 5 years in a grain maize/winter wheat rotation: effect of maize residue treatments on *Fusarium graminearum* infection and deoxynivalenol contamination in wheat. *Mycotoxin Res* 2011;**27**:81–96.
75. Edwards SG. Influence of agricultural practices on fusarium infection of cereals and subsequent contamination of grain by trichothecene mycotoxins. *Toxicol Lett* 2004;**153**:29–35.
76. Mehl HL, Jaime R, Callicott KA, Probst C, Garber NP, Ortega-Beltran A, et al. *Aspergillus flavus* diversity on crops and in the environment can be exploited to reduce aflatoxin exposure and improve health. *Ann N Y Acad Sci* 2012;**1273**:7–17.
77. Bucheli P, Meyer I, Vuataz A, Viani R. Industrial storage of green robusta coffee under tropical conditions and its impact on raw material quality and ochratoxin A content. *J Agric Food Chem* 1998;**46**:4507–11.
78. Denziel T, Jarvis B, Rolfe EJ. A field survey of pistachio (*Pistacia vera*) nut production and storage in Turkey with particular reference to aflatoxin contamination. *J Sci Food Agric* 1976;**27**:1027–34.
79. Landers KE, Davis ND, Diener UL. Influence of atmospheric gases on aflatoxin production by *Aspergillus flavus* in peanuts. *Phytopathology* 1967;**57**:1086–90. Cited in: Andersen B, Thrane U. Food borne fungi in fruit and cereals and their production of mycotoxins. In: Hocking AD, Thrane U, Pitt JI, Samson RA, editors. *Advances in food mycology*, New York: Springer; 2006.
80. EN 149:2009–08. Respiratory protective devices—filtering half masks to protect against particles—requirements, testing, marking; EN 149:2001+A1:2009.

Chapter 20

Mycotoxins: Genotoxicity Studies and Methodologies

Carina Ladeira

Environmental Health Research Group & Genetic and Metabolism Research Group, Escola Superior de Tecnologia da Saúde de Lisboa, IPL, Portugal

INTRODUCTION

Human exposure to mycotoxins is determined by environmental or biological monitoring. In environmental monitoring, mycotoxins are measured in food, air, or other samples; in biological monitoring, the presence of residues, adducts, and metabolites are assayed directly in tissues, fluids, and excreta.[1] Human exposure to mycotoxins may result from consumption of plant-derived foods that are contaminated with toxins, the carryover of mycotoxins and their metabolites in animal products such as meat and eggs, or exposure to air and dust containing toxins.[2,3] Examples of mycotoxins of greatest public health and agro-economic significance include aflatoxins (AFs), ochratoxins, trichothecenes, zearalenone, fumonisins (FUMs), tremorgenic toxins, and ergot alkaloids.[2,4,5]

A challenge in the field of exposure assessment of mycotoxins is to develop accurate and reliable biomarkers for human studies. The biomarker approach may be a promising tool for directly measuring toxin-mediated biological perturbations or to directly measure the amounts of mycotoxins present in the organism.[6]

In molecular epidemiology, it is possible to demonstrate with more certainty the association between putative carcinogens and specific cancers.[1] Biological markers of AF, ochratoxin A (OTA), and FUM exposure have attracted attention for mycotoxin biomonitoring studies. However, while AF and ochratoxin biomarkers have been successfully applied and validated over the past decade, large drawbacks remain in finding a suitable FUM biomarker.[6]

The toxicity of mycotoxins differs depending on the kind of toxin. It was observed in animals that the toxicity was related to the species, the dose ingested, the duration of the exposure, and sex and age.[7]

Biomonitoring of AFs can be done by analyzing the presence of AF metabolites in blood, milk, and urine; moreover, excreted DNA adducts and blood protein adducts can be monitored. The AFB1-N^7-guanine adduct represents the most reliable urinary biomarker for AF exposure but reflects only recent exposure.

Environmental Mycology in Public Health. http://dx.doi.org/10.1016/B978-0-12-411471-5.00020-X

Numerous studies have shown that carcinogenic potency is highly correlated with the extent of total DNA adducts formed in vivo.[1]

CYTOKINESIS-BLOCK MICRONUCLEUS ASSAY

The cytokinesis-block micronucleus (CBMN) assay is a comprehensive system for measuring DNA damage; cytostasis and cytotoxicity-DNA damage events are scored specifically in once-divided binucleated cells. End points possible to be measured are micronuclei (MN), a biomarker of chromosome breakage and/or whole chromosome loss; nucleoplasmic bridges (NPB), a biomarker of DNA misrepair and/or telomere end-fusions; and nuclear buds (NBUD), a biomarker of elimination of amplified DNA and/or DNA repair complexes. Cytostatic effects are measured via the proportion of mono-, bi-, and multinucleated cells and cytotoxicity via necrotic and/or apoptotic cell ratios.[8,9]

The CBMN assay has become one of the most commonly used methods for assessing chromosome breakage and loss in human lymphocytes both in vivo and ex vivo.[10] In the CBMN assay, once-divided cells are recognized by their binucleated appearance after blocking cytokinesis with cytochalasin-B, an inhibitor of microfilament ring assembly required for the completion of cytokinesis.[9] The restriction of scoring just MN in binucleated cells prevents confounding effects caused by suboptimal or altered cell division kinetics. The biological meaning of the presence of MN in mononucleated cells indicates DNA damage that was present in the cells before they were put into culture with cytochalasin-B, whereas binucleated cells may contain preexisting MN as well as MN expressed during culture as a result of chromosome breaks accumulated during G0 phase in vivo.[11,12] Because of its reliability and good reproducibility the CBMN assay,[9,13] has been extensively used to evaluate the presence and extent of chromosome damage in human populations exposed to genotoxic agents in various occupational settings, in the environment, or as a consequence of lifestyle.[14]

The CBMN assay is visualized as a cytome concept, which implies that every cell in the system studied is scored cytologically for its viability status (necrosis and apoptosis), its mitotic status (mononucleated, binucleated, and multinucleated), and its chromosomal damage or instability status (presence of MN, NPB, NBUD, and a number of centromere probe signals among nuclei or MN of binucleated cells if such molecular tools are combined with the assay).[9]

Use of the CBMN assay in in vitro genetic toxicology testing is well established; in fact, it has become an accepted standard method to assess the genotoxic hazard of chemicals and has led to the development of the Organization for Economic Cooperation and Development (OECD) guideline for this purpose, the OECD 487 guideline.[15]

The CBMN assay is an effective tool for studying cellular and nuclear dysfunction caused by in vitro or in vivo aging, micronutrient deficiency or excess, genotoxin exposure, and genetic defects in genome maintenance. It is also fruitful in the emerging fields of nutrigenomics and toxicogenomics and their combinations, because it has become increasingly clear that nutrient status also affects sensitivity to exogenous genotoxins.[9,16] Many results obtained indicate the potential predictive value of the CBMN assay with respect to cancer risk and validate its use as a test for detecting nutritional, environmental, and genetic factors that are potentially carcinogenic.[9]

Concerning human biomonitoring, the CBMN assay is widely used for in vivo exposure to genotoxins and has become a standard biodosimetry method endorsed by the International Atomic Energy Agency and the World Health Organization to measure exposure to ionizing radiation.[17] The assay measures MN and other nuclear anomalies in ex vivo mitogen-stimulated lymphocytes from in vivo systemic exposed persons, in this way integrating in vivo systemic exposure of lymphocytes and in vivo/ex vivo response to the genotoxic stress. Its predictivity for the detection of genetic risks is supported by the fact that it allows measurement at the single-cell level of both structural and numerical chromosome aberrations.[15]

In summary, the CBMN assay is a robust assay for genetic damage with applications in ecotoxicology, nutrition, radiation sensitivity testing both for cancer risk assessment and optimization of radiotherapy, biomonitoring of human populations, and importantly, testing of new pharmaceuticals and other chemicals. In the future, an automated system is expected that can reliably score the various end points possible with the CBMN assay.[9]

MICRONUCLEUS

MN originate from chromosome fragments or whole chromosomes that lag behind anaphase during nuclear division and are not included in the main nuclei.[13,18,19] MN are small extranuclear bodies that arise in dividing cells from acentric chromosome/chromatid fragments or whole chromosome/chromatid that lag behind in anaphase and are not included in the daughter nuclei in telophase.[20] At telophase, a nuclear envelope forms around the lagging chromosomes and fragments, which then uncoil and gradually assume the morphology of an interphase nucleus, except that they are smaller than the main nuclei in the cell (hence, the term "micronucleus").[19] MN harboring chromosomal fragments may result from direct double-strand DNA breakage, conversion of single-strand breaks into double-strand breaks after cell replication, or inhibition of DNA synthesis.[20]

MN can be formed via different pathways: namely, from acentric chromosome or chromatid fragments. A small proportion of acentric chromosome fragments may simply arise from unrepaired double-stranded DNA breaks. Other mechanisms that could lead to MN formation from acentric fragments include simultaneous excision repair of damaged (e.g., 8-oxo-deoxyguanosine) or inappropriate bases incorporated in DNA (e.g., uracil) that are in proximity to and on opposite complementary DNA strands.[21] Another mechanism that may lead to MN from chromosome loss events is hypomethylation of cytosine in centromeric and pericentromeric repeat sequences such as classical satellite repeats at pericentromeric regions and higher-order repeats of satellite DNA in centromeric DNA.[21] Because of the central role of kinetochore proteins in engaging chromosomes with the spindle, it is probable that mutations leading to defects in kinetochore and microtubule interaction dynamics could cause of MN formation owing to chromosome loss at anaphase. Other variables likely to increase MN from chromosome loss are defects in mitotic spindle assembly, mitosis check point defects, and abnormal centrosome amplification.[21]

The fate of MN after their formation in the micronucleated cell is poorly understood. Their postmitotic fate includes: (1) elimination of the micronucleated cell as a consequence of apoptosis; (2) expulsion from the cell (when the DNA within the MN is not expected to be functional or capable of replication owing to the absence of the necessary cytoplasmic components); (3) reincorporation into the main nucleus (when reincorporated chromosome may be indistinguishable from those of the main nucleus and might resume normal biological activity); and (4) retention within the cell's cytoplasm as an extranuclear entity (when MN may complete one or more rounds of DNA/chromosome replication).[20,22]

The key advantage of the CBMN assay lies in its ability to detect both clastogenic and aneugenic events, leading to structural and numerical chromosomal aberrations, respectively.[20] Clastogens induce MN by breaking the double helix of DNA, forming acentric fragments that are incapable of adhering to the spindle fibers, and integrating in the daughter nuclei, and are thus left out during mitosis. The same occurs to whole chromosomes with damaged kinetochores; they cannot attach to the microtubules that pull the chromatids toward the daughter cells during mitosis and thus they remain outside the new nuclei. This damage could be generated by chemicals reacting with proteins forming the kinetochores.[23,24]

Aneugens are chemicals that prevent the formation of the spindle apparatus during mitosis. These agents generate not only whole chromatids that are left out of the nuclei, thus forming MN, but also the formation of multinucleated cells, in which each nucleus would contain a different number of chromosomes. These agents are also likely to induce an increase in mitotic figures that are clearly seen in the same slides. With the CBMN assay is possible to distinguish between MN originating from whole chromosomes and those originating from acentric fragments, as well as to determine whether malsegregation of chromosomes is occurring between nuclei in a binucleated cell that may not contain MN, by using centromeric probes.[8,10,19]

Pancentromeric DNA probes are used to distinguish between MN originating from any whole chromosome loss event and MN containing acentric chromosome fragments. The use of chromosome-specific centromeric DNA probes allows both the determination of specific chromosome loss events resulting in MN and unequal segregation of specific chromosomes among daughter nuclei even in the absence of MN formation.[21] Pancentromeric probes should be used only to distinguish between MN originating from chromosome breaks (centromere negative) and chromosome loss (centromere positive). Chromosome-specific centromere probes should be used only to measure malsegregation (owing to nondisjunction or chromosome loss) involving unique chromosomes.[8,10,19,21,25] Evaluation of the mechanistic origin of individual MN by centromere and kinetochore identification contributes to the high sensitivity and specificity of the method.[20]

Important factors influence baseline MN frequency in human lymphocytes. Age and gender are the most important demographic variables affecting the MN index; frequencies in females are greater than those in males by a factor of 1.2–1.6, depending on the age group.[26] MN frequency was significantly and positively correlated with age in males and females and is affected by dietary factors such as folate deficiency and plasma levels of vitamin B12 and homocysteine. It was also proposed that the MN index can be influenced by the propensity of an individual's cells to undergo apoptosis, and genetic factors such as genetic polymorphisms.[10,20,27]

In a general form, the formation of MN is attributed to a variety of insults to genetic material, which could be classified as exogenous and endogenous factors. Exogenous factors include radiation, chemical agents, and microorganism invasion. Endogenous factors include genetic defects, pathological changes, deficiency of essential nutritional ingredients (e.g., folic acid), and injuries induced by deleterious metabolic products (such as reactive oxygen species).[28]

The hypothesis of a predictive association between the frequency of MN in CBMN assay in lymphocytes and cancer development is supported by a number of findings: (1) an association between MN frequency and cancer risk was inferred from mechanistic similarities with chromosomal aberrations, which were shown to be predictive for cancer; (2) in vitro, high concordance is observed between chromosomal aberrations and MN; (3) an increase in MN frequency is observed in lymphocytes of cancer patients and in patients with syndromes that make them cancer-prone, such as Bloom syndrome and ataxia telangiectasia; (4) MN frequency is significantly associated with the blood concentration of vitamins such as folate, whose deficiencies are associated with increased risk for some cancers; (5) there is a direct link between MN frequencies and early stages of carcinogenesis: namely, a significant association between increased MN frequencies and low-grade and high-grade diagnostic categories of cervical carcinogenesis in women.[20]

Formation of nuclear anomalies such as MN, chromosomal rearrangements, and anaphase bridges (leading to breakage–fusion–bridge cycles and generation of more MN) are events commonly seen in the early stages of carcinogenesis. Elevated levels of MN indicate defects in DNA repair and chromosome segregation that could result in generation of daughter cells with altered gene dosage, or deregulation of gene expression that could lead to evolution of the chromosome instability phenotype often seen in cancer. These considerations give mechanistic support to a possible causal association between MN frequency and the risk of cancer. A study by Bonassi et al.[29] observed an association between MN frequency and cancer risk in non-hematological malignancies, which suggests that genome damage events in lymphocytes may be correlated with cancer-initiating events in other tissues via a common genetic, dietary, or environmental factor.

NUCLEOPLASMIC BRIDGES

NPB occur when centromeres of dicentric chromosomes are pulled to opposite poles of the cell at anaphase. In the absence of breakage of the anaphase bridge, the nuclear membrane eventually surrounds the daughter nuclei and the anaphase bridge; in this manner, an NPB is formed.[21] Various mechanisms could lead to NPB formation after DNA misrepair of strand breaks in DNA. Typically, a dicentric chromosome and an acentric chromosome fragment are formed that result in the formation of an NPB and an MN, respectively.[8,9,13,19] Misrepair of DNA strand breaks could also lead to the formation of dicentric ring chromosomes and concatenated ring chromosomes, which could also result in the formation of NPB. An alternative mechanism for dicentric chromosome and NPB formation is telomere end fusion caused by telomere shortening, loss of telomere capping proteins, or defects in telomere cohesion.[9,30] The study of Rudolph et al.[31] in models of rodent and human intestinal cancer in vivo correlates with telomere length, which indicates that NPB formation may also be used as a surrogate measure of critically short telomeres.[9,31]

The two mechanisms of NPB formation can be distinguished in binucleated cytokinesis-blocked cells using telomere probes. NPB arising from telomere end fusions are expected to be telomere positive if they retain telomere dysfunction caused by loss of telomere-binding proteins without telomere attrition. In contrast, an NPB caused by misrepair of DNA breaks has a low probability of occurring within the telomeric sequences and is therefore likely to be telomere negative. Furthermore, NPB arising from misrepair of DNA breaks are also likely to be associated with an MN originating from the acentric fragment generated during misrepair.[21] NPB can break and form MN.[8,32] For about 40% of MN, two or more arise from a single NPB. When two or more MN are observed after an NPB resolution, normally an MN in each daughter cell remains.[33]

Study from Umegaki and Fenech validated the use of NPB as a biomarker of DNA damage in human WIL2-NS cells treated with hydrogen peroxide or superoxide, or after co-incubation with activated human neutrophils. Therefore, the importance of scoring NPB should not be underestimated because it provides direct evidence of genome damage resulting from misrepaired DNA breaks or telomere end fusions, which is otherwise not possible to deduce by scoring MN only.[8] NPB formation has been shown increase by a wide range of exposures including endogenous oxidants, ionizing radiation, polycyclic aromatic hydrocarbons, the cigarette smoke carcinogen 4-(methylnitrosamino)-1-(3-pyridyl)-1-butanone, vanadium pentoxide, and deficiencies in folate and selenium.[21]

NUCLEAR BUDS

Nuclear buds (NBUD) are biomarkers of elimination of amplified DNA and/or DNA repair complexes. The nuclear budding process has been observed in cultures grown under strong selective conditions that induce gene amplification as well as under moderate folic acid deficiency.[9] Gene amplification has a crucial role in the malignant transformation of human cells because it mediates the activation of oncogenes or the acquisition of drug resistance.[34] Studies conducted by Shimizu et al.[35–37] showed that amplified DNA is selectively localized to specific sites at the periphery of the nucleus and eliminated via nuclear budding to form MN during S-phase of mitosis. Amplified DNA may be eliminated through recombination between homologous regions within amplified sequences forming mini-circles of acentric and atelomeric DNA (double minutes), which localize to distinct regions within the nucleus, or through the excision of amplified sequences after segregation to distinct regions of the nucleus. The process of nuclear budding occurs during S-phase and the NBUD are characterized by having the same morphology as an MN, except that they are linked to the nucleus by a narrow or wide stalk of nucleoplasmic material, depending on the stage of the budding process. Excess of DNA may in general be expelled from the nucleus by the formation of NBUD and subsequent micronucleation.[32] The duration of the nuclear budding process and the extrusion of the resulting MN from the cell remain largely unknown,[8,9,13] although a study by Utani et al.[34] demonstrated that at least some of the cytoplasmic MN may be eliminated from the cell by extrusion.

NBUD are also classified as tentative precursors of MN. They are morphologically similar to MN in shape, structure, and size, except that they are connected to the nucleus by a narrow or wide stalk of nucleoplasmic material depending on the stage of the budding process.[21,23,32] NBUD may also be explained by the conventional model of MN formation, assuming they derive from anaphase laggards that independently form a nuclear envelope in telophase before fully integrating into the nucleus or from remnants of broken anaphase bridges.[32] The DNA in these buds is replicated and can subsequently be released as MN in the cytoplasm. NBUD have also been shown to form when an NPB between two nuclei breaks and the remnants shrink back toward the nuclei.[21]

NBUD originate from interstitial or terminal acentric fragments. Such NBUD may possibly represent nuclear membrane entrapment of DNA that has been left in cytoplasm after nuclear division or from excess DNA that is extruded from the nucleus. Whether NBUD are also a mechanism to eliminate excess chromosomes in a hypothesized process known as aneuploidy rescue remains unclear because there is only limited evidence for this possibility.[21]

In conclusion, the CBMN assay has evolved into an efficient cytome assay of DNA damage and misrepair, chromosomal instability, mitotic abnormalities, cell death, and cytostasis, enabling direct and/or indirect measurement of various aspects of cellular and nuclear dysfunction, such as unrepaired chromosome breaks fragments and asymmetrical chromosome rearrangement (MN or NPB accompanied by MN originating from acentric chromosomal fragments); telomere end fusions (NPB with telomere signals in the middle of the bridge and possibly without accompanying MN); malsegregation of chromosomes caused by spindle or kinetochore defects or cell-cycle checkpoint malfunction (MN containing whole chromosomes or asymmetrical distribution of chromosome-specific centromere signals in the nuclei of BN cells); nuclear elimination of amplified DNA and/or DNA repair complexes (NBUD); chromosomal instability phenotype and breakage–fusion–bridge cycles (simultaneous expression of MN, NPB, and NBUD); DNA hypomethylation (specific elevation in the frequency of MN containing whole chromosomes 1, 9, and 16); and altered mitotic activity and/or cytostasis (NDI) and cell death by necrosis or apoptosis (ratios of necrotic and apoptotic cells).[9]

ASSESSING GENOTOXIC EFFECTS OF MYCOTOXINS BY CBMN

DNA single strain breaks in *Escherichia coli* and increased number of MN in PK15 cells, HepG2 cells, human lymphocytes, and V79 cells has been reported in relation with mycotoxins exposure (Klaric et al., 2013).

Citrinin (CTN) induced MN in human-derived liver (HepG2) cells at concentrations of ≥10 μM or more and reduced the percentage of binucleated cells in a dose-dependent manner.[38] The distribution of centromere-positive and -negative MN in experiments with CIT was similar to that seen with potent aneugens.[38,39] In addition, in PK15 cells a significantly higher number of MN and NBUD were noted when applied at highest concentration, compared with control cells.[40]

CTN genotoxicity was established in the study of Flajs and Peraica,[41] in which various cell cultures exposed to CTN showed a significant increase in MN frequency. A study by Dönmez-Altuntas et al.[42] showed that CTN in increasing concentrations induced MN frequency and reduced the percentage of binucleated/mononucleated cells in cultured human lymphocytes. This indicates that CTN has genotoxic potential to cause genetic damage in human cells.

Patulin (PAT) is a toxic contaminant present at significant concentrations in some food products. There is a particular concern regarding apples and apple products. This mycotoxin has been described as a clastogen and also as an aneugen in mammalian cells. In a study performed by Alves et al.,[43] the three studied doses revealed an increase in a dose-dependent manner in an CBMN assay in lymphocytes.

Results from the study of Dönmez-Altuntas et al.[46] demonstrated that PAT at concentrations of 2.5, 5.0, and 7.5 μM caused a significant dose-dependent increase in the frequency of apoptotic cells and of necrotic cells, an increase in the number cells of NPBs, and a significant dose-dependent decrease in NDI rates in human lymphocytes. The data indicate that PAT has the ability to induce DNA damage (at 5 μM or more). However, the cytotoxic and cytostatic effects were more pronounced in human lymphocytes.

A study by Glaser and Stopper[44] revealed the dose-dependent formation of MN. After staining with an antibody against centromeres, kinetochore-positive and -negative cells were present, which is in agreement with the results of

Pfeiffer et al.[39] Also, a striking number of NPB were detected shortly after PAT treatment. In addition, results from a study by Zhou et al.[45] revealed PAT as capable of inducing the formation of MN in HepG2 cells.

Work performed by Dönmez-Altuntas et al.[46] and Robbiano et al.[47] suggested that a high concentration of OTA induced MN frequency in cultured human lymphocytes, and provided additional evidence for the genotoxicity of OTA in various in vitro systems. In addition, in PK15 cells a significantly higher number of MN and NBUD was noted after 12 h exposure to OTA.[40]

A study by Degen et al.[48] found a dose-dependent MN in ovine seminal vesicle cell cultures. That study also performed kinetochore analysis and concluded that OTA-induced MN revealed a pattern consistent with the idea of a mixed mode of action: In addition to its clastogenic activity, OTA may interfere with chromosomal distribution during cell division. This result was also achieved by a study by Knasmüller et al.[38] in which MN formation by OTA in human cells partly resulted from chromosome-breaking effects.

A study from Ouanes et al.[49] clearly indicated that zearaleone (ZEN) induced MN in dividing Vero cells and in bone marrow cells of treated mice in a dose-dependent manner. The authors explained these results by showing that ZEN caused damage by covalent binding (DNA adducts).

COMET ASSAY

Rydberg and Johanson were the first researchers to directly quantitate levels of DNA damage in individual cells by embedding them in agarose on slides and lysing them under mild alkali conditions to allow for the partial unwinding of DNA.[52]

Östling and Johanson developed a microgel electrophoresis technique to detect DNA damage at the level of a single cell. In their technique, cells embedded in agarose were placed on a microscope slide, the cells were lysed by detergents and high salt, and the liberated DNA electrophoresed under neutral conditions. Cells with an increased frequency of DNA double-strand breaks displayed increased migration of DNA toward the anode. The migrating DNA was quantitated by staining with ethidium bromide and by measuring the intensity of fluorescence at two fixed positions within the migration pattern using a microscope photometer. The neutral conditions used greatly limited the general utility of the assay.[50,51]

Subsequently, Singh et al. introduced a microgel technique involving electrophoresis under alkaline (pH >13) conditions to detect DNA damage in single cells. At this pH, increased DNA migration is associated with incomplete excision repair sites and alkali labile sites.[50,52] Because almost all genotoxic agents induce orders of magnitude more single-strand breaks and/or alkali labile sites than double-strand breaks, this version of the assay offered greatly increased sensitivity for identifying genotoxic agents.[51] Two years later, Olive and colleagues introduced another alkaline version of this assay in which DNA is electrophoresed at a pH of ≈12.3. Since the introduction of the alkaline (pH >13) comet assay in 1988, the breadth of applications and the number of investigators using this technique have increased almost exponentially. Compared with other genotoxicity assays, the advantages of the technique include (1) its demonstrated sensitivity for detecting low levels of DNA damage; (2) the requirement for small numbers of cells per sample; (3) flexibility; (4) low costs; (5) ease of application; and (6) the ability to conduct studies using a relatively short time period (a few days) needed to complete an experiment.[50]

The comet assay or single-cell gel electrophoresis is a simple, sensitive method for detecting DNA-strand breaks. Cells embedded in agarose on a microscope slide are lysed with detergent and 2.5 M NaCl and fresh Triton X-100 to remove membranes and soluble cell constituents, including most histones, leaving the DNA supercoiled and attached to a nuclear matrix as a nucleoid. A break in one strand of a DNA loop is enough to release the supercoiling, and during electrophoresis the relaxed loops are able to extend toward the anode.[53–57] Electrophoresis causes DNA loops containing breaks to move toward the anode, forming "comets" when stained and visualized by fluorescence microscopy. The relative content of DNA in the tail indicates the frequency of breaks.[54,58,59]

The DNA strand breaks can originate from the direct modification of DNA by chemical agents or their metabolites; from the processes of DNA excision repair, replication, and recombination; or from the process of apoptosis. Direct breakage of DNA strands occurs when reactive oxygen species interact with DNA alkaline labile sites that can be generated by depurination of an adducted base of the nucleotide and subsequent conversion of the abasic site to a strand break detected by alkaline treatment (pH >13.1).[55]

This assay was adapted to measure oxidized purines and oxidized pyrimidines by incubation of the nucleoids with bacterial DNA repair enzymes[56] including formamidopyrimidine DNA glycosylase (FPG), which recognizes the oxidized purine 8-oxo-Gua, endonuclease III, to detect oxidized pyrimidines, T4 endonuclease V to detect UV-induced pyrimidines dimmers, AlkA (3-methyladenine DNA glycosylase) for alkylated bases, or uracil DNA glycosylase, which removes misincorporated uracil from DNA.[57]

The comet assay has become a standard method for assessing DNA damage, with a wide range of applications in genotoxicity testing, human biomonitoring and molecular epidemiology, ecogenotoxicology, and fundamental research in DNA damage and repair[60,61]; studying the mechanisms of action of genotoxic chemicals; investigating oxidative damage as a factor in disease; monitoring oxidative stress in animals or human subjects resulting from exercise, diet, or exposure to environmental agents; studying the effects of dietary antioxidants; and monitoring environmental pollution by studying sentinel organisms.[56,59] This assay is useful for evaluating xenobiotic impacts based on its use of small cell samples and its ability to evaluate DNA damage in nonproliferation cells such as lymphocytes. In addition, the ability to obtain sufficient numbers of cells for analysis from different tissues (for instance, lymphocytes and buccal cells) provides a relatively noninvasive procedure for analysis.[51] Most studies assayed human blood cells because they circulate in the body, and the cellular, nuclear, and metabolic state of the blood cells can reflect the overall extent of body exposure.[62] Although like all tissues, lymphocytes are highly specialized, they can be seen to reflect the overall state of the organism insofar as they circulate through the whole body.[63] In addition, in biomonitoring studies, nasal epithelial cells and buccal cells have drawn the most attention because they are cells from tissues that come into direct contact with ingested or inhaled compounds.[55]

The congruence of results between the comet assay and other end points such MN or sister chromatid exchanges has been a principal reason to increase use of the comet assay as a biomarker for hazard assessment, particularly in monitoring the effects of occupational hazards.[51,62] Biological monitoring has been an important tool for the surveillance of medical health programs in European countries and to monitor occupational hazards in the United States.[62]

Formamidopyrimidine DNA Glycosylase

Measuring DNA strand breaks gives limited information. Breaks may represent the direct effect of some damaging agent, but generally they are quickly rejoined. They may in fact be apurinic/apyrimidinic (AP) sites baseless sugars, which are alkali labile and therefore appear as breaks. Or they may be intermediates in cellular repair because both nucleotide and base excision–repair processes cut out damage and replace it with sound nucleotides.[62,66] AP sites are alkali labile, so in principle they are expected to appear among the strand breaks, detected in the standard alkaline comet assay. However, it has not been convincingly demonstrated that all AP sites are converted under these conditions.[56,64]

To make the assay more specific and sensitive, an extra step was introduced of digesting the nucleoids with an enzyme that recognizes a particular kind of damage and creates a break. FPG detects the major purine oxidation product 8-OHG as well as other altered purines.[55,60,63,65] This enzyme was named for its ability to recognize imidazole-ring–opened purines, or formamidopyrimidines: namely, 8-oxo-G, 2,6-diamino-4-hydroxy-5-formadopyrimidine and 4,6-diamino-5-formamidopyrimidine, which occur during the spontaneous breakdown of damaged purines; however, a major substrate in cellular DNA is 8-OHG.[56,59,61,66]

A mammalian analogue of FPG, OGG1, has been applied in the Comet assay; however, studies performed comparing FPG and OGG1 revealed the ineffectiveness of OGG1.[56] For that reason, FPG continues to be the enzyme of choice for oxidized purines.

Image Analysis

Three scoring methods are available: visual scoring, semi-automated image analysis, and automated image analysis. Results from a study[67] verified that all three approaches can be regarded as trustworthy and interchangeable to a large extent.

The most important parameters to measure in the comet assay are the tail length, the relative fluorescence intensity of the head and tail (normally expressed as the percentage of DNA in the tail), and the tail moment.[70]

The percentage of DNA in the tail is considered the parameter that can best be compared between laboratories. The consensus in the International Workshop on Genotoxicity Test Procedures was that image analysis is preferred but not required and that the parameter percent tail DNA appeared to be the most linearly related to dose and the easiest to understand intuitively.[58,68]

The percent tail DNA values are constrained to a maximum of 100 and a minimum of 0 with no variability at the extremes and maximum variability at intermediate values such as 50%. The percent tail DNA has an advantage in that it can be standardized over studies, whereas tail length and moment, although consistent within the study, may not be comparable across studies.[69] Therefore, relative tail intensity is the most useful parameter because it bears a linear relationship to break frequency, is relatively unaffected by threshold settings, and allows discrimination of damage over the widest possible range. It also gives a clear indication of what the comets actually looked like.[69,70] A satisfactory condition for the assay is that untreated control cells should have a background level of breaks (i.e., about 10% DNA in the tail) and there are suggestions that negative control cells should have between 0% and 20% DNA in tail, which would obviate statistical problems.[69]

ASSESSING GENOTOXIC EFFECTS OF MYCOTOXINS BY COMET ASSAY

A study by Gao et al.[71] showed that ZEA caused DNA damage in HEK293 cells measured by comet assay. The damage measured by this technique was DNA strand breaks and oxidative stress on DNA strand breaks, both induced by ZEA.

FB1 caused an increase in MN formation in different cell lines and DNA strand breaks in rat liver and kidney, as measured by alkaline and FPG-modified comet assay.[72,73]

Some studies reported an increase in DNA damage in in vitro treatment with the mycotoxin OTA, in kidney,[38,47,74–77] confirming the genotoxic potential of OTA. A study by Domijan et al.[72] confirmed by comet assay standard and FPG modification that the measured end points suggested that oxidative stress was responsible for OTA-induced DNA damage; these results were confirmed even with lower OTA doses.

A study by Gursoy-Yuzugullu et al.[78] detected single- and double-strand DNA breaks using alkaline and neutral comet assay, respectively, by AF.

A study by Anderson et al.[79] did not show significantly higher results between African subjects from Gambia and those from the United Kingdom, owing to possible AF exposure.

Flajs and Peraica[41] showed an increase in DNA damage using the comet assay in Vero cells exposed for 24 h to CTN. However, the same method gave negative results in human-derived liver cells (HepG2) and human embryonic kidney cells (HEK293) no matter whether FPG was present. The latter was also achieved in a study by Liu et al.[80] This suggests that CTN-induced oxidative stress did not affect DNA. In addition, results obtained in a study by Knasmüller et al.[38] revealed negative findings by comet assay.

PAT demonstrated genotoxic potential and the ability to induce oxidative DNA damage in human cells, because treatment of HEK293[80] and in HepG2[81] cells with this mycotoxin significantly increased the values of tail moments in a concentration-dependent manner; this clearly indicated that PAT is a clastogen able to induce DNA strand breaks in human cells.

Studies on Interactions between Mycotoxins

The human population is probably exposed to multiple mycotoxins. Exposure can occur in different environments and food might be contaminated by several mycotoxins. Multiple exposure may lead to additive, synergistic, or antagonistic effects.[82] The combined toxicity of mycotoxins is hard to predict based on the toxic effect of a single mycotoxin.

AFB1 and OTA are some of the most relevant mycotoxins because of their toxic effects and demonstrated human exposure. The International Agency for Research on Cancer classified AFB1 as class 1 (human carcinogen) and OTA as class 2B (possible human carcinogen).[83–85]

The potential role of oxidative stress in AFB1 and OTA combined genotoxicity was tested by the FPG-modified comet assay in hepatocellular carcinoma epithelial cells (HepG2). Interestingly, the combination provoked a significant decrease in DNA damage compared with treatment with AFB1 alone.[82]

OTA and AFB1 are nephrotoxic mycotoxins that contaminate various commodities, and humans are constantly exposed to low levels of these mycotoxins. OTA is genotoxic after oxidative metabolism. This activity is through to have a central role in OTA-mediated carcinogenesis and may be divided into direct (covalent DNA adduction) and indirect (oxidative DNA damage) mechanisms of action.[86]

FB1 acts as carcinogen as well as a promoter of carcinogenesis and could potentiate OTA genotoxicity and carcinogenicity. AF levels in Europe are increasing; because of their carcinogenic activity, their interactions with OTA might pose a serious threat to humans and animals.[87]

Combined OTA and FB1 treatment, measured either by the standard comet or FPG-modified comet assay showed a synergistic increase in tail intensity and olive tail moment in kidney cells, even at doses corresponding to daily human exposure in Europe.[72]

The combined effect of CIT and OTA may cause oxidative stress by increasing the tail length and tail intensity of kidney and liver cells measured with the hOGG1 comet assay.[87]

A study by Klaric et al.[88] that assessed the individual and combined genotoxic effects of FB1, beauvericin (BEA), and OTA showed that combined treatment with two or all of the three mycotoxins increased MN frequency, mostly in an additive manner. A marked increase in MN frequency was observed with combinations of BEA plus OTA, and FB1 plus BEA plus OTA. In a general form, FB1, BEA, and OTA are genotoxic to PK15 and have predominantly clastogenic effects. All three mycotoxins induce MN and NPBs in PK15 cells in dose-dependent manner. The same author studied the combined effect of OTA and CTN in PK15 and concluded that both toxins exert genotoxic activity, measured by the formation of MN and NBUD. In addition, some OTA and CTN combinations significantly increased the frequency of NPB.[40]

A study by Knasmuller et al.[89] concluded that three *Fusarium* toxins (vomitoxin, moniliformin, and fumonisin B1) are clastogenic under in vitro conditions at very low concentrations.

Final Remarks

Human exposure to mycotoxins may result from a wide range of sources such as food or air. Thus, mycotoxins are of great concern in the contexts of public health and agro-economics. A challenge in the field of exposure assessment of mycotoxins and their effects is to develop accurate and reliable biomarkers for human studies. The CBMN assay and comet assay are sensitive and reliable methods that allow less invasive measurement of genotoxic effects potentially caused by mycotoxin exposure. Molecular epidemiology instruments such as a biomarker approach may be a promising tool for directly measuring toxin-mediated biological perturbations or to directly measure the amounts of mycotoxins present in the organism. Although AF and ochratoxin biomarkers have been successfully applied and validated over the past decade, further studies are needed to engage other mycotoxins.

REFERENCES

1. Bennett JW, Klich M. Mycotoxins. *Clin Microbiol Rev* 2003;**16**:497–516.
2. Zain ME. Impact of mycotoxins on humans and animals. *J Saudi Chem Soc* 2011;**15**:129–44.
3. da Rocha MEB, Freire FCO, Maia FMF, Guedes MIF, Rondina D. Mycotoxins and their effects on human and animal health. *Food Control* 2014;**36**:159–65.
4. Hussein HS, Brasel JM. Toxicity, metabolism, and impact of mycotoxins on humans and animals. *Toxicology* 2001;**167**:101–34.
5. Wang JS, Groopman JD. DNA damage by mycotoxins. *Mutat Res* 1999;**424**:167–81.
6. Marin S, Ramos AJ, Cano-Sancho G, Sanchis V. Mycotoxins: occurrence, toxicology, and exposure assessment. *Food Chem Toxicol Int J Publ Br Ind Biol Res Assoc* 2013;**60**:218–37.
7. Sirot V, Fremy J-M, Leblanc J-C. Dietary exposure to mycotoxins and health risk assessment in the second French total diet study. *Food Chem Toxicol Int J Publ Br Ind Biol Res Assoc* 2013;**52**:1–11.
8. Fenech M. Cytokinesis-block micronucleus assay evolves into a "cytome" assay of chromosomal instability, mitotic dysfunction and cell death. *Mutat Res* 2006;**600**:58–66.
9. Fenech M. Cytokinesis-block micronucleus cytome assay. *Nat Protoc* 2007;**2**:1084–104.

10. Fenech M, Crott J, Turner J, Brown S. Necrosis, apoptosis, cytostasis and DNA damage in human lymphocytes measured simultaneously within the cytokinesis-block micronucleus assay: description of the method and results for hydrogen peroxide. *Mutagenesis* 1999;**14**:605–12.
11. Fenech M, Chang WP, Kirsch-Volders M, Holland N, Bonassi S, Zeiger E. HUMN project: detailed description of the scoring criteria for the cytokinesis-block micronucleus assay using isolated human lymphocyte cultures. *Mutat Res* 2003;**534**:65–75.
12. Kirsch-Volders M, Fenech M. Inclusion of micronuclei in non-divided mononuclear lymphocytes and necrosis/apoptosis may provide a more comprehensive cytokinesis block micronucleus assay for biomonitoring purposes. *Mutagenesis* 2001;**16**:51–8.
13. Fenech M, Crott JW. Micronuclei, nucleoplasmic bridges and nuclear buds induced in folic acid deficient human lymphocytes—evidence for breakage–fusion-bridge cycles in the cytokinesis-block micronucleus assay. *Mutat Res Mol Mech Mutagen* 2002;**504**:131–6.
14. Bonassi S, Fenech M, Lando C, Lin YP, Ceppi M, Chand WP, et al. HUman MicroNucleus project: international database comparison for results with the cytokinesis-block micronucleus assay in human lymphocytes: I. Effect of laboratory protocol, scoring criteria, and host factors on the frequency of micronuclei. *Environ Mol Mutagen* 2001;**37**:31–45.
15. Kirsch-Volders M, Bonassi S, Knasmueller S, Holland N, Bolognesi C, Fenech MF. Commentary: critical questions, misconceptions and a road map for improving the use of the lymphocyte cytokinesis-block micronucleus assay for in vivo biomonitoring of human exposure to genotoxic chemicals-a HUMN project perspective. *Mutat Res* 2014. http://dx.doi.org/10.1016/j.mrrev.2013.12.001.
16. Fenech M. The Genome Health Clinic and Genome Health Nutrigenomics concepts: diagnosis and nutritional treatment of genome and epigenome damage on an individual basis. *Mutagenesis* 2005;**20**:255–69.
17. Vral A, Fenech M, Thierens H. The micronucleus assay as a biological dosimeter of in vivo ionising radiation exposure. *Mutagenesis* 2011;**26**:11–7.
18. Fenech M. The advantages and disadvantages of the cytokinesis-block micronucleus method. *Mutat Res* 1997;**392**:11–8.
19. Fenech M. The in vitro micronucleus technique. *Mutat Res* 2000;**455**:81–95.
20. Mateuca R, Lombaert N, Aka PV, Decordier I, Kirsch-Volders M. Chromosomal changes: induction, detection methods and applicability in human biomonitoring. *Biochimie* 2006;**88**: 1515–31.
21. Fenech M, Kirsch-Volders M, Natarajan AT, Surralles J, Crott JW, Parry J, et al. Molecular mechanisms of micronucleus, nucleoplasmic bridge and nuclear bud formation in mammalian and human cells. *Mutagenesis* 2011;**26**:125–32.
22. Shimizu N. Molecular mechanisms of the origin of micronuclei from extrachromosomal elements. *Mutagenesis* 2011;**26**:119–23.
23. Serrano-García L, Montero-Montoya R. Micronuclei and chromatid buds are the result of related genotoxic events. *Environ Mol Mutagen* 2001;**38**:38–45.
24. Utani K, Kohno Y, Okamoto A, Shimizu N. Emergence of micronuclei and their effects on the fate of cells under replication stress. *PloS One* 2010;**5**:e10089.
25. Norppa H, Falck GC-M. What do human micronuclei contain? *Mutagenesis* 2003;**18**:221–33.
26. Bolognesi C, Lando C, Forni A, Landini E, Scarpato R, Migliore L, et al. Chromosomal damage and ageing: effect on micronuclei frequency in peripheral blood lymphocytes. *Age Ageing* 1999;**28**:393–7.
27. Fenech M. Important variables that influence base-line micronucleus frequency in cytokinesis-blocked lymphocytes—a biomarker for DNA damage in human populations. *Mutat Res Mol Mech Mutagen* 1998;**404**:155–65.

28. Huang Y, Fenech M, Shi Q. Micronucleus formation detected by live-cell imaging. *Mutagenesis* 2011;**26**:133–8.
29. Bonassi S, Znaor A, Ceppi M, Lando C, Chang WP, Holland N, et al. An increased micronucleus frequency in peripheral blood lymphocytes predicts the risk of cancer in humans. *Carcinogenesis* 2007;**28**:625–31.
30. Thomas P, Umegaki K, Fenech M. Nucleoplasmic bridges are a sensitive measure of chromosome rearrangement in the cytokinesis-block micronucleus assay. *Mutagenesis* 2003;**18**: 187–94.
31. Rudolph KL, Millard M, Bosenberg MW, DePinho RA. Telomere dysfunction and evolution of intestinal carcinoma in mice and humans. *Nat Genet* 2001;**28**:155–9.
32. Lindberg HK, Wang X, Järventaus H, Falck GC-M, Norppa H, Fenech M. Origin of nuclear buds and micronuclei in normal and folate-deprived human lymphocytes. *Mutat Res* 2007;**617**: 33–45.
33. Hoffelder DR, Luo L, Burke NA, Watkins SC, Gollin SM, Saunders WS. Resolution of anaphase bridges in cancer cells. *Chromosoma* 2004;**112**:389–97.
34. Utani K, Kawamoto J, Shimizu N. Micronuclei bearing acentric extrachromosomal chromatin are transcriptionally competent and may perturb the cancer cell phenotype. *Mol Cancer Res MCR* 2007;**5**:695–704.
35. Shimizu N, Itoh N, Utiyama H, Wahl GM. Selective entrapment of extrachromosomally amplified DNA by nuclear budding and micronucleation during S phase. *J Cell Biol* 1998;**140**: 1307–20.
36. Shimizu N, Shimura T, Tanaka T. Selective elimination of acentric double minutes from cancer cells through the extrusion of micronuclei. *Mutat Res* 2000;**448**:81–90.
37. Shimizu N, Shingaki K, Kaneko-Sasaguri Y, Hashizume T, Kanda T. When, where and how the bridge breaks: anaphase bridge breakage plays a crucial role in gene amplification and HSR generation. *Exp Cell Res* 2005;**302**:233–43.
38. Knasmüller S, Cavin C, Chakraborty A, Mersch-Sundermann V, Gelderblom W, Zöhrer E, et al. Structurally related mycotoxins ochratoxin A, ochratoxin B, and citrinin differ in their genotoxic activities and in their mode of action in human-derived liver (HepG2) cells: implications for risk assessment. *Nutr Cancer* 2004;**50**:190–7.
39. Pfeiffer E, Gross K, Metzler M. Aneuploidogenic and clastogenic potential of the mycotoxins citrinin and patulin. *Carcinogenesis* 1998;**19**:1313–8.
40. Klarić MS, Zelježić D, Rumora L, Peraica M, Pepeljnjak S, Domijan A-M. A potential role of calcium in apoptosis and aberrant chromatin forms in porcine kidney PK15 cells induced by individual and combined ochratoxin A and citrinin. *Arch Toxicol* 2012;**86**:97–107.
41. Flajs D, Peraica M. Toxicological properties of citrinin. *Arh Hig Rada Toksikol* 2009;**60**: 457–64.
42. Dönmez-Altuntas H, Dumlupinar G, Imamoglu N, Hamurcu Z, Liman BC. Effects of the mycotoxin citrinin on micronucleus formation in a cytokinesis-block genotoxicity assay in cultured human lymphocytes. *J Appl Toxicol JAT* 2007;**27**:337–41.
43. Alves I, Oliveira NG, Laires A, Rodrigues AS, Rueff J. Induction of micronuclei and chromosomal aberrations by the mycotoxin patulin in mammalian cells: role of ascorbic acid as a modulator of patulin clastogenicity. *Mutagenesis* 2000;**15**:229–34.
44. Glaser N, Stopper H. Patulin: mechanism of genotoxicity. *Food Chem Toxicol Int J Publ Br Ind Biol Res Assoc* 2012;**50**:1796–801.
45. Zhou S, Jiang L, Geng C, Cao J, Zhong L. Patulin-induced genotoxicity and modulation of glutathione in HepG2 cells. *Toxicon Off J Int Soc Toxinol* 2009;**53**:584–6.
46. Dönmez-Altuntaş H, Hamurcu Z, Imamoglu N, Liman BC. Effects of ochratoxin A on micronucleus frequency in human lymphocytes. *Nahrung* 2003;**47**:33–5.

47. Robbiano L, Baroni D, Carrozzino R, Mereto E, Brambilla G. DNA damage and micronuclei induced in rat and human kidney cells by six chemicals carcinogenic to the rat kidney. *Toxicology* 2004;**204**:187–95.
48. Degen GH, Gerber MM, Obrecht-Pflumio S, Dirheimer G. Induction of micronuclei with ochratoxin A in ovine seminal vesicle cell cultures. *Arch Toxicol* 1997;**71**:365–71.
49. Ouanes Z, Abid S, Ayed I, Anane R, Mobio T, Creppy EE, et al. Induction of micronuclei by Zearalenone in Vero monkey kidney cells and in bone marrow cells of mice: protective effect of Vitamin E. *Mutat Res* 2003;**538**:63–70.
50. Tice RR, Agurell E, Anderson D, Burlinson B, Hartmann A, Kobayashi H, et al. Single cell gel/comet assay: guidelines for in vitro and in vivo genetic toxicology testing. *Environ Mol Mutagen* 2000;**35**:206–21.
51. Valverde M, Rojas E. The comet assay in human biomonitoring. In: Anderson D, Dhawan A, editors. *The comet assay in toxicology*. Royal Society of Chemistry; 2009. p. 227–51.
52. Hartmann A, Agurell E, Beevers C, Brendler-Schwaab S, Burlinson B, Clay P, et al. Recommendations for conducting the in vivo alkaline comet assay. 4th international comet assay workshop. *Mutagenesis* 2003;**18**:45–51.
53. Fairbairn DW, Olive PL, O'Neill KL. The comet assay: a comprehensive review. *Mutat Res* 1995;**339**:37–59.
54. Collins A, Dusinská M, Franklin M, Somorovská M, Petrovská H, Duthie S, et al. Comet assay in human biomonitoring studies: reliability, validation, and applications. *Environ Mol Mutagen* 1997;**30**:139–46.
55. Moller P, Knudsen LE, Loft S, Wallin H. The comet assay as a rapid test in biomonitoring occupational exposure to DNA-damaging agents and effect of confounding factors. *Cancer Epidemiol Biomark Prev Publ Am Assoc Cancer Res Cosponsored Am Soc Prev Oncol* 2000;**9**:1005–15.
56. Azqueta Amaya, Shaposhnikov Sergey, Collins Andrew. Detection of oxidised DNA using DNA repair enzymes. In: Anderson D, Dhawan A, editors. *The comet assay in toxicology*. Royal Society of Chemistry; 2009. p. 58–63.
57. Collins Andrew, Dusinska Maria. Applications of the comet assay in human biomonitoring. In: Anderson D, Dhawan A, editors. *The comet assay in toxicology*. Royal Society of Chemistry; 2009: 201–202.
58. Kumaravel TS, Jha AN. Reliable comet assay measurements for detecting DNA damage induced by ionising radiation and chemicals. *Mutat Res* 2006;**605**:7–16.
59. Dusinska M, Collins AR. The comet assay in human biomonitoring: gene–environment interactions. *Mutagenesis* 2008;**23**:191–205.
60. Collins AR. The comet assay for DNA damage and repair: principles, applications, and limitations. *Mol Biotechnol* 2004;**26**:249–61.
61. Collins AR. Investigating oxidative DNA damage and its repair using the comet assay. *Mutat Res* 2009;**681**:24–32.
62. Valverde M, Rojas E. Environmental and occupational biomonitoring using the comet assay. *Mutat Res* 2009;**681**:93–109.
63. Collins AR, Oscoz AA, Brunborg G, Gaivão I, Giovannelli L, Kruszewski M, et al. The comet assay: topical issues. *Mutagenesis* 2008;**23**:143–51.
64. Johansson C, Møller P, Forchhammer L, Loft S, Godschalk R, Langie S, et al. An ECVAG trial on assessment of oxidative damage to DNA measured by the comet assay. *Mutagenesis* 2010;**25**:125–32.
65. Collins AR, Dusinská M, Horváthová E, Munro E, Savio M, Stětina R. Inter-individual differences in repair of DNA base oxidation, measured in vitro with the comet assay. *Mutagenesis* 2001;**16**:297–301.

66. Smith CC, O'Donovan MR, Martin EA. hOGG1 recognizes oxidative damage using the comet assay with greater specificity than FPG or ENDOIII. *Mutagenesis* 2006;**21**:185–90.
67. Azqueta A, Meier S, Priestley C, Gutzkow K, Brunborg G, Sallette J, et al. The influence of scoring method on variability in results obtained with the comet assay. *Mutagenesis* 2011;**26**:393–9.
68. Hartmann A, Speit Guenter. Comet assay—protocols and testing strategies. In: Anderson D, Dhawan A, editors. *The comet assay in toxicology*. Royal Society of Chemistry; 2009. p. 378.
69. Lovell DP, Omori T. Statistical issues in the use of the comet assay. *Mutagenesis* 2008;**23**: 171–82.
70. Collins A. Comet assay. Principles, applications, and limitations. In: Didenko VV, Totowa NJ, editors. *From methods in molecular biology—in situ detection of DNA damage: methods and protocols*. Humana Press Inc.; 2004.
71. Gao F, Jiang L, Chen M, Geng CY, Yang G, Ji F, et al. Genotoxic effects induced by zearalenone in a human embryonic kidney cell line. *Mutat Res* 2013;**755**:6–10.
72. Domijan A-M, Zeljezić D, Kopjar N, Peraica M. Standard and Fpg-modified comet assay in kidney cells of ochratoxin A- and fumonisin B(1)-treated rats. *Toxicology* 2006;**222**:53–9.
73. Domijan A-M, Peraica M, Vrdoljak AL, Radić B, Zlender V, Fuchs R. The involvement of oxidative stress in ochratoxin A and fumonisin B1 toxicity in rats. *Mol Nutr Food Res* 2007;**51**:1147–51.
74. Kamp HG, Eisenbrand G, Schlatter J, Würth K, Janzowski C. Ochratoxin A: induction of (oxidative) DNA damage, cytotoxicity and apoptosis in mammalian cell lines and primary cells. *Toxicology* 2005;**206**:413–25.
75. Russo A, La Fauci L, Acquaviva R, Campisi A, Raciti G, Scifo C, et al. Ochratoxin A-induced DNA damage in human fibroblast: protective effect of cyanidin 3-O-beta-d-glucoside. *J Nutr Biochem* 2005;**16**:31–7.
76. Zeljezić D, Domijan A-M, Peraica M. DNA damage by ochratoxin A in rat kidney assessed by the alkaline comet assay. *Braz J Med Biol Res Rev Bras Pesqui Médicas E Biológicas Soc Bras Biofísica Al* 2006;**39**:1563–8.
77. Cosimi S, Orta L, Mateos S, Cortés F. The mycotoxin ochratoxin A inhibits DNA topoisomerase II and induces polyploidy in cultured CHO cells. *Toxicol Vitro Int J Publ Assoc BIBRA* 2009;**23**:1110–5.
78. Gursoy-Yuzugullu O, Yuzugullu H, Yilmaz M, Ozturk M. Aflatoxin genotoxicity is associated with a defective DNA damage response bypassing p53 activation. *Liver Int Off J Int Assoc Study Liver* 2011;**31**:561–71.
79. Anderson D, Yu T-W, Hambly RJ, Mendy M, Wild CP. Aflatoxin exposure and DNA damage in the comet assay in individuals from The Gambia, West Africa. *Teratog Carcinog Mutagen* 1999;**19**:147–55.
80. Liu B-H, Yu F-Y, Wu T-S, Li SY, Su MC, Wang MC, et al. Evaluation of genotoxic risk and oxidative DNA damage in mammalian cells exposed to mycotoxins, patulin and citrinin. *Toxicol Appl Pharmacol* 2003;**191**:255–63.
81. Ayed-Boussema I, Abassi H, Bouaziz C, Hlima WB, Ayed Y, Bacha H. Antioxidative and antigenotoxic effect of vitamin E against patulin cytotoxicity and genotoxicity in HepG2 cells. *Environ Toxicol* 2013;**28**:299–306.
82. Corcuera LA, Arbillaga L, Vettorazzi A, Azqueta A, López de Cerain A. Ochratoxin A reduces aflatoxin B1 induced DNA damage detected by the comet assay in Hep G2 cells. *Food Chem Toxicol Int J Publ Br Ind Biol Res Assoc* 2011;**49**:2883–9.
83. Overall evaluations of carcinogenicity: an updating of IARC Monographs volumes 1 to 42. *IARC Monogr Eval Carcinog Risks Hum Suppl World Health Organ Int Agency Res Cancer* 1987;**7**:1–440.

84. IARC. IARC monographs on the evaluation of carcinogenic risks to humans. *Some naturally occurring substances: food items and constituents, heterocyclic aromatic amines and mycotoxins*, vol. 26. 1997.
85. IARC Working Group on the Evaluation of Carcinogenic Risks to Humans. Some traditional herbal medicines, some mycotoxins, naphthalene and styrene. *IARC Monogr Eval Carcinog Risks Hum World Health Organ Int Agency Res Cancer* 2002;**82**:1–556.
86. Pfohl-Leszkowicz A, Manderville RA. Ochratoxin A: an overview on toxicity and carcinogenicity in animals and humans. *Mol Nutr Food Res* 2007;**51**:61–99.
87. Klarić MS, Rašić D, Peraica M. Deleterious effects of mycotoxin combinations involving ochratoxin A. *Toxins* 2013;**5**:1965–87.
88. Klarić MS, Rumora L, Ljubanović D, Pepeljnjak S. Cytotoxicity and apoptosis induced by fumonisin B(1), beauvericin and ochratoxin A in porcine kidney PK15 cells: effects of individual and combined treatment. *Arch Toxicol* 2008;**82**:247–55.
89. Knasmüller S, Bresgen N, Kassie F, Mersch-Sundermann V, Gelderblom W, Zöhrer E, et al. Genotoxic effects of three *Fusarium* mycotoxins, fumonisin B1, moniliformin and vomitoxin in bacteria and in primary cultures of rat hepatocytes. *Mutat Res* 1997;**391**:39–48.

Chapter 21

Mycotoxin Analytical Methods

Magdalena Twarużek, Anna Błajet-Kosicka, Robert Kosicki, Jan Grajewski

Kazimierz Wielki University, Faculty of Natural Science, Institute of Experimental Biology, Department of Physiology and Toxicology, Bydgoszcz, Poland

EXTRACTION AND ANALYTICAL TECHNIQUES

Several extraction techniques are used for mycotoxin isolation, including shaking, blending, pressurized liquid extraction (PLE), supercritical fluid extraction (SFE), microwave-assisted extraction (MAE), ultrasound-assisted extraction (UAE), and matrix solid phase dispersion (MSPD). The solid phase extraction (SPE) technique, including immunoaffinity columns (IACs), is commonly applied to purify extracts. The most frequently used instrumental analytical techniques for mycotoxin detection and determination are gas chromatography (GC) with electron capture (ECD), flame ionization (FID), or single mass spectrometric (MS) detection; high-performance liquid chromatography (HPLC) coupled to ultraviolet (UV) or diode array detection (DAD); fluorescence (FLD), MS, or tandem mass spectrometric (MS/MS) detectors; as well as immunochemical methods including enzyme-linked immunosorbent assay (ELISA) and rapid tests. Special sampling devices (e.g., RCS air sampler, SpinCon bioaerosol sampler, Andersen GPS-1 polyurethane foam (PUF) high-volume air sampler) are used for air and dust sample collection. Besides immunoassays, a variety of biological methods have been used for mycotoxin analysis including cytotoxicity tests, mainly 3-(4,5-dimethylthiazol-2-yl)-2,5-diphenyltetrazolium bromide (MTT) assay.

FOOD AND FEED OF CEREAL ORIGIN

Both the complexity of the matrix and the detection method influence the choice of sample treatment applied. For extraction, different organic solvents (mainly acetonitrile, methanol, acetone, and ethyl acetate) and their mixtures as well as water have been used, depending on the physicochemical properties of the mycotoxin of interest. The mixture of acetonitrile–water (84:16, v/v) is commonly used to extract trichothecenes from cereals[1] but it also can be applied for simultaneous isolation of aflatoxins, ochratoxin A (OTA), deoxynivalenol, T-2 and HT-2 toxins,

Environmental Mycology in Public Health. http://dx.doi.org/10.1016/B978-0-12-411471-5.00021-1

zearalenone, and fumonisins B_1 and B_2 from wheat flour.[2] Bertuzzi et al.[3] examined different extraction solvent mixtures—methanol–water (80:20, v/v) and acetone–water (85:15, v/v; 70:30, v/v; 60:40, v/v; 50:50, v/v)—to extract aflatoxins from maize. The best process efficiency for all analytes was for two acetone–water mixtures (70:30, v/v; 60:40, v/v). For the simultaneous determination of aflatoxins and OTA in baby food, a mixture of methanol–water (80:20, v/v) was applied for the extraction of mycotoxins.[4] This combination was also the most efficient mixture for aflatoxin B_1 extraction from pig feed matrix.[5] Scott et al.[6] evaluated different solvent mixtures to extract fumonisins from rice, corn-based food, and beans. Results showed that acetonitrile–methanol–water (25:25:50, v/v/v) was generally a better extraction solvent than methanol–water (75:25 or 80:20, v/v) for recovery of mycotoxins. The efficiency of different extraction mixtures for the extraction of fumonisins from maize was also compared by Marschik et al.[7] The authors confirmed that extraction efficiency differs when spiked and naturally contaminated samples are investigated, and the most appropriate extractant turned out to be acetonitrile–methanol–water (25:25:50, v/v/v). Considering multi-mycotoxin extractions from cereals, mixtures of acetonitrile–water result in higher process efficiency compared with methanol–water.[8] The usually extraction procedure is carried out by shaking or blending, although other methods have been applied. Alternative techniques used for mycotoxin extraction from cereal-based matrices include PLE, SFE, MAE, UAE, and MSPD. In PLE, both temperature and pressure are used to extract organic compounds from solid or semisolid matrices. The use of elevated pressures allows solvents to be used above their atmospheric boiling points to increase solvation power and extraction kinetics. Increased temperatures can also disrupt strong solute–matrix interactions. These increase the extraction efficiency and rate and reduce the consumption of organic solvents and operation time.[9] This technique was applied to extract *Fusarium* mycotoxins from maize,[10] zearalenone, and α-zearalenol from cereals and swine feed,[11] fumonisins from corn-based baby food,[12] OTA from organic and conventional bread,[13] ochratoxin A from rice,[14] and 17 mycotoxins (including aflatoxins, OTA, and *Fusarium* mycotoxins) from cereal-based commodities.[15] Use of the PLE technique may become laborious and time consuming because obstruction of extraction cells often occurs owing to swelling of starches in cereals. Moreover, a thorough cleanup of PLE extract typically needs to be performed because of more co-extracted impurities compared with traditional methods.[16] SFE is a process of separating one component (the extractant) from another (the matrix) using supercritical fluids as the extracting solvent. Extraction is usually from a solid matrix, but it can also be from liquids. Carbon dioxide (CO_2) is the most used supercritical fluid, sometimes modified by co-solvents such as ethanol or methanol. Extraction conditions for supercritical CO_2 are above the critical temperature of 31 °C and critical pressure of 74 bar, but the addition of modifiers may slightly alter this.[17] Extraction of macrocyclic lactone mycotoxins from maize flour was performed using SFE[18,19]; however, this technique is expensive and rarely used to prepare samples for mycotoxin determination.[20] MAE is the process of using microwave energy to heat solvents in contact with a sample to partition analytes from the sample matrix into the solvent. The ability to heat the sample solvent mixture rapidly is inherent in MAE and the main advantage of this technique.[21] In MAE, extraction occurs as the result of changes in the sample structure caused by electromagnetic waves, and process acceleration as well as high extraction yield may result from a synergistic combination of heat and mass gradients.[22] In principle, only samples or solvents containing dipolar materials or microwave absorbents can be affected by microwaves, which heat the extraction body from inside to outside in a short time and are much different from common heating methods. Acceleration results from fast and uniform heating.[9] Regarding mycotoxin determination, the MAE technique was used to extract zearalenone from wheat and corn[23] as well as for aflatoxin isolation from grain and grain products.[24] Another way to enhance extraction efficiency is to use ultrasonication during the process. In UAE, several extractions can be performed at the same time, no advanced laboratory equipment is required, and the technique is relatively inexpensive compared with previously listed methods.[25] However, in the case of mycotoxin extraction, no significant differences in process efficiency were observed when UAE was compared with conventional extraction techniques used to isolate zearalenone from corn[26] and enniatins, beauvericin, and fusaproliferin from pasta.[27] Li et al.[28] applied UAE to extract fumonisins B_1 and B_2 from corn with methanol–water (3:1, v/v). MSPD is a patented analytical approach for the extraction of solid and/or viscous biological samples. In MSPD, the sample is mixed with standard SPE bonded-phase solid support materials. In this process, the bonded-phase support acts as both an abrasive to produce disruption of sample architecture and a bound solvent that assists in accomplishing sample disruption. The sample is dispersed over the surface of the bonded-phase support material, producing a unique mixed-character phase for conducting target analyte isolation.[25,29] MSPD was validated for different cereal samples analyzed for aflatoxins and OTA.[30] Cereal (wheat, rice, maize, rye, barley, oats, spelt, and sorghum) and cereal products (snacks, pasta, soup, biscuits, and flour) were prepared with MSPD for the determination of 14 mycotoxins.[31] This process was also applied to isolate 32 mycotoxins from barley[32]; however, selected analytes were not adequately extracted and the recoveries were not suitable enough.

The next step of sample preparation is extract cleanup and mycotoxin enrichment. The technique of widest application at this stage is SPE. SPE involves partitioning a liquid phase (sample matrix or extract with analytes) and a solid (sorbent) phase. The sample extract is loaded onto the SPE solid phase, undesired components are washed away, and the analytes are eluted with another solvent into a collection vial. The SPE technique can also be used to retain impurities from the sample extract.[33] The SPE multifunctional columns, containing alumina, active charcoal, and Celite polymers, are commonly used for cereal-based sample extract cleanup. Multifunctional columns designed by Romer Labs® (MultiSep® and MycoSep®) have been widely applied for food and feed matrix extract cleanup. MultiSep®/MycoSep® 226 columns, developed for aflatoxins and zearalenone, have been used for determination of trichothecenes in barley and barley products,[34] deoxynivalenol, zearalenone, T-2, HT-2,

OTA, and fumonisins (B_1, B_2) in vegetable animal feed.[35] Two-stage cleanup (MultiSep® 226 and 227, designed for trichothecenes A and B) was performed during evaluation of trichothecene B, their derivatives, and precursor content in wheat.[36] MultiSep® 227 was also used in the procedure of quantifying deoxynivalenol, T-2, and HT-2 toxins in bread.[37] Scarpino et al.[38] used MycoSep® 240 MON in maize extract purification for the determination of moniliformin. The use of polymeric reversed-phase SPE columns (Oasis HLB) has been described for 26 mycotoxins in maize silage[39] and nine mycotoxins (aflatoxins, OTA, deoxynivalenol, zearalenone, fumonisins, and T-2 toxin) in maize.[40] IACs are a special form of SPE technique in which the column is filled with anti-mycotoxin antibodies coupled covalently to an appropriate carrier and stored in phosphate-buffered saline. The mycotoxin binds to the specific antibodies; after the matrix components are removed, it is eluted with an appropriate organic solvent that denatures the antibodies. IACs provide clean extracts and are applicable to complex matrices. However, they lack reusability and have higher costs compared with other materials. IACs were developed for all major mycotoxins. Corn-based food samples have been prepared by means of IACs for the determination of fumonisins.[41,42] Aflatoxins, OTA, zearalenone, deoxynivalenol, T-2, and HT-2 toxins have been isolated from cereal food and feed sample extracts with IACs, as well.[43–47] Application of IACs is limited in the case of multi-mycotoxin methods, but columns for simultaneous determination of several mycotoxins have been developed and applied to cereals[48] and animal feed samples.[49] A promising tool for mycotoxin isolation is molecularly imprinted polymers (MIPs). MIPs are highly cross-linked polymers capable of selective molecular recognition.[50] Several attempts were made to synthesize and use MIPs for mycotoxins present in cereals, such as OTA,[51] fusaric acid,[52] fumonisins,[53,54] ergot alkaloids,[55] zearalenone,[56] and citrinin.[57] The Quick, Easy, Cheap, Effective, Rugged, and Safe (QuEChERS) sample preparation method, developed for pesticide residue analysis, was applied to samples analyzed for mycotoxins. The procedure is based on the extraction of mycotoxins using a mixture of water–acetonitrile with the addition of buffering inorganic salts. This approach has been applied to cereal sample preparation for the determination of OTA in bread,[58] 17 mycotoxins,[59] 14 mycotoxins in rice,[60] 22 mycotoxins in sorghum,[61] trichothecenes, zearalenone, and patulin in milled grain-based products.[62]

OTHER FOOD MATRICES

Because different non-cereal food can be also contaminated with mycotoxins, several methodologies are applied for sample preparation of these matrices. Non-cereal food can be either solid (e.g., dried fruit, spices, cheese, meat) or liquid (milk, wine, beer, juices). For solid matrices the extraction techniques are generally the same as for cereal samples. PLE was used to extract aflatoxins from nuts[63]; aflatoxins and OTA from animal-derived foods, for example, swine, bovine and sheep muscle, liver, and kidney, as well as hen eggs[64];

and 35 mycotoxins from various traditional Chinese medicines.[65] Chinese medicine (*Zizyphi fructus*) samples were prepared for aflatoxin evaluation using the SFE technique.[66] UAE was applied to aflatoxins and OTA extraction from hazelnuts, pork, and nutmeg samples.[67–69] Extraction of aflatoxins from liquid samples (milk or oil) can be assisted by ultrasound.[70,71] For this group of mycotoxins, several procedures engaging the MSPD technique were developed for sample preparation: for example, peanuts, chili powder, green beans, black sesame, and olive oil (aflatoxin B_1, B_2, G_1, and G_2), and aflatoxin M_1 in cheese.[72–75] Also macrocyclic lactone mycotoxins were extracted with this technique from fish tissue.[76] Extraction techniques used for liquid food samples are liquid–liquid extraction (LLE) and SPE. Fabiani et al.[77] compared three different cleanup methods (MycoSep, IACs, and LLE with dichloromethane under acidic conditions) in sample preparation to determine OTA in wine. The use of IACs and MycoSep showed good recovery whereas LLE recovery was worse. Multifunctional columns (MycoSep® 228) were also compared with homemade polyvinylpolypyrrolidone-florisil columns to analyze patulin in apple and hawthorn juices and jams; both approaches showed good recovery.[78] For the determination of aflatoxin M_1 in milk, both IACs and Oasis HLB columns have been used for analyte isolation.[79,80] IACs were also used for aflatoxin M_1 isolation from other dairy products such as cheese and yogurt.[81] The use of MIPs for mycotoxin investigation in liquid samples has been also described. OTA was determined in red wine, beer, and grape juice samples cleaned up with MIP sorbents.[82,83] Zhao et al.[84] developed and used MIPs for patulin evaluation in apple juice (recovery > 90%). Apple juice samples for patulin determination as well as apple–pear juice and puree have been prepared based on QuEChERS methodology.[85,86] A QuEChERS-like method was developed to identify and quantify 15 mycotoxins in beer-based drinks.[87] Solid non-cereal sample extracts have been purified using these techniques as well. Multifunctional columns MycoSep® 226 and MycoSep® 229 were used for the determination of aflatoxins in peanuts and OTA in spices, respectively.[88,89] SPE with Oasis HLB columns was applied to animal-derived food extracts before quantification of aflatoxins and OTA,[64] whereas MIPs were used for OTA determination in coffee samples.[82] Preparation of a QuEChERS-based sample was employed in multi-mycotoxin analyses of fruits, spices, oil seeds, and nuts.[90–92] Deoxynivalenol, T-2, and HT-2 toxins were isolated from paprika and chili pepper,[93] OTA from cocoa and chocolate[94] and fumonisins from medicinal herbs and spices samples[95] using IACs.

SEPARATION TECHNIQUES FOR IDENTIFICATION AND DETERMINATION OF MYCOTOXINS

Gas and liquid chromatography are the most popular separation techniques used for the determination of mycotoxins. GC is commonly applied to quantify trichothecenes, especially of the A group compounds, because they are not fluorescent and do not absorb the UV–Vis light strongly. Because mycotoxins are not volatile compounds, they require derivatization before injection into a GC column. The chemical conversion of trichothecenes leads to trimethylsilyl, trifluoroacetyl, pentafluoropropionyl, and heptafluorobutyryl derivatives.[96,97] Poor linearity of calibration curves and repeatability as well as memory effects from previous injections are some of the main problems connected with GC methods.[98] The most popular GC detectors, such as ECD, FID, and mass spectrometers (MS), are used in mycotoxin quantification. Trichothecene B was determined in poultry feed mixtures with GC-ECD after sample extraction with acetonitrile–water (84:16, v/v) on a rotary shaker and cleanup with MycoSep® 225 columns. For chemical conversion, trimethylsilylimidazole–*bis*-(trimethylsilyl)trifluoroacetamide–trimethyl-chlorosilane, 3:3:2, was used.[99] GC-ECD was also the technique for the determination of T-2 and HT-2 derivatized with heptafluoropropionic anhydrite in paprika and chili pepper samples.[93] Deoxynivalenol, diacetoxyscirpenol, and T-2 toxin were determined in popcorn samples with GC-FID, with TFA as the derivatization agent, after extraction with methanol and purification with dichloromethane and aluminum oxide/charcoal columns.[100] Coupling GC with MS has been applied in quantification of OTA in wine and beer[101]; patulin in apple juice[102]; 10 trichothecenes in oats[103]; and OTA, zearalenone, deoxynivalenol, and T-2 toxin in grain, feed premixes, and meat.[104] The natural fluorescence of aflatoxins, OTA, and zearalenone makes them usable for fluorescence detection, a common method used in HPLC. The fluorescence of aflatoxins depends on the composition of the mobile phase. In reverse-phase chromatography, aflatoxins are eluted with methanol–water or acetonitrile–water mobile phases, but the fluorescence of aflatoxin B_1 and G_1 is reduced considerably and derivatization is necessary to enhance detection.[105] Enhancement can be achieved by pre-column addition of trifluoroacetic acid to form hemiacetals[106,107] or after chromatographic separation by a reaction with bromine (KBr and HNO_3 introduced to the mobile phase) using electrochemical cells.[108,109] The detection of fumonisins with FLD also needs a derivatization step, usually performed by reaction with *o*-phthalaldehyde in the presence of 2-mercaptoethanol.[95] Different derivatizing agents have been used for trichothecene A (T-2 and HT-2 toxins), for example, coumarin-3-carbonyl chloride, 1- and 2-anthroylnitrile, 1- and 2-naphthoyl chloride, and pyrene-1-carbonyl cyanide for fluorescence detection.[1] Ochratoxin A has often been detected using excitation and emission wavelengths of 333 and 460 nm, respectively,[110–112] whereas for zearalenone excitation of 274 nm and emission wavelengths of 446–450 nm have been a common choice.[113–115] A diode array or UV detectors have been used to quantify trichothecene B as well as patulin[116–118]; their use for OTA has been also described.[119] Aresta et al.[120] determined OTA and cyclopiazonic, mycophenolic, and tenuazonic acids in cornflakes using HPLC with diode array UV detection. UV detection has been also applied in quantification of *Alternaria* mycotoxins (alternariol and alternariol monomethyl ether) in wine and grape and cranberry juices.[121] Application of mass spectrometry (MS) as a detection method in liquid chromatography caused interest and the development of procedures for mycotoxin determination. Coupling of HPLC with MS or tandem mass spectrometry (MS/MS) is a universal, selective, and sensitive technique that enables simultaneous detection of a large number of compounds. The lack of a derivatization procedure makes this technique more popular than GC-MS. Possible ion suppression or enhancement caused by matrix components, which affects the accuracy of measurements, is one of the LC-MS/MS disadvantages. Monbaliu et al.[122,123] developed a multi-method using LC-MS/MS for the determination of 23 mycotoxins in sweet pepper, after extraction with ethyl acetate–formic acid (99:1, v/v), followed by splitting the extract and cleanup with aminopropyl columns and octadecyl columns or by using strong anion-exchange columns. Mycotoxins were analyzed in positive electrospray ionization using selected reaction monitoring. Limits of detection with the multi-mycotoxin method varied from 0.3 to 42.5 ng/g. A simple and sensitive LC-MS method for the determination of patulin in fruit juice and dried fruit samples was developed by Kataoka et al.[124] Electrospray ionization conditions in the negative ion mode were optimized for MS detection and the pseudo-molecular ion $[M-H]^-$ was used to detect patulin in selected ion monitoring mode. As an internal standard,[13] C-patulin was used and the detection limit was 23.5 ng/L. The LC-MS/MS method has been applied in determination of OTA in cheese samples, using OTB as the internal standard.[125] Rasmussen et al.[126] focused on the detection method of 27 secondary fungi metabolites in silages using LC-MS/MS. The authors applied the QuEChERS approach for sample preparation. The method was successfully validated for the determination of eight analytes qualitatively and 19 quantitatively, but the validation results for citrinin and fumonisins were unsatisfying. The limit of detection for the quantitatively validated analytes ranged from 1 to 739 ng/g. An LC-MS/MS method for the simultaneous determination of aflatoxins, dyes, and pesticides in spices was developed by Ferrer Amate et al.[127] This technique has been used for the determination of OTA, mycophenolic acid, and fumonisin B_2 in meat products,[128] deoxynivalenol, zearalenone, T-2, and HT-2 toxins in breakfast cereals and baby food[129]; 25 mycotoxins in cassava flour, peanut cake, and maize samples,[130] aflatoxins, ochratoxin A, deoxynivalenol, zearalenone, T-2 and HT-2 toxins in cereal-based foods[131,132]; trichothecenes in grains[133]; 29 mycotoxins in distillers dried grains with solubles; common ingredients for animal feed[134]; aflatoxin M_1 in milk[70,135]; aflatoxins in nonalcoholic beer[136]; 11 mycotoxins in beer and wine[137]; seven mycotoxins in vegetable animal feed[35]; moniliformin in maize[38]; deoxynivalenol, zearalenone, T-2, and HT-2 toxins in malting barley[138]; ergot alkaloids in cereals[139]; and aflatoxins and OTA in bee pollen.[140] MALDI-TOF mass spectrometers have been used to identify mycotoxin-producing fungi.[141] Liquid and gas chromatography are the most frequently used separation techniques for mycotoxin determination. In addition, the application of capillary electrophoresis (CE) has been described. Maragos and Appell[142] developed a CE method of zearalenone determination in maize using cyclodextrin-enhanced fluorescence. Macrocyclic lactone mycotoxins were also evaluated by CE with amperometric detection in flour samples.[19] This technique has been applied for OTA determination in wine.[143,144]

IMMUNOASSAY AND OTHER METHODS

The ELISA is a popular method used for monitoring the most important mycotoxins. This technique is based on the reaction of a mycotoxin and a specific antibody, usually performed on microtiter plates. Main advantages of this technique are simple sample preparation and inexpensive equipment, but cross-reactivity with related mycotoxins and other matrix components may lead to false positive or negative results, which should be confirmed by chromatographic methods.[98] This technique has been applied in determination of aflatoxins in herbal medicine[145] and deoxynivalenol in rice and corn silage.[146] ELISA seems to be a popular technique for the determination of aflatoxin M_1 in milk.[147–149] Lateral flow devices (LFDs) are rapid immunoassays based on the interaction between specific antibodies present on a membrane strip and antibody-coated dyed receptors, for example, latex or colloidal gold, that react with an analyte to form an analyte–receptor complex.[98] These assays are usually qualitative and give a yes/no answer regarding the presence of a contaminant or semiquantitative. LFDs for the quantitative determination of aflatoxins in maize were developed by Anfossi et al.[150] LFDs have been applied in the analysis of grains and feeds for aflatoxin B_1, deoxynivalenol, fumonisins, T-2 toxin, and zearalenone.[5,151–154] Another technique, based on the interaction between a fluorescently labeled antigen and a specific antibody in solution, is fluorescence polarization immunoassay (FPIA). Use of the FPIA test has been described for the detection of OTA in red wine, rice, and wheat,[155–157]; OTA and zearalenone in grains[158]; and trichothecenes in wheat.[159,160] Near-infrared spectroscopy (NIR) in connection with principal component analysis and cluster analysis is a methodology that allows the determination of mycotoxins without the extraction procedure. NIR has been applied for the rapid detection of aflatoxin B_1 in maize and barley,[161] deoxynivalenol in durum and common wheat,[162] and fumonisins in maize.[163] In surface plasmon resonance (SPR) biosensors refractive index changes are used to detect mass changes at metal (usually gold) sensor surface interfaces. Biosensors based on SPR have been developed to detect OTA in cereals and beverages,[164] T-2 and HT-2 toxin in cereals and maize-based baby food,[165] nivalenol and deoxynivalenol in wheat,[166] and aflatoxin B_1 in corn.[167]

ENVIRONMENTAL SAMPLES

The detection of mycotoxins is a difficult task requiring careful work by analytical chemists with knowledge of fungal metabolites plus access to state-of-the-art instrumentation. Building materials represent specific matrices with an infinite number of combinations of materials (e.g., wallpaper, paint, dust) that may interfere with analytical methods. Identification of specific mycotoxins is further complicated by the fact that fungi produce so many different metabolites.[168]

Many studies concerned with indoor health problems and mycotoxins focus on trichothecenes, although this group of metabolites is interesting only when growth of *Stachybotrys* and perhaps *Memnoniella* has occurred. Buildings materials, indoor air, and air conditioning systems may also be reservoirs of other fungi such as *Aspergillus*, *Penicillium*, and *Fusarium*. In this regard, the occurrence of aflatoxins, sterigmatocystin, citrinin, and OTA in homes and buildings has been reported.[169]

Regarding sampling, ventilation ducts where dust had previously been found to be culture-positive for *Stachybotrys chartarum* were cleaned using a vacuum cleaner.[170] Dust samples were collected on cotton swabs.[171] In water-damaged Danish buildings, material was collected onto a 0.45-μm filter by a sampling device attached to a vacuum cleaner.[172] Airborne dust samples were collected using an RCS (Biotest Diagnostics, USA) air sampler that collects airborne microorganisms quantitatively onto a culture medium according to the impaction principle.[173] Two samplers were employed to collect airborne trichothecene mycotoxins: a SpinCon PAS 450-10 bioaerosol sampler (Sceptor Industries, Inc., USA) and an Andersen GPS-1 PUF high-volume air sampler (Thermo Electron Corporation, USA). The SpinCon sampler has been evaluated in the outdoor environment and has been determined to be a highly effective air-sampling device.[174,175] High-volume samplers incorporating PUF are generally designed to collect airborne pesticides and organic pollutants in the outdoor environment.[176–178] Airborne particulates were collected either by area or by personal sampling. Air was drawn through 25-mm glass fiber filters without binders (Gelman Science, USA) by means of Aircheck pumps.[179] Air sampling was performed with an Enerfluid® pump with an airflow of 23 L/min. It was composed of a Teflon filter with a pore diameter of 1 μm.[180] Air samples were collected from an indoor laboratory using the SASS 2000, a wet-walled cyclone collector (Research International, USA). This system uses a high-efficiency blower to pull air at a rate of 265 L/min into a cyclonic cup where water is injected in a fine spray. Submicron-sized particles and molecular species are partitioned into the aqueous phase, with recapture and recirculation of the aqueous phase.[181] Air samples were taken in two ways: using a cassette filter or a BioSampler (SKC, 84, USA) connected to an AirCon-2 (Gilian Instrument Co., Clearwater, USA) high-flow sampler pump.[182]

An important step in the analysis of mycotoxins is extraction, and especially sample cleanup. For instance, samples taken from water-damaged dwellings were extracted with can–H_2O (84:16, v/v), centrifuged for 10 min at 6000 g, and cleaned by passing through a Bond Elut Mycotoxin column.[183] Water-damaged gypsum boards heavily infested with *S. chartarum* were found in a school and in a domestic residence in Copenhagen. Samples were extracted with dichloromethane and cleaned up on Sep-Pak® C_{18} modules.[172] Krysińska-Traczyk et al.[184]

applied a mixture of can–H_2O (75:25) to extract mycotoxins from grain dust. The filtrate was divided into three parts: The first part was degreased and cleaned in a column containing Florisil for moniliformin, the second part was cleaned in Ochratest Vicam columns for OTA, and the third part of the extract was cleaned in columns containing activated carbon, Celite, and Al_2O_3 neutral (1:1:1) for trichothecenes. To analyze aflatoxins in dust samples IACs (Aflatest, Vicam) were used and aflatoxins were extracted with methanol–water (80:20, v/v).[179] In other research,[185] IACs were not used and aflatoxins were extracted with methanol–acetonitrile (50:50, v/v). When possible, partial purification of samples with an immunoaffinity column,[186] or some other method employing a phase different from the one used in the final analytical column, is desirable.

In the case of analysis of samples of the indoor environment, many different extraction methods are described, depending on the sample type (air, dust, mycelium, or building materials). Several extraction solvents, such as acetonitrile, methanol–water, dichloromethane–ethyl acetate, and ethyl acetate, are used, followed by defatting, filtration, centrifugation, and/or evaporation steps.[170,171,187–190]

Many studies have been conducted in which mycotoxins were measured in dust collected in occupational environments. For instance, Smoragiewicz et al.[191] based identification of trichothecenes in dust from a ventilation system on false-positive results originating from the use of nonspecific thin-layer chromatographic (TLC) and liquid chromatographic methods. The most common fungi collected from dust samples were analyzed by TLC analysis that revealed that isolates produced known mycotoxins.[192,193] TLC is not specific enough to detect trochothencens.[168] In studies performed by Krysińska-Traczyk et al.[184] in grain dust, the level of moniliformin as well as OTA was determined by HPLC. The levels of trichothecenes of the B group, deoxynivalenol (DON) and nivalenol were determined using gas chromatography combined with a mass spectrometer (GC-MS). Production characteristics and climatic data were studied as determinants of trichothecenes in settled dust samples obtained during the production of barley, oats, and spring wheat. Trichothecenes were also determined by GC-MS.[194] The micro-method by HPLC was performed for aflatoxins and OTA in dust samples from different workplaces in Italy, where three of the most susceptible foodstuffs (cocoa, coffee, and spices) are processed.[179] To detect OTA in building materials collected from homes in flood areas, HLPC with fluorescence detection was used.[195] A similar HPLC-FLD method for the determination of particulate aflatoxins and OTA in air samples collected during the usual production process in a number of workplaces in a coffee factory was used by Tarin et al.[185] Also, OTA in airborne dust from Norwegian cow farms was analyzed by HPLC-FLD.[196] Dust samples were collected from heating ducts in a household where symptoms resembling ochratoxin poisoning in animals occurred and determination was made by HPLC and confirmed by LC-MS.[186]

However, many researchers have preferred to use MS-based methods, especially tandem MS (MS/MS), because of the high analytical specificity offered. HPLC identified by tandem mass spectrometry and quantified using electrospray ionization on a quadrupole ion trap mass analyzer was used to analyze bulk samples of moldy interior finishes, including samples of wallpaper, cardboard, wood, plywood, plasterboard, paper-covered gypsum board, mineral wool, plaster, sand, soil, linoleum, polyurethane insulation, pipe insulation, and paint.[187] LC-MS/MS was also used to demonstrate sterigmatocystin in carpet dust from damp indoor environments[188] and in dust samples collected in New Orleans homes mold-contaminated because of flooding after Hurricane Katrina.[197] GC-MS and GC-MS/MS were used to detect verrucarol and trichodermol, hydrolysis products of macrocyclic trichothecenes and trichodermin, respectively, of *S. chartarum* in mold-affected building materials[170,198,199] and settled house dust.[170] Also, GC-MS/MS was used to identify verrucarol in dust samples collected in homes associated with Hurricane Katrina.[197] GC ion trap mass spectrometry was used to detect mycotoxins in water-damaged gypsum board heavily infested with *S. chartarum* found in a school and in a domestic residence in Copenhagen. The trichothecenes were detected as heptafluorobutyrlated derivatives. Extracts of samples from both locations yielded verrucarol after hydrolysis strongly indicated the presence of toxic macrocyclic trichothecenes (probably satratoxin H and G). In addition the sample from the domestic residence yielded trichodermol.[172] Nielsen et al.[200] used the same method to detect mycotoxins produced on artificially inoculated building materials.

Specificity is extremely important; ideally, a combination of LC (or GC) and MS should be used. LC with DAD can be used with caution and only for the analysis of compounds with highly characteristic UV spectra, not for components with end absorption or a single UV maximum. However, even when analyses are performed by LC with MS or MS/MS, trichothecene analysis can be difficult owing to the lability of these metabolites and their tendency to form adducts.[201] For instance, the report by Tuomi et al.[187] of deoxynivalenol, verrucarol, 3-acetyl-deoxynivalenol, diacetoxyscirpenol, and T-2 tetraol in moldy building materials may have resulted from false positives because no trichothecene-producing fungi were isolated, no T-2 toxin was detected in samples containing T-2 tetraol, and no deoxynivalenol was found in samples containing 3-acetyl-deoxynivalenol. Such odd results, in which no apparent producer is present and in which there is inconsistency in the list of detected compounds, must be verified using different analytical methods.[168] In direct analysis, without prior culture, LC-MS/MS has been used to demonstrate satratoxins and satratoxin G in indoor building material samples,[171,173,187,202] satratoxin G in carpet dust,[188] and satratoxin G and satratoxin H in air samples.[183] Rolle-Kampczyk et al.[203] developed a new method based on CE to detect mycotoxins: OTA, sterigmatocystin, citrinin, gliotoxin, and patulin in collected house dust samples.

The macrocyclic trichothecenes level was measured using a monoclonal, antibody-based ELISA test. The levels of these mycotoxins were measured on wall surfaces, floor dust, and ambient air of moldy dwellings.[180] Brasel et al.[178,204] also investigated the presence of airborne macrocyclic trichothecene mycotoxins in indoor environments with known *S. chartarum* contamination, using ELISA. The sensitivity and accuracy of the competitive ELISA described for roridin A by Märtlbauer et al.[205] seem to be useful in environmental analytical chemistry. This method was applied to analyze macrocyclic trichothecens.[206–208]

Ngundi et al.[181] developed and used a competitive immunoassay technique to detect DON in indoor air samples using an array biosensor.

Besides immunoassays, other biological methods have been used for mycotoxin analysis, including a cytotoxicity test mainly by MTT.[206,208–212] Cytotoxicity assays based on application of extracts to cell cultures have been useful in studies involving potential exposure to mycotoxins and demonstrate adequate sensitivity for air sampling or building materials or for determinations in settled dust. Trichothecene concentrations can also be estimated via their inhibitory effect on protein translation.[213] The spectrum of analytical methods applied for mycotoxin determination is broad and based on chemical and biological techniques.

REFERENCES

1. Lattanzio VM, Pascale M, Visconti A. Current analytical methods for trichothecene mycotoxins in cereals. *Trend Anal Chem* 2009;**28**:758–68.
2. Škrbić B, Godula M, Đurišić-Mladenović N, Živančev J. Multimycotoxin analysis by UHPLC-HESI-MS/MS: a preliminary survey of Serbian wheat flour. *Agron Res* 2011;**9**(Special issue II):461–8.
3. Bertuzzi T, Rastelli S, Mulazzi A, Pietri A. Evaluation and improvement of extraction methods for the analysis of aflatoxins B_1, B_2, G_1 and G_2 from naturally contaminated maize. *Food Anal Method* 2012;**5**:512–9.
4. Beltrán E, Ibáñez M, Sancho JV, Cortés MA, Yusà V, Hernàndez F. UHPLCMS/MS highly sensitive determination of aflatoxins, the aflatoxin metabolite M_1 and ochratoxin A in baby food and milk. *Food Chem* 2011;**126**:737–44.
5. Delmulle BS, De Saeger S, Sibanda L, Barna-Vetro I, Van Peteghem CH. Development of an immunoassay-based lateral flow dipstick for the rapid detection of aflatoxin B_1 in pig feed. *J Agric Food Chem* 2005;**53**:3364–8.
6. Scott PM, Lawrence GA, Lombaert GA. Studies on extraction of fumonisins from rice, corn-based foods and beans. *Mycotoxin Res* 1999;**15**:50–60.
7. Marschik S, Hepperle J, Lauber U, Schnaufer R, Maier S, Kühn C, et al. Extracting fumonisins from maize: efficiency of different extraction solvents in multi-mycotoxin analytics. *Mycotoxin Res* 2013;**29**:119–29.
8. Sulyok M, Berthiller F, Krska R, Schuhmacher R. Development and validation of a liquid chromatography/tandem mass spectrometric method for the determination of 39 mycotoxins in wheat and maize. *Rapid Commun Mass Spectrom* 2006;**20**:2649–59.
9. Chen Y, Guo Z, Wang X, Qiu C. Sample preparation. *J Chromatogr A* 2008;**1184**:191–219.
10. Royer D, Humpf HU, Guy PA. Quantitative analysis of Fusarium mycotoxins in maize using accelerated solvent extraction before liquid chromatography/atmospheric pressure chemical ionization tandem mass spectrometry. *Food Addit Contam A* 2004;**21**:678–92.

11. Urraca JL, Marazuela MD, Moreno-Bondi MC. Molecularly imprinted polymers applied to the clean-up of zearalenone and alpha-zearalenol from cereal and swine feed sample extracts. *Anal Bioanal Chem* 2006;**385**:1155–61.
12. D'Arco G, Fernández-Franzón M, Font G, Damiani P, Mañes J. Analysis of fumonisins B(1), B(2) and B(3) in corn-based baby food by pressurized liquid extraction and liquid chromatography/tandem mass spectrometry. *J Chromatogr A* 2008;**1209**:188–94.
13. González-Osnaya L, Soriano JM, Moltó JC, Mañes J. Dietary intake of ochratoxin A from conventional and organic bread. *Int J Food Microbiol* 2007;**118**:87–91.
14. Juan C, Zinedine A, Idrissi L, Mañes J. Ochratoxin A in rice on the Moroccan retail market. *Int J Food Microbiol* 2008;**126**:83–5.
15. Desmarchelier A, Oberson JM, Tella P, Gremaud E, Seefelder W, Mottie P. Development and comparison of two multiresidue methods for the analysis of 17 mycotoxins in cereals by liquid chromatography electrospray ionization tandem mass spectrometry. *J Agric Food Chem* 2010;**58**:7510–9.
16. Hajslova J, Zachariasova M, Cajka T. Analysis of multiple mycotoxins in food. In: Zweigenbaum J, editor. *Mass spectrometry in food safety: methods and protocols*. Wilmington, Delaware: Food Safety. Agilent Technologies Inc.; 2011. p. 233–58.
17. Sapkale GN, Patil SM, Surwase US, Bhatbhage PK. Supercritical fluid extraction-a review. *Int J Chem Sci* 2010;**8**:729–43.
18. Zougagh M, Téllez H, Sánchez A, Chicharro M, Rios A. Validation of a screening method for rapid control of macrocyclic lactone mycotoxins in maize flour samples. *Anal Bioanal Chem* 2008;**391**:709–14.
19. Arribas AS, Bermejo E, Zapardiel A, Téllez H, Rodriguez-Flores J, Zougagh M, et al. Screening and confirmatory methods for the analysis of macrocyclic lactone mycotoxins by CE with amperometric detection. *Electrophoresis* 2009;**30**:499–506.
20. Kralj Cigić I, Prosen H. An overview of conventional and emerging analytical methods for the determination of mycotoxins. *Int J Mol Sci* 2009;**10**:62–115.
21. Eskilsson CS, Björklund E. Analytical-scale microwave-assisted extraction. *J Chromatogr A* 2000;**902**:227–50.
22. Veggi PC, Martinez J, Meireles MAA. Fundamentals of microwave extraction. In: Chemat F, Cravotto G, editors. *Microwave-assisted extraction for bioactive compounds: theory and practice. Food engineering series 4*. New York: Springer; 2013. p. 15–52.
23. Pallaroni L, von Holst C, Eskilsson S, Björklund E. Microwave-assisted extraction of zearalenone from wheat and corn. *Anal Bioanal Chem* 2002;**374**:161–6.
24. Chen S, Zhang H. Development of a microwave-assisted-extraction-based method for the determination of aflatoxins B_1, G_1, B_2, and G_2 in grains and grain products. *Anal Bioanal Chem* 2013;**405**:1623–30.
25. Ridgway K, Lalljie SP, Smith RM. Sample preparation techniques for the determination of trace residues and contaminants in foods. *J Chromatogr A* 2007;**1153**:36–53.
26. Pallaroni L, Von Holst C. Comparison of alternative and conventional extraction techniques for the determination of zearalenone in corn. *Anal Bioanal Chem* 2003;**376**:908–12.
27. Serrano AB, Font G, Mañes J, Ferrer E. Comparative assessment of three extraction procedures for determination of emerging *Fusarium* mycotoxins in pasta by LC-MS/MS. *Food Control* 2013;**32**:105–14.
28. Li C, Wu YL, Yang T, Huang-Fu WG. Rapid determination of fumonisins B_1 and B_2 in corn by liquid chromatography-tandem mass spectrometry with ultrasonic extraction. *J Chromatogr Sci* 2012;**50**:57–63.
29. Barker SA. Matrix solid phase dispersion (MSPD). *J Biochem Biophys Meth* 2007;**70**: 151–62.

30. Rubert J, Soler C, Mañes J. Optimization of matrix solid-phase dispersion method for simultaneous extraction of aflatoxins and OTA in cereals and its application to commercial samples. *Talanta* 2010;**82**:567–74.
31. Serrano AB, Font G, Ruiz MJ, Ferrer E. Co-occurrence and risk assessment of mycotoxins in food and diet from Mediterranean area. *Food Chem* 2012;**135**:423–9.
32. Rubert J, Dzuman Z, Vaclavikova M, Zachariasova M, Soler C, Hajslova J. Analysis of mycotoxins in barley using ultra high liquid chromatography high resolution mass spectrometry: comparison of efficiency and efficacy of different extraction procedures. *Talanta* 2012;**99**:712–9.
33. Żwir-Ferenc A, Biziuk M. Solid phase extraction technique—trends, opportunities and applications. *Pol J Environ Stud* 2006;**15**:677–90.
34. Barthel J, Gottschalk C, Rapp M, Berger M, Bauer J, Meyer K. Occurrence of type A, B and D trichothecenes in barley and barley products from the Bavarian market. *Mycotoxin Res* 2012;**28**:97–106.
35. Åberg AT, Solyakov A, Bondesson U. Development and in-house validation of an LC-MS/MS method for the quantification of the mycotoxins deoxynivalenol, zearalenone, T-2 and HT-2 toxin, ochratoxin A and fumonisin B_1 and B_2 in vegetable animal feed. *Food Addit Contam A* 2013;**30**:541–9.
36. Kadota T, Kimura M, Hirano S, Tajima O, Nakajima T, Kamata Y, et al. Development of a simultaneous liquid chromatography/tandem mass spectrometric method for the determination of type B trichothecenes, their derivatives, and precursors in wheat. *Rapid Commun Mass Spectrom* 2011;**25**:3481–90.
37. Monaci L, De Angelis E, Visconti A. Determination of deoxynivalenol, T-2 and HT-2 toxins in a bread model food by liquid chromatography-high resolution-orbitrap-mass spectrometry equipped with a high-energy collision dissociation cell. *J Chromatogr A* 2011;**1218**:8646–54.
38. Scarpino V, Blandino M, Negre M, Reyneri A, Vanara F. Moniliformin analysis in maize samples from North-West Italy using multifunctional clean-up columns and the LC-MS/MS detection method. *Food Addit Contam A* 2013;**30**:876–84.
39. Van Pamel E, Verbeken A, Vlaemynck G, De Boever J, Daeseleire E. Ultrahigh-performance liquid chromatographic-tandem mass spectrometric multimycotoxin method for quantitating 26 mycotoxins in maize silage. *J Agric Food Chem* 2011;**59**:9747–55.
40. Wang Y, Xiao C, Guo J, Yuan Y, Wang J, Liu L, et al. Development and application of a method for the analysis of 9 mycotoxins in maize by HPLC-MS/MS. *J Food Sci* 2013;**78**:M1752–6.
41. Muscarella M, Magro SL, Nardiello D, Palermo C, Centonze D. Determination of fumonisins B_1 and B_2 in maize food products by a new analytical method based on high-performance liquid chromatography and fluorimetric detection with post-column derivatization. *Method Mol Biol* 2011;**739**:187–94.
42. Solfrizzo M, De Girolamo A, Gambacorta L, Visconti A, Stroka J, Van Egmond HP. Determination of fumonisins B_1 and B_2 in corn-based foods for infants and young children by LC with immunoaffinity column cleanup: interlaboratory validation study. *J AOAC Int* 2011;**94**:900–8.
43. Manova R, Mladenova R. Incidence of zearalenone and fumonisins in Bulgarian cereal production. *Food Control* 2009;**20**:362–5.
44. Zaied C, Abid S, Zorgui L, Bouaziz C, Chouchane S, Jomaa M, et al. Natural occurrence of ochratoxin A in Tunisian cereals. *Food Control* 2009;**20**:218–22.
45. Bensassi F, Zaied C, Abid S, Hajlaoui MR, Bacha H. Occurrence of deoxynivalenol in durum wheat in Tunisia. *Food Control* 2010;**21**:281–5.
46. Muscarella M, Iammarino M, Nardiello D, Magro SL, Palermo C, Centonze D. Simultaneous determination of aflatoxins B_1, B_2, G_1, and G_2 in foods and feed materials. *Method Mol Biol* 2011;**739**:203–10.

47. Pascale M, Panzarini G, Visconti A. Determination of HT-2 and T-2 toxins in oats and wheat by ultra-performance liquid chromatography with photodiode array detection. *Talanta* 2012;**89**:231–6.
48. Vaclavikova M, MacMahon S, Zhang K, Begley TH. Application of single immunoaffinity clean-up for simultaneous determination of regulated mycotoxins in cereals and nuts. *Talanta* 2013;**117**:345–51.
49. Senyuva HZ, Gilbert J, Türköz G, Leeman D, Donnelly C. Analysis of deoxynivalenol, zearalenone, T-2, and HT-2 toxins in animal feed by LC/MS/MS—a critical comparison of immunoaffinity column cleanup with no cleanup. *J AOAC Int* 2012;**95**:1701–8.
50. Vasapollo G, Sole RD, Mergola L, Lazzoi MR, Scardino A, Scorrano S, et al. Molecularly imprinted polymers: present and future prospective. *Int J Mol Sci* 2011;**12**:5908–45.
51. Ali WH, Derrien D, Alix F, Pérollier C, Lépine O, Bayoudh S, et al. Solid-phase extraction using molecularly imprinted polymers for selective extraction of a mycotoxin in cereals. *J Chromatogr A* 2010;**1217**:6668–73.
52. Appell M, Jackson MA, Wang LC, Ho CH, Mueller A. Determination of fusaric acid in maize using molecularly imprinted SPE clean-up. *J Sep Sci* 2014;**37**:281–6.
53. De Smet D, Dubruel P, Van Peteghem C, Schacht E, De Saeger S. Molecularly imprinted solid-phase extraction of fumonisin B analogues in bell pepper, rice and corn flakes. *Food Addit Contam A* 2009;**26**:874–84.
54. Bryła M, Jędrzejczak R, Roszko M, Szymczyk K, Obiedziński MW, Sękul J, et al. Application of molecularly imprinted polymers to determine B_1, B_2, and B_3 fumonisins in cereal products. *J Sep Sci* 2013;**36**:578–84.
55. Lenain P, Diana Di Mavungu J, Dubruel P, Robbens J, De Saeger S. Development of suspension polymerized molecularly imprinted beads with metergoline as template and application in a solid-phase extraction procedure toward ergot alkaloids. *Anal Chem* 2012;**84**:10411–8.
56. Lucci P, Derrien D, Alix F, Pérollier C, Bayoudh S. Molecularly imprinted polymer solid-phase extraction for detection of zearalenone in cereal sample extracts. *Anal Chim Acta* 2010;**672**:15–9.
57. Guo BY, Wang S, Ren B, Li X, Qin F, Li J. Citrinin selective molecularly imprinted polymers for SPE. *J Sep Sci* 2010;**33**:1156–60.
58. Paíga P, Morais S, Oliva-Teles T, Correia M, Delerue-Matos C, Duarte SC, et al. Extraction of ochratoxin A in bread samples by the QuEChERS methodology. *Food Chem* 2012;**135**:2522–8.
59. Zhang JM, Wu YL, Lu YB. Simultaneous determination of carbamate insecticides and mycotoxins in cereals by reversed phase liquid chromatography tandem mass spectrometry using a quick, easy, cheap, effective, rugged and safe extraction procedure. *J Chromatogr B* 2013;**915–916**:13–20.
60. Koesukwiwat U, Sanguankaew K, Leepipatpiboon N. Evaluation of a modified QuEChERS method for analysis of mycotoxins in rice. *Food Chem* 2014;**153**:44–51.
61. Oueslati S, Blesa J, Moltó JC, Ghorbel A, Mañes J. Presence of mycotoxins in sorghum and intake estimation in Tunisia. *Food Addit Contam A* 2014;**31**:307–18.
62. Rodríguez-Carrasco Y, Moltó JC, Berrada H, Mañes J. A survey of trichothecenes, zearalenone and patulin in milled grain-based products using GC-MS/MS. *Food Chem* 2014;**146**:212–9.
63. Campone L, Piccinelli AL, Aliberti L, Rastrelli L. Application of pressurized liquid extraction in the analysis of aflatoxins B(1), B(2), G(1) and G(2) in nuts. *J Sep Sci* 2009;**32**:3837–44.
64. Chen D, Cao X, Tao Y, Wu Q, Pan Y, Huang X, et al. Development of a sensitive and robust liquid chromatography coupled with tandem mass spectrometry and a pressurized liquid extraction for the determination of aflatoxins and ochratoxin A in animal derived foods. *J Chromatogr A* 2012;**1253**:110–9.

65. Han Z, Ren Y, Zhu J, Cai Z, Chen Y, Luan L, et al. Multianalysis of 35 mycotoxins in traditional Chinese medicines by ultra-high-performance liquid chromatography-tandem mass spectrometry coupled with accelerated solvent extraction. *J Agric Food Chem* 2012;**60**:8233–47.
66. Liau BC, Jong TT, Lee MR, Chang CM. Supercritical fluid extraction and quantification of aflatoxins in *Zizyphi fructus* by liquid chromatography/atmospheric pressure chemical ionization tandem mass spectrometry. *Rapid Commun Mass Spectrom* 2007;**21**:667–73.
67. Bacaloni A, Cavaliere C, Cucci F, Foglia P, Samperi R, Laganà A. Determination of aflatoxins in hazelnuts by various sample preparation methods and liquid chromatography-tandem mass spectrometry. *J Chromatogr A* 2008;**1179**:182–9.
68. Duarte SC, Lino CM, Pena A. Novel IAC-LC-ESI-MS(2) analytical set-up for ochratoxin A determination in pork. *Food Chem* 2013;**138**:1055–61.
69. Kong WJ, Liu SY, Qiu F, Xiao XH, Yang MH. Simultaneous multi-mycotoxin determination in nutmeg by ultrasound-assisted solid-liquid extraction and immunoaffinity column clean-up coupled with liquid chromatography and on-line post-column photochemical derivatization-fluorescence detection. *Analyst* 2013;**138**:2729–39.
70. Wang H, Zhou XJ, Liu YQ, Yang HM, Guo QL. Determination of aflatoxin M_1 in milk by triple quadrupole liquid chromatography-tandem mass spectrometry. *Food Addit Contam A* 2010;**27**:1261–5.
71. Wang X, Li P, Yang Y, Zhang W, Zhang Q, Fan S, et al. Determination of aflatoxins in cereals and oils by liquid chromatography-triple quadrupole tandem mass spectrometry. *China J Chromatogr* 2011;**29**:517–22.
72. Blesa J, Soriano JM, Moltó JC, Marin R, Mañes J. Determination of aflatoxins in peanuts by matrix solid-phase dispersion and liquid chromatography. *J Chromatogr A* 2003;**1011**:49–54.
73. Hu YY, Zheng P, Zhang ZX, He YZ. Determination of aflatoxins in high-pigment content samples by matrix solid-phase dispersion and high-performance liquid chromatography. *J Agric Food Chem* 2006;**54**:4126–30.
74. Cavaliere C, Foglia P, Guarino C, Marzioni F, Nazzari M, Samperi R, et al. Aflatoxin M_1 determination in cheese by liquid chromatography-tandem mass spectrometry. *J Chromatogr A* 2006;**1135**:135–41.
75. Cavaliere C, Foglia P, Guarino C, Nazzari M, Samperi R, Laganà A. Determination of aflatoxins in olive oil by liquid chromatography-tandem mass spectrometry. *Anal Chim Acta* 2007;**596**:141–8.
76. Laganà A, Bacaloni A, Castellano M, Curini R, De Leva I, Faberi A, et al. Sample preparation for determination of macrocyclic lactone mycotoxins in fish tissue, based on on-line matrix solid-phase dispersion and solid-phase extraction cleanup followed by liquid chromatography/tandem mass spectrometry. *J AOAC Int* 2003;**86**:729–36.
77. Fabiani A, Corzani C, Arfelli G. Correlation between different clean-up methods and analytical techniques performances to detect ochratoxin A in wine. *Talanta* 2010;**83**:281–5.
78. Zhou Y, Kong W, Li Y, Logrieco AF, Xu J, Yang M. A new solid-phase extraction and HPLC method for determination of patulin in apple products and hawthorn juice in China. *J Sep Sci* 2012;**35**:641–9.
79. Hussain I, Anwar J, Asi MR, Munawar MA, Kashif M. Aflatoxin M_1 contamination in milk from five dairy species in Pakistan. *Food Control* 2010;**21**:122–4.
80. Wang Y, Liu X, Xiao C, Wang Z, Wang J, Xiao H, et al. HPLC determination of aflatoxin M_1 in liquid milk and milk powder using solid phase extraction on OASIS HLB. *Food Control* 2012;**28**:131–4.
81. Iha MH, Barbosa CB, Favaro RM, Trucksess MW. Chromatographic method for the determination of aflatoxin M_1 in cheese, yogurt, and dairy beverages. *J AOAC Int* 2011;**94**:1513–8.

82. Lee TP, Saad B, Khayoon WS, Salleh B. Molecularly imprinted polymer as sorbent in micro-solid phase extraction of ochratoxin A in coffee, grape juice and urine. *Talanta* 2012;**88**:129–35.
83. Cao J, Kong W, Zhou S, Yin L, Wan L, Yang M. Molecularly imprinted polymer-based solid phase clean-up for analysis of ochratoxin A in beer, red wine, and grape juice. *J Sep Sci* 2013;**36**:1291–7.
84. Zhao D, Jia J, Yu X, Sun X. Preparation and characterization of a molecularly imprinted polymer by grafting on silica supports: a selective sorbent for patulin toxin. *Anal Bioanal Chem* 2011;**401**:2259–73.
85. Desmarchelier A, Mujahid C, Racault L, Perring L, Lancova K. Analysis of patulin in pear- and apple-based foodstuffs by liquid chromatography electrospray ionization tandem mass spectrometry. *J Agric Food Chem* 2011;**59**:7659–65.
86. Kharandi N, Babri M, Azad J. A novel method for determination of patulin in apple juices by GC-MS. *Food Chem* 2013;**141**:1619–23.
87. Tamura M, Uyama A, Mochizuki N. Development of a multi-mycotoxin analysis in beer-based drinks by a modified QuEChERS method and ultra-high-performance liquid chromatography coupled with tandem mass spectrometry. *Anal Sci* 2011;**27**:629–35.
88. Fu Z, Huang X, Min S. Rapid determination of aflatoxins in corn and peanuts. *J Chromatogr A* 2008;**1209**:271–4.
89. Bononi M, Gallone F, Tateo F. Validation data for HPLC/FLD determinations of ochratoxin A in red paprika and black pepper adopting a one-step clean-up procedure. *Food Addit Contam A* 2010;**27**:249–54.
90. Lacina O, Zachariasova M, Urbanova J, Vaclavikova M, Cajka T, Hajslova J. Critical assessment of extraction methods for the simultaneous determination of pesticide residues and mycotoxins in fruits, cereals, spices and oil seeds employing ultra-high performance liquid chromatography-tandem mass spectrometry. *J Chromatogr A* 2012;**1262**:8–18.
91. Arroyo-Manzanares N, Huertas-Pérez JF, Gámiz-Gracia L, Garcia-Campaña AM. A new approach in sample treatment combined with UHPLC-MS/MS for the determination of multi-class mycotoxins in edible nuts and seeds. *Talanta* 2013;**115**:61–7.
92. Yogendrarajah P, Van Poucke C, De Meulenaer B, De Saeger S. Development and validation of a QuEChERS based liquid chromatography tandem mass spectrometry method for the determination of multiple mycotoxins in spices. *J Chromatogr A* 2013;**1297**:1–11.
93. Valle-Algarra FM, Mateo EM, Mateo R, Gimeno-Adelantado JV, Jiménez M. Determination of type A and type B trichothecenes in paprika and chili pepper using LC-triple quadrupole-MS and GC-ECD. *Talanta* 2011;**84**:1112–7.
94. Turcotte AM, Scott PM. Ochratoxin A in cocoa and chocolate sampled in Canada. *Food Addit Contam A* 2011;**28**:762–6.
95. Kong W, Xie T, Li J, Wei J, Qiu F, Qi A, et al. Analysis of fumonisins B_1 and B_2 in spices and aromatic and medicinal herbs by HPLC-FLD with on-line post-column derivatization and positive confirmation by LC-MS/MS. *Analyst* 2012;**137**:3166–74.
96. Valle-Algarra FM, Medina A, Gimeno-Adelantado JV, Llorens A, Jiménez M, Mateo R. Comparative assessment of solid-phase extraction clean-up procedures, GC columns and perfluoroacylation reagents for determination of type B trichothecenes in wheat by GC-ECD. *Talanta* 2005;**66**:194–201.
97. Jeleń HH, Wasowicz E. Determination of trichothecenes in wheat grain without sample cleanup using comprehensive two-dimensional gas chromatography-time-of-flight mass spectrometry. *J Chromatogr A* 2008;**1215**:203–7.
98. Pascale M. Detection methods for mycotoxins in cereal grains and cereal products. *Matica Srpska Proc Nat Sci* 2009;**117**:15–25.

99. Labuda R, Parich A, Berthiller F, Tančinovà D. Incidence of trichothecenes and zearalenone in poultry feed mixtures from Slovakia. *Int J Food Microbiol* 2005;**105**:19–25.
100. Oliveira AQ, Soares LMV, Sawazaki E. Survey of deoxynivalenol, diacetoxyscirpenol and T2 toxin in popcorn hybrids planted in the state of Sao Paulo and in popcorn commercialized in the city of Campinas, SP. *Food Sci Technol (Campinas)* 2001;**21**:330–3.
101. Soleas GJ, Yan J, Goldberg DM. Assay of ochratoxin A in wine and beer by high-pressure liquid chromatography photodiode array and gas chromatography mass selective detection. *J Agric Food Chem* 2001;**49**:2733–40.
102. Marks HS. Rapid gas chromatography/mass spectrometry determination and confirmation of patulin in apple juice. *J AOAC Int* 2007;**90**:879–83.
103. Edwards SG. Fusarium mycotoxin content of UK organic and conventional oats. *Food Addit Contam A* 2009;**26**:1063–9.
104. Amelin VG, Karaseva NM, Tretyakov AV. Simultaneous determination of trichothecene mycotoxins, ochratoxin A and zearalenone in grain and products of its processing, feed premixes and meat by gas chromatography. *J Anal Chem* 2013;**68**:61–7.
105. Romero-González R, Martínez Vidal JL, Garrido Frenich A. *Food science and technology. Liquid chromatography for the determination of mycotoxins in foods*. New York: Nova Science Publisher Inc.; 2010. p. 63.
106. Akiyama H, Goda Y, Tanaka T, Toyoda M. Determination of aflatoxins B_1, B_2, G_1 and G_2 in spices using a multifunctional column clean-up. *J Chromatogr A* 2001;**932**:153–7.
107. Hernández Hierro JM, Garcia-Villanova RJ, Rodríguez Torrero P, Toruño Fonseca IM. Aflatoxins and ochratoxin A in red paprika for retail sale in Spain: occurrence and evaluation of a simultaneous analytical method. *J Agric Food Chem* 2008;**56**:751–6.
108. O'Riordan MJ, Wilkinson MG. Comparison of analytical methods for aflatoxin determination in commercial chilli spice preparations and subsequent development of an improved method. *Food Control* 2009;**20**:700–5.
109. Zhu Z, Liu G, Chen Y, Cheng J. Assessment of aflatoxins in pigmented rice using a validated immunoaffinity column method with fluorescence HPLC. *J Food Compos Anal* 2013;**31**: 252–8.
110. Ariño A, Herrera M, Estopañan G, Juan T. High levels of ochratoxin A in licorice and derived products. *Int J Food Microbiol* 2007;**114**:366–9.
111. Vecchio A, Mineo V, Planeta D. Ochratoxin A in instant coffee in Italy. *Food Control* 2012;**28**:220–3.
112. Kong W, Wei R, Logrieco AF, Wei J, Wen J, Xiao X, et al. Occurrence of toxigenic fungi and determination of mycotoxins by HPLC-FLD in functional foods and spices in China markets. *Food Chem* 2014;**146**:320–6.
113. Majerus P, Graf N, Krämer M. Rapid determination of zearalenone in edible oils by HPLC with fluorescence detection. *Mycotoxin Res* 2009;**25**:117–21.
114. Zaied C, Zouaoui N, Bacha H, Abid S. Natural occurrence of zearalenone in Tunisian wheat grains. *Food Control* 2012;**25**:773–7.
115. Iqbal SZ, Asi MR, Jinap S, Rashid U. Detection of aflatoxins and zearalelone contamination in wheat derived products. *Food Control* 2014;**35**:223–6.
116. Klinglmayr C, Nöbauer K, Razzazi-Fazeli E, Cichna-Markl M. Determination of deoxynivalenol in organic and conventional food and feed by sol-gel immunoaffinity chromatography and HPLC-UV detection. *J Chromatogr B* 2010;**878**:187–93.
117. Aoyama K, Akashi H, Mochizuki N, Ito Y, Miyashita T, Lee S, et al. Interlaboratory study of LC-UV and LC-MS methods for the simultaneous determination of deoxynivalenol and nivalenol in wheat. *J Food Hyg Soc Jpn* 2012;**53**:152–6.

118. Guo Y, Zhou Z, Yuan T, Yue T. Survey of patulin in apple juice concentrates in Shaanxi (China) and its dietary intake. *Food Control* 2013;**34**:570–3.
119. Csutorás CS, Rácz L, Rácz K, Fűtő P, Forgó P, Kiss A. Monitoring of ochratoxin A during the fermentation of different wines by applying high toxin concentrations. *Microchem J* 2013;**107**:182–4.
120. Aresta A, Cioffi N, Palmisano F, Zambonin CG. Simultaneous determination of ochratoxin A and cyclopiazonic, mycophenolic, and tenuazonic acids in cornflakes by solid-phase microextraction coupled to high-performance liquid chromatography. *J Agric Food Chem* 2003;**51**:5232–7.
121. Scott PM, Lawrence GA, Lau BP. Analysis of wines, grape juices and cranberry juices for Alternaria toxins. *Mycotoxin Res* 2006;**22**:142–7.
122. Monbaliu S, Van Poucke C, Van Peteghem C, Van Poucke K, Heungens K, De Saeger S. Development of a multi-mycotoxin liquid chromatography/tandem mass spectrometry method for sweet pepper analysis. *Rapid Commun Mass Spectrom* 2009;**23**:3–11.
123. Monbaliu S, Van Poucke K, Heungens K, Van Peteghem C, De Saeger S. Production and migration of mycotoxins in sweet pepper analyzed by multimycotoxin LC-MS/MS. *J Agric Food Chem* 2010;**58**:10475–9.
124. Kataoka H, Itano M, Ishizaki A, Saito K. Determination of patulin in fruit juice and dried fruit samples by in-tube solid-phase microextraction coupled with liquid chromatography-mass spectrometry. *J Chromatogr A* 2009;**1216**:3746–50.
125. Zhang X, Cudjoe E, Vuckovic D, Pawliszyn J. Direct monitoring of ochratoxin A in cheese with solid-phase microextraction coupled to liquid chromatography-tandem mass spectrometry. *J Chromatogr A* 2009;**1216**:7505–9.
126. Rasmussen RR, Storm IM, Rasmussen PH, Smedsgaard J, Nielsen KF. Multi-mycotoxin analysis of maize silage by LC-MS/MS. *Anal Bioanal Chem* 2010;**397**:765–76.
127. Ferrer Amate C, Unterluggauer H, Fischer RJ, Fernàndez-Alba AR, Masselter S. Development and validation of a LC-MS/MS method for the simultaneous determination of aflatoxins, dyes and pesticides in spices. *Anal Bioanal Chem* 2010;**397**:93–107.
128. Sørensen LM, Mogensen J, Nielsen KF. Simultaneous determination of ochratoxin A, mycophenolic acid and fumonisin B(2) in meat products. *Anal Bioanal Chem* 2010;**398**:1535–42.
129. Romagnoli B, Ferrari M, Bergamini C. Simultaneous determination of deoxynivalenol, zearalenone, T-2 and HT-2 toxins in breakfast cereals and baby food by high-performance liquid chromatography and tandem mass spectrometry. *J Mass Spectrom* 2010;**45**:1075–80.
130. Ediage EN, Di Mavungu JD, Monbaliu S, Van Peteghem C, De Saeger S. A validated multi-analyte LC-MS/MS method for quantification of 25 mycotoxins in cassava flour, peanut cake and maize samples. *J Agric Food Chem* 2011;**59**:5173–80.
131. Lattanzio VM, Gatta SD, Godula M, Visconti A. Quantitative analysis of mycotoxins in cereal foods by collision cell fragmentation-high-resolution mass spectrometry: performance and comparison with triple-stage quadrupole detection. *Food Addit Contam A* 2011;**28**:1424–37.
132. Lattanzio VM, Gatta SD, Suman M, Visconti A. Development and in-house validation of a robust and sensitive solid-phase extraction liquid chromatography/tandem mass spectrometry method for the quantitative determination of aflatoxins B_1, B_2, G_1, G_2, ochratoxin A, deoxynivalenol, zearalenone, T-2 and HT-2 toxins in cereal-based foods. *Rapid Commun Mass Spectrom* 2011;**25**:1869–80.
133. Lim CW, Tai SH, Lee LM, Chan SH. Analytical method for the accurate determination of tricothecenes in grains using LC-MS/MS: a comparison between MRM transition and MS3 quantitation. *Anal Bioanal Chem* 2012;**403**:2801–6.

134. Jackson LC, Kudupoje MB, Yiannikouris A. Simultaneous multiple mycotoxin quantification in feed samples using three isotopically labeled internal standards applied for isotopic dilution and data normalization through ultra-performance liquid chromatography/electrospray ionization tandem mass spectrometry. *Rapid Commun Mass Spectrom* 2012;**26**:2697–713.
135. Biancardi A, Piro R, Dall'asta C, Galaverna G. A simple and reliable liquid chromatography-tandem mass spectrometry method for the determination of aflatoxin M_1 in milk. *Food Addit Contam A* 2013;**30**:381–8.
136. Khan MR, Alothman ZA, Ghfar AA, Wabaidur SM. Analysis of aflatoxins in nonalcoholic beer using liquid-liquid extraction and ultraperformance LC-MS/MS. *J Sep Sci* 2013;**36**:572–7.
137. Al-Taher F, Banaszewski K, Jackson L, Zweigenbaum J, Ryu D, Cappozzo J. Rapid method for the determination of multiple mycotoxins in wines and beers by LC-MS/MS using a stable isotope dilution assay. *J Agric Food Chem* 2013;**61**:2378–84.
138. Běláková S, Benešová K, Čáslavský J, Svoboda Z, Mikulikovà R. The occurrence of the selected fusarium mycotoxins in Czech malting barley. *Food Control* 2014;**37**:93–8.
139. Arroyo-Manzanares N, Malysheva SV, Vanden Bussche J, Vanhaecke L, Diana Di Mavungu J, De Saeger S. Holistic approach based on high resolution and multiple stage mass spectrometry to investigate ergot alkaloids in cereals. *Talanta* 2014;**118**:359–67.
140. Xue XF, Selvaraj JN, Zhao L, Dong H, Liu F, Liu Y, et al. Simultaneous determination of aflatoxins and ochratoxin a in bee pollen by low-temperature fat precipitation and immunoaffinity column cleanup coupled with LC-MS/MS. *Food Anal Method* 2014;**7**:690–6.
141. Chalupová J, Raus M, Sedlářová M, Sebela M. Identification of fungal microorganisms by MALDI-TOF mass spectrometry. *Biotechnol Adv* 2014;**32**:230–41.
142. Maragos CM, Appell M. Capillary electrophoresis of the mycotoxin zearalenone using cyclodextrin-enhanced fluorescence. *J Chromatogr A* 2007;**1143**:252–7.
143. Spadaro D, Ciavorella A, Lore A, Garibaldi A, Gullino ML. Low levels of ochratocin A in wines from Piedmont. *Commun Agric Appl Biol Sci* 2007;**72**:327–32.
144. Almeda S, Arce L, Valcárcel M. Combined use of supported liquid membrane and solid-phase extraction to enhance selectivity and sensitivity in capillary electrophoresis for the determination of ochratoxin A in wine. *Electrophoresis* 2008;**29**:1573–81.
145. Shim WB, Kim K, Ofori JA, Chung YC, Chung DH. Occurrence of aflatoxins in herbal medicine distributed in South Korea. *J Food Prot* 2012;**75**:1991–9.
146. Hiraoka H, Yamamoto K, Mori Y, Asao N, Fukunaka R, Deguchi K, et al. Modified use of a commercial ELISA kit for deoxynivalenol determination in rice and corn silage. *Mycotoxin Res* 2013;**29**:79–88.
147. Bilandžić N, Božić Đ, Đokić M, Sedak M, Kolanović BS, Varenina I, et al. Seasonal effect on aflatoxin M_1 contamination in raw and UHT milk from Croatia. *Food Control* 2014;**40**:260–4.
148. Hassan HF, Kassaify Z. The risks associated with aflatoxins M_1 occurrence in Lebanese dairy products. *Food Control* 2014;**37**:68–72.
149. Kos J, Lević J, Đuragić O, Kokić B, Miladinović I. Occurrence and estimation of aflatoxin M_1 exposure in milk in Serbia. *Food Control* 2014;**38**:41–6.
150. Anfossi L, D'Arco G, Calderara M, Baggiani C, Giovannoli C, Giraudi G. Development of a quantitative lateral flow immunoassay for the detection of aflatoxins in maize. *Food Addit Contam A* 2011;**28**:226–34.
151. Kolosova AY, De Saeger S, Sibanda L, Verheijen R, Van Peteghem C. Development of a colloidal gold-based lateral-flow immunoassay for the rapid simultaneous detection of zearalenone and deoxynivalenol. *Anal Bioanal Chem* 2007;**389**:2103–7.

152. Shim WB, Yang ZY, Kim JS, Kim JY, Kang SJ, Woo GJ, et al. Development of immunochromatography strip-test using nanocolloidal gold-antibody probe for the rapid detection of aflatoxin B_1 in grain and feed samples. *J Microbiol Biotechnol* 2007;**17**:1629–37.
153. Molinelli A, Grossalber K, Fuhrer M, Baumgartner S, Sulyok M, Krska R. Development of qualitative and semi-quantitative immunoassay-based rapid strip tests for the detection of T-2 toxin in wheat and oat. *J Agric Food Chem* 2008;**56**:2589–94.
154. Molinelli A, Grossalber K, Krska R. A rapid lateral flow test for the determination of total type B fumonisins in maize. *Anal Bioanal Chem* 2009;**395**:1309–16.
155. Park JH, Chung DH, Lee IS. Application of fluorescence polarization immunoassay for the screening of ochratoxin A in unpolished rice. *J Life Sci* 2006;**16**:1006–13.
156. Zezza F, Longobardi F, Pascale M, Eremin SA, Visconti A. Fluorescence polarization immunoassay for rapid screening of ochratoxin A in red wine. *Anal Bioanal Chem* 2009;**395**: 1317–23.
157. Lippolis V, Pascale M, Valenzano S, Porricelli ACR, Suman M, Visconti A. Fluorescence polarization immunoassay for rapid, accurate and sensitive determination of ochratoxin A in wheat. *Food Anal Method* 2014;**7**:298–307.
158. Bondarenko AP, Eremin SA. Determination of zearalenone and ochratoxin A mycotoxins in grain by fluorescence polarization immunoassay. *J Anal Chem* 2012;**67**:790–4.
159. Lippolis V, Pascale M, Visconti A. Optimization of a fluorescence polarization immunoassay for rapid quantification of deoxynivalenol in durum wheat-based products. *J Food Prot* 2006;**69**:2712–9.
160. Lippolis V, Pascale M, Valenzano S, Pluchinotta V, Baumgartner S, Krska R, et al. A rapid fluorescence polarization immunoassay for the determination of T-2 and HT-2 toxins in wheat. *Anal Bioanal Chem* 2011;**401**:2561–71.
161. Fernández-Ibañez V, Soldado A, Martínez-Fernández A, de la Roza-Delgado B. Application of near-infrared spectroscopy for rapid detection of aflatoxin B_1 in maize and barley as analytical quality assessment. *Food Chem* 2009;**113**:629–34.
162. De Girolamo A, Lippolis V, Nordkvist E, Visconti A. Rapid and non-invasive analysis of deoxynivalenol in durum and common wheat by fourier-transform near infrared (FT-NIR) spectroscopy. *Food Addit Contam A* 2009;**26**:907–17.
163. Della Riccia G, Del Zotto S. A multivariate regression model for detection of fumonisins content in maize from near infrared spectra. *Food Chem* 2013;**141**:4289–94.
164. Yuan J, Deng D, Lauren DR, Aguilar MI, Wu Y. Surface plasmon resonance biosensor for the detection of ochratoxin A in cereals and beverages. *Anal Chim Acta* 2009;**656**:63–71.
165. Meneely JP, Sulyok M, Baumgartner S, Krska R, Elliott CT. A rapid optical immunoassay for the screening of T-2 and HT-2 toxin in cereals and maize-based baby food. *Talanta* 2010;**81**:630–6.
166. Kadota T, Takezawa Y, Hirano S, Tajima O, Maragos CM, Nakajima T, et al. Rapid detection of nivalenol and deoxynivalenol in wheat using surface plasmon resonance immunoassay. *Anal Chim Acta* 2010;**673**:173–8.
167. Park JH, Kim YP, Kim IH, Ko S. Rapid detection of aflatoxin B_1 by a bifunctional protein crosslinker-based surface plasmon resonance biosensor. *Food Control* 2014;**36**:183–90.
168. Nielsen KF. Mycotoxin production by indoor molds. *Fungal Genet Biol* 2003;**39**:103–17.
169. Jarvis BB, Miller JD. Mycotoxins as harmful indoor air contaminants. *Appl Microbiol Biotechnol* 2005;**66**:367–72.
170. Bloom E, Bal K, Nyman E, Larsson L. Optimizing a GC-MS method for screening of *Stachybotrys* mycotoxins in indoor environments. *J Environ Monit* 2007;**9**:151–6.

171. Bloom E, Bal K, Nyman E, Must A, Larsson L. Mass spectrometry-based strategy for direct detection and quantification of some mycotoxins produced by *Stachybotrys* and *Aspergillus* spp. in indoor environments. *Appl Environ Microbiol* 2007;**73**:4211–7.
172. Nielsen KF, Hansen MO, Larsen TO, Thrane U. Production of trichothecene mycotoxins on water damaged gypsum boards in Danish buildings. *Int Biodeterior Biodegradation* 1998;**42**:1–7.
173. Bloom E. *Mycotoxins in indoor environments determination using mass spectrometry* [Doctoral thesis]. Department of Laboratory Medicine Division of Medical Microbiology Lund University Sweden; 2008. p. 81.
174. Barnes C, Schreiber K, Pacheco F, Landuyt J, Hu F, Portnoy J. Comparison of outdoor allergenic particles and allergen levels. *Ann Allergy Asthma Immunol* 2000;**84**:47–54.
175. Cage BR, Schreiber K, Barnes C, Portnoy J. Evaluation of four bioaerosol samplers in the outdoor environment. *Ann Allergy Asthma Immunol* 1996;**77**:401–6.
176. Martin JW, Muir DC, Moody CA, Ellis DA, Kwan WC, Solomon KR, et al. Collection of airborne fluorinated organics and analysis by gas chromatography/chemical ionization mass spectrometry. *Anal Chem* 2002;**74**:584–90.
177. Shoeib M, Harner T. Characterization and comparison of three passive air samplers for persistent organic pollutants. *Environ Sci Technol* 2002;**36**:4142–51.
178. Brasel TL, Martin JM, Carriker CG, Wilson SC, Straus DC. Detection of airborne *Stachybotrys chartarum* macrocyclic trichothecene mycotoxins in the indoor environment. *Appl Environ Microbiol* 2005;**71**:7376–88.
179. Brera C, Caputi R, Miraglia M, Iavicoli I, Salerno A, Carelli G. Exposure assessment to mycotoxins in workplaces: aflatoxins and ochratoxin A occurrence in airborne dusts and human sera. *Microchem J* 2002;**73**:167–73.
180. Charpin-Kadouch C, Maurel G, Felipo R, Queralt J, Ramadour M, Dumon H, et al. Mycotoxin identification in moldy dwellings. *J Appl Toxicol* 2006;**26**:475–9.
181. Ngundi MM, Quadr SA, Wallace EV, Moore MH, Lassman ME, Shriver-Lake LC, et al. Detection of deoxynivalenol in foods and indoor air using an array biosensor. *Environ Sci Technol* 2006;**40**:2352–6.
182. Vesper S, Dearborn DG, Yike I, Allan T, Sobolewski J, Hinkley SF, et al. Evaluation of *Stachybortrys chartarum* in the house of an infant with pulmonary hemorrhage: quantitative assessment before, during and after remediation. *J Urban Health* 2000;**75**:68–85.
183. Gottschalk CH, Bauer J, Meyer K. Detection of satratoxin G and H in indoor air from a water-damaged building. *Mycopathologia* 2008;**166**:103–7.
184. Krysińska-Traczyk E, Kiecana I, Perkowski J, Dutkiewicz J. Levels of fungi and mycotoxins in samples of grain and grain dust collected on farms in eastern Poland. *Ann Agric Environ Med* 2001;**8**:269–74.
185. Tarin A, Rosell MG, Guardino X. Use of high-performance liquid chromatography to assess airborne mycotoxins. Aflatoxins and ochratoxin A. *J Chromatogr A* 2004;**1047**:235–40.
186. Richard JL, Plattner RD, Mary J, Liska SL. The occurrence of ochratoxin A in dust collected from a problem household. *Mycopathologia* 1999;**146**:99–103.
187. Tuomi T, Reijula K, Johnsson T, Hemminki K, Hintikka EL, Lindroos O, et al. Mycotoxins in crude building materials from water-damaged buildings. *Appl Environ Microbiol* 2000;**66**:1899–904.
188. Engelhart S, Loock A, Skutlarek D, Sagunski H, Lommel A, Färber H, et al. Occurrence of toxigenic *Aspergillus versicolor* isolates and sterigmatocystin in carpet dust from damp indoor environments. *Appl Environ Microbiol* 2002;**68**:3886–90.
189. Delmulle B, De Saeger S, Adams A, De Kimpe N, Van Peteghem C. Development of a liquid chromatography/tandem mass spectrometry method for the simultaneous determination of 16 mycotoxins on cellulose filters and in fungal cultures. *Rapid Commun Mass Spectrom* 2006;**20**:771–6.

190. Hintikka EL, Holopainen R, Asola A, Jestoi M, Peitzsch M, Kalso S, et al. Mycotoxins in the ventilation systems of four schools in Finland. *World Mycotoxin J* 2009;**2**:369–79.
191. Smoragiewicz W, Cossette B, Boutard A, Krzystyniak K. Trichothecene mycotoxins in the dust ventilation system in office buildings. *Int Arch Occup Environ Health* 1993;**65**:113–7.
192. Abdel-Hafez SII, Shoreit AAM. Mycotoxins producing fungi and mycoflora of air-dust from Taif, Saudi Arabia. *Mycopathologia* 1985;**92**:65–71.
193. Abdel-Hafez SII, Shoreit AAM, Abdel-Hafez AII, el Maghraby OM. Mycoflora and mycotoxin-producing fungi of air-dust particles from Egypt. *Mycopathologia* 1986;**93**:25–32.
194. Nordby KCH, Straumfors Halstensen A, Elen O, Clasen PE, Langseth W, Kristensen P, et al. Trichothecene mycotoxins and their determinants in settled dust related to grain production. *Ann Agric Environ Med* 2004;**11**:75–83.
195. Twarużek M, Grajewski J, Składanowska B, Janinska B, Fischer G. Untersuchungen unterschiedlicher Baumaterialien auf das Vorkommen von Schimmelpilzen und deren Mykotoxinen. *Mycotoxin Res* 2001;**17A**:137–41.
196. Skaug MA, Eduard W, Størmer FC. Ochratoxin A in airborne dust and fungal conidia. *Mycopathologia* 2000;**151**:93–8.
197. Bloom E, Grimsley LF, Pehrson C, Lewis J, Larsson L. Molds and mycotoxins in dust from water-damaged homes in New Orleans after hurricane Katrina. *Indoor Air* 2009;**19**:153–8.
198. Hinkley SF, Jarvis BB. Chromatographic method for *Stachybotrys* toxins. *Methods Mol Biol* 2001;**157**:173–94.
199. Nielsen KF. *Mold growth on building materials: secondary metabolites, mycotoxins and biomarkers* [Ph.D. thesis]. Lyngby, Denmark: BioCentrum-DTU, Technical University of Denmark; 2002. p. 122.
200. Nielsen KF, Thrane U, Larsen TO, Nielsen PA, Gravesen S. Production of mycotoxins on artificially inoculated building materials. *Int Biodeterior Biodegradation* 1998;**42**:9–16.
201. Razzazi-Fazeli E, Rabus B, Cecon B, Böhm J. Simultaneous quantification of A-trichothecene mycotoxins in grains using liquid chromatography–atmospheric pressure chemical ionization mass spectrometry. *J Chromatogr A* 2002;**968**:129–42.
202. Tuomi T, Saarinen L, Reijula K. Detection of polar and macrocyclic trichothecene mycotoxins from indoor environments. *Analyst* 1998;**123**:1835–41.
203. Rolle-Kampczyk U, Köller G, Popp P, Popp P. Mycotoxins in house dust problems and analytical approaches. *Mycotoxin Res* 2002;**18**:198–202.
204. Brasel TL, Douglas DR, Wilson SC, Straus DC. Detection of airborne *Stachybotrys chartarum* macrocyclic trichothecene mycotoxins on particulates smaller than conidia. *Appl Environ Microbiol* 2005;**71**:114–22.
205. Märtlbauer E, Gareis M, Terplan G. Enzyme immunoassay for the macrocyclic trichothecene roridin A: production, properties and use of rabbit antibodies. *Appl Environ Microbiol* 1988;**54**:225–30.
206. Johanning E, Gareis M, Yang CHS, Hintikka E-L, Nikulin M, Jarvis B, et al. Toxicity screening of materials from buildings with fungal indoor air quality problems (*Stachybotrys chartarum*). *Mycotoxin Res* 1998;**14**:60–73.
207. Johanning E, Gareis M. Mycotoxin screening of indoor environments in sentinel health investigations. In: *Bioaerosols—6th international scientific conference on bioaerosols, fungi, bacteria, mycotoxins in indoor and outdoor environments and human health*. Fungal Research Group Foundation; 2012. p. 195–200.
208. Twarużek M, Grajewski J, Gareis M. Toxicity study of field samples from water damaged houses in flooded areas in Poland. In: *Bioaerosols—6th international scientific conference on bioaerosols, fungi, bacteria, mycotoxins in indoor and outdoor environments and human health*. Fungal Research Group Foundation; 2012. p. 185–94.

209. Gareis M. Cytotoxicity testing of samples originating from buildings. In: Johanning E, Yang CS, editors. *Fungi and bacteria in indoor air environments. Proceeding of the international conference, Saratoga springs, New York*. Albany, NY: Eastern New York Occupational Health Program; October 6–7, 1994. p. 139–44.
210. Johanning E, Gareis M, Yang ChS. Umweltmedizinische Inhalationsrisiken bei toxischen Bioaerosolen-Exposition Untersuchung und MTT Cytotoxizitäts-Screening von Feldproben. In: *20. Mykotoxin Workshop, Detmold, 8–10 Juni*. 1998a. p. 32–40.
211. Ruotsalainen M, Hirvonen MR, Hyvärinen A, Meklin T, Savolainen K, Nevalainen A. Cytotoxicity, production of reactive oxygen species and cytokines induced by different strains of *Stachybotrys* sp. from mouldy buildings in RAW264.7 macrophages. *Environ Toxicol Pharmacol* 1998;**6**:193–9.
212. Smith JE, Anderson JG, Lewis CW, Murad YM. Cytotoxic fungal spores in the indoor atmosphere of damp domestic environment. *FEMS Microbiol Lett* 1992;**100**:337–43.
213. Yike I, Allan T, Sorenson WG, Dearborn DG. Highly sensitive protein translation assay for trichothecene toxicity in airborne particulates. Comparison with cytotoxicity assays. *Appl Environ Microbiol* 1999;**65**:88–94.

Chapter 22

Indoor Microbial Volatile Organic Compound (MVOC) Levels and Associations with Respiratory Health, Sick Building Syndrome (SBS), and Allergy

Xi Fu

Occupational and Environmental Medicine, Department of Medical Sciences, Uppsala University, University Hospital, Uppsala, Sweden

INTRODUCTION

Microorganisms can produce specific volatile organic compounds (VOCs) during their primary and secondary metabolism. These compounds are called microbial volatile organic compounds (MVOC).[1,2] MVOCs are a special type of VOC. As a subgroup of VOC, they have the same range of boiling points as VOC, 240–260 °C.[3]

A database of MVOC was established by Lemfack et al.[4] This database contains around 1200 MVOC that have been identified up to March 2014. There are around 530 fungal MVOC (from mold). Around 250 MVOC from fungi emission have been measured in indoor environmental studies.[5]

MVOCs are produced when mold consumes sugars and amino acids as the carbon source. Moisture is the key factor for indoor mold growth. The World Health Organization (WHO) has listed commonly identified mold species occurring in damp buildings.[6] Around 20 mold species require an equilibrium relative humidity (ERH) over 80%, including *Cladosporium*, *Mucor circinelloides*, *Rhizopus*, *Sporobolomyces*, *Alternaria*, *Exophiala*, *Epicoccum*, *Fusarium*, *Trichoderma*, *Ulocladium*, *Rhodotorula*, and *Stachybotrys*.[6] However, a number of species from *Aspergillus* or *Penicillium* genus can live at an ERH lower than 80%. Microbial volatile metabolites from primary and secondary metabolism of mold include alcohols, aldehydes, hydrocarbons, acids, ethers, esters, ketones, lactones, phenols, terpenoids, and sulfur and nitrogen compounds[7,8] (Table 1). There are a large number of studies measuring MVOC emission patterns from commonly detected mold. For instance, emission patterns of *Aspergillus* species

Environmental Mycology in Public Health. http://dx.doi.org/10.1016/B978-0-12-411471-5.00022-3

TABLE 1 Commonly Detected MVOC

Alcohols 1-Butanol 2-Butanol 2-Methyl-1-butanol 2-Methyl-1-propanol 3-Methyl-1-butanol 3-Methyl-2-butanol 3-Octanol 1-Octen-3-ol 2-Octen-1-ol 2-Pentanol	**Esters** Isobutyl acetate Ethyl isobutyrate Ethyl-2-methylbutyrate **Ketones** 2-Heptanone 2-Hexanone 3-Octanone **Terpenoids** Geosmin 2-Methylisoborneol
Ethers 3-Methylfuran 2-Pentylfuran	**Sulfur and nitrogen compounds** Dimethyl disulfide 2-Isopropyl-3-methoxy pyrazine

Refs. 9–12.

include 2-heptanone, 2-hexanone, 3-octanol, and α-terpineol.[13] Cultivation studies investigating MVOC emission patterns for different mold have reported that the emission depends on the type of species, the substrates, and the growth condition including type of nutrients, moisture condition, light, and CO_2 and O_2 levels.[1,14]

MEASUREMENT AND ANALYSIS OF MVOC

Sampling Methods

MVOCs can be adsorbed in porous material such as charcoal by pumped air sampling or diffusion sampling. Two main types of adsorbents are commonly used in current sampling tubes or filters: charcoal and Tenax® material. Charcoal is a material of good adsorption capability. The adsorbed MVOC can be desorbed by solvent extraction using carbon disulfide with 1% methanol,[13] or methylene chloride.[15] Tenax® TA is one type of Tenax® material, which is a polymer resin based on 2,6-diphenylene oxide[16] Tenax® GR is another type of Tenax® material, which is composed of 70% Tenax® TA and 30% graphite.[16] The compounds sampled by Tenax® TA and Tenax® GR can both be desorbed by thermal desorption.[17,18] The desorption efficiency from a certain type of material varies between different compounds.

MVOCs are commonly sampled by active sampling, pumping air through the adsorbent tube. A commonly used air flow for activated charcoal tubes is 200 mL/min and the sampling time varies from 4 to 8 h.[12,15,19,20] The pumped Tenax® sampling method requires a lower total sampling volume, typically an airflow rate of 0.05–0.1 L/min for a maximum of 2 h. The advantage of active sampling method is that the sampling volume is well-defined (Table 2). In some occupational exposure studies, personal active sampling in the breathing zone during one workday has been performed.[15]

Passive sampling with either charcoal or Tenax® diffusion samplers were also commonly applied in indoor environment studies.[21–24] Diffusion samplers can be of a badge-like design. They are usually placed 1–2 m above the floor level and at least 0.5–1 m away from walls or furniture (Table 2). Diffusion samplers can also be used in personal sampling, by placing the sampler on the participant's collar.

Analytical Methods

Gas chromatography/mass spectrometry with selected ion monitoring is currently widely used for MVOC analysis.[10,12,20,23] Selected ion monitoring provides good specificity and sensitivity because only selected mass fragments are monitored. In some earlier studies from the 1990s, MVOC was analyzed by gas chromatography using flame ionization detectors or electron capture detection.[13,25]

Electronic nose, or e-nose, is a technique mimicking human olfaction to detect and recognize odor. The method was originally used in the food industry. It was developed and applied in indoor environment studies to detect MVOC and fungal contamination in the beginning of the twenty-first century.[26] Experimental studies have shown that it is possible for the electronic nose to recognize fungi species according to their emitted MVOC patterns.[27]

MVOC LEVELS IN INDOOR AND OCCUPATIONAL ENVIRONMENTS

MVOC levels have been measured in a number of indoor environment studies, mostly to study associations with building dampness and microbial contamination.[3] Moreover, some studies measured MVOC in occupational environments such as the indoor environment in aircraft.[9,15] Wilkins et al. demonstrated that indoor levels of certain MVOC, such as 2-methyl-1-propanol, dimethyl disulfide, dimethyl trisulfide, dimethyl tetrasulfide, and 2-ethyl-1-hexanol, were associated with a history of mold dampness in buildings.[8] Elke et al. measured MVOC by diffusion sampling and reported that 3-methylbutan-1-ol, 2-hexanone, 2-heptanone, and 3-octanol were main indicators of mold growth.[13] Lorenz et al. concluded that 3-methylfuran, 1-octen-3-ol, and dimethyl disulfide were main indicators of microbial contamination and concluded that if any of these three compounds were above 50 ng/m^3, this indicated a microbial source.[28] Fischer et al. reported that a slight moldy odor may occur at an MVOC concentration in the air of 50–1720 ng/m^3 and a strong moldy odor can occur if the MVOC concentration is 160–12,300 ng/m^3.[29]

TABLE 2 Examples for Various Sampling Methods for MVOC Used in Different Studies

	Adsorbent	Sampler	Procedure	References
Active	Charcoal	Anasorb 747, SKC Inc., Eighty four, PA, USA	2.0 L/min, 1 h	19
		Anasorb 747, SKC Inc., Eighty four, PA, USA	0.2 L/min, 6 h	12
		Anasorb 747, SKC Inc., Eighty four, PA, USA	0.2 L/min, 4 h	20
		Anasorb 747, SKC Inc., Eighty four, PA, USA	0.2 L/min, 8 h (personal sampling)	15
		GK 26-16, SKC, Eighty four, PA, USA	0.025 L/min	13
	Tenax	Tenax TA	0.1 L/min, 20 min	19
		Tenax TA	0.05 L/min, 1.5 h	17
Passive	Charcoal	3500 OVM, 3M, Neuss, Germany	1.5 m height, 1 m from wall, 4 weeks	13
		3500 OVM, 3M, Neuss, Germany	1.5–2 m height, 0.5 m to next furniture, 4 weeks	22
		Supelco VOC-SD tube-type passive sampler, Sigma–Aldrich, MO, USA	1–1.5 m high, 1 m from wall, 2 days	23,24
	Tenax	Tenax GR	10 days	21

There is no standard method for MVOC analysis. Reported indoor MVOC levels vary a lot and it is difficult to compare different studies because of differences in sampling and analytical methods. Total MVOC has been reported, calculated by mass summation of identified individual compounds. However, different studies included different compounds in the total MVOC.[20,23,30] Table 3 summarizes levels of individual MVOC in indoor and outdoor environments and occupational settings in North Europe.

MVOCs IN INDOOR ENVIRONMENT AS INDICATOR OF HIDDEN MICROBIAL GROWTH

Measurement of MVOC has been applied to indicate microbial growth in different types of environment. This originated in the food industry in 1970s, when MVOC were analyzed to identify the source of unpleasant odors from various food products.[1] During 1990s, MVOC levels in indoor air started to be implemented in certain countries, especially Sweden[11] and Germany,[13] as an indicator of hidden microbial growth in building construction. In a cold climate, microbial growth is often hidden inside building construction. Modern buildings in a cold climate usually contain a water-vapor barrier in building construction to avoid the moisture, generated inside the building, to condense in the cold part of the construction. This dampness barrier is usually a plastic film in the inner part of the wall and roof construction. It has been demonstrated experimentally that MVOC can migrate through this plastic film, and reach the indoor air. Therefore, detection of MVOC was implemented as a way to find hidden microbial contamination without the need to open the building structure.[2,11] Measurement of MVOC levels emit from microbial contaminants demonstrated that certain MVOC are main indicators of microbial growth, including 3-methylfuran and 1-octen-3-ol.[8,13]

Although MVOCs are indicators for mold growth, the association between the levels of MVOC and mold status in a building can be weak.[19] In a recent study, MVOC were measured in indoor air in homes, and some association were found between specific MVOC and a history of microbial growth or dampness in the building. However, the study concluded that the sum of MVOC sometimes used by consultant companies to indicate microbial growth was not associated with the status of dampness of the building.[29]

Most indoor studies on MVOC have investigated the concentration of individual MVOC. However, researchers in France instead chose to investigate patterns of MVOC and compare the pattern in the indoor environment with patterns achieved in cultivation laboratory studies. A fungal MVOC index was developed by the French Scientific and Technical Center for Buildings, which can help identify microbial growth in buildings.[31]

HEALTH EFFECT AND SICK BUILDING SYNDROME

Associations between indoor levels of MVOC and health concerns have been investigated in cross-sectional studies in northern Europe (Sweden, Estonia, and Iceland), France, and Japan, including sick building syndrome (SBS), asthma, allergic symptoms, chronic bronchitis, and rhinitis. SBS refers to certain

TABLE 3 Indoor, Outdoor, and Occupational Levels of MVOC in Studies From Northern Europe

Compound Name	Nordic Home Exposure (ng/m³)	Exposure in Occupational Environment			Outdoor Air	References
		House Painters	School/ Office	Aircraft Cabin		
1-Butanol (ng/m³)	650–78,000		440–3520	80–4090		9,10
2-Butanol	190–14,000		94–1380	540–14,440		9,10,12
2-Methyl-1-butanol	10–940		<1–37	260–9410	<1–210	9,10
3-Methyl-1-butanol	<1–5320		<1–560	1110–31,320	<1–3800	1,9,10
1-Octen-3-ol	7–460	220–8040	<1–210	11–280	<1–1900	1,9,10,12,15
2-Pentanol	1–220	35–1120	<1–320	1–1098	<1–630	9,10,12,15
3-Methylfuran	1–370	11–460	<1–160	1–670	<1–110	9,10,12,15
2-Pentylfuran	<1–46,000	160–7400	13–180	1–120		9,10,12,15
Isobutyl acetate	1–2530	130–850	<1–2680	1–3000		9,10,12,15
Ethyl isobutyrate	1–72	<1–780	<1	1–93		9,10,12,15
Ethyl-2-methylbutyrate	<1–2390	<1–33	<1–89	<1–46		9,10,15
2-Heptanone	40–3170	110–5380	18–750	18–450	<1–1100	9,10,12,15
2-Hexanone	11–300	110–5380	7–140	7–210	<1–800	9,10,12,15
3-Octanone	13–170	0–29	<1–24	<1–38	<1–2	9,10,12,15
Dimethyl disulfide	1–4080	<1–56	<1–710	1–220	<1–260	9,12,15

nonspecific symptoms, including eye, nose, skin, throat symptoms, and general symptoms such as tiredness and headache.[32,33] Building dampness is considered to be a major cause of SBS.[33]

One study from Japanese homes found an association between SBS and the indoor concentration of MVOC, especially for the compound 1-octen-3-ol.[23] In the Japanese home environment study, positive associations were also found between the indoor concentration of 1-octen-3-ol and allergic rhinitis and conjunctivitis.[24] Another study from homes included three countries in northern Europe (Sweden, Estonia, and Iceland). There were positive associations between any SBS symptom and levels of 2-pentanol, 2-hexanone, 2-pentylfuran, 1-octen-3-ol, and 3-methylfuran, especially for mucosal symptoms. One study among pupils in Swedish elementary schools found associations between higher indoor concentrations of total MVOC and nocturnal breathlessness and doctor-diagnosed asthma. Moreover, there were positive associations between nocturnal breathlessness and specific MVOC, including 3-methylfuran, 3-methyl-1-butanol, dimethyl disulfide, 2-heptanone, 1-octen-3-ol, and 3-octanone.[10] One epidemiological study from homes in France reported positive associations between the fungal index, developed by researchers in France, and current asthma and chronic bronchitis, especially in rural areas.[34] Finally, one occupational study measured personal exposure to MVOC by personal air sampling in the breathing zone of house painters. An association was found between personal exposure to 1-octen-3-ol and clinical signs from the nasal mucosa, including reduced nasal patency and increased levels of myeloperoxidase in nasal lavage fluid.[15] MPO is an indicator of neutrophilic inflammation.

Finally, acute effects of exposure to some MVOC have been studied in exposure chamber studies using MVOC concentrations much higher than the levels in indoor environments. Experimental exposure to 3-methylfuran (1 mg/m^3 for 2 h) increased blinking frequency and levels of biomarkers of inflammation in nasal lavage (MPO and lysozyme). Moreover, lung function (forced vital capacity) was decreased.[35] Experimental exposure to 1-octen-3-ol (10 mg/m^3 for 2 h) induced eye and nose symptoms, increased blinking frequency, and increased levels of biomarkers of nasal inflammation. Levels of eosinophil cationic protein, MPO, and lysozyme increased in nasal lavage. Ratings of headache and nausea were also increased.[35] In contrast, experimental exposure to 3-methyl-1-butanol (1 mg/m^3 for 2 h) had almost no effect on biomarkers or medical symptoms, except a slight increase in eye irritation.[36]

CONCLUSIONS

Some MVOC are associated with medical symptoms and clinical signs, but further longitudinal, epidemiological studies on health effects of exposure to MVOC are needed. The most consistent health associations so far have been found for 1-octen-3-ol. Further investigations are needed to confirm that indoor air measurement of MVOC is a suitable method to identify hidden microbial growth. International standardization of sampling and analytical methods for

MVOC is needed. Moreover, more MVOC studies are needed in different countries and different climate zones in the world.

REFERENCES

1. Korpi A, Jarnberg J, Pasanen AL. Microbial volatile organic compounds. *Crit Rev Toxicol* 2009;**39**:139–93.
2. Wessen B, Ström G, Palmgren U. Microbial problem buildings-analysis and verification. In: *Proc indoor air*, vol. 4. 1999. p. 875–9.
3. Douwes J. *Building dampness and its effect on indoor exposure to biological and non-biological pollutants* 2009.
4. Lemfack MC, Nickel J, Dunkel M, Preissner R, Piechulla B. MVOC: a database of microbial volatiles. *Nucleic Acids Res* 2014;**42**:D744–8.
5. Morath SU, Hung R, Bennett JW. Fungal volatile organic compounds: a review with emphasis on their biotechnological potential. *Fungal Biol Rev* 2012;**26**:73–83.
6. WHO. Guidelines for indoor air quality: dampness and mold. World Health Organization.
7. Larsen TO, Frisvad JC. Characterization of volatile metabolites from 47 penicillium taxa. *Mycol Res* 1995;**99**:1153–66.
8. Wilkins K, Larsen K. Variation of volatile organic compound patterns of mold species from damp buildings. *Chemosphere* 1995;**31**:3225–36.
9. Fu X, Lindgren T, Guo M, Cai GH, Lundgren H, Norback D. Furry pet allergens, fungal DNA and microbial volatile organic compounds (MVOCs) in the commercial aircraft cabin environment. *Environ Sci Process Impacts* 2013;**15**:1228–34.
10. Kim JL, Elfman L, Mi Y, Wieslander G, Smedje G, Norback D. Indoor molds, bacteria, microbial volatile organic compounds and plasticizers in schools–associations with asthma and respiratory symptoms in pupils. *Indoor air* 2007;**17**:153–63.
11. Ström G, West J, Wessen B, Palmgren U. *Quantitative analysis of microbial volatiles in damp Swedish houses*. Amsterdam: Elsevier Science; 1994.
12. Wieslander G, Norback D, Venge P. Changes of symptoms, tear film stability and eosinophilic cationic protein in nasal lavage fluid after re-exposure to a damp office building with a history of flooding. *Indoor Air* 2007;**17**:19–27.
13. Elke K, Begerow J, Oppermann H, Kramer U, Jermann E, Dunemann L. Determination of selected microbial volatile organic compounds by diffusive sampling and dual-column capillary GC-FID—a new feasible approach for the detection of an exposure to indoor mould fungi? *J Environ Monit: JEM* 1999;**1**:445–52.
14. Claeson A-S. *Volatile organic compounds from microorganisms-identification and health effects*. Umea University; 2006.
15. Wieslander G, Norback D. Ocular symptoms, tear film stability, nasal patency, and biomarkers in nasal lavage in indoor painters in relation to emissions from water-based paint. *Int Arch Occup Environ Health* 2010;**83**:733–41.
16. Scientific Instrument Services I. *Tenax® ta adsorbent resin physical properties* 2014.
17. Betancourt DA, Krebs K, Moore SA, Martin SM. Microbial volatile organic compound emissions from *Stachybotrys chartarum* growing on gypsum wallboard and ceiling tile. *BMC Microbiol* 2013;**13**:283.
18. Cao X, Nicholas Hewitt C. Evaluation of tenax-GR adsorbent for the passive sampling of volatile organic compounds at low concentrations. *Atmospheric Environment Part A. General Topics* 1993;**27**:1865–72.
19. Schleibinger H, Laussmann D, Bornehag CG, Eis D, Rueden H. Microbial volatile organic compounds in the air of moldy and mold-free indoor environments. *Indoor Air* 2008;**18**:113–24.

20. Wessen B, Schoeps KO. Microbial volatile organic compounds–what substances can be found in sick buildings? *Analyst* 1996;**121**:1203–5.
21. Fischer G, Schwalbe R, Moller M, Ostrowski R, Dott W. Species-specific production of microbial volatile organic compounds (MVOC) by airborne fungi from a compost facility. *Chemosphere* 1999;**39**:795–810.
22. Matysik S, Herbarth O, Mueller A. Determination of microbial volatile organic compounds (MVOCs) by passive sampling onto charcoal sorbents. *Chemosphere* 2009;**76**:114–9.
23. Araki A, Kawai T, Eitaki Y, Kanazawa A, Morimoto K, Nakayama K, et al. Relationship between selected indoor volatile organic compounds, so-called microbial VOC, and the prevalence of mucous membrane symptoms in single family homes. *Sci Total Environ* 2010;**408**:2208–15.
24. Araki A, Kanazawa A, Kawai T, Eitaki Y, Morimoto K, Nakayama K, et al. The relationship between exposure to microbial volatile organic compound and allergy prevalence in single-family homes. *Sci Total Environ* 2012;**423**:18–26.
25. Begerow J, Jermann E, Keles T, Koch T, Dunemann L. Screening method for the determination of 28 volatile organic compounds in indoor and outdoor air at environmental concentrations using dual-column capillary gas chromatography with tandem electron-capture-flame ionization detection. *J Chromatogr A* 1996;**749**:181–91.
26. Kuske M, Romain A-C, Nicolas J. Microbial volatile organic compounds as indicators of fungi. Can an electronic nose detect fungi in indoor environments? *Build Environ* 2005;**40**:824–31.
27. NSaN M. Fungal volatile fingerprints: discrimination between dermatophyte species and strains by means of an electronic nose. In: *International Symposium on Olfaction and Electronic Noses*, vol. 131. 2008. p. 117–20.
28. Lorenz W, Diederich T, Conrad M. Practical experiences with MVOC as an indicator for microbial growth. In: *9th International Conference on Indoor Air Quality and Climate – Indoor Air 2002*. Rotterdam (Netherlands): in-house publishing; 2002. p. 341–6.
29. Fischer G, Dott W. Relevance of airborne fungi and their secondary metabolites for environmental, occupational and indoor hygiene. *Arch Microbiol* 2003;**179**:75–82.
30. Sahlberg B, Gunnbjornsdottir M, Soon A, Jogi R, Gislason T, Wieslander G, et al. Airborne molds and bacteria, microbial volatile organic compounds (MVOC), plasticizers and formaldehyde in dwellings in three north European cities in relation to sick building syndrome (SBS). *Sci Total Environ* 2013;**444**:433–40.
31. Moularat S, Hulin M, Robine E, Annesi-Maesano I, Caillaud D. Airborne fungal volatile organic compounds in rural and urban dwellings: detection of mould contamination in 94 homes determined by visual inspection and airborne fungal volatile organic compounds method. *Sci Total Environ* 2011;**409**:2005–9.
32. Joshi SM. The sick building syndrome. *Indian J Occup Environ Med* 2008;**12**:61–4.
33. Norback D. An update on sick building syndrome. *Curr Opin Allergy Clin Immunol* 2009;**9**: 55–9.
34. Hulin M, Moularat S, Kirchner S, Robine E, Mandin C, Annesi-Maesano I. Positive associations between respiratory outcomes and fungal index in rural inhabitants of a representative sample of French dwellings. *Int J Hyg Environ Health* 2013;**216**:155–62.
35. Walinder R, Ernstgard L, Johanson G, Norback D, Venge P, Wieslander G. Acute effects of a fungal volatile compound. *Environ Health Perspect* 2005;**113**:1775–8.
36. Ernstgard L, Norback D, Nordquist T, Wieslander G, Walinder R, Johanson G. Acute effects of exposure to vapors of 3-methyl-1-butanol in humans. *Indoor Air* 2013;**23**:227–35.

Part IV

Methods in Environmental Mycology

Section I

Environmental Sampling

Chapter 23

Air, Surface and Water Sampling

Ana Catarina Pinheiro[1], Carla Viegas[2]
[1]*Pharmacy Department, Centro Hospitalar do Algarve, Faro Unit, Portugal;* [2]*Environment & Health RG, Lisbon School of Health Technology, Polytechnic Institute of Lisbon, Environmental Health Institute, Faculty of Medicine from Lisbon University, Lisbon, Portugal*

INTRODUCTION

Air sampling collection for fungal contamination analysis can be performed using different methods. The choice of method depends on whether a quantitative or qualitative analysis is desired. This chapter presents some of the most frequent techniques.[1,2]

The positive correlation between spore level and the risk of infection is widely accepted; the usual method to determine this risk is air sampling.[3–6] This approach was chosen from a variety of settings.[3,7–18]

Different methods can be applied to assess air contamination: into passive (gravity plates) or active (impact, impingers, and filters).

PASSIVE METHODS

Exposed-Plate Gravitational Method or Deposition Samplers

The results obtained with these techniques depend on the dispersion ability of each fungal genera and the different conditions of the surrounding air (turbulent vs calm atmosphere, for instance). In this case, sampling can be done by exposing media-containing plates for 30 min.[19] This method may be used especially in settings such as health care facilities with lower contamination.[20] Passive methods do not allow quantitative results, only colony-forming units (CFU) per plate, because it is almost impossible to calculate the exact volume of air sampled. As such, the result for each plate cannot be compared with any other sample because conditions cannot easily be replicated.

ACTIVE METHODS

These methods involve sampling a known volume of air per unit time; the samplers, named volumetric samplers, are equipped with a suction device that allow fungal quantification (CFU per cubic meter) when collected volumes are

Environmental Mycology in Public Health. http://dx.doi.org/10.1016/B978-0-12-411471-5.00023-5

previously defined. It allows comparison with outdoor data, the main source of indoor contamination.[2,21]

Impact

According to Predicala et al.,[22] this methodology presents the best results compared with other sampling methods. Also called the slit-to-agar method, it has been chosen by many authors[23,24] and by the American Conference of Industrial Hygienists,[25] and it is recommended by the Canadian Health Organization.[26] Fleischer et al.[27] compared gravity plates with the impact method and obtained statistically significant differences between them. A higher amount of CFU per cubic meter was obtained using the active method because particles that transport fungi depend on airflow, which keeps them airborne. In addition, despite limitations to achieve comparable data, this method is widely applied in hospitals,[28–30] schools,[31,32] and other environments.[17,33]

(Liquid) Impingers

When spore mortality is an issue, this type of samplers is the best choice.[2] Formerly known as scrubbers, they are often necessary in occupational settings with higher fungal loads because this approach allows dilution of the sample before plate incubation. This is not generally possible with samplers that employ impaction on solid media (see above).[34] It requires the use of a tampon solution such as sterile phosphate-buffered saline (PBS) and is particularly relevant when using molecular biology protocols for fungal identification[3,35,36] and to research fungal metabolites such as mycotoxins.

Suspending dust samples in liquid and making further dilutions before plating normally leads to higher concentration estimates of culturable fungi in house dust.[37] Dilution may help disperse aggregates of fungal and actinomycetes spores and culturable fungal fragments. Consequently, such microorganisms would constitute a larger proportion of total culturable isolate. On the downside, microorganisms are present in small numbers and as single units they may be less represented.[37] Moreover, impingers cannot operate for long periods because liquid evaporation can hamper the microorganism's viability.[2]

Filters

It is possible to perform both direct and indirect analysis using this method. Air passing through a porous and homogeneous smooth surface, such as a membrane filter, allows for particulate deposition and the filter can be removed and placed directly under a microspore for analysis.[2]

The filter membrane can also be placed on culture media and incubated to allow fungal growth or even digested with a tampon solution such as sterile PBS and then inoculated in the selected media.[20] Filter samples can also be dispersed in a liquid before cultivation. The resulting suspension can be diluted before culture and subsamples can be cultivated under different conditions, allowing different organisms to be detected. As mentioned for air samples, aggregates may disperse when collected or re-suspended in a liquid, leading to higher colony counts than methods with direct cultivation of the collection plates.[35] Because of the risk of dehydration—because the surface where the particles are collected is completely dry—this method is suited only for resistant microorganisms or fungal spores.[2]

Spore Counting Method

Another method to achieve fungal load is one involving counting propagules impacted onto a sticky surface and identifying them according to morphological features. This method involves sampling a known volume of air per unit time.[18] However, identification is restricted because it is possible to confirm only the presence of fungus and, when possible, to identify the fungal genera.

Electrostatic and Thermal Precipitation

Both methods are used to collect small particles that are deviated, owing to electrostatic charges or differences in the temperature inside the apparatus.[2]

STRATEGY

Given the possible methods presented, which should we choose? Several criteria must be considered[2]:

- the possibility of a health risk for the individuals performing the study
- whether there is a specific species to be monitored
- how fast the results must be presented
- the size of the particles to be studied and monitored
- sampling type, considering whether it is preferable to define an area (and the number of samplers to be placed) or to proceed with personal sampling.

When health reasons are behind a given study, it is relevant to know the inhalable fraction; then, personal samplers are more adequate and few bioaerosol samplers collect the inhalable fraction or can be used for breathing zone sampling, which is a further requirement.[38]

If fungal contamination is not expected to experience significant changes during a season or even a day, the sampling period may not cover all of the exposure hours of any given setting, but in this situation, a detailed observation and analysis of all activities not included in the assessed period must have been performed previously or the sampling period could be selected taking into account the most critical scenario.[39] Nevertheless, exposure variability of microbial agents can be substantially greater than that commonly found for chemical agents. The number of samples must then be increased to obtain exposure estimates with sufficient precision.[35]

A basic problem in quantitative exposure assessment of fungi, as any other biologic agent, is the variability of exposure. As an example, stationary measurements of fungal spores with a sampling time of 8 h, at a fixed location in a sorting plant for timber, showed day-to-day variations from $< 0.1 \times 10^6$ to 20×10^6 spores/m^3.[35] However, the exposure variability of microbial agents can be even greater because microorganisms may proliferate rapidly under favorable conditions.[35] Day-to-day variability represents fewer problems in epidemiological studies of short-term effects, in which the mean exposure during the entire study period that is related to the observed effects can be measured. However, exposure measurement of all exposed time—in occupational epidemiological and hygiene studies of longer duration—is not feasible, and exposure has to be estimated from a limited number of measurements. A group-based strategy is usually chosen to limit costs. This strategy assumes that workers performing similar jobs also have similar exposure levels and that measurements performed on a subset of the workers are valid for the other workers within the group. Then, the measurements of all workers can be combined and the mean exposure of the whole group is determined with better precision than the mean exposure of the individual workers, because the group mean is based on a larger number of measurements. The efficiency of this strategy depends on the similarity of exposure levels among workers, because combining workers with diverse exposure levels in groups may lead to poor precision of exposure–response associations in the epidemiological analysis.[40]

SURFACE SAMPLING

The biological materials found on any given surface may not correspond to those encountered in the air samples. Several studies corroborate the importance of coupling surface and air sampling to achieve a more complete fungal characterization from an indoor environment.[28,41–50]

Surface sampling is inexpensive and easy to perform, and in the case of direct analysis, it can give immediately results.

Direct Analysis

This method allows the identification of morphologically distinguishable cells[37] but does not allow discrimination between viable and nonviable spores or hyphae. A direct microscopic examination of a surface has the advantage of showing exactly what is there, revealing indoor reservoirs of spores that have not yet become airborne. This analysis can be performed using different approaches[37]:

- Using tape and applying pressure onto the surfaces and then onto a slide for microscopic observation. Through this technique it is possible to confirm fungal colonization and, when possible, to identify fungal genera. It does not, however, allow for fungal quantification.
- Material sampling to confirm fungal colonization. This allows observation directly into the microscope to confirm colonization and allows fungal identification. As before, fungal quantification remains impossible in most cases.

Indirect Analysis

Indirect analysis requires the use of growth media and standard incubation periods and temperatures. Surface dust culture, swabs, or contact plates are substantially interchangeable and their use should be tailored to the situation. All are able to detect fungi on surfaces but are selective for viable components.

- Dust collection

Concentrations of house dust mite, cat, dog, and cockroach antigens in dust are measured by immunoassay.[51] However, culture-based and microscope-based methods are used to identify the potential presence of microbial allergens and toxins in air and dust samples.[37,52] Although the active biological agents are not themselves detected by culture methods, their presence is assumed if the organism with which they are associated is detected.[37] This method allows fungal flora characterization in an indoor space. Some species generally associated with infiltration problems, such as *Acremonium*, *Chaetomium*, *Stachybotrys*, and *Ulocladium* can be identified this way because, owing to their intrinsic features, they are not readily airborne.

- Surface scrapings. Samples from surfaces are collected and placed in growth media. Through this technique it is possible to confirm fungal colonization and, when possible, to identify fungal genera. It does not, however, allow for fungal quantification.
- Surface swabbing. This methodology allows fungal quantification (CFU per square meter) when the swabbed area is predetermined. Surface samples can be executed by swabbing the surfaces using a $10 \times 10\text{-cm}^2$ stencil disinfected with 70% alcohol solution between samples according to International Standard 18593–2004.
- Contact plates. The media plate is pressed against a surface and spores can be quantified and species identified after incubation.

WATER SAMPLING

For water sampling, sterile glass bottles can be used and water can be filtered through sterile filters (pore size, 0.45 mm). The number of CFU of different species can be noted as CFU/500 ml. To analyze drinking water, cold tap water should be obtained after allowing it to run until it reaches a steady temperature, and after flaming.[53]

REFERENCES

1. Horner W. Assessment of the indoor environment: evaluation of mold growth indoors. *Immunol Allergy Clin N Am* 2003;**23**(3):519–31.
2. De Nuntiis P, Maggi O, Mandrioli P, Ranalli G, Sorlini C. Monitoring the biological aerossol. In: Mandrioli P, Caneva G, Sabbioni C, editors. *Cultural heritage and aerobiology – methods and measurement techniques for biodeterioration monitoring*. Dordrecht: Kluwer Academic Publishers; 2003. p. 107–44.

3. Cabral JPS. Can we use indoor fungi as bioindicators of indoor air quality? Historical perspectives and open questions. *Sci Total Environ* 2010;**408**:4285–95.
4. Dales RE, Cakmaic S, Burnett RT, Judek S, Coateas F, Brooic JR. lnfluence of ambient fungal spores on emergency visits for asthma to a regional chiidren's hospital. *Am J Respir Crit Care Med* 2000;**162**:2087–90.
5. Hay RI, Clayton YM, Goodiey JM. Fungal aerobiology: how, when and where? *J Hosp lnfect* 1995;**30**:352–7.
6. Manuel RJ, Kibbler CC. The epidemiology and prevention of invasive aspergillosis. *J Hosp Infect* 1998;**39**:95–109.
7. Solans X, Alono RM, Constans A. Exposici6n laboral a hongos y bactérias ambientales en una planta de selecci6n de residuos de envases. *Rev lberoam Micol* 2007;**24**(2):131–5.
8. Lugauskas A, Jaskelevicius B. Micromycetes hazardous to human health in buildings of various age and use in Vilnius. *Indoor Built Environ* 2007;**16**(4):358–70.
9. Kim KY, Park JB, Jang GY. Assessment of bioaerosols in the public buildings of Korea. *Indoor Built Environ* 2007;**16**(5):465–71.
10. Zorman T, Jersek B. Assessment of bioaerosol concentrations in different indoor environments. *Indoor Built Environ* 2008;**17**(2):155–63.
11. Ozkutuk A, Ceylan E, Ergor G. The relationship between moulds isolated from indoor air and features of the house environment. *Indoor Built Environ* 2008;**17**(3):269–73.
12. Crawford J, Rosebaum P, Anagnost S. Home conditions and indoor airborne fungal levels in a cohort of infants at risk for ashma. In: *Healthy buildings, 9th international conference & exhibition, Syracuse, NY, September 13–17, 2009*. p. 694.
13. Homna Y, Hayashi M, Hasegawa K. Field measurement of airborne fungal spores of detached houses with insulated crawl space foundation in Japan. In: *Healthy buildings, 9th international conference & exhibition, Syracuse, NY, September 13–17, 2009*. p. 237.
14. Must A, Bloom E, Sandberg G. Determination of mycoflora and mycotoxins in concealed constructions and in the indoor occupational space. In: *Healthy buildings, 9th international conference & exhibition, Syracuse, NY, September 13–17, 2009*. p. 284.
15. Ando N, Yoshino H, Takaki R. Field survey on indoor microorganism in twenty-five residences in japan. In: *Healthy buildings, 9th international conference & exhibition, Syracuse, NY, September 13–17, 2009*. p. 708.
16. Sánchez M, Muñoz M, González A. Fungal indoor air characterization in an old and in a modern building of Madrid, Spain. In: *Healthy buildings, 9th international conference & exhibition, Syracuse, NY, September 13–17, 2009*. p. 706.
17. Pinheiro AC, Macedo MF, Jurado V, Saiz-Jimenez C, Viegas C, Brandão J, et al. Mould and yeasts identification in archival settings: preliminary results on the use of traditional methods and molecular biology options in Portuguese archives. *Int Biodeterior Biodegrad* 2011;**65**(4):619–27.
18. Tormo_Molina R, Garijo MA, Rocirı́guez SF, Palacios IS. Monitoring the occurrence of indoor fungi in a hospital. *Rev lberoam Micol* 2012;**29**(4):227–34.
19. Sudharsanam S, Srikanth P, Krishnamurthy S, Steinberg R. Microorganisms in bioaerosols in indoor air of hospital and non-hospital settings. *Sri Ramachandra J Med* 2009;**2**:52.
20. Sudharsanam S, Swaminathan S, Ramalingam A, Thangavel G, Annamalai R, Steinberg R, et al. Characterization of indoor bioaerosols from a hospital ward in a tropical setting. *Afr Health Sci* 2012;**12**(2):217–25.
21. Viegas C. Exposição profissional a fungos: o caso dos trabalhadores dos ginásios com piscina. *Caminhos do Conhecimento* 2012:318.

22. Predicala BZ, Urban JE, Maghirang RG. Assessment of bioaerosols in swine barns by filtration and impaction. *Curr Microbiol* 2002;**44**(2):136–40.
23. Blomquist G, Palmgren U, Ström G. Improved techniques for sampling airborne fungal particles in highly contaminated environments. *Scand J Work, Environ Health* 1984;**10**(4):253–8.
24. Smid T, Schokkin E, Boleij JS. Enumeration of viable fungi in occupational environments: a comparison of samplers and media. *Am Industrial Hyg Assoc J* 1989;**50**(5):235–9.
25. Verhoeff AP, Van Wijnen JH, Brunekreef B. Presence of viable mould propagules in indoor air in relation to house damp and outdoor air. *Allergy* 1992;**47**(2):83–91.
26. *Health Canada – indoor air quality in office buildings: a technical guide*. Vancouver: Health Canada; 1993. Available in. http://www.hc-sc.gc.ca/ewh-semt/alt_formats/hecs-sesc/pdf/pubs/air/office_buildingimmeubles_bureaux/93ehd-dhm166-eng.pdf.
27. Fleisher M, Bober-Gheek B, Bortkiewicz O. Microbiological control of airborne contamination in hospitals. *Indoor Built Environ* 2006;**15**(1):53–6.
28. Brenier-Pinchart MP, Bernadette Lebeau B, Quesada JL, et al. Influence of internal and outdoor factors on filamentous fungal flora in hematology wards. *Am J Infect Control* 2009;**37**:631–7.
29. Vacova M, Buchta V, Prymula R, et al. The occurrence of microscopic fungi in air samples from a transplant intensive care unit. *Indoor Built Environm* 2006;**15**:115–8.
30. Gniadek A, Macura AB. Intensive care unit environment contamination with fungi. *Adv Med Sci* 2007;**52**:283–7.
31. Canha N, Ribeiro M, Freitas MC, Almeida M, Almeida SM, Cabo S, et al. Fungi, bacteria and pollens seasonally quantified at 3 basic schools in Lisbon, *ASHRAE IAQ 2010 Conference: Airborn Infection Control - Ventilation, IAQ & Energy, Kuala Lumpur, Malaysia, November 10–12, 2011*. ISBN 978-1-936504-04-6.
32. Madureira J, Pereira C, Paciência I, Teixeira JP, Oliveira Fernandes E. Identification and levels of airborne fungi in Portuguese primary schools. *J Toxicol Environ Health, Part A* 2014;**77**:816–26.
33. Pinheiro AC, Viegas C, Viegas S, Veríssimo C, Brandão J, Macedo MF. Indoor air quality in Portuguese archives: a snapshot on exposure levels. *J Toxicol Environ Health* 10/2012;**75**(22–23):1359–70.
34. Thorne PS, Heederik D. Assessment methods for bioaerosols. In: Salthammer T, editor. *Organic indoor air pollutants—occurrence, measurement, evaluation*. Weinheim, Germany: Wiley/VCH; 1999. p. 85–103.
35. Eduard W, Halstensen A. Quantitative exposure assessment of organic dust. *Scandinavian Journal of Work, Environ Health* 2009;**7**:30–5.
36. Viegas C, Malta-Vacas J, Sabino R, Viegas S, Veríssimo C. Accessing indoor fungal contamination using conventional and molecular methods in Portuguese poultries. *Environ Monit Assess* 2014;**186**(3):1951–9. ISSN 0167-6369. http://dx.doi.org/10.1007/s10661-013-3509-4.
37. Macher JM. Evaluation of a procedure to isolate culturable microorganisms from carpet dust. *Indoor Air* 2001;**11**:134–40. http://dx.doi.org/10.1034/j.1600-0668.2001.110208.x.
38. Eduard W, Heederik D, Duchaine C, Green BJ. Bioaerosol exposure assessment in the workplace: the past, present and recent advances. *Environ Monit* 2012;**14**:334.
39. Piteira C. *A qualidade do ar interior em instalações hospitalares*. Lisboa: Lidel; 2007.
40. Burdorf A. Analysis and modelling of personal exposure. In: Nieuwenhuijsen MJ, editor. *Exposure assessment in occupational and environmental epidemiology*. Oxford: Oxford University Press; 2003. p. 85–102.
41. Klánová K, Hollerová J. Hospital indoor environment: screening for micro-organisms and particulate matter. *Indoor Built Environ* 2003;**12**(1–2):61–7.

42. Lugauskas A, Krikstaponis A. Microscopic fungi found in the libraries of Vilnius and factors affecting their development. *Indoor Built Environ* 2004;**13**(3):169–82.
43. Kordbacheh P, Zaini F, Kamali P. Study on the sources of nosocomial fungal infections at intensive care unit and transplant wards at a teaching hospital in Tehran. *Iran J Public Health* 2005;**34**(2):1–8.
44. Srikanth P, Sudharsanam S, Steinberg R. Bio-aerosols in indoor environment: composition, health effects and analysis. *Indian J Med Microbiol* 2008;**26**(4):302–12.
45. Hayashi M, Osawa H. A field study on mould, mites and environmental factors in houses. In: *Healthy buildings, 9th international conference & exhibition, Syracuse, NY, September 13-17, 2009*. p. 262.
46. Lu Z, Lu WZ, ZHANG JL. Microorganisms and particles in AHU systems: measurement and analysis. *Build Environ* 2009;**44**(4):694–8.
47. Reboux R, Houdrouge K, Gloaguidi-haore Z. Usefulness of environmental fungal surveillance in the prevention of invasive aspergillosis in haematology patients. *Mycoses* 2009;**52**(1):27.
48. Viegas C, Alves C, Carolino E, Pinheiro C, Rosado L, Silva-Santos C. Assessment of fungal contamination in a group of Lisbon's gymnasiums with a swimming pool. *Italian J Occup Environ Hyg* 2011;**2**(1):15–20.
49. Viegas C, Carolino E, Sabino R, Viegas S, Veríssimo C. Fungal contamination in swine: a potential occupational health threat. *J Toxicol Environ Health, Part A* 2013;**76**(4–5):272–80. http://dx.doi.org/10.1080/15287394.2013.757205. Available from http://hdl.handle.net/10400.21/2459.
50. Viegas C, Almeida-Silva M, Gomes AQ, Wolterbeek HT, Almeida SM. Fungal contamination assessment in Portuguese elderly care centers. *J Toxicol Environ Health, Part A: Curr Issues* 2014;**77**(1–3):14–23. http://dx.doi.org/10.1080/15287394.2014.861336.
51. Trudeau WL, Caldas EF. Identifying and measuring indoor biologic agents. *J Allergy Clin Immunol* 1994;**94**:393–400.
52. American Industrial Hygiene Association (AIHA). *Field guide for the determination of biological contaminants in environmental samples*. Fairfax, VA: AIHA; 1996.
53. Engelhart S, Simon A, Exner M. Wachstumsverhalten und Persistenz von *Aspergillus fumigatus* und *Fusarium solani* in Trinkwasser. *Hyg Med* 2009;**34**:128–31.

Section II

Chapter 24

Sand and Soil Sampling

João Brandão[1], Valerie J. Harwood[2]

[1]*Reference Unit for Parasitic and Fungal Infections, Department of Infectious Diseases, National Institute of Health Doutor Ricardo Jorge, Lisboa, Portugal;* [2]*Department of Integrative Biology, University of South Florida, Tampa, FL, USA*

Beach sand includes habitats in which viruses, bacteria, protozoa, and fungi thrive (reviewed in Whitman et al.[1]). Pathogens, including fungi, may present a significant and underappreciated risk to beach users.[2–4] One study conducted in Portugal found potentially pathogenic fungi in 48% of beach sand samples.[4]

Sand habitats are characterized by nonhomogeneous, patchy distribution of microorganisms, For example, a study in Minnesota, USA, found highly variable levels of *Escherichia coli* across beach transects,[5] and heterogeneous spatial distribution was also observed for enterococci in Florida sand.[6] Accurate conclusions drawn about the risk of infectious disease associated with exposure to beach sand depend on representative sampling, but this is difficult to achieve in water,[7,8] let alone sand.

An ideal sampling scheme would take into account all factors that could contribute to variability; however, that is logistically impossible and a compromise must be made that includes considerations of time and expense, as well as the extent of sampling coverage. Sabino et al.[4] explained that since the year 2000 in Portugal, a standard three-point composite sample of the beach, collected from the center and middle two halves, is used to generate a sample that is potentially more representative of the entire beach than a single sample. This approach is designed to improve sampling effort while minimizing cost. The consensus reached at a meeting in Lisbon in 2012 (Microareias) of several beach quality professionals was that sand samples (one set of wet sand and one of dry sand) are to be collected at three equidistant points along the beach, attempting to represent the beach sand as a whole.[9] The frequency of sampling did not generate any particular recommendation; however, in terms of timing it was agreed that water and sand samples should be collected simultaneously to avoid wasting human and financial resources.

The strategy of sampling as many points as possible while carrying out individual analysis is particularly useful during a research study that aims to categorize areas of the beach and fluctuation of microbial levels alongside its

Environmental Mycology in Public Health. http://dx.doi.org/10.1016/B978-0-12-411471-5.00024-7

length. Bearing in mind that often beach sand can be a source of contamination of recreational waters,[10] particularly in climates where rain is frequent, it might be necessary to characterize a beach to determine which points are most relevant for microbiological contamination of both beach users and nearby water. In such cases, the cost of time and materials necessary for multiple sampling and analysis should be included in the project funding.

Great variability is also attributed to the near-shore (wet sand) versus backshore (dry sand) position on the beach. For example, a study in Hawai'i found higher enterococci levels in backshore sand than in foreshore sand,[11] and similar results were found for *E. coli* in Florida.[7] Such findings call into question the generalization that wet sand should contain more microorganisms than dry sand,[9] but they support the recommendation that both wet and dry sand should be sampled to provide a better understanding of microbial distribution.

The methodology used to collect composite samples from both wet and dry sand, as described in Sabino et al.,[9] is identical. Sample collection for wet sands was defined as an area up to 1 or 2 m from the sea during low tide, that is, the inter-tidal zone. Microbial levels in these samples should be influenced by water quality in the water body and by contributions from runoff, the degradation of organic material, and the influence of beach users.[12]

A great deal remains to be learned about the most appropriate methodology for sampling sand in general. Even less is known about the distribution of fungi than about bacteria in sand. Researchers and regulatory agencies, whenever possible, are encouraged to carry out sampling transects on beaches and to analyze the samples separately to generate more data on the distribution of total and potentially pathogenic fungi in beach sand.

Concerns regarding sandboxes are minute compared with those involving a beach and do not require such attention by the authors of this book. However, applying the same procedure of three samples into one batch will always enhance representativeness, and standardization will lead to comparability.

REFERENCES

1. Whitman R, Harwood VJ, Edge TA, Nevers M, Byappanahalli M, Vijayavel K, et al. Microbes in beach sands: integrating environment. *Ecol Public Health Rev Environ Sci Biotechnol* 2014;**13**:329–68.
2. Heaney CD, Sams E, Dufour AP, Brenner KP, Haugland RA, Chern E, et al. Fecal indicators in sand, sand contact, and risk of enteric illness among beachgoers. *Epidemiol* 2012;**23**:95–106.
3. Pereira E, Figueira C, Aguiar N, Vasconcelos R, Vasconcelos S, Calado G, et al. Microbiological and mycological beach sand quality in a volcanic environment: Madeira archipelago. *Port Sci Total Environ* 2013;**461–462**:469–79.
4. Sabino R, Veríssimo C, Cunha MA, Wergikoski B, Ferreira FC, Rodrigues R, et al. Pathogenic fungi: an unacknowledged risk at coastal resorts? New insights on microbiological sand quality in Portugal. *Mar Pollut Bull* 2011;**62**:1506–11.
5. Ishii S, Hansen DL, Hicks RE, Sadowsky MJ. Beach sand and sediments are temporal sinks and sources of *Escherichia coli* in Lake Superior. *Environ Sci Technol* 2007;**41**:2203–9.

6. Bonilla TD, Nowosielski K, Cuvelier M, Hartz A, Green M, Esiobu N, et al. Prevalence and distribution of fecal indicator organisms in South Florida beach sand and preliminary assessment of health effects associated with beach sand exposure. *Mar Pollut Bull* 2007;**54**:1472–82.
7. Enns AA, Vogel LJ, Abdelzaher AM, Solo-Gabriele HM, Plano LR, Gidley ML, et al. Spatial and temporal variation in indicator microbe sampling is influential in beach management decisions. *Water Res* 2012;**46**:2237–46.
8. Whitman RL, Nevers MB. *Escherichia coli* sampling reliability at a frequently closed Chicago Beach: monitoring and management implications. *Environ Sci Technol* 2004;**38**:4241–6.
9. Sabino R, Rodrigues R, Costa I, Carneiro C, Cunha M, Duarte A, et al. Routine screening of harmful microorganisms in beach sands: implications to public health. *Sci Total Environ* 2014;**472**:1062–9.
10. MC1 P, Solo-Gabriele HM, Piggot AM, Klaus JS, Zhang Y. Relationships between sand and water quality at recreational beaches. *Water Res* 2011;**45**(20):6763–9.
11. Cui H, Yang K, Pagaling E, Yan T. Spatial and temporal variation in enterococcal abundance and its relationship to the microbial community in Hawaii beach sand and water. *Appl Environ Microbiol* 2013;**79**:3601–9.
12. Brandão J, Wergikosky B, Rosado C, Noronha G, Rosado L, Veríssimo C, et al. *Relatório final do projecto "Qualidade Microbiológica de areias de Praias Litorais"*. Associação Bandeira Azul da Europa; 2002.

Chapter 25

Processing Methodologies

Carla Viegas[1], Ana Catarina Pinheiro[2]
[1]*Environment & Health RG, Lisbon School of Health Technology, Polytechnic Institute of Lisbon, Environmental Health Institute, Faculty of Medicine from Lisbon University, Lisbon, Portugal;*
[2]*Pharmacy Department, Centro Hospitalar do Algarve, Faro Unit, Portugal*

Microorganisms, and more specifically fungi, can be identified and/or quantified by several methods, presented in the following.

MICROSCOPY

Fluorescence, phase-contrast, and electron microscopy are some possible options for counting microorganisms and collecting data for identification.[1] The first involves the use of fluorescent dyes and the staining of fungal structures that respond after the fluorochrome is excited by the wavelength used[2] in this type of microscope.

With the phase-contrast method, the internal details of microorganisms can be detected and observed because a three-dimensional image is created using parallel beams of light that respond separately when different density structures are encountered.[2]

A much shorter wavelength is used in the electron microscope; as such, magnification and resolution are much greater when using this technique. Samples have to be coated with metal ions to create the necessary contrast.[2]

CLASSICAL CULTURING METHODS

Environmental fungi typically are grown at room temperature (25 °C) and in the dark or with alternating light and dark.[3] Data on culturable microorganisms are expressed as colony-forming units derived from plate counts in which each colony results from the multiplication of a single cell or an aggregate of cells in a homogeneous or heterogeneous cluster.[3]

Some culture methods collect particles directly by impaction on semisolid nutrient plates that are cultivated under standardized conditions. As before, the colonies that grow on the plates are counted and results are given as colony-forming units. Indirect surface samples and most air sampling methods require conventional analysis for fungal identification.

Environmental Mycology in Public Health. http://dx.doi.org/10.1016/B978-0-12-411471-5.00025-9

One of the most important advantages of culture methods is their high sensitivity potential to identify cultivable organisms. Viability is crucial in the study of infectious organisms.[4,5]

Besides providing information on a microorganism's viability, and approximate number, culture methods allow fungal spores to be identified. This is particularly important in studies of fungal contamination of indoor environments where fungal concentrations are low and the presence of species such as *Aspergillus versicolor* indicates indoor fungal growth. A culture method is less harmful to the microorganism because it is not subjected to dehydration, as happens when collecting on filters and dry surfaces.[6]

The main disadvantages of culture methods are poor precision and a highly variable underestimation of exposure. Underestimation depends on sampling strain, microbial robustness, and the size of aggregates that may grow into only one colony. Furthermore, aggregates may break up when dispersed in a liquid before samples are cultivated, and the number of colony-forming units depends on culture conditions, the nutrient medium, and the presence of other species. Results based on cultivation are therefore semiquantitative at best.[4]

The impaction method allows only conventional methods to be applied regarding fungal identification, and it is known that the exclusive use of conventional methods for fungal quantification (fungal culture) may underestimate results for different reasons. The incubation temperature chosen will not be the most suitable for every fungal species, which may result in the inhibition of some species and favoring of others.[7] Differences in fungi growth rates may also result in data underestimation, because fungal species with higher growth rates may inhibit others species' growth. Finally, underestimated data can result from nonviable fungal particles that may have been collected or fungal species that do not grow in the culture media used, although these species may have clinical relevance in the context.[8,9]

Investigators may also wish to use analytical methods not based on microorganism culturability to assess the total number of bacteria and fungi present, detect selected biological agents (e.g., specific antigens, endotoxin, or glucans), or measure marker compounds (e.g., muramic acid or ergosterol).[3,10]

Although there is a great diversity of methodologies for assessing fungal contamination, it is also essential to complement findings from environment surveillance with data from health surveillance,[11] to achieve the real exposure to fungi and their metabolites.

BIOCHEMICAL METHODS

Specially designed to determine the total biomass of a given sample, biochemical methods are advantageous in terms of time consumption because samples can be directly analyzed upon retrieval. The most common method is based on adenosine triphosphate luminescence and is particularly intriguing in terms of fungal quantification because most studies where it is applied find much higher contamination than those using culture methods, a clear indication that the conditions provided for culture analysis may hamper development of all viable components.[1] Detractors of this theory claim that this method is not suited for fungal detection and consider it only a preliminary screening test for biological activity.[1]

IMMUNOLOGICAL ASSAYS

Enzyme immunoassay or enzyme-linked immunoassay (ELISA) is a technique used to assess atmospheric allergen load.[1] This type of assay is based on binding of an antibody to an antigen, a mycotoxin in the case of fungal assessment.

ELISA is the most frequently applied type of assay, and the substrate for the enzyme is usually a chromogenic substance. The developed color after incubation time is usually measured spectrophotometrically, but the most simple detection method is the visual comparison of color intensity, which provides either semiquantitative results or a yes/no response at a certain concentration level or concentration range.[12]

MOLECULAR BIOLOGY APPROACHES

The use of molecular methods requires prior knowledge of the genome of the target species or strains.[13] Highly significant advances have been achieved in this area in the past few years. Nevertheless, available genetic information is still scarce and scattered for many species,[14] essentially regarding species complexes. In addition, costs associated with molecular biology techniques are high, which reinforces the use of conventional methodologies. Therefore, and in parallel with molecular biology, the use of conventional methods is still recommended to confirm fungal species.[15] The other reason to couple both approaches is that most molecular biology methods show the viable and the unviable spectra of the fungal contamination, which makes it difficult to determine whether the contamination is active or which species or genera are behind a given problem.

Molecular biology methods are based on the amplification of DNA (structural genomics) or RNA (functional genomics), or on the study of proteins (proteomics). The DNA analysis begins with the nucleic acid's extraction from the cell and proceeds with amplification, required to obtain a high number of copies. Polymerase chain reaction (PCR) is the most common method of amplification. Comparing with the basic procedure, real-time PCR has the advantage of providing a certain degree of quantification because the method provides the user with the initial and final number of copies and these data can be (carefully) extrapolated to estimate the level of contamination in a given setting.

When the analysis is species-specific, positive DNA detection means a positive result. When the search is broader—using universal primers for fungi, for instance—sequencing of the resulting DNA is the most common route. Comparing the result with the available international databases provides a relatively confident result. The study of biodiversity and microbial community profiling demands the use of other techniques before sequencing. Denaturing Gel Gradient Electrophoresis (DGGE), Denaturing High Performance Liquid Chromatography (DHPLC), Single Strand Conformation Polymorphism (SSCP), or Terminal Restriction Fragment Length Polymorfism (t-RFLP) are some of these techniques. Common to all of them are the existence of more than one species in a given sample and the need to isolate each of them molecularly to be able to sequence each separately. Using temperature or eluent gradients, all of these techniques are constantly evolving for better performance. They are the most commonly used in environmental profiling.

The advent of bioinformatics has been proven invaluable for the study of large amounts of data. DNA microarrays or DNA chips can analyze thousands of genes on a microslide.[16] One future possibility for this technique is the tailored manufacture of a chip containing a set of probes for every microorganism of interest: in the case of environmental health, one could have a series of pathogenic fungal species in a single chip and quickly determine whether they are present in a given setting.

One problem with using DNA or RNA analysis is the lack of specificity for living organisms. Another is the bias inflicted by the sometimes rigid cell walls of fungal hyphae, which can hamper the DNA/RNA extraction process and the entire analysis that follows this first step.

Fluorescent in situ hybridization (FISH) is set to overcome this problem, because no extraction is needed. This is a powerful method that can detect metabolically active fungi directly in the environment without cultivation when RNA is present. A probe of a nucleic acid short sequence, normally designed to target the 18S or 28S rRNA gene, is labeled with a fluorochrome and incorporated in biological materials or environmental samples. After incorporation and incubation, the probe is hybridized to the DNA or RNA in biological materials to form a double-stranded molecule and the sites of hybridization are detected and visualized. A washing step performed after incubation results in a fluorescent signal of exclusively target organisms. The signal is correlated with the ribosome content, and therefore increases in cells with higher metabolic activity. Examples of situations in which the method has been used are varied, from mines to sewage filters. Several factors can influence the efficiency of FISH; limitations include fungal and substrate inherent autofluorescence, non-specific binding of probes, and low ribosome content.[17]

REFERENCES

1. Favalli MA, Gallo F, Maggi O, Mandrioli P, Pacini E, Pasquariello G, et al. Analysis of the biological aerosol. In: Mandrioli P, Caneva G, Sabbioni C, editors. *Cultural heritage and aerobiology—methods and measurement techniques for biodeterioration monitoring*. Dordrecht: Kluwer Academic Publishers; 2003. p. 145–72.
2. Murray PR, Rosenthal KS, Pfaller MA. *Medical microbiology*. 7th ed. Elsevier; 2013.
3. Macher JM. Evaluation of a procedure to isolate culturable microorganisms from carpet dust. *Indoor Air* 2001;**11**:134–40. http://dx.doi.org/10.1034/j.1600-0668.2001.110208.x.

4. Eduard W, Halstensen A. Quantitative exposure assessment of organic dust. *Scand J Work Environ Health* 2009;**7**:30–5.
5. Samson R, Hoekstra E, Frisvad J. *Introduction to food and airborne fungi*. 6th ed. Utrecht: Centraal Bureau Voor Schimmelcultures; 2000.
6. De Nuntiis P, Maggi O, Mandrioli P, Ranalli G, Sorlini C. Monitoring the biological aerossol. In: Mandrioli P, Caneva G, Sabbioni C, editors. *Cultural heritage and aerobiology—methods and measurement techniques for biodeterioration monitoring*. Dordrecht: Kluwer Academic Publishers; 2003. p. 107–44.
7. Zorman T, Jeršek B. Assessment of bioaerosol concentrations in different indoor environments. *Indoor Built Environ* 2008:155–63.
8. Strachan D, Flannigan B, Mccabe E. Quantification of airborne moulds in the homes of children with and without wheeze. *Thorax* 1990;**45**(5):382–7.
9. Bartlett K, Kennedy S, Brauer M. Evaluation and a predictive model of airborne fungal concentrations in school classrooms. *Ann Occup Hyg* 2004;**48**(6):547–54.
10. American Industrial Hygiene Association (AIHA). *Field guide for the determination of biological contaminants in environmental samples*. Fairfax, VA: AIHA; 1996.
11. Prista J, Uva AS. Exposição profissional a agentes químicos: os indicadores biológicos na vigilância da saúde dos trabalhadores. *Rev Saúde Trab* 2003;**4**:5–12.
12. Cigić IK, Prosen H. An overview of conventional and emerging analytical methods for the determination of mycotoxins. *Int J Mol Sci* January 2009;**10**(1):62–115.
13. Douwes J, Thorne P, Pearce N. Bioaerosol health effects and exposure assessment: progress and prospects. *Ann Occup Hyg* 2003;**47**(3):187–200.
14. Horner WE. Assessment of the indoor environment: evaluation of mold growth indoors. *Immunol Allergy Clin North Am* 2003;**23**(3):519–31.
15. Borman A. Conventional methods versus molecular biology. In: *4th trends in medical mycology, Athens*. October 18–21, 2009.
16. Gonzalez JM. Overview on existing molecular techniques with potential interest in cultural heritage. In: Saiz-Jimenez C, editor. *Molecular biology and cultural heritage*. Lisse: Balkema; 2003. p. 3–13.
17. Tsui CKM, Woodhall J, Chen W, Lévesque CA, Lau A, Schoen CD, Baschien C, Najafzadeh MJ, de Hoog GS. Molecular techniques for pathogen identification and fungus detection in the environment. *IMA Fungus* 2011;**2**(2):177–89. http://dx.doi.org/10.5598/imafungus.2011.02.02.09.

Chapter 26

Molecular Approaches to Detect and Identify Fungal Agents in Various Environmental Settings

Raquel Sabino[1], Ana Catarina Pinheiro[2]
[1]*National Institute of Health Dr. Ricardo Jorge, Infectious Diseases Department, Lisbon, Portugal;*
[2]*Pharmacy Department, Centro Hospitalar do Algarve, Faro Unit, Portugal*

The evaluation of health risks associated with exposure to biological agents is a very complex issue, not only because of differences in individual susceptibility, but also because it must take into consideration microorganisms' biological, physiological and genetic diversity. Therefore, the adoption of universal guidelines for detection of fungi and quantification of fungal exposure and its impact on human health is still a challenge.[1] And although limit values for fungal air contamination have already been proposed, these are also not consensual because of the lack of uniformity in the environmental monitoring procedures[2] and laboratory methodologies.

Conventional methods for detecting and identifying fungal species are based on isolation, cultivation, and subsequent observation of morphological traits and/or cultural characteristics of colonies, such as their color, texture, topography, conidial size, conidiophore structure, diffusible pigments, exudates, growth area, aerial and submerged hyphae, growth rate, and thermotolerance.

It is known that the exclusive use of conventional methods of fungal quantification (fungal culture) may underestimate the results for various reasons.[3,4] In fact, the use of molecular analysis in the study of environmental settings has led to the conclusion that only 1% of the total number of prokaryotic species is revealed by conventional methods.[5] Although no values have been determined for fungi, one can expect a high percentage for them as well.

Intra- and interspecific variations in morphological characteristics due to differences in growth and culture conditions can pose serious problems in identification. The chosen incubation temperature can also result in the inhibition of some species and the fostering of others.[6] For example, the ideal temperature for the detection of thermophilic species, such as *Aspergillus fumigatus,* is 40 °C[7,8] because other saprophytic fungal species will not grow

Environmental Mycology in Public Health. http://dx.doi.org/10.1016/B978-0-12-411471-5.00026-0

at this temperature. Environmental samples are incubated at room temperature (±25 °C) and, therefore, data about the presence of thermophilic fungi in these settings is underestimated with conventional methodologies. Different growth rates may also result in data underestimation, because the fungal species with the fastest growth rates may inhibit others. Finally, underestimated data can result from nonviable organisms that may have been collected or fungal species that do not grow in the culture medium used, although these species may have clinical relevance in the context.[9,10] Culture methods are time-consuming and laborious, days or even weeks may be required to isolate the fungi in culture. Several environmental fungi, such as the genera *Fusarium* and *Aspergillus*, are classified into "species complexes" and each section includes many related species, termed as "cryptic species."[11] These are species that present very similar morphological features but with very significant differences at the molecular level. These differences can sometimes result in different susceptibilities to several antifungal substances[12] and, as such, are important to identify. Thus, for species delimitation, a polyphasic approach is suggested as the "gold standard," and this includes morphological, physiological, and ecological data coupled with the understanding of a combination of multilocus sequences. Specific primers and probes for genes encoding mycotoxins have been described. One of the target genes for mycotoxin production is the nor-1 gene, which encodes a reductase enzyme (norsolorinic acid reductase), a protein that takes part in the aflatoxin biosynthetic pathway.[13] Whereas some species belonging to the section *Flavi* complex present a complete functional nor-1 gene, and are therefore aflatoxin producers, others are atoxigenic because they either lack this gene or have an incomplete gene sequence.[14]

To decipher cryptic species, assess mycotoxin production potential, and overcome most of the limitations found when using the traditional culturing methods, new diagnostic tools based on polymerase chain reaction (PCR) methods have been developed. Their use in the detection and quantification of fungi has been increasing as they become an important part of any environmental evaluation. PCR-based techniques have been widely used in specific and direct detection of DNAs and RNAs of microorganisms from clinical and environmental samples to determine, fast, accurately, and quantitatively, the composition of microbial communities. They have also been applied directly to environmental samples to circumvent the time required for culture and subculture and to increase the sensitivity of detection. Compared to conventional methods, recognition of fungal DNA directly from environmental samples offers two additional advantages: (1) fungal DNA can be amplified a million-fold or more by using the PCR methodology and, as such, the presence of the microbes can be detected from as little as one or a few target cells[15] and (2) DNA derived from dead as well as from living fungal cells can be amplified in this system. Nonculturable and/or nonviable spores may still be allergenic and cause health problems.[15]

After DNA/RNA extraction multiple methodologies can be pursued (Figure 1).

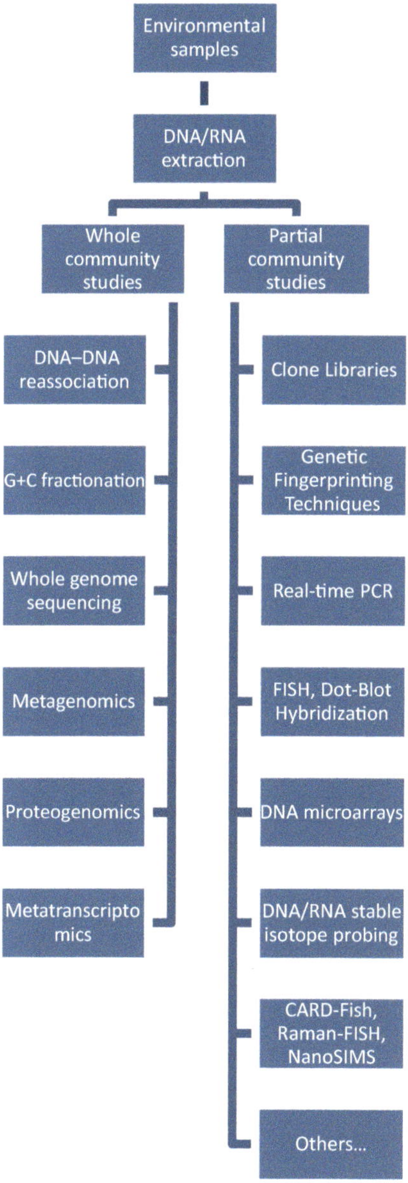

FIGURE 1 Culture-independent methods based on the amplification of DNA/RNA from microorganisms. *Adapted From Rastogi and Sani.*[25]

Partial community studies are the still the most commonly used, although great advances have been achieved in the "omics" and whole genome sequencing strategies. Because of their complexity and ever-growing diversity only a brief description of some of the methods is offered here.

The straightforward option to analyze environmental PCR products is to clone them, sequence the individual gene fragments, and compare them with international databases such as GenBank or KNAW-CBS. Clone libraries can be used alone or combined with genetic fingerprinting techniques. These include various methods such as DGGE/TTGE, DHPLC, SSCP, RAPD, ARDRA, or T-RFLP and produce a community fingerprint based on either sequence polymorphism or length polymorphism. DGGE/TGGE (denaturing- or temperature-gradient gel electrophoresis) and DHPLC (denaturing high-performance liquid chromatography) can be used to separate multiple DNA sequences according to their mobility under increasingly denaturing conditions. In single-strand conformation polymorphism (SSCP), the PCR products are denatured followed by electrophoretic separation of single-stranded DNA fragments on a nondenaturing polyacrylamide gel. In all of them separation is based on subtle differences in sequences (often a single base pair), which results in measurable differences in mobility.

In DGGE, TGGE, and SSCP the DNA bands can be excised from the gel, reamplified, and sequenced. In DHPLC they can be collected for the same purpose. In RAPD, or random amplified polymorphic DNA, the use of short primers generates PCR amplicons of various lengths, which can be separated on a gel or used as a distinctive pattern. A pattern is also what is obtained when the PCR product is digested with restriction endonucleases (ARDRA, or amplified ribosomal DNA restriction analysis). ARDRA is also known as restriction fragment length polymorphism (RFLP), but when a fluorescently labeled primer is used, then the method is termed T-RFLP. These terminal restriction fragments are separated and analyzed on an automated sequencer.

Real-time PCR was developed to address the need for quantification, and for this particular kind of amplification intercalating dyes or fluorescent probes are used to measure the amount (or accumulation) of amplicons for each PCR cycle. This approach brings us closer to determining the original amount of DNA in a given sample.

To fully comprehend a fungal community—or even a single colony—whole-genome analysis is rapidly becoming the right choice. Microbial genomes are sequenced using a shotgun cloning method and, once obtained, the sequences are aligned and assembled by computer programs. Prokaryotes are still the main focus of these complete-genome analyses, but given recent improvements (pyrosequencing, for instance), it should be a matter of time until we get to know full genomes for the fungi as well.

Metagenomics (environmental genomics or community genomics) is, basically, the same principle but applied to a whole community. Microbial genomes are retrieved directly from environmental samples but, instead of using pure cultures (as above), the entire community DNA is brought under analysis without prior isolation. This incredible step makes it possible to advance to the post-genomic era, in which metatranscriptomics step in to reveal the link between genetic potential and function.

As far as fungal communities and/or isolated cultures are concerned, the most commonly used target for fungal diagnostic PCR primers is the ribosomal RNA gene operon (Figure 2), encoding the 18S, 5.8S, and 28S RNA subunits,

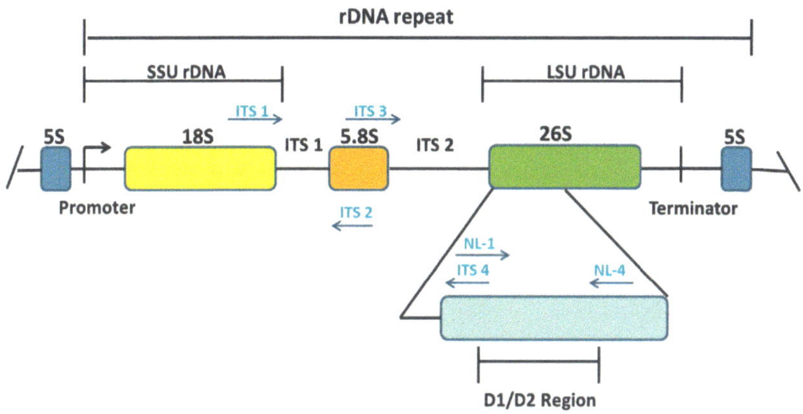

FIGURE 2 RNA gene operon.

of which there are hundreds of copies encoded in the fungal genome, improving the sensitivity of the diagnosis. The nuclear-encoded ribosomal RNA genes (rDNA) of fungi exist as a multiple-copy gene family comprising highly similar DNA sequences.

These genes have conserved and divergent regions that are very amenable to the design of panfungal and species-specific primers.[16,17]

For most yeasts and dematiaceous fungi, especially the human pathogenic species, the D1/D2 domains of the large subunit ribosomal DNA (LSU rDNA) are very useful for identification.[18] Although rRNA genes are highly conserved, the internal transcribed spacer (ITS) regions are distinctive.[19]

The ITS region contains two variable noncoding regions that are nested within the rDNA repeat between the highly conserved small subunit 5.8S and large subunit rRNA genes. The ITS region is one of the most widely sequenced DNA regions in fungi; the entire region is often between 600 and 800 bp and is amplified using the universal primers ITS1 and ITS4.[17] Because of its higher degree of variation compared to other genic regions of rDNA (small subunit, SSU, and LSU), it is frequently used to identify genera or species, depending on the taxon. The universal fungal oligonucleotide primer pair ITS3 and ITS4 amplifies portions of the 5.8S and 28S rDNA subunits, and the ITS2 region, allowing species identification.

To determine the fungal barcoding region of excellence, a multinational multilaboratory consortium analyzed and compared three subunits from the nuclear ribosomal RNA cistron: the ITS and the nuclear ribosomal SSU and LSU. The nuclear ribosomal SSU showed poor species-level resolution in fungi and was discarded. The nuclear ribosomal LSU showed superior species resolution in some taxonomic groups but was otherwise slightly inferior to the ITS, with the highest probability of successful identification for the broadest range of fungi. As such, the ITS was formerly proposed for adoption as the primary fungal bar-code marker to the Consortium for the Barcode of Life.

Supplementary bar codes may be developed for particular narrowly circumscribed taxonomic groups as it is unlikely that a single marker–bar code system will be capable of identifying every specimen to a species level in a kingdom as old and diverse as fungi.[20]

Species identification based on molecular methods can be a cost-effective, fast, good discriminatory approach for delineating species.[21] In fact, only molecular biology made it possible to detect potentially toxigenic/pathogenic fungi in different occupational settings.[22,23]

Identification based upon comparative sequencing of the ribosomal ITS region to the species complex level and of a protein-encoding locus, such as the β-tubulin region, elongation factor α, calmodulin, and others, for identification of species within complexes has been recommended for species identification.[11] The sequences of those protein-encoding regions are highly discriminatory, allowing their use in identification to the species level.

Conventional culture method setbacks have justified the need for modern molecular biology approaches. But these are not problem-free solutions. As cautioned, the extremely simple methodology descriptions may be misleading, as each one really follows a very strict recipe for success. And the use of still new technologies comes with a price sometimes too high for routine use.[24]

Specific methodologies can also carry some problems,[25] starting with the first step: DNA/RNA extraction. There can be incomplete or preferential cell lysis, which influences the final result by either distorting community composition or impairing quantitative studies (such as real-time PCR). Next there can be problems with the amplification process: inhibition compounds (such as humic acids when dealing with soil samples) generally extracted alongside DNA or RNA can make it extremely difficult to perform a PCR successfully. When trying to eliminate these contaminants one can be also eliminating part of the genomic material and this decreases PCR efficiency (and, again, impairs quantification). Preferential amplification can also be a cause of quality and quantity misinterpretation. Finally, the formation of PCR artifacts can also impair a correct reading of a fungal community. And these are the general problems associated with any PCR. Fungal colonies or communities present additional ones.

A prior knowledge of the genome of the target species/strains is needed[26] and available genetic information is still scarce and scattered for many species,[27] essentially regarding species complexes. Despite being less sensitive, less specific, and more time consuming,[28] conventional methods are still needed when it is necessary to characterize the fungal distribution in poorly studied settings. In the case of a possible fungal occupational exposure through inhalation, only conventional methods offer the advantage of enabling identification and quantification of only the viable fraction, which is the most likely to represent a health risk.[3,29] In environmental exposure, following the ALARA principle ("as low as reasonably achievable") in terms of the intention to keep the exposure to carcinogenic substances at the lowest achievable level, the simple presence of a potential mycotoxin producer strain is determinant to employ corrective measures.[4,30]

Given these scenarios, a growing number of studies suggest the need for both molecular and conventional methods to be employed in the identification and quantification of fungi.[4,30,31] As mentioned by Rastogi and Sani,[25] "Culture-based and culture-independent molecular techniques are neither contradictory nor excluding and should be considered as complementary." Using both cultural and molecular methodologies we could quantify viable microorganisms and simultaneously identify potentially toxigenic species, resulting in complementary information useful in the adoption of strategies to minimize exposure to mycotoxins.

REFERENCES

1. Sigler L, Abbott S, Gauvreau H. Assessment of worker exposure to airborne molds in honeybee overwintering facilities. *Am Ind Hyg Assoc J* 1996;**57**:484–90.
2. Green J, Tovey E, Sercombe J. Airborne fungal fragments and allergenicity. *Med Mycol* 2006;**44**:S245–55.
3. Eduard W, Halstensen AS. Quantitative exposure assessment of organic dust. *SJWEH* 2009;**7**:30–5.
4. Viegas S, Veiga L, Malta-Vacas J, Sabino R, Figueredo P, Almeida A, et al. Occupational exposure to aflatoxin (AFB1) in poultry production. *J Toxicol Environ Health A* 2012;**75**:1330–40.
5. Muyzer G. Genetic fingerprinting of microbial communities – present status and future perspectives. Methods of microbial community analysis, microbial biosystems: New Frontiers proceedings of the 8th international symposium on microbial ecology. In: Bell CR, Brylinsky M, Johnson-Green P, editors. *Atlantic Canada society for microbial ecology*. Halifax: Canada; 1999.
6. Zorman T, Jersek B. Assessment of bioaerosol concentrations in different indoor environments. *Indoor Built Environ* 2008;**17**:155–63.
7. Kosalec I, Pepeljnjak S, Jandrlie M. Gliotoxin production by *Aspergillus fumigatus* strains. *Arh Hig Rada Toksikol* 2005;**56**:269–73.
8. Fulleringer SL, Seguin D, Warin S, Bezille A, Desterque C, Arné P, et al. Evolution of the environmental contamination by thermophilic fungi in a turkey confinement house in France. *Poult Sci* 2006;**85**:1875–80.
9. Bartlett K, Kennedy S, Brauer M. Evaluation and a predictive model of airborne fungal concentrations in school classrooms. *Ann Occup Hyg* 2004;**48**:547–54.
10. Strachan D, Flannigan B, Mccabe E. Quantification of airborne moulds in the homes of children with and without wheeze. *Thorax* 1990;**45**:382–7.
11. Balajee SA, Houbraken J, Verweij PE, Hong S-B, Yaghuchi T, Varga J, et al. *Aspergillus* species identification in the clinical setting. *Stud Mycol* 2007;**59**:39–46.
12. Balajee SA, Gribskov JL, Hanley E, Nickle D, Marr KA. *Aspergillus lentulus* sp. nov., a new sibling species of A fumigatus. *Eukaryot Cell* 2005;**4**:625–32.
13. Mayer Z, Bagnara A, Färber P, Geisen R. Quantification of the copy number of nor-1, a gene of the aflatoxin biosynthetic pathway by real-time PCR, and its correlation to the cfu of *Aspergillus flavus* in foods. *Int J Food Microbiol* 2003;**82**:143–51.
14. Cary J, Ehrlich K. Aflatoxigenicity in *Aspergillus*: molecular genetics, phylogenetic relationships and evolutionary implications. *Mycopathologia* 2006;**162**:167–77.
15. Wu Z, Wang X, Blomquist G. Evaluation of PCR primers and PCR conditions for specific detection of common airborne fungi. *J Environ Monit* 2002;**4**:377–82.

16. Bruns TD, Vilgalys R, Barns SM, Gonzalez D, Hibbett DS, Lane DJ, et al. Evolutionary relationships within the fungi: analyses of nuclear small subunit rRNA sequences. *Mol Phylogenet Evol* 1992;**1**:231–41.
17. White TJ, Bruns T, Lee S, Taylor J. Amplification and direct sequencing of fungal ribosomal RNA genes for phylogenetics. In: Innis MA, Gelfand DH, Sninsky JJ, White TJ, editors. *PCR protocols: a guide to methods and applications*. New York: Academic Press, Inc.; 1990. p. 315–22.
18. Abliz P, Fukushima K, Takizawa K, Nishimura K. Identification of pathogenic dematiaceous fungi and related taxa based on large subunit ribosomal DNA D1/D2 domain sequence analysis. *FEMS Immunol Med Mic* 2004;**40**:41–9.
19. Reiss E, Morrison CJ. Nonculture methods for diagnosis of disseminated candidiasis. *Clin Microbiol Rev* 1993;**6**:311–23.
20. Schoch CL, Seifert KA, Huhndorf S, Robert V, Spouge JL, Levesque CA, Fungal Barcoding Consortium; Fungal Barcoding Consortium Author List, et al. Nuclear ribosomal internal transcribed spacer (ITS) region as a universal DNA barcode marker for fungi. *Proc Natl Acad Sci USA* 2012;**109**:6241–6.
21. Deak E, Balajee SA. Molecular methods for identification of *Aspergillus* species. *Aspergillosis: from diagnosis to prevention* 2010;**1**:75–85.
22. Bellanger AP, Reboux G, Murat JB, Bex V, Millon L. Detection of *Aspergillus fumigatus* by quantitative polymerase chain reaction in air samples impacted on low-melt agar. *Am J Infect Control* 2010;**38**:195–8.
23. Malta-Vacas J, Viegas S, Sabino R, Viegas C. Fungal and microbial volatile organic compounds exposure assessment in a waste sorting plant. *J Toxicol Environ Health A* 2012;**75**:1410–7.
24. Borman A. Conventional methods versus molecular biology. In: *4th Trends in Medical Mycology, Athens*. 2009.
25. Rastogi G, Sani R. Molecular techniques to assess microbial community structure, function, and dynamics in the environment. In: Ahmad I, et al., editor. *Microbes and microbial technology. Agricultural and environmental applications*. Springer Science; 2011, ISBN: 978-1-4419-7930-8. p. 29–57.
26. Douwes J, Thorne P, Pearce N, Heederik D. Bioaerosol health effects and exposure assessment: progress and prospects. *Ann Occup Hyg* 2003;**47**:187–200.
27. Horner W. Assessment of the indoor environment: evaluation of mold growth indoors. *Immunol Allergy Clin* 2003;**23**:519–31.
28. Stetzenbach L, Buttner M, Cruz P. Detection and enumeration of airborne biocontaminants. *Curr Opin Biotech* 2004;**15**:170–4.
29. Samson R, Hoekstra E, Frisvad J. *Introduction to food and airborne fungi*. 6th ed. Utrecht: Centraalbureau voor Schimmelcultures; 2000.
30. Viegas S, Veiga L, Figueredo P, Almeida A, Carolino E, Sabino R, et al. Occupational exposure to aflatoxin B1 in swine production and possible contamination sources. *J Toxicol Environ Health A* 2013;**76**:944–51.
31. Pitkäranta M, Meklin T, Hyvärinen A, Paulin L, Auvinen P, Nevalainen A, et al. Analysis of fungal flora in indoor dust by ribosomal DNA sequence analysis, quantitative PCR, and culture. *Appl Environ Microbiol* 2008;**74**:233–44.

Index

Note: Page numbers followed by "f" and "t" indicate figures and tables respectively.

A

E

F

H

Kolibri 360 GmbH
Ferdinand-Jühlke-Straße 7
99095 Erfurt, Deutschland
produktsicherheit@kolibri360.de